W0043938

Positron Emission Tomography:
A Critical Assessment of Recent Trends

NATO ASI Series

Advanced Science Institute Series

A Series presenting the results of activities sponsored by the NATO Science Committee, which aims at the dissemination of advanced scientific and technological knowledge, with a view to strengthening links between scientific communities.

The Series is published by an international board of publishers in conjunction with the NATO Scientific Affairs Division

A Life Sciences	Plenum Publishing Corporation
B Physics	London and New York
C Mathematical and Physical Sciences	Kluwer Academic Publishers
D Behavioural and Social Sciences	Dordrecht, Boston and London
E Applied Sciences	
F Computer and Systems Sciences	Springer-Verlag
G Ecological Sciences	Berlin, Heidelberg, New York, London,
H Cell Biology	Paris and Tokyo
I Global Environment Change	

PARTNERSHIP SUB-SERIES

1. **Disarmament Technologies**	Kluwer Academic Publishers
2. **Environment**	Springer-Verlag / Kluwer Academic Publishers
3. **High Technology**	Kluwer Academic Publishers
4. **Science and Technology Policy**	Kluwer Academic Publishers
5. **Computer Networking**	Kluwer Academic Publishers

The Partnership Sub-Series incorporates activities undertaken in collaboration with NATO's Cooperation Partners, the countries of the CIS and Central and Eastern Europe, in Priority Areas of concern to those countries.

NATO-PCO-DATA BASE

The electronic index to the NATO ASI Series provides full bibliographical references (with keywords and/or abstracts) to about 50,000 contributions from international scientists published in all sections of the NATO ASI Séries. Access to the NATO-PCO-DATA BASE is possible via a CD-ROM "NATO Science and Technology Disk" with user-friendly retrieval software in English, French, and German (©WTV GmbH and DATAWARE Technologies, Inc. 1989). The CD-ROM contains the AGARD Aerospace Database.

The CD-ROM can be ordered through any member of the Board of Publishers or through NATO-PCO, Overijse, Belgium.

3. High Technology – Vol. 51

Positron Emission Tomography: A Critical Assessment of Recent Trends

edited by

Balázs Gulyás

Department of Neuroscience,
Karolinska Institute, Stockholm, Sweden
and
Hungarian PET Foundation,
Debrecen, Hungary

and

Hans W. Müller-Gärtner

Institute for Medicine,
Forschungszentrum Jülich GmbH, Jülich, Germany
and
Department of Nuclear Medicine,
Heinrich-Heine-University of Düsseldorf,
Düsseldorf, Germany

Springer Science+Business Media, B.V.

Proceedings of the NATO Advanced Research Workshop on
Positron Emission Tomography: A Critical Assessment of Recent Trends
Debrecen, Hungary
1–5 October, 1996

A C.I.P. Catalogue record for this book is available from the Library of Congress.

ISBN 978-94-010-6097-4   ISBN 978-94-011-4996-9 (eBook)
DOI 10.1007/978-94-011-4996-9

Printed on acid-free paper

All Rights Reserved
© 1998 Springer Science+Business Media Dordrecht
Originally published by Kluwer Academic Publishers in 1998
Softcover reprint of the hardcover 1st edition 1998

No part of the material protected by this copyright notice may be reproduced or utilized in any form or by any means, electronic or mechanical, including photocopying, recording or by any information storage and retrieval system, without written permission from the copyright owner.

TABLE OF CONTENTS

PREFACE

Under the auspices of the Scientific Affairs Division of the North Atlantic Treaty Organisation (NATO), a NATO Advanced Research Workshop (ARW) was held in Hungary's PET Center in Debrecen between 1 and 5 October 1996. The workshop, entitled: Positron Emission Tomography: A Critical Assessment of Recent Trends, focused on the recent advances of PET with special regard to its clinical and research applications. The workshop was attended by 96 participants from 17 different countries. The great majority of the student attendants came from Central and Eastern European countries, indicating the surge of interest in this region of Europe in the application of PET.

The present volume contains the papers delivered at the workshop by the invited speakers. The first part of the book concentrates on the technical and administrative aspects of establishing and running a PET center. The second part focuses on novel technical developments regarding camera designs, radiochemistry, data acquisition, kinetic modelling, image processing, registration and analysis. In the further parts of the book the reader is guided through some recent clinical and research applications of PET, including, among others, the fields of oncology, neurology, psychiatry, and neurosciences. The volume is also completed with two informative chapters on the recent status of PET in the world: First, there is a short overview of PET centers and PET projects in Central and Eastern Europe where PET has in recent years been gaining an appropriate status in the field of functional imaging. Second, there is a database containing basic information on PET centers world-wide.

It should be emphasised here that the present volume also contains two chapters authored by two student participants of the meeting: Pascale Schumann and József Lövey, who proved to be the best student speakers at the meeting. In addition to their presentation, thirteen student participants delivered papers or exhibited posters at the workshop; the high standard of their presentations was appreciated by the participants.

By sponsoring a seminal workshop on PET in Hungary in 1996, the Scientific Affairs Division of NATO has made an outstanding contribution to the promotion of scientific exchange in the field of functional imaging and to the future perspectives of the applications of positron emission tomography in the Central and Eastern European countries.

Stockholm and Düsseldorf, 1 May 1997.

Balázs Gulyás
and
Hans W. Müller Gärtner
editors

Part one:
How to establish and run a PET Center?

Planning a Proposal for a PET Centre

Andy Holley
CTI Europe Ltd.
15 Knowles Avenue, Crowthorne, Berkshire, UK. RG45 6DU.

This is a guide to the major items to be considered when putting together a proposal for establishing a new PET Centre. PET is a unique tool offering researchers unrivalled access to in vivo quantitative measurements not available by any other technique. As a clinical tool it has been shown to increase diagnostic accuracy and hence ensure cost efficient use of scarce patient care facilities. However, it is also expensive and any potential funding body is going to need to be absolutely convinced of the benefits and the ability of the applicant to make effective use of this powerful tool. Seeking funding is a major undertaking, likely to take 3-5 years from inception to birth so the first prerequisite is to have the commitment and determination needed to overcome the various obstacles and hurdles.

Clear objectives.
The first rule is to have clearly defined and achievable objectives. The centre will either be research, clinical or some combination of both. A research centre must propose projects which are well thought out and relevant. A clinical centre must propose diagnostic benefits which are demonstrably relevant (and cost effective) to identified diseases. Funding is unlikely to be available for general topics such as "Neurology Research" or "clinical oncology". The success or failure of the search for funding is likely to hinge on this crucial section.

Sources of funding.
The government research councils and national health care organisation are unlikely to have sufficient funding to start up a new PET centre. However it is always worth an application. There are sometimes exceptional circumstances which do motivate a government to fund new facilities. The government may be persuaded to fund the cyclotron part of the project as a regional source for distribution of positron emitting isotopes. As a minimum they should be asked to fund the on going running costs, if the initial capital comes from another source.

Many European PET centres are actually funded by private benefactors, wealthy individuals who would like to see some of their money used for medical research. This is especially appealing if the alternative is to give it to the taxman.

B. Gulyás and H.W. Müller-Gärtner (eds.),
Positron Emission Tomography: A Critical Assessment of Recent Trends, 3–9.
© 1998 Kluwer Academic Publishers.

Site Planning.

A typical PET centre covers around 400-600 sqm. Rooms will include,

> Cyclotron Room
> Radiochemistry Laboratory
> Scanner room
> Scanner operation room
> Patient Preparation
> Patient WC

A blood laboratory is also needed if quantitative studies are to be performed.
A room for off-line analysis of data is recommended for a clinical centre and essential for a research centre.

Before submitting an application for funding you are advised to request a specific layout of how the equipment could fit within the space which you have available.

Equipment.

A typical PET centre will include a cyclotron, chemistry equipment and a PET scanner. The particular requirements will depend on the objectives of the project.

A modern low energy cyclotron is quite easily capable of producing prodigious amounts of all the commonly required PET isotopes for either clinical or research use. The use of automated chemistry modules facilitates the routine production of common tracers. Options such as dual extraction allow simultaneous production of two different isotopes, facilitating operation in a busy research centre.

A clinical PET centre will also need a basic kit of chemistry equipment for quality control. A sample list should be available from manufacturers. However a research centre will need a much more extensive range of chemistry equipment according to the aims of the project.

PET scanners come in a variety of price performance bands. A low cost system is available for routine clinical work and some clinical research. For advanced research it will be necessary to invest in a premium scanner with state of the art resolution and sensitivity. Some people plan to start off with a second-hand PET scanner with the intention to upgrade later. The problem is that once you have got a scanner it is difficult to get funds to upgrade. If you are not able to work properly with the second-hand scanner then the project is seen to be a failure and not worth supporting. If you are able to work properly then why do you need a new one ?! Since the ongoing running costs of isotope production etc. are largely independent of whether the scanner is new or second-hand I would advise pushing strongly for a new scanner from the start.

Pharmaceutical companies will sometimes invest in part of the cost of a PET centre. You will need to be able to convince them that there are real benefits in investing in your project instead of working with the existing major centres. This is not likely unless you have some significant advantage such as unrivalled access to a particular patient group.

Sales of isotope are a good way to raise funds to cover the ongoing running costs of the cyclotron. A modern small cyclotron can easily produce sufficient quantities of isotope for distribution as well as meeting your internal requirements. Once you have covered the initial investment and basic running costs, the incremental cost of making additional isotope for external use is minimal. The use of a dual extraction means that the system can make F-18 for distribution at the same time as making other isotopes for internal use. The main hurdle is to ensure that you have qualified staff who can meet the regulations for commercial sale of the radiopharmaceutical.

Experience.

It is important that at least some of the key people have hands on experience of using PET to demonstrate that they have the skills required for PET. It is therefore strongly recommended that they spend a period working at an existing PET centre. A period of 6-12 months is recommended to get actively involved in contributing to the work of that centre. A shorter visit is likely to be superficial and not enable the worker to become an active contributor. Funding for such visits is available from a variety of sources. Your own National Research council may be able to help. The European Molecular Biology Organisation(EMBO) [i] has some funds for working visits as does NATO[ii]. The Royal Society[iii] in England will fund working visits to UK centres. The DAAD[iv] funds working visits to Germany and the National Science Foundation can fund working visits to USA. The EEC concerted action[v] program provides an excellent list of European PET centres together with contact details and a summary of their research program. The Institute of Clinical PET (ICP) [vi] can provide lists of American PET centres.

Location.

The key to a successful project is in choosing the right location. A research centre must be situated close to the support infrastructure including chemists, biologists and a major medical research establishment. A clinical centre needs to be easily accessible for referring physicians and their patients. Some people have tried to situate new PET centres at existing nuclear research centres so as to share an existing accelerator. This is fraught with difficulties due to the conflicting priorities of the other existing users of the accelerator. This option should only be considered if the research centre has significant amounts of freely available beam time as well as an established tradition of medical research, including on-site patient handling facilities.

The cost of site preparation will vary enormously according to individual circumstances. The introduction of self shielded cyclotrons has eliminated the need for a dedicated bunker and greatly reduced the cost of adapting a building.

Budgetary prices are as follows:

PET Scanner
   Clinical
   Mid range
   Premium
	$ 1,000,000 - 1,200,000
	$ 1,600,000 - 1,800,000
	$ 2,000,000 - 2,600,000

Cyclotron
   Clinical
   Extended
	$ 1,400,000
	$ 1,800,000

Chemistry Equipment
   Basic
   Research
	$ 300,000
	$600,000 upwards

Site preparation	??

Running Costs.

This is a major concern for any funding body. The capital investment, whilst large, is a one off payment. The ongoing running costs stretch interminably into the future. The main part of the running costs are in two areas, staffing and maintenance.

A modern cyclotron is easily controlled by a graphical workstation and does not require a dedicated operator. Furthermore, the use of automated chemistry modules to produce standard compounds greatly reduces the number of chemists required. A routine PET centre will need a staff of 4-6 people as follows:

      Radio Chemist (also operates cyclotron)
      Chemistry Technician
      Nurse/Radiographer (s) for patient preparation
           and scanner operation.
      Administrator
      Doctor(s)

In addition you will need part time support from people in other departments.

      Radiopharmacist for quality control
      Radiation Protection Officer
      Computer Scientist for system management
      Physicist for scanner calibration and QC

A major PET research centre which is developing new compounds and pioneering new techniques will have a much larger team to include research chemists, physicists, computer scientists etc. The larger European PET centres have more than 50 people supporting their PET programs.

Maintenance.
All manufacturers offer a full maintenance contract covering preventative maintenance and on call service to correct any faults which may arise. If you have significant in house technical expertise you should ask about training one of your own staff to provide service with telephone support and on site support from the manufacturer as required.

Other running costs include chemicals, consumable, such as swabs and syringes, and stationery.

It is strongly recommended that the budget also includes a travel budget to allow co operative visits to other centres.

8

A typical budget is as follows:

Staff.
Clinical (4-5)
  Radiochemist, Chemistry Technician
  Nurse/Radiographer, Administrator
  Doctor
Research (more !!)
  More chemists, Physicists
  Computer Scientists, Mathematicians.
	Actual cost will vary from country to country.

Travel	$30,000
Maintenance Scanner, Cyclotron	8-12 % price. Can be reduced if own staff are trained for first line.
Consumables. Chemicals, target materials swabs etc.	Typically < $50 per patient. Budget $40,000 for 800 patients.
Miscellaneous Telephone, Stationery etc.	$15,000

Summary.
The cost of PET is high, but so are the benefits. In order to secure funding you will need to submit a strong proposal with defined objectives which clearly demonstrate that the benefits more than justify the costs. The proposal should include

> Clearly defined objectives
> Competence of key personnel
> Proposed location
> Capital Investment
> Running costs.

This, together with a lot of perseverance and dedication, should enable you to establish a new PET centre in your own country.

[i] EMBO. Tel + (49) 6221 383 031, FAX + (49) 6221 384 879

[ii] NATO Science Affairs Division, B-1110 Brussels
Tel(32) 2 728 4111, Fax: ((32) 2 728 4232

[iii] Royal Society, 6 Carlton Terrace, London SW1Y 5AG
Atn. Sonja Kvesic, Tel (44) 171 839 5561

[iv] DAAD. Kennedy Allee 50, Bonn, Germany
Tel: +(49) 228 882 0

[v] EEC Concerted Action
c/o Prof. D Comar, CERMEP, hopital Neurocardilogique, 59 Bd pinel, 69003 Lyon,
Tel (33) 72 68 86 00, FAX +(33) 72 68 86 10

[vi] Institute of Clinical PET, 11781 Lee Jackson Memorial highway, Suite 360
Fairfax, VA 22033, USA. Tel" +(1) 703 691 2255, FAX +(1) 703 691 2259

THE USE AND REGISTRATION OF PET-RADIOPHARMACEUTICALS

- EUROPEAN AND WORLD TRENDS -

Geerd-J. Meyer
Abteilung Nuklearmedizin und spezielle Biophysik
Medizinische Hochschule Hannover
D - 30623 Hannover

1. Abstract

The development of regulatory schemes, for the production of PET radiopharmaceuticals can be understood only, in view of the development of PET as a new diagnostic tool. It seems at the moment that PET starts to cross the border from a research tool to a real clinical utility. Although some extreme expectations of the expansion have not been ful- filled in the last two years, the steady increase in PET investments is a strong indicator for its success. However, the increasing availability in the USA, as well as in Japan and especially in Germany was not matched by regulatory actions, to back up the legal foun- dations for its clinical use. Reasons for the hampered development have been identified as being regulatory deficits, educational deficits, and struggles between interest groups, as well as the complexity of the method itself.

New ways of defining quality control demands and the qualification for responsibilities had to be developed within conservative bodies, as national drug licensing agencies, pharmacopoeial commissions, professional interest groups, and education departments.

Although regulatory approaches differ to quite some extent in many countries, harmoni- zation efforts are observed not only in Europe, but also between Europe and the USA.

Nevertheless progress has been made in several of these fields during the last five years, especially within the field of production and quality control. It is hoped that in the com- ing two to three years many of the remaining issues can be solved.

A relatively unrestricted availability of fluorinated radiopharmaceuticals for PET will au- tomatically lead to an enormous demand. This can be derived from the high demand for

B. Gulyás and H.W. Müller-Gärtner (eds.),
Positron Emission Tomography: A Critical Assessment of Recent Trends, 11–24.
© 1998 Kluwer Academic Publishers.

double headed SPECT gamma cameras which can be operated in coincidence mode. On the other hand this will further trigger an increase the of number of operational PET camera locations. In summary diagnostic PET procedures should become available as a clinical routine tool all over Europe, - at least in all locations close to universities and in larger cities -.

2. Development of PET Radiopharmaceuticals.

Positron emission tomography (PET) has evolved over more than 20 years by now. With regard to this relative long history, its clinical use and availability is still quite limited, when compared with other modern diagnostic tools, as e.g. Ultrasound or NMR, which both have gone through a similar time of development. The reason for the hampered development of PET as a clinical utility is mainly due to the extreme complexity of the method, which besides a cost intensive investment requires a continuous interdisciplinary teamwork of physicists, radiochemists and physicians.

On the other hand PET has established itself as a powerful research tool in life sciences. In this area, interdisciplinary teamwork is rather a stimulating co-factor, than a negative aspect of the method.

We therefore have two aspects of PET development, which need clearly to be distinguished, in order to evaluate the current status of PET in both its health-political and health-economical importance, as well as in terms of its scientific relevance and impact.

The scientific and research and development ("R&D") areas are located within research centres and universities. In general all operations performed in such an environment can be regarded as activities, which employ radioactive labelled materials and tracers which do not necessarily fall under the laws governing the use of pharmaceuticals. As long as all applications are performed under the shield of research, special laws on investigational drugs will usually cover their use in humans.

On the other hand, the routine use of PET as a medical utility in nuclear medicine requires that the tracers are definitively recognized as pharmaceuticals.

With respect to these different aspects, the PET tracers have to be classified according to their use. There is, however, no criterion which can be used unambiguously to classify a PET tracer as a routine radiopharmaceutical or a radioactive research drug. The use is often changing from investigational applications to routine use and even often back again to mere research applications.

The drug laws, however, do not provide for such ambivalent use and therefore require some action by the PET community in order to establish equivocal criteria for the designation of PET tracers.

Since European harmonization has influenced the drug laws of EU member states to quite some extent, a European consensus has to be found for the criteria.

Although individual nations may choose their own designations for national implementation, the best choice on a European level was the evaluation of PET tracer production and use by an EANM task-group. The results of a European survey have been published recently. The impact of this evaluation on the European bodies is indicated by the fact that the results have been used by the Eur. Pharmacopoeial commission for its priority rating.

In order to judge the current status of PET tracer use and production, it is useful to map the PET locations.

The national distribution of PET locations is given in table 1.

TABLE 1. National distribution of PET installations. (The numbers include installation sites which are currently under construction, - they should be regarded as values with a small uncertainty. - Planned sites amount in nearly all countries to about the same number of those which are currently in operation already.)

Country	PET Installations
USA	73
Germany	25
Japan	12
Belgium	6
Italy	5
UK	5
France	4
Canada	4
Sweden	3
Switzerland	3
Australia	3
Netherlands	2
Spain	2
Austria	2
Denmark	2
Finland	2
Russia	2
Hungary	1
others	6

14

Except for those countries which have installed a relative large number of systems, PET is predominantly regarded as a clinical research tool.

With respect to this application the radiosynthetic capacity is the most important factor which determines the efficacy of the PET methodology.

In order to demonstrate the enormous potential of PET in the clinical research area, the increasing number of compounds, which have been labelled with short-lived positron emitting radionuclides is shown in fig 1.

FIGURE 1. Number of labelled compounds that have been prepared up to the years 1986, 1990, 1995. (1986 Data taken from A.P. Wolf. Additional data extracted from the three most important journals: i.e. J. Labelled Comp. Radiopharm., Appl.Radiat.Isot., and J.Nucl.Med.)

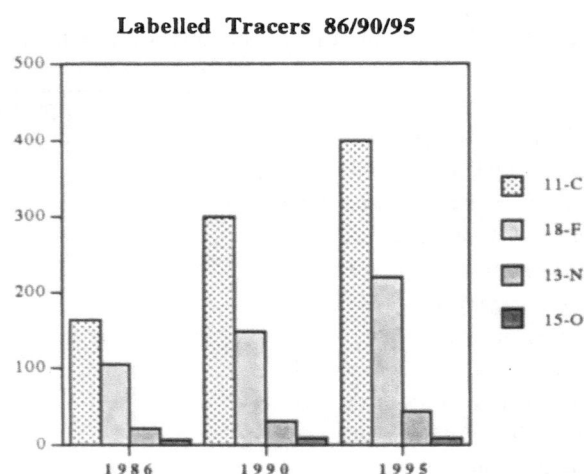

As can be seen from fig 1, there is an overwhelming variety of tracers, especially within the pool of ^{11}C-compounds. Nevertheless ^{18}F-labelled tracers have become available in an very large number also.

The main motivation for their development stems from the longer halflife of ^{18}F, which guarantees a slightly longer shelf-life of the labelled product. This helps to reduce the logistic difficulties associated with clinical PET studies. In some cases the longer halflife is necessary also in order study a relatively slow biochemical process. On the other hand, the fluorine-label may alter the biochemical behaviour of a compound, which in its native substrate form does not contain a fluorine atom.

Despite the rapid increase of labelled products which can be used in PET, their use is relatively limited.

3. Current Use of PET Radiopharmaceuticals.

The 1993 European survey on the use of PET Radiopharmaceuticals has revealed, that only few products are used on a broad scale. The data are summarized in figs. 2 and 3. Although the survey missed the data from a few centres, the general conclusions that can been drawn are valid, since the production and application activities of those centres, who did not respond in time are known from their publication activities.. Furthermore it can be assumed that the relative importance of PET tracers in terms of their frequency of production and use are valid also for the USA and Japan.

FIGURE 2. The graph displays the number of PET centres which produce a certain PET tracer on a regular basis.

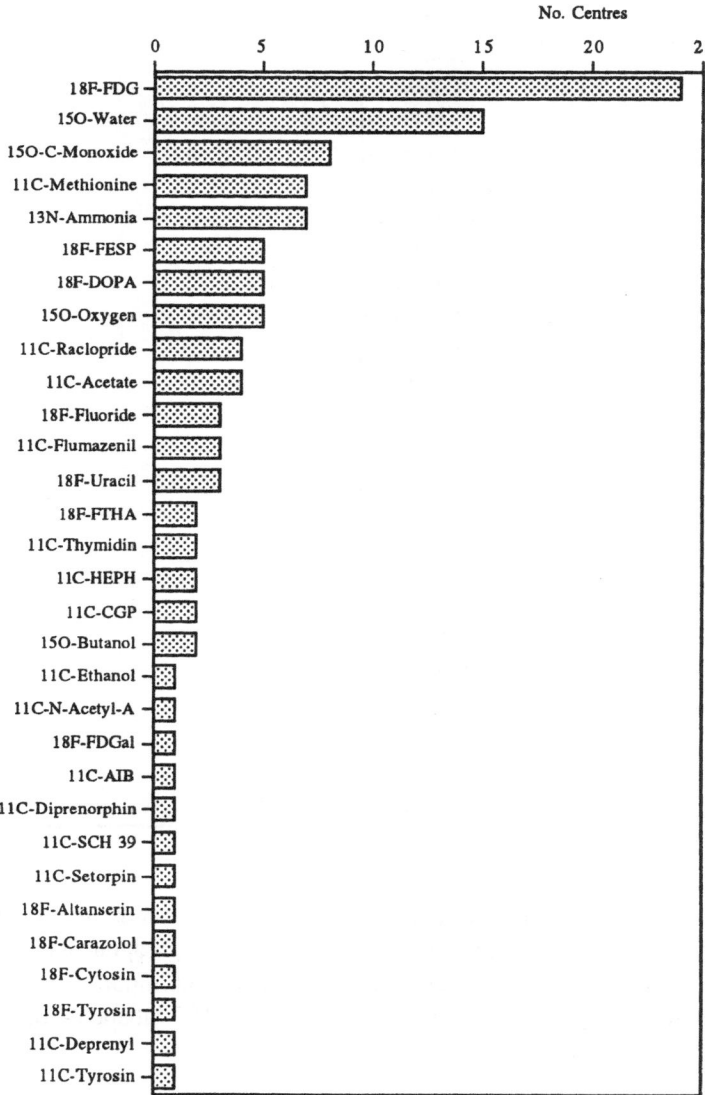

16

FIGURE 3.
The graph displays the number of Individual PET applications per week which are performed by the PET centres

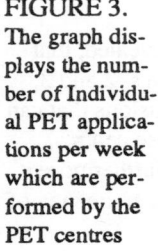

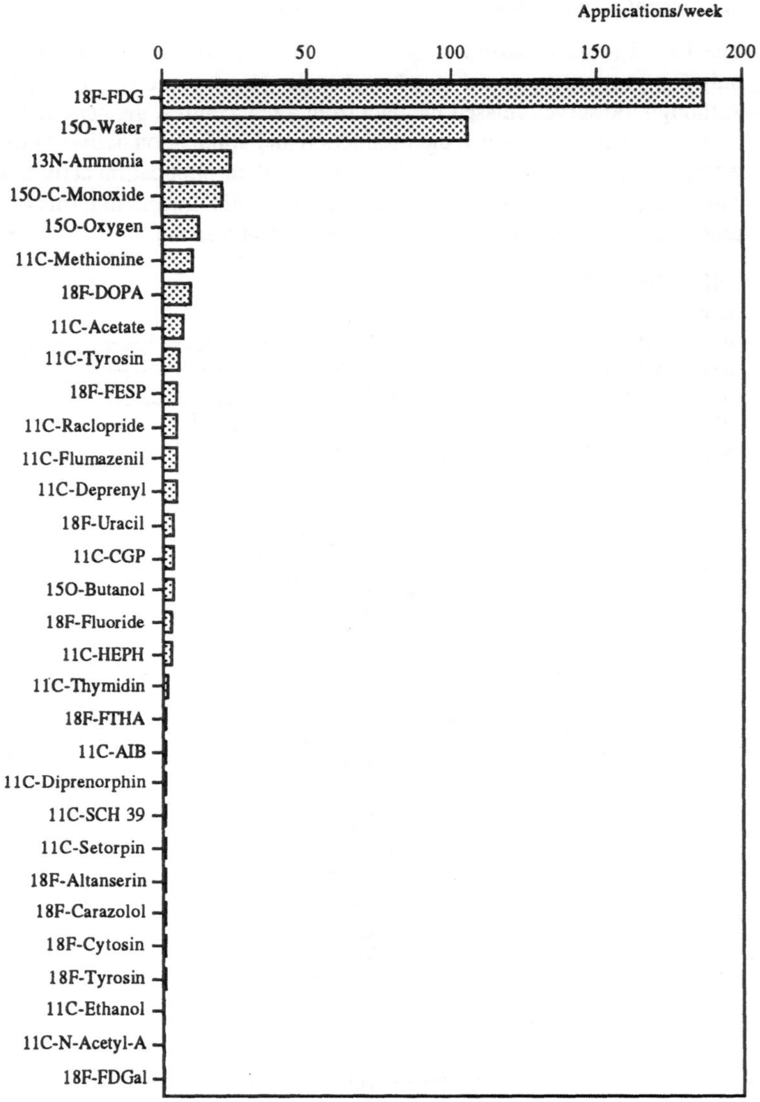

It is shown very clearly that ^{18}FDG is by far the most important tracer, followed by ^{15}O-labelled water. The latter has reached this high application frequency because it is used quite much in stimulation studies. Its clinical importance is not really matched by the frequency of its use. Interestingly no other ^{18}F-labelled product has yet reached any broad importance.

Since it is not easy to judge the relative importance of other tracers, a simple multiplication of the data shown in fig 2 and 3 has been performed, in order to gain a relative rating which covers the frequency of use as well as the spread of production. The graph illustrating this relative importance is given in fig. 4. It has been used to establish the order for the preparation of monographs in the Eur Pharm. at the priority rating.

FIGURE 4. Relative importance of PET tracers indicated by a frequency factor which has been obtained by multiplying the applications per week by the number of centres preparing the radiopharmaceutical on a routine schedule. The result requires a logarithmic scale.

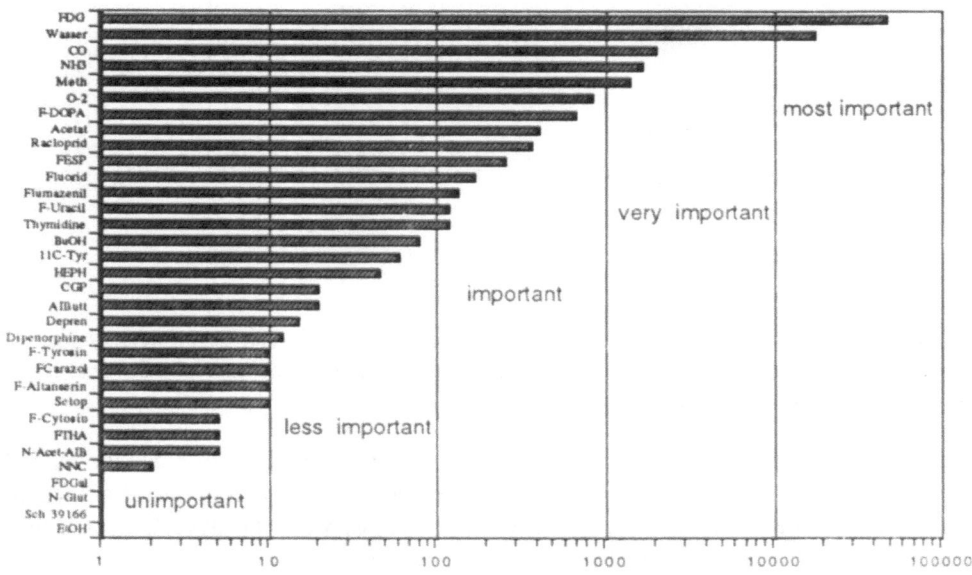

It is very probable that in the future the overall application frequency will correlate even less with the tracer production capability, when further clinical demand will concentrate the use of FDG especially in such PET locations, which depend on deliveries from outside.

4. Clinical usefulness of PET Radiopharmaceuticals

It remains difficult to judge the clinical usefulness of PET applications by the data which are currently available. Future follow-up of the use as provided by the last European sur-

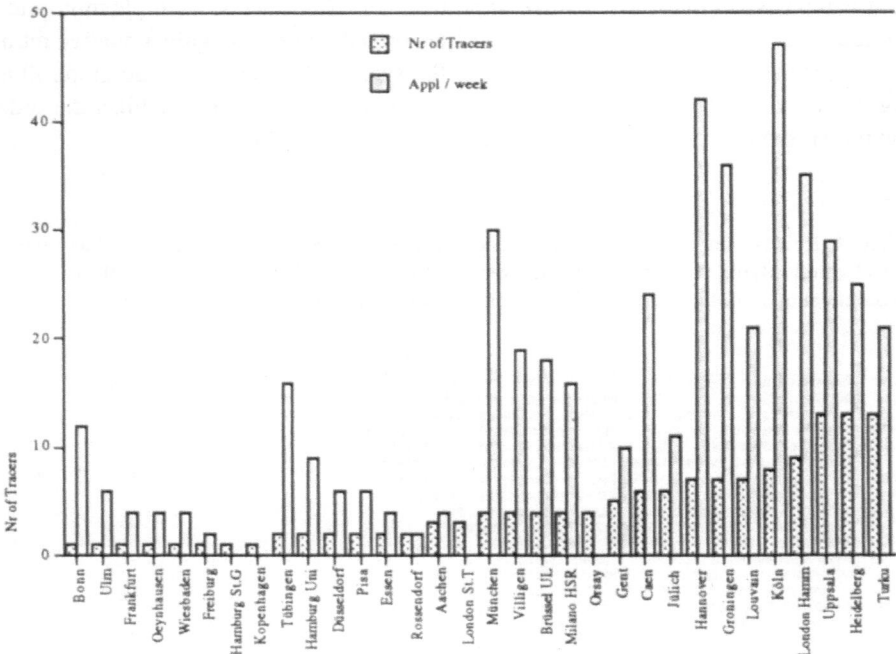

FIGURE 5. The graph displays a combination of the data from figs 2 and 3. This overlay illustrates that centres with a broad variety of tracers tend to be very busy, but that the number of regularly produced tracers is not necessarily an indicator.

vey needs to be planned in order to obtain data, which can be used to judge the overall acceptance of the methodology. Current trends can be seen only from partial aspects of the methodology or from geographically limited areas, were the development is fast enough, in order to see statistically relevant changes on a year to year basis.

One indicator has been presented from year to year by Henry Wagner during the SNM meeting. As shown in fig. 6 a,b, the yearly increase in PET presentations has been very steady.From these figures it can be seen, that the rising increase stems mainly from the presentations with ^{18}F-labelled tracers. It is clear that this is solely due to the number of FDG studies, which represent over 85% of all papers dealing with ^{18}F-labelled compounds.

Another indicator for the acceptance of the methodology can be derived from the development of PET in countries with a high technological standard and economic strength in the health care sector, like the USA, and Germany.

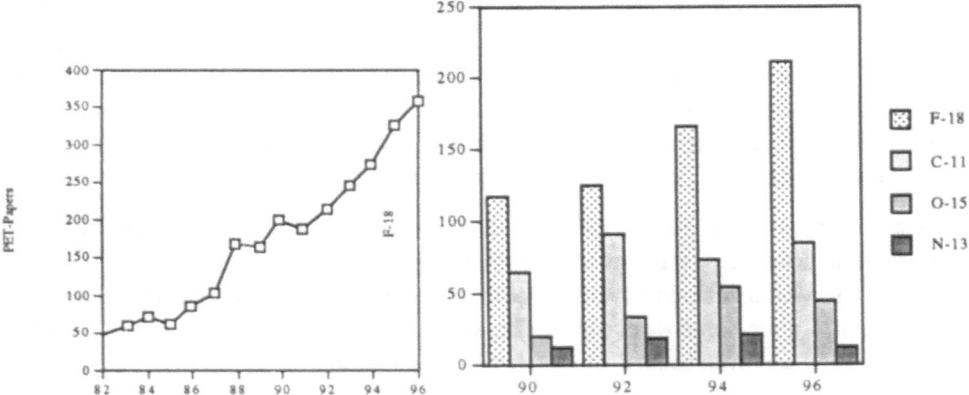

FIGURE 6 a. Number of PET presentations at SNM congresses.

FIGURE 6 b Number of papers dealing with PET tracers labelled with different PET-radionuclides

The development in Germany is shown in fig 7.

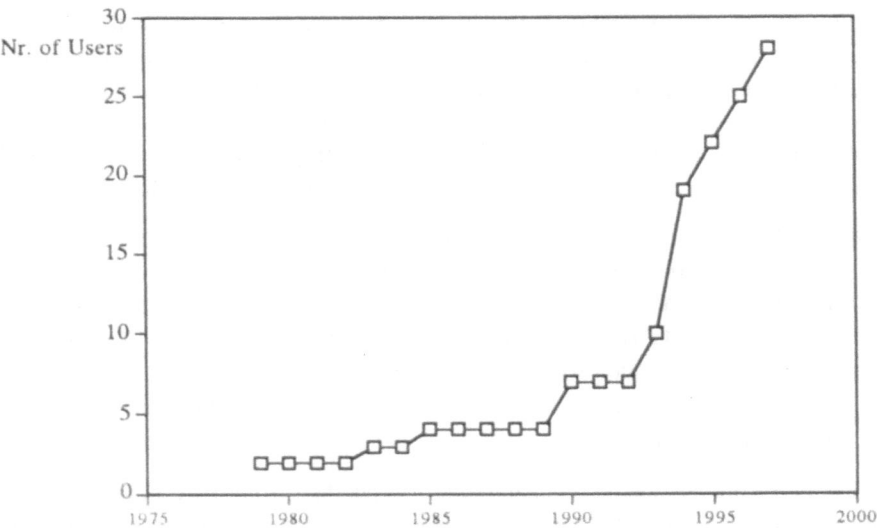

FIGURE 7. Development of the number of PET users in Germany during the past 12 years.

Although this trend is not that pronounced in the USA, a steady increase has been observed there also over the last 15 years.

It must be considered, however, that a broad clinical use of the PET methodology is not only heavily dependent on the availability of short-lived positron emitting radionuclides, i.e. cyclotron operations at large, but also from the economical and legal surrounding, governing the crucial aspects of reimbursement and the drug laws.

The complex interaction of these factors, which are quite different from country to country have influenced the development in all countries, and especially in the USA. It is clear by now, that the hampered development of meaningful regulatory systems, which has been caused to quite some extent by fights of different interest groups, has caused an enormous delay and even some decrease in the amount of investment in PET.

5. Legal Aspects

5.1 HISTORICAL DEVELOPMENT OF INITIATIVES FOR THE LEGALIZATION OF PET RADIOPHARMACEUTICAL PRODUCTION

Radiotracers for PET have been developed at special research centres, around cyclotron locations, were even chemists were usually hard to find. The radiochemists who actually started a new field of application for radioisotopes were not trained in pharmaceutical sciences at all. Their background was favourably in classical radiochemistry with an additional interest in organic chemistry. Further more any work in this new area was technically so demanding, that people needed a special interest in engineering also. This aspect, in combination with the limited marketing prospects, lead to a broad neglect of the PET tracer field by pharmaceutical scientists and the corresponding legal bodies.

Nevertheless with the spread of the method the awareness about the associated problems arose amongst the radiochemists themselves. In 1987 engaged radiochemists approached the USP in order to get an approval to prepare a monograph on FDG for the USP. The request was approved, and a first draft was prepared by 1988. In parallel a first international workshop on harmonization needs of the quality aspects of radiopharmaceuticals labelled with short lived radioisotopes was organized by the IAEA in Heidelberg (Germany) in 1989 (Vera Ruiz et al. 1990).

The outcome of this workshop indicated, that independent from the pharmacopoeial approach, more discussion on the quality issues was necessary. The reason for this can be summarized shortly as follows:

The quality of short-lived radiopharmaceuticals depends on the reliability and built-in quality assurance systems during synthesis, since conventional testing methods become inefficient or even obsolete. Although this principle was not totally new in the pharmacopoeial world, it starts only now to gain more respect also in conventional production of pharmaceuticals. For example, "parametric release", is a relatively new aspect in modern pharmacy, and was thought a "quite impossible approach" some years ago only.

In order to formulate the quality needs for PET radiopharmaceuticals in more detail, a task-group was formed within the Eur. Concerted Action on PET (ECA), which during 1990 - 1992 established practical guidelines for the quality control of PET radiopharmaceuticals. (Meyer et al. 1993).

In parallel to this action, a European Harmonization of Drug Laws became effective in 1991, which influenced the market of conventional radiopharmaceuticals substantially. The ECA quality control task group was quite concerned about the effects, which the European Harmonization could have on radiopharmaceuticals for PET.

In response to a memo, sent by the ECA task group to the European regulatory body (CPMP) in 1991 about the concerns regarding PET radiopharmaceutical regulations, the CPMP indicated, that it was interested in regulating only such products, which would require a marketing authorization, and that it would not see any needs to cover PET radiopharmaceuticals in the next future. Instead it was suggested to cover the regulatory needs by the European Pharmacopoeia.

Triggered additionally by national authorities and legislative bodies, this lead in 1992 to the formation of an expert group for PET radiopharmaceuticals within Expert Group 14 of the Eur. Pharm. Commission. This was ordered to produce monographs for ^{18}F-FDG, and ^{15}O-Water.

After the termination of the ECA task group in 1993 the Eur. Assoc. Nucl. Med. (EANM) formed a task group for the licensing of PET radiopharmaceuticals, which first continued and finished some paperwork of the ECA task group. Additionally the EANM task group organized a survey on the current use of PET radiopharmaceuticals in Europe and published the material in combination with dosimetry and toxicity data which are necessary for the formulation of specific product characteristics. (Meyer et al. 1995)

Preliminary data from this survey were used in 1994 by the Eur. Pharm Comm. for the priority rating and led to the authorization to produce four more monographs for PET radiopharmaceuticals, namely for ^{13}N-Ammonia, ^{15}O-Carbonmonoxide, ^{18}F-Fluoro-DOPA, and ^{11}C-Methionine.

The monograph for ^{18}F-FDG is now under revision for nearly three years. Although this seems to be a terribly long time, it is a quite normal processing time, especially in view of the difficulties which the expert group for PET radiopharmaceuticals was faced with, when it started to formulate a type of monograph which differed in many respects from the current scheme of radiopharmaceuticals. It is close to be finished now and will be published hopefully in the beginning of 1997.

After this pioneering PET-monograph, it is hoped that other PET radiopharmaceuticals will require a much shorter processing time.

5.2 LICENSING STRATEGIES

Besides the formulation of officially recognized quality standards for PET radiopharmaceuticals, a crucial legal issue remains the appointment of responsibilities in the preparation of these products. The issue is quite differently dealt with within different countries. Besides the fact, that most of the experience is with radiochemists in research institutions, few countries have recognized radiopharmacy as a profession only. A European harmonization in this area would be very desirable. A workshop for collecting the essential data, was held in 1990 already, which clearly formulated the future needs (Cox et al 1990). Political action, however, was difficult to initiate until 1996, when a postgraduate training system was established within the Eur. School of Nucl. Med.

Because of the complexity of the preparations, some national authorities tend to regard the preparation of PET radiopharmaceuticals under the legislative aspects of manufacturing, rather than pharmacy practice. The legal implications and difficulties, which stem from this standpoint have been reviewed recently (Cox and Meyer, 1995). The main focus has been the preparation of non-PET radiopharmaceuticals, which can not be prepared by hospital pharmacies, although they may be covered by the Eur. Pharmacopoeia. However this legal standpoint of the licensing bodies, hampers also that an "Orphant Drug Status" can be established, which could be helpful for PET radiopharmaceuticals as well.

The current legal status implies that all routinely used PET radiopharmaceuticals need to be licensed. The new European licensing authority, however, will only be interested in products, which are meant to cross inner European borders. This is not necessarily the case for many producers of ^{18}F-FDG even, - not to speak of other PET radiopharmaceuticals labelled with shorter lived radionuclides. Therefore, some national authorities tend to cover the preparation of PET radiopharmaceuticals under the laws of pharmacy practice. This controversial handling of the legal status of PET-radiopharmaceuticals is likely to persist for a couple of years.

National licensing is therefore the preferred route of legalization. This has not jet occurred in any country in the EU, except for Switzerland, where the situation is relatively simple, due to the fact that there is only one producer with two production sites. However, the licensing process is quite far advanced in Germany also, and will hopefully be settled within 1997.

The strategy developed so far includes the formation of a non-profit working group of PET-radiopharmaceutical producers within existing scientific bodies. Standard application files will be elaborated by some groups of producers, who will then after licensing of the standard procedure, modify the files to meet the requirements of other individual production sites. Producers with commercial interests will have to pay for the services of the working group.

In order to obtain a general production authorization for PET radiopharmaceuticals, federal authorities have to issue the necessary permits. A European harmonization in this respect is not in sight jet. The German and Swiss PET-radiopharmaceutical producers have, however, initiated actions on both a national and European level to harmonize production authorization. This initiative intends to set up general rules for the recognition of qualification in radiopharmaceutical chemistry for both radiochemists and pharmacists and to offer appropriate postgraduate training.

5.3   DIFFERENCES BETWEEN THE USA AND EUROPE

Although many of the problems discussed above have been observed in the USA as well as in Europe, some issues are distinctively different. The main difference stems from the corporate status of the US pharmacopoeia. Whereas the European Pharmacopoeia as well as several national ones present legally binding quality standards also for drug manufacturing , the USP presents references in pharmacy practice only.

Since nuclear pharmacy is a specialty in the US, it is understandable, that this specialty would like to obtain responsibility in the field of PET radiopharmaceutical production. On the other hand the US licensing authority (FDA) would like to see PET radiopharmaceutical production under manufacturing rules, i.e. under its own control. Since USP monographs on PET radiopharmaceuticals have been formulated quite simple and formally, the FDA has had a reason to argue about the quality control of PET radiopharmaceuticals. An end of this struggle is not jet in sight. Chances are good that the FDA will win this issue, especially since the Institute of Clinical PET (ICP) a professional lobbying association of PET-users and PET-Industry favours an arrangement with FDA standards.

The severe problem which continues to exist through all these years in the USA is, that reimbursement of PET procedures is linked to the acknowledgement of the PET radiopharmaceuticals by the FDA. This kind of link is less pronounced in many of the European countries and especially in Germany, were PET procedures are reimbursed regularly, even without a definitive general regulation.

5.4 EUROPEAN STATUS QUO AND FUTURE TRENDS

The latest advancements in gamma camera design and performance have led to high expectations regarding the future use of ^{18}F-FDG and possibly other ^{18}F-labelled radiopharmaceuticals. All gamma camera producers are currently developing double headed systems which allow coincidence measurement of 511 keV photons. Virtually all orders for high end equipment ask for this kind of option. These promising expectations will, however, only be met by a real increase of the clinical impact of nuclear medicine, when ^{18}F-FDG becomes readily available.

For a limited time it seems possible that the necessary production can be performed by research centres and university institutions. If these production facilities can launch the expected demand, industrial enterprises and joint ventures will follow automatically. Furthermore the research institutions will have to concentrate on the development of new tracers, leaving the routine production to commercially based cyclotrons.

Time must show, if ^{18}F-labelled radiopharmaceuticals can play a successful role in clinical nuclear medicine. As shown in this review, the obstacles are different from all earlier problems, which nuclear medicine has encountered in the past. Scientific and clinical evidence of the usefulness of these products is nearly overwhelming in such important areas as oncology, cardiology, and neurology. If the legal and logistic problems can be solved, ^{18}F-labelled radiopharmaceuticals will help nuclear medicine to establish itself a much sounder base than it has at the moment.

6.   References

Vera Ruiz, H., Marcus, C.S., Pike, V.W., Coenen, H.H., Fowler, J.A., Meyer, G.-J., Cox, P.H., Vaalburg, W., Cantineau, R., Lambrecht, R.M. (1990) Quality Control of Cyclotron Produced Radiopharmaceuticals. *Nuclear Medicine and Biology* 17, 445-456

Cox, P.H., Coenen, H.H., Deckart, H., Fueger, G.F., Hesslewood, S.R., Kristensen, K., Komarek P. Meyer, G.-J., Stöcklin , G., Schubiger, P.A. (1990)
Report and recommendations on the requirements for postgraduate training in radiopharmacy and radiopharmaceutical chemistry. *Eur. J. Nucl. Med.* 17, 203-211

Meyer, G.-J., Coenen H.H., Waters, S.L., Langström, B., Cantineau, R., Strijkmans, K., Vaalburg, W., Halldin, C., Crouzel, C., Maziere, B., Luxen, A. (1993) Quality Assurance and Quality Control of Short-Lived Radiopharmaceuticals for PET.
in: G. Stöcklin, V. Pike (eds) *Radiopharmaceuticals for Positron Emission Tomography - Methodological Aspects.* Kluwer Acad. Press, Dordrecht 1993, p 91 - 150

Cox, P.H., Meyer, G.-J. (1995) Radiopharmaceuticals 1994 *Nil desperandum*
Eur. J. Nucl. Med. 22, 563-570

Meyer, G.-J., Coenen, H.H., Waters S.L., Langström, B., Maziere, B., Luxen, A. (1995) PET Radiopharmaceuticals in Europe: Current use and data relevant for the formulation of specific product characteristics (SPCs) *Eur. J. Nucl. Med.* 22, 1420-1432

FDG DISTRIBUTION - A NEW MARKET TREND

S.R.LINDBÄCK
GEMS PET Systems AB
Husbyborg
752 29 Uppsala
Sweden

1. FDG in PET

The history of fluorine-18 labelled deoxyglucose (FDG) in positron emission tomography PET is almost as old as PET itself. The deoxyglucose method was developed by Sokoloff and colleagues[1] already in 1977 using ^{14}C-deoxyglucose in animal autoradiographic studies. Reivich and associates[2] first applied it in human studies with ^{18}F-FDG. FDG was found to have the attractive feature of being able to penetrate the blood -brain barrier while after metabolising, the compound could no longer diffuse through the cell membrane but was trapped in the cell and, hence, could be imaged by PET coincidence technique. In addition FDG has turned out to be one of the rather limited number of compounds in PET for which a good kinetic model has been developed which allows quantitative measurements to be made.

Today, almost 20 years later, FDG is still by far the most commonly used PET radio-tracer. In 1995 there were 44 PET centers in operation in Europe. One of the task forces reporting at the European Nuclear Medicine meeting in 1995 investigated the frequency of use of PET radio-tracers at 24 of these PET centres. The result showed that FDG was used almost twice as frequent as the second compound on the list ^{15}O-Water and almost 10 times more frequent than the third one on the list, ^{13}N-Ammonia. This clearly demonstrates the unique position of FDG in PET.

FDG has found very important applications in oncology, neurology and cardiology. It seems to be generally accepted by now that the future predominant application for clinical PET will be in oncology. Examples of applications include identification of primary tumour, detection of tumour recurrence, monitoring of treatment effectiveness, patient staging, avoidance of unnecessary surgery just to mention the most important ones.

2. A New Market for FDG?

Widespread use of PET has long been hampered by the big capital investment needed to acquire a cyclotron and/or a PET scanner and by the non-availability of PET tracers. Work in recent years on gamma cameras upgraded with dual head collimators and/or coincidence electronics (CD) for use with FDG, seems however, to present a part solution to these problems. However, poor spatial resolution in the case of dual collimators only will probably limit the use of this approach. CD on the other hand with a typical resolution of 6-8 mm will probably allow adequate wholebody scans, in spite of low sensitivity, in approximately one hour. This opens the avenue for applications in clinical oncology. Most major equipment suppliers can now provide or will soon be able to provide gamma cameras for use with FDG

Today most medium sized and large hospitals around the world have already gamma cameras that are upgradeable or they will probably be in a position to buy new ones as the price is only a fraction of that of a dedicated PET scanner.

The results from retrofitted or redesigned gamma cameras have also been judged enough encouraging for the radiopharmaceutical distribution industry to enter into a

B. Gulyás and H.W. Müller-Gärtner (eds.),
Positron Emission Tomography: A Critical Assessment of Recent Trends, 25–31.
© 1998 *Kluwer Academic Publishers.*

26

number of deals with existing or planned cyclotron sites for regional distribution of FDG, especially in the USA.

3. How to produce FDG

Production of FDG takes place in two subsequent steps. In the first step fluorine-18 is produced in a target connected to a small accelerator, by irradiating ^{18}O enriched water with protons. In the second step the irradiated target water is transferred to a chemistry module where the synthesis takes place that labels ^{18}F to deoxyglucose. Cyclotrons with proton energies ranging from 10-18 MeV are by far the most common accelerator for this application today but also lower energy linear accelerators of both RF and DC type have appeared on the market or have been proposed.

Typical target irradiation times range from one to two hours at 20-40 µamp beam current while FDG syntheses times range from 30-50 minutes.

Figure 1 shows the nuclear cross-section[3] for the commonly used (p,n) reaction. The resulting yield of ^{18}F as obtained by integration of the cross-section over energy is shown in figure 2 and has been expressed as percentage of the maximum possible yield which occurs at around 18 MeV.

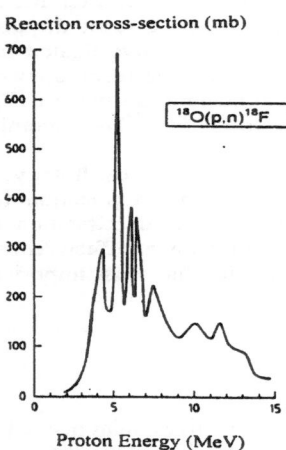

Reaction cross-section (mb)

^{18}O(p.n)^{18}F

Proton Energy (MeV)

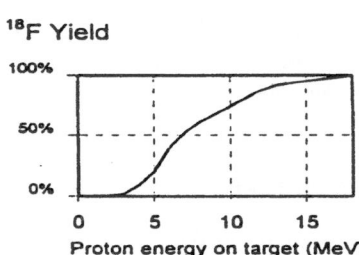

^{18}F Yield

Proton energy on target (MeV)

Figure 1. Nuclear cross-section for the reaction ^{18}O(p,n)^{18}F

Figure 2. Relative ^{18}F yield vs proton energy on target

Several commercial proton accelerators with FDG modules are available on the market today with varying capacity for ^{18}F production as summarised in Table 1.

TABLE 1. Summary of commercial accelerators for ^{18}F / FDG production

Type	MeV	Company	Yield ^{18}F EOB 1 hr	
Cyclotron	18	IBA	1500 mCi	56 GBq
Cyclotron	16.5	GE	2500 mCi	93 GBq
Cyclotron	13	EBCO	1000 mCi	37 GBq
Cyclotron	12	Oxford	750 mCi	28 GBq
Cyclotron	11	CTI	1000 mCi	37 GBq
RF Linac	7	AccSys	700 mCi	26 GBq

About 40-60% of the ^{18}F radioactivity subjected to FDG conversion comes out as labelled FDG at the end of synthesis. The conversion efficiency depends on choice of synthesis, input radioactivity level, synthesis time etc., to mention some of the factors involved.

Figure 3 below shows the General Electric PETtrace as an example of a dedicated cyclotron for large scale commercial production of ^{18}F, while figure 4 depicts the GE FDG MicroLab, a fully automated unit for production of FDG which makes use of a disposable kit.

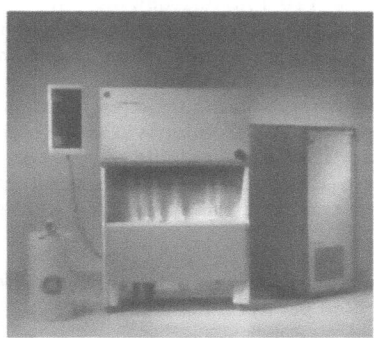

Figure 3. The PETtrace cyclotron from GE. Picture shows targets with functional support panel. The ^{18}F-target is the one at the lowest position in the picture

Figure 4. FDG MicroLab from GE for production of FDG using a disposable kit.

4. Production site layout

The logistics of production and distribution of FDG including quality control, dispensing and packing for shipment is shown schematically in figure 5. The cyclotron can either be placed in a concrete bunker or be installed with its own radiation shield. After irradiation the target water is pushed through a small bore tube, by applying helium overpressure, into the FDG chemistry module which is placed in a hot cell in the radiochemistry lab. At the end of the synthesis a small sample is drawn for quality control while the bulk of the production is brought to a laminar flow hot cell for dispensing of the prescribed number and size of doses to be delivered.

After dispensing the dose vials are packed and moved out to a pickup area for transport. Since time is of essence the transport will start even if the result of the quality control is not available. If a quality problem occurs the shipment has to be cancelled.

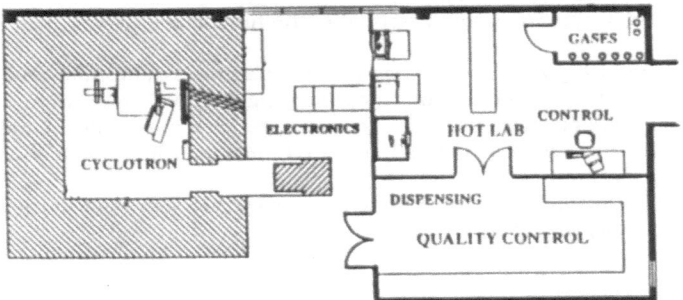

Figure 5. Schematic facility layout showing cyclotron bunker, hot lab and quality control room

5. Capacity of a dedicated distribution centre

In order to calculate the number of satellite sites that can be handled by a dedicated FDG distribution cyclotron certain assumptions have to be made. We will assume that the host site has its own dedicated PET scanner requiring doses for 6 patients per day. Each patient will need an injected dose of 10 mCi (370 Mbq) of FDG and the patient flow is set at one patient every 1.5 hr.

For the satellite sites we will assume only four patients (doses) per day but the same injected dose per patient and patient flow rate as for the host site. A simple calculation taking the 110 minutes half-life of ^{18}F into consideration gives the amount of FDG required at time of arrival to a satellite site, viz. 121 mCi (4.48 Gbq).

The following assumptions have been made for the preparation of doses and handling time on arrival to the site:

- Dispensing/packing time: 40 minutes
- Handling time from arrival to the site to injection: 10 minutes

In Table 2 we have calculated the corresponding amount of FDG needed at end of synthesis (EOS) for some different site distances as well as the approximate number of sites that can be supplied vs distance if 2000 mCi of FDG is available.

TABLE 2. Amount of FDG needed at EOS vs site distance and number of sites that can be supplied per 2000 mCi batch of FDG.

Distance	EOS	Time of Arrival	Number of Sites
(hr)	(mCi)	(mCi)	
0.5	188	121	10
1.0	227	121	9
1.5	274	121	7
2.0	331	121	6
3.0	484	121	4
4.0	706	121	3

Remotely located sites clearly put very tough demands on production capability.

In order to make a complete distribution scenario we also need to make some assumptions regarding the production of ^{18}F and FDG. The following assumptions have been made:

- Dual target irradiation will be utilised (two targets simultaneously). This simply doubles the yield of ^{18}F but not necessarily of FDG as the conversion efficiency tends to get lower at higher input levels of radioactivity.
- FDG synthesis time 35 minutes
- ^{18}F to FDG effective conversion rate at 5 Ci input: 40% EOS
- Start of first irradiation at 6:00 a.m.
- Start of a second irradiation immediately after the first one.

The scenario shown in figure 6 was created using the above assumptions and assuming satellite sites arbitrarily distributed at distances varying from 0.5 hour up to 3 hours. In order not to delay patient schedules unduly at remote sites those sites should receive delivery from batch number one. The in-house PET scanner will receive one dose only from the first batch.

As can be seen, all sites have received their doses by 11:15 a.m. and no patient study needs to take place after 5:25 p.m.

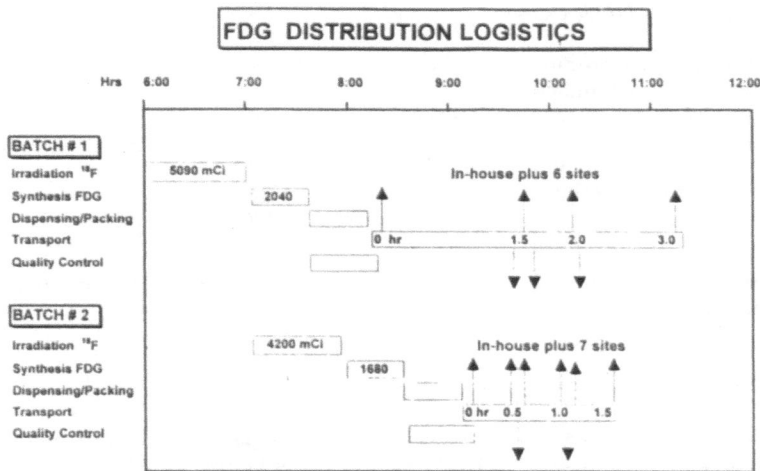

Figure 6. Example of FDG distribution from a dedicated production centre. All satellite sites get 121 mCi of FDG at time of arrival.

6. Production cost estimates.

The following estimates are to be seen as guidelines only.

6.1 STAFFING

With the assumption of an in-house PET scanner program the following staff must be considered a minimum:
- 1   Physician
- 1   Radiochemist (PhD)
- 1   Technologist (Chemistry)
- 1   Technologist (Electronics)
- 1/3 Administration

6.2 CAPITAL INVESTMENT

• Building	0.3 - 1.0 M$
• Cyclotron with Hot Lab	1.5 - 1.9 M$
• PET Scanner	1.3 - 2.4 M$
Total	3.1 - 5.3 M$

The cost range for the building indicated above is meant to reflect all from modifications of an existing building to a new building. Cost variations of cyclotron/hot lab depend on size of cyclotron, number of tracers/precursors purchased and number of hot cells and instruments for the hot lab. Finally, the choice of PET scanner will to a large extent depend on whether the in-house program is primarily clinical or research oriented.

6.3 FDG PRODUCTION COST

Assuming one batch production every day or 230 productions per year the following annual costs can be estimated:

- O-18 target water and FDG chemicals   50 k$
- Electric power consumption   7 k$
- Quality control consumables   5 k$
- Packing material   <u>25 k$</u>

Total   87 k$

In addition come transportation costs which are very site related so no attempt will be made here to estimate those costs.

7. Distribution Sites 1996/97

As mentioned earlier distribution of FDG has already started at least on a smaller scale in many places in Europe and the USA. Table 3 and 4 below list these places as they are known to the author today.

TABLE 3. European FDG Distribution sites in operation/planned.

In Operation 1996	Planned/Operation 1997-
• KFK Karlsruhe • KFA Jülich • USZ Zürich • Rigshospitalet Copenhagen • DESY Hamburg	• EuroPet Freiburg • Bonn • Amsterdam • Seibersdorf

TABLE 4. FDG Distribution sites in the USA in operation/planned.

In Operation 1996	Planned/Operation 1997 - 2000
• Tampa • Phoenix • New York • Chicago • CTI/Syncor at a number of existing cyclotron sites	• CTI/Syncor at a number of existing/new cyclotron sites

8. Summary

Regional distribution of FDG is becoming a reality. The ultimate success or failure depends on how well the new generation of gamma cameras equipped with coincidence electronics will meet clinical protocol requirements. If they do there is potential for a big growth in clinical PET and we will see a growing number of FDG distribution sites world-wide.

9. References

1. Sokoloff L., Reivich M., Kennedy C. et al (1977): The [^{14}C] deoxyglucose method for the measurement of local cerebral glucose utilisation: Theory, procedure and normal values in the conscious and anaesthetised albino rat. J.Neorochem 28:897.
2. Reivich M., Kuhl D.E., Wolf A. et al (1979): The [^{18}F]-fluorodeoxyglucose method for the measurement of local cerebral glucose utilisation in man. Circ. Res. 44:127.
3. Stöcklin G., Pike V.W.(eds.) (1993), Radiopharmaceuticals for Positron Emission Tomography, Kluwer Academic Publishers. Printed in the Netherlands, 1-43.

STATUS OF CLINICAL PET IN THE USA AND THE ROLE AND ACTIVITIES OF THE INSTITUTE FOR CLINICAL PET

PAUL D. SHREVE
University of Michigan Medical Center
Ann Arbor, Michigan
USA

Modern Positron Emission Tomography (PET) scanners were developed in the early 1970s by Kuhl (l, 2), Ter-Pogossian (3) and Phelps (4). The principal motivation for the development of the coincidence imaging technology was to facilitate the study of cerebral physiology using positron emitting tracers such as ^{15}O oxygen and ^{15}O water. Studies of cerebral blood flow and glucose metabolism were almost exclusively the focus of PET imaging until the 1980s when PET tomographs capable of accommodating the torso permitted extension of positron based tracers to cardiology. By the early 1990s studies of glucose metabolism in malignant neoplasms throughout the body emerged. Because PET was developed principally to answer scientific questions in an academic setting rather than provide clinical patient care, as was largely the case with CT and MRI, PET imaging remained in the province of academic institutions, considered primarily a research tool rather than a clinically useful technology. Potential clinical applications were nevertheless recognized and developed in neurology in the early 1980s, cardiology in the mid 1980s and oncology in the early 1990's. Localization of epileptic foci, assessment of brain tumors, and differentiation of recurrent primary brain neoplasms from radiation necrosis were seen as useful applications of PET in clinical neurology. In cardiology, measurement of myocardial blood flow and determination of myocardial viability in zones of non-contractile myocardium were recognized as suitable clinical applications. Finally, identification and characterization of malignant neoplasms throughout the body gained recognition as a key clinical application. The growing roster of clinical applications of PET were so encouraging to many working in the field that in 1991, an entire issue of the Journal of Nuclear Medicine was dedicated to Clinical PET, with a cover title announcing: "Clinical PET - It's Time Has Come" (5). The number of PET installations in the USA continued an upward trend, but then in the mid 1990s began to level off (Fig. 1). In contrast, the pace of new PET installations

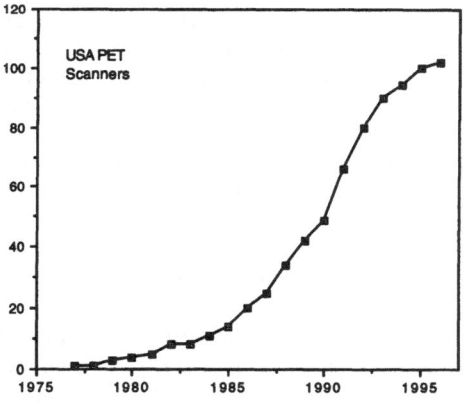

Figure 1. Number of active PET scanners in USA

33

B. Gulyás and H.W. Müller-Gärtner (eds.),
Positron Emission Tomography: A Critical Assessment of Recent Trends, 33–42.
© 1998 *Kluwer Academic Publishers.*

in Europe and Asia are accelerating. Given these recent developments, one is inclined to ask, "Why hasn't the 'time for clinical PET' fully arrived in the USA?"

The recent drop off in new PET installations in the USA does not reflect a lack of enthusiasm for clinical PET, indeed articles in the scientific literature dealing with PET applications in clinical oncology continue to expand both in numbers and scope. Rather, several impediments to clinical PET in the USA have slowed its growth and dissemination. These impediments include: the complexity and cost of establishing and operating a PET center, the convoluted and ever changing regulatory environment in the USA, the resistance among indemnity insurance plans to new technologies, the indifference within the radiological community to nuclear medicine in general, and PET in particular, and the insular nature of some aspects of academic medicine.

Complexity and Cost of PET

The complexity and cost of establishing and operating a PET center has often been cited as a major impediment to clinical PET. In the 1980's the initial investment, which necessarily included a cyclotron, radiosynthesis laboratory, and PET scanner with required site work was in the 5 to 7 million US dollars range. This, in comparison to the purchase and siting of a CT or MRI scanner, at that time in the range of 0.5 to 1.5 million dollars, was daunting. More so than even the steep initial capitalization, the operating expenses for a PET center were unusually high due to personnel requirements. At that time, it was customary for a PET center to have a cyclotron operator, a chemist, physicist, two technicians as well as a physician and often a unit manager on the personnel roster. These expenses became even more extreme considering typical PET facilities performed two to four scans per day. Low exam volume translated into high unit costs per scan, insuring PET the stigma of being the most expensive of imaging technologies. Early PET centers thus were confined almost exclusively to academic medical centers or affluent acute care hospitals with academic aspirations. Nearly all PET centers in the USA had some component of research with only a handful of centers dedicated to purely clinical operation. In many instances there was an anticipation that the initial mixed academic research and clinical patient load of a PET center would evolve into predominately or exclusively a clinical operation with the advent of reimbursement for the clinical studies. This however did not materialize due to an evolving regulatory environment and major changes in the structure and economics of the health care in the USA in the late 1980's.

Regulatory Impediments

A key impediment to approval of payment for clinical PET studies has been the regulations surrounding the PET radiopharmaceuticals themselves. Government approval for routine clinical use of CT and MRI rested on the safety and efficacy of the imaging devices themselves. Approval for PET, in contrast, has required first government regulatory approval of each PET radiopharmaceutical to be employed in a clinical application of PET. Uncertainty regarding whether and how the federal government would be regulating the production and use of PET radiopharmaceuticals stalled evaluations for payment of specific PET studies as both the federal government agencies including HCFA (Health Care Financing Administration) and private insurance carriers refused to make decisions concerning payment for clinical PET studies until the federal Food and Drug Administration (FDA) clarified whether it would regulate such radiopharmaceuticals and approve them for clinical use. Radiopharmaceuticals did not exist when the legislation creating the FDA was written in 1906 and further amended in 1934, but because both reactor byproduct radiopharmaceuticals and cyclotron produced radiopharmaceuticals (^{201}Tl, ^{67}Ga, ^{111}In) involved interstate commerce in their manufacture and sale, the FDA had authority to regulate them as new drugs introduced into interstate commerce. In fact, the FDA declared in 1973 that all radiopharmaceuticals would be considered drugs subject to FDA regulation. Due to the short physical half-life, in many instances PET radiopharmaceuticals would be manufactured and compounded at the site of use, and not directly involve interstate commerce. Some in the USA PET community advanced an argument that the production and compounding of PET radiopharmaceuticals falls within the practice of pharmacy to be regulated by individual state pharmacy board, which is exempt from FDA regulation. Bureaucratic and legal maneuvering consequently dominated the late 1980s regarding which government agencies, if any, would or could regulate PET radiopharmaceuticals even while active efforts to obtain reimbursement for clinical PET scans were underway.

As early as 1987 the Society of Nuclear Medicine and the American College of Nuclear Physicians (6) as well as The Committee on Advanced Cardiac Imaging and Technology of the American Heart Association (7) determined that PET was clinically useful for assessing epileptic foci, for the differentiation of recurrent primary brain neoplasms from radiation necrosis, for early diagnosis of Alzheimer's and related dementias, for assessment of myocardial viability, and based on a 1989 workshop sponsored by the National Cancer Institute, the diagnosis of malignant neoplasms. The health care finance administration of the US Federal Government was petitioned in 1989 for reimbursement of selected PET clinical studies. HCFA however deferred its decision and requested the office of Health Technology Assessment to review the available data and assess the clinical efficacy of PET. The office of Health Technology Assessment however would not release its assessment until radiopharmaceuticals had been approved by the FDA. The FDA had approved a ^{82}Sr/^{82}Rb generator and HCFA approval payment for myocardial perfusion PET scans performed with ^{82}Rb, however FDG, the principal tracer for the clinical PET studies, had not been approved. Meanwhile in 1990 the Health Insurance Association of America held forums on clinical PET and some private insurance carrier reimbursement for PET scans was emerging. The absence of federal government reimbursement through Medicare and Medicaid and of FDA approval for FDG was nonetheless considered by many insurance carriers *ipso facto* evidence that PET scans were still investigational, and therefore would not be considered for reimbursement.

Various professional societies in the United States with interests in clinical PET such as the Society of Nuclear Medicine, American College of Nuclear Physicians, American College of Cardiology, American College of Radiology as well as corporations with a commercial interest in clinical PET were approaching the regulatory difficulties from different directions. For example, Syncor Corporation, a major distributor of radiopharmaceuticals with an interest in the potential business of distributing PET radiopharmaceuticals, filed a lawsuit against the FDA challenging the FDA's jurisdiction over the regulation of FDG. The lawsuit was joined by The American College of

Nuclear Physicians, but not other key professional societies. Despite the suite, the FDA ruled in April 1995 that it had jurisdiction over PET radiopharmaceuticals and would move to regulate them. The FDA further stipulated that every type of PET radiopharmaceutical made at each cyclotron production site required a New Drug Application (NDA) and all cyclotron facilities would have to conform to Good Manufacturing Practice standards (GMP). This created a great deal of consternation within the PET community because initially the FDA had stated, recognizing the unique nature of PET radiopharmaceutical production and use, the onerous regulatory requirements and record keeping associated with GMP (originally to deal with large pharmaceutical production facilities) would not be directly applicable to PET cyclotron facilities. Such exceptions were not specified in the FDA announcement, which further stipulated that compliance must be achieved within 18 months. There was widespread fear in the USA PET community that clinical PET could be buried under an avalanche of regulatory burdens, and the anticipated added regulatory related expenses would only further elevate the cost of clinical PET.

Indemnity Resistance

To further complicate the impasse of reimbursement through government health programs Medicare and Medicaid, reimbursement through private insurance carriers was variable and in some cases unpredictable. In the late 1980's and early 1990's private medical insurance remained primarily on a fee for service, cost based, system known as indemnity plans. Due to the explosion in medical procedures, not incidentally including clinical imaging procedures, there was great resistance among indemnity insurance carriers to any new imaging technology. Imaging technologies were seen by the insurance carriers, in some respects erroneously, as a principal source of health care cost inflation at a time when health care costs in the United States were spiraling at rates as high as 20% annually. Actually, a major underlying contribution to the rapid escalation in the total cost of diagnostic imaging was not the technology itself, but a rapid escalation in the 1980s of self referral practice patterns for imaging procedures. For example, MRI was not the highest diagnostic imaging expense for indemnity based government health care, rather cardiac echo procedures, which proliferated with the surging tide of cardiology specialists who performed them, became the single most expensive category. Nevertheless, high technology imaging procedures were a more visible target than questionable physician practice patterns. Accordingly, the indemnity insurance community feared a potential repeat of the notorious 1980s MRI scanner proliferation if PET imaging were unleashed on the practicing medical community. Both a pattern of clinical "piling on" and the "woodwork" effect seemed inevitable if payments for PET scans became routine. "Piling on" refers to the tendency of new diagnostic tests to be used as a supplement to existing tests rather than a substitute. This was observed, for example, when MRI scans were performed in addition to CT scans rather than replacing CT scans. The "woodwork effect" is a phenomena insurance plans or entitlement programs experience when a new benefit becomes available; people come "out of the woodwork" to claim the new benefit. Again the MRI experience was instructive. Initially approved for the imaging diagnosis of multiple sclerosis, indications for an MRI scan rapidly unraveled in the 1980s as people with headaches or any vague neurologic symptom increasingly demanding MRI scans. Thus the standard for proof of the diagnostic utility and clinical value of PET repeatedly was edged upward by payers understandably weary of new imaging technologies slipping by cursory or even non-existent effectiveness assessments (8,9).

Radiology Community Indifference

In addition to a climate characterized by regulatory uncertainty and payment resistance by both private insurance carriers and the federal government, there was to some extent an indifference to the possibilities of clinical PET in the radiology community. In the USA, most nuclear imaging procedures are performed by diagnostic radiologists. The vast majority of academic nuclear medicine is within a diagnostic radiology department. Nuclear medicine was an exciting discipline which attracted the scientifically oriented academic radiologists in the 1950s and 1960s. Beginning in the 1970s, wave after wave of powerful anatomic imaging technologies, CT, ultrasound, and MRI, fostered rapid growth in the numbers and status of diagnostic radiologists. These new developments in anatomic imaging eclipsed and even obviated many radiotracer based imaging procedures, gradually isolating nuclear medicine within the diagnostic radiology community to the fading world of "unclear" medicine. PET represented a powerful new extension of the tracer method to diagnostic imaging, and increasingly proved capable of providing substantially superior diagnostic accuracy over anatomic imaging methods in several important clinical applications. Because PET was largely developed for research purposes, however many radiologists continued to view PET as a research tool for neuroscientists and the like rather than a practical clinical imaging modality. Likewise, aided by effective promotion by manufacturers there were high expectations for MRI in the late 1980's. Many radiologists believed MRI would answer nearly any clinical question, allow for "knifeless biopsies", render CT obsolete, and the like. Little enthusiasm existed for heavy financial outlays for PET scanner/cyclotron when echo-planar MRI and spiral CT emerged as the next iteration for the legions of young radiologists trained in these modalities. In many academic and private practice settings those with an interest in radiotracer imaging simply could not persuade their colleges much of a future existed for Nuclear Medicine and in particular for clinical PET.

Academic Isolation

Finally, in part because PET was developed primarily in academic centers as a tool for basic scientific investigation, clinical PET imaging suffered from the academic isolation. CT and MRI technology were rapidly developed by industry for clinical applications. Propelled by potent industry sales and promotion efforts and nurtured in the cost plus reimbursement milieu of the 1970s and 1980s, these technologies rapidly diffused throughout the practice setting. In contrast, at the same time PET remained isolated in academic departments, willingly plied as an esoteric research tool by those largely involved with PET. This in part reflected patterns of interest among those in academic medicine. Many of the leaders in nuclear medicine had entered the field in the 1950s and 1960s because the tracer method was central to so much of basic biomedical research. In some respects nuclear medicine was populated by the scientifically rather than a clinically oriented radiologist and internist. Diagnostic radiology in general however evolved as a service academic department, with clinical diagnostic imaging seen as the predominant focus. An evolving outsider status of nuclear medicine within diagnostic radiology was further exacerbated by an institutional schism between nuclear medicine as a specialty, and nuclear radiology as subspecialty of diagnostic radiology, which had existed since the American Board of Nuclear Medicine was founded in 1973. Clinically oriented diagnostic radiologists viewed PET as an expensive scientific instrument their nuclear medicine colleges used for research which was often perceived as remote from practical patient care. Moreover, in some cases those who used PET for such esoteric research tended to view clinical PET with suspicion: clinical PET scans would occupy scanner time, potentially crowding out productive scientific work. Academic fiefdoms formed at many institutions with PET viewed by those working in the field as a source of academic productivity, something to be kept isolated from diagnostic radiologists with clinical interests. Manufacturers of PET technology consequently found few PET users in the USA with broad interest or expertise in clinical patient care to advise and encourage them to develop PET related technology best suited for clinical applications, a process essential for reducing the cost and expanding the availability of clinical PET to patients.

Formation of ICP

Due to the slow pace of development of clinical PET and the various impediments described above, the Institute of Clinical PET (ICP) was founded in January 1991. The ICP was founded as a non-profit educational foundation with the explicit purpose of making clinical effective positron imaging imaging available to patients. Its function was to bring physicians from different specialty backgrounds, scientists, academic institutions and industry with an interest in clinical PET together. While the initial purpose was to move clinical PET forward in the USA, the ICP has gown into an international organization for clinical PET and currently has over 300 institutional and individual members worldwide.

ICP Role and Activities

The ICP is a private organization, funded by individual, institutional, and corporate member dues as well as corporate and institutional grants. Making clinically effective positron imaging available to patients is the overall mission of the ICP. The term positron imaging is used to be inclusive of the rapidly evolving configurations of positron radiotracer imaging technology. The ICP promotes clinical applications of positron imaging by developing educational programs for professionals and by increasing public awareness of the value of positron based radiotracer imaging. The ICP also works to obtain government approval for the clinical use of positron radionuclide tracers and reimbursement for clinical positron imaging. To this end the ICP represents USA PET centers before US government agencies and coordinates multi-center clinical trials employing clinical PET.

In addition to hosting an annual scientific meeting, the ICP acts as a clearing house to disseminate information to both professionals and the public on the clinical utility and cost effectiveness of PET imaging. Some examples of ICP publications include the full NDA (new drug application) with Appendix for PET tracers such as FDG, the Drug Master File for FDG, the costs of clinical PET, FDA PET workshop transcript, FDA public hearing transcript, medical imaging advisory committee transcripts, a videotape of cardiac PET applications and a series of monographs concerning the clinical application and economic implications of PET. For individual members the ICP provides a quarterly newsletter describing the status of active ICP initiatives, up to date information on current activity of reimbursement for clinical PET in the USA, notice of upcoming meetings of interest to the PET community, and proceedings of the international PET conference held annually in October. There is also an International Council to share clinical and regulatory information amongst countries.

Presently the ICP address is:

> 11781 Lee Jackson Memorial Highway, Suite 360
> Fairfax, Virginia 22033 USA
> Voice 703-691-2255
> FAX 703-691-2259

The ICP address for E-Mail is: peticp@aol.com

There is also an award winning WEB site at http://www.icppet.org.

ICP Activities and the Complexity of Cost of PET

Since its inception, the ICP has worked to lower impediments to clinical PET on a variety of fronts. For example, to address the complexity and cost of clinical PET, the ICP has engaged industry members to develop more efficient and lower cost PET technology. Already these efforts have resulted in a new generation of clinically oriented PET technology. Linear accelerator based methods of positron radionuclide production are under evaluation. A new generation of cyclotrons, dramatically smaller and inexpensive to operate, and highly automated for routine production of the key PET tracers such as FDG have been introduced. Collaboration between cyclotron manufactures and radiopharmaceutical distributors has resulted in an impending network of production/distribution sites in the USA to provide FDG to hospitals and clinics, eliminating the need for on site cyclotron/production for FDG imaging. This brings the initial investment for PET well into the range of other diagnostic imaging modalities, particularly as newer PET scanners dedicated to clinical applications are available for roughly one million US dollars. In fact as shown in Figure 2, the purchase price range of a dedicated PET tomograph is similar to MRI, fluoroscopy suites, and even overlaps high end CT scanners. Further, a new series of dual purpose hybrid SPECT/PET devices, based on conventional dual head Anger

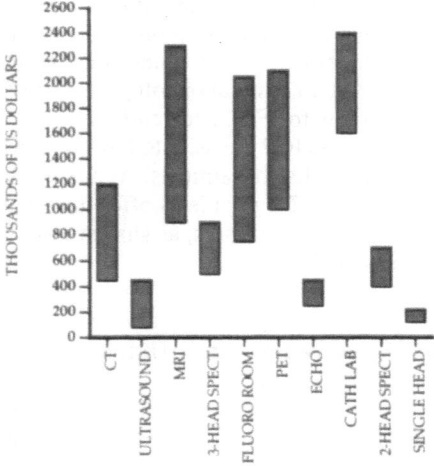

Figure 2. Comparative Costs of Imaging Equipment

gamma cameras, which will sell for less than seven hundred thousand US dollars will bring to positron imaging a broad range of prices and capabilities. Nuclear medicine equipment manufacturers, radiopharmaceutical distributors and established PET companies have formed an industry committee which meets quarterly with the ICP Board of Directors in cooperation with the ICP Business Plan. The underlying goal of the ICP business plan: make clinically effective positron imaging available to patients.

ICP Activities and Regulatory Impediments

Since regulatory approval of PET radiopharmaceuticals has been among the most important impediments to clinical PET, the ICP has worked to coordinate academic and industry efforts to deal with the FDA, HCFA and other US federal government agencies. ICP has served to represent the diverse practice, academic and industry interests as one voice to these government agencies, providing a unified front and demonstrating to the federal agencies a substantial constituency for clinical PET exists. For example the ICP spearheaded work on a Drug Master File for FDG, standardizing and codifying FDG production and formulation in anticipation of FDA regulations. In part due to ICP efforts, FDG has been approved for clinical use in brain PET imaging for localization of epileptic foci, and the ICP is working vigorously to extend FDA approval of FDG for cancer imaging. To facilitate better communication and understanding between the PET community and the federal regulators, the ICP has sponsored multiple workshops, seminars and joint training sessions, usually at the ICP International PET conference. At the recent ICP International PET Conference, the ICP organized a ICP/FDA Joint Regulatory Session where PET/cyclotron operators could interact directly with ranking FDA regulators and discuss the implications of the newly promulgated FDA rules and the timetable for compliance (transcript available to ICP members). In response to the 1995 decision by the FDA to regulate PET radiopharmaceuticals, the ICP initiated development of drug master files for PET tracers of clinical interest and is coordinating the submition of group amended new drug applications (ANDAs) by clinical PET centers. The ICP has undertaken surveys to determine current usage of the different radiopharmaceuticals employed in clinical applications of PET. The ICP Clinical Validation Task Force is coordinating the submission of appropriate data concerning FDG PET in clinical oncology to the FDA in an effort to broaden the clinical indications for FDG, currently approved only for the detection of brain epileptic foci. Further, the ICP has assisted industry with approval/reimbursement of specific devices or PET radiopharmaceuticals. While the focus on the regulatory front has been in the USA, the ICP's goal is to offer the materials used in conforming to US regulations to international members, as similar issues are being raised by regulatory agencies in Europe and Asia.

ICP Activities and Indemnity Resistance

To meet the indemnity resistance, the ICP has coordinated clinical trials to demonstrate the cost effectiveness of clinical PET and spearheaded education campaigns for patients, unions and company benefits offices, and insurance carriers in general. Several publications detailing the cost effectiveness of clinical PET have been completed and are available through the ICP publications office. It is of particular interest that in the

USA, managed care organizations have been more receptive to clinical PET than the indemnity insurance carriers or the federal government, presumably reflecting the ability of such organizations to control utilization of new technologies and integrate innovative patient care algorithms quickly and relatively uniformly in their patient population. Recognizing this, the ICP established a Managed Care Task Force, in a focused effort to educate managed care entities and the employers who purchase health plans regarding how clinical PET can be used to both reduce costs and improve patient management. Individuals have been recruited to work directly with specific managed care plans, corporate health care plans and trade unions. In addition sourcebooks containing information on PET providers nationwide, and current data and journal articles relevant to different clinical circumstances where PET is clinically useful, and cost information have been compiled. Again while the ICP's efforts have been chiefly focused on US health care providers and purchasers, the information will be largely germane to international members working to convince their government and private payers of PET's value.

ICP Activities and Radiology Community Indifference

The ICP from its inception was formed by various specialty groups, including diagnostic radiology. Many prominent members of the ICP are diagnostic radiologists, are faculty in academic radiology departments, or are in private practice radiology groups. As an organization, the ICP has acted to harmonize activities among major US professional organizations with an interest in clinical PET, including The Society of Nuclear Medicine, American College of Nuclear Physicians, American College of Radiology, Amercian Academy of Neurology and the American College of Cardiology. Further, ICP members have worked to promote results of clinical PET studies in the radiological literature and at radiology meetings and conferences to educate radiologists regarding the clinical utility of PET. The same databases used to convince government and payers of PET's usefulness and superiority over anatomic based imaging methods in appropriate clinical circumstances have been advanced to the radiology community. Many of the indications for clinical PET, such as the solitary pulmonary nodule, are well known in the radiologic community to be inadequately characterized by anatomic based imaging methods such as CT or MR. In targeting such applications of PET, and providing cost effectiveness and outcomes data, the ICP has facilitated a realization among radiologists that PET is not dead, or just for research, but has come of age as an important part of the diagnostic imaging armamentarium.

ICP Activities and Academic Isolation

Since making clinically effective positron imaging available to patients is the overall focus of the ICP, its function is effectively to bring PET technology out of academic isolation. Bringing academic scientists and physicians together with practicing radiologists, nuclear physicians, cardiologists and neurologists and focusing on issues related to clinical use of PET has broken down many perceptual barriers relating the complexity and expense of performing PET scans. By promoting industry efforts to improve PET technology, resulting in less expensive and simpler imaging and positron radiopharmaceutical production hardware, and disseminating expertise and strategies for reimbursement, ICP efforts have rendered academic fiefdoms less relevant; the growing prospect of a substantial clinical revenue stream from clinical PET likely will reorder priorities in many academic centers.

Summary

Clinical PET in the USA has not grown as rapidly as initially expected due to both anticipated and unanticipated impediments. Through efforts directed by organizations such as the Institute for Clinical PET, these impediments are now diminishing, and sustained growth in positron based radiotracer medical imaging in US clinical practice appears imminent.

References

1. Kuhl, D.E.and Edwards, R.Q. (1963) Image separation radioisotope scanning, *Radiology* **30**, 653-661.

2. Kuhl, D.E., Edwards, R.Q., Ricci, A.R., *et. al.* (1973) Quantitative section scanning using orthogonal tangent correction, *J Nucl Med* **14**, 196-200.

3. Ter-Pogossian, M.M., Phelps, M.E., and Hoffman, E.J. (1975) A positron emission transaxial tomograph for nuclear medicine imaging (PETT), *Radiology* **114**, 89-98.

4. Phelps, M.E., Hoffman, E.J., Mullani, N.A., *et. al.* (1975) Application of annihilation coincidence detection to transaxial reconstruction tomography, *J Nucl Med* **16**, 210-223.

5. Wagner, H.N., Jr. (1991) Clinical PET: Its time has come, *J Nucl Med* **32**, 561-564.

6. ACNP/SNM Task Force on Clinical PET (1987) Positron emission tomography: Clinical status in the United States in 1987, *J Nucl Med* **29**,1136-1143.

7. Bonow, R.O., Berman, D.D.S., Gibbons, R.J., *et. al.* (1991) Cardiac positron emission tomography. A report for health professions from the Committee on Advanced Cardiac Imaging and Technology of the Council on Clinical Cardiology, American Heart Association, *Circulation* **84**,447-454.

8. Cooper, L.S., Chalmers, T.C., McCally, M., *et. al.* (1988) The poor quality of early evaluations of magnetic resonance imaging, *JAMA* **259**,3277-3280.

9. Chalmers, T.C. (1988) PET scans and technology assessment, *JAMA* **260**,2713-2715.

Part two:
PET: Novel methodological approaches

RECENT TRENDS IN PET CAMERA DESIGNS

D.L. BAILEY
MRC Cyclotron Unit, Hammersmith Hospital
London. UK

1. Introduction

The development that has had the greatest influence on PET camera design in the last five years has undoubtedly been the introduction of full volume, septa-less acquisition and reconstruction. 3D PET, as it is now commonly referred to, was developed in the search for increased sensitivity and grew out of large area gas detector PET systems [1]. It was based on the original work of 3D reconstruction from electron micrographs of biomacromolecules by Vanstein and Orlov [2,3]. The translation of these ideas to multiring bismuth germanate (BGO) tomographs was first implemented at the Hammersmith Hospital in a collaboration with groups from Geneva and Brussels, encouraged by the manufacturer CTI [4], and was quickly followed by others [5,6]. This has realised an increase in sensitivity of around fivefold for most multiring systems. It has allowed the development of full-time 3D systems based on BGO or NaI(Tl) detectors. The impact of 3D PET has been dramatic. In neuroscience, statistical mapping of 3D rCBF "activation" studies in single subjects, rather than groups, became possible. The duration over which ^{11}C-labelled radiotracers could be studied was greatly increased and the quality of each datum was improved, which aids in better modelling and functional mapping results. At this point in time, while application of 3D PET to other regions of the body has been more protracted, it is nevertheless making slow progress [7,8,9].

Apart from this major initiative, PET cameras have continued to evolve slowly rather than change dramatically. Much focus has been placed on the clinical market, and systems have become far more reliable and the data more accessible. This is not to say that dramatic events are not unfolding, rather that, at present, many exciting new developments have not reached commercial production. New scintillators, new light collection devices, animal scanners, faster electronics and computing, and higher resolution detectors are all on the near horizon.

2. Current Status of PET Camera Developments

Today, the vast majority of BGO based multiring systems are capable of both 2D and 3D acquisition. The is achieved by having automatically retractable collimation, or

B. Gulyás and H.W. Müller-Gärtner (eds.),
Positron Emission Tomography: A Critical Assessment of Recent Trends, 45–56.
© 1998 *Kluwer Academic Publishers.*

septa. This in itself is a complication in terms of gantry design, engineering, and reliability. It has, though, allowed users to ease into, rather than leap into, 3D operation. Logistical concerns about the increased acceptance of scattered coincidences, and data size and management had to be addressed, but, using specialized hardware (usually i-860 based) reconstruction times have been made acceptable. In parallel, the detectors in block systems have gradually become smaller, which increases the number of lines of response and hence data size, for the same active volume being scanned. Unfortunately, decreasing detector size also decreases the packing fraction in block detectors as more sawcuts required to section the blocks, and this decreases to some extent the overall sensitivity of the detector.

As experience with 3D PET has mounted, a number of issues have arisen as a consequence of the move to 3D systems which require further attention. The biggest impact that 3D PET has had is on the detector and electronics performance. Random coincidences increase as the square of the singles rates on the detectors and therefore 3D performance is maximal at the lowest counting rates practical. In neurological "activation" studies using either $H_2^{15}O$ or ^{15}O-butanol the total dose that could be delivered in any one injection had to be drastically reduced [10,11,12,13]. This can be exploited by allowing the same *total* dose of radiotracer to be delivered over a greater number of scans, with less dose being injected in each scan. From a purely measurement point of view this can be advantageous, but may introduce other physiological/cognitive variables such as attention, learning and habituation effects. The alternative, of course, is to give less total radioactivity. In neurotransmitter, receptor, drug binding and pharmacokinetic studies (all unique to investigation with radiotracers) 3D PET affords better quality data and the potential to study the time course over a longer duration, due to the improved sensitivity [14,15,16,17,18].

Experience with 3D PET outside of the head has been less encouraging. The recent trend has been to extend the axial field of view and decrease the diameter of the ring of PET systems. When these systems are operated in septa-retracted 3D mode they have a very large field of view for single photons, and to a lesser extent, scattered coincidence events. The decreased ring diameter leaves less room for shielding at the axial ends of the detectors and this, combined with the extended axial field of view, both cause the system to be more "open" to an unwanted larger single photon acceptance field of view. Increased single photons from outside of the coincidence field of view increases both system deadtime and random coincidence rates, thus effectively reducing the live time of the tomograph. The situation is exemplified when scanning the thorax or abdomen using a radiotracer that has significant cerebral uptake and is excreted via the kidneys; the brain and the bladder may be outside the field of view of the tomograph, but both will contribute a large photon flux on the detectors. This is the case with ^{18}FDG, and it has been found that whole body or thorax/abdomen scanning using this tracer necessitates lower doses than are typically used in 2D. This has implications for data quality and patient throughput, but nevertheless is compatible with lowering the radiation dose.

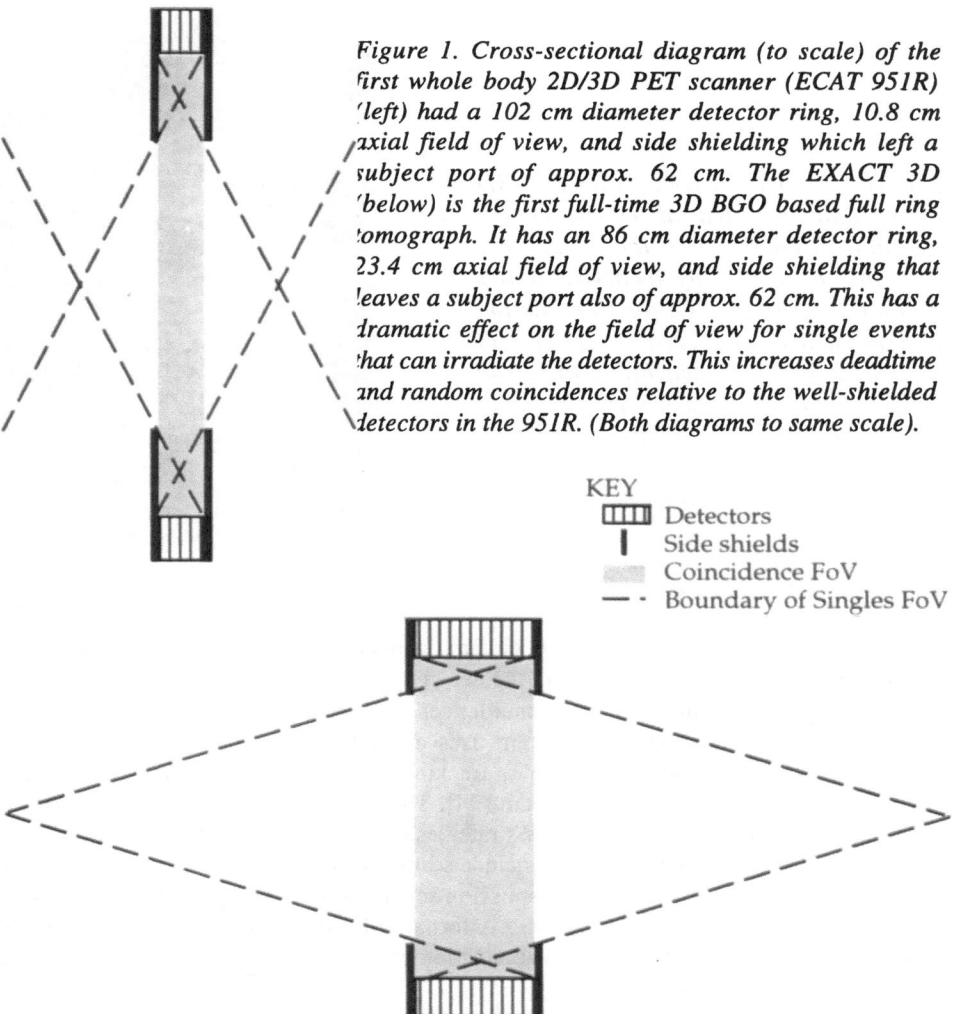

Figure 1. Cross-sectional diagram (to scale) of the first whole body 2D/3D PET scanner (ECAT 951R) (left) had a 102 cm diameter detector ring, 10.8 cm axial field of view, and side shielding which left a subject port of approx. 62 cm. The EXACT 3D (below) is the first full-time 3D BGO based full ring tomograph. It has an 86 cm diameter detector ring, 23.4 cm axial field of view, and side shielding that leaves a subject port also of approx. 62 cm. This has a dramatic effect on the field of view for single events that can irradiate the detectors. This increases deadtime and random coincidences relative to the well-shielded detectors in the 951R. (Both diagrams to same scale).

KEY
▥▥ Detectors
| Side shields
▥ Coincidence FoV
— · Boundary of Singles FoV

Perhaps the most ambitious BGO tomograph that has been constructed to date is the 24 cm axial field of view whole body full-time 3D tomograph at the Hammersmith Hospital in London (ECAT EXACT 3D) [19] developed by CTI (Knoxville, TN, USA). This tomograph uses the latest standard high resolution block detector (8x8 detectors per block) with individual detector dimensions 4.25 mm x 4.0 mm and a greatly simplified gantry design due to the lack of any moving parts. In all, the tomograph has nearly 30,000 individual detectors. This device has a number of innovative features which have been necessitated by designing a "3D only" system, and provides a useful example of changes that future machines will use.

48

Figure 2. The ECAT EXACT 3D tomograph installed at Hammersmith Hospital in London is a large (23.4 cm axial field of view), fully 3D device. It uses a point source of ^{137}Cs for transmission measurements. The great simplicity of design achieved by removing rotating rod sources and retractable septa is easily appreciated in this photograph. In addition, the increasing trend to modularity of detector design is readily seen.

Transmission scanning is poor in a 3D system using coincidence techniques such as rod sources of ^{68}Ge. The local deadtime of the detectors near the source is in general so great that the effective count rate for coincidences is severely limited. The alternative employed on the Hammersmith system uses a single photon emitting point source [20,21,22,23] of ^{137}Cs and a scintillation tracking approach to localise the point source to provide a pseudo "coincidence" event [24]. The single photon approach will have a large scatter fraction and therefore new strategies to deal with the scatter in transmission data are required. Potential candidates include scatter correction using any of the currently implemented approaches used in 3D emission scanning (dual energy windows, integral transform/subtraction, model-based). Alternatively the attenuation data can be reconstructed and segmented and the attenuation coefficients reassigned to the expected narrow-beam values [25].

Finally, vying with the Hammersmith system in terms of ambition, although for different reasons, is the "Dual PET" system developed in Akita, Japan [26]. The system uses BGO scintillators and can operate in both 2D and 3D. It is essentially two independent PET cameras on a common mounting track. While one PET camera is recording data from the head, the other records data from the heart to provide an input function for dynamic on-line modelling. This is a camera designed by modellers, for modelling, and as such should provide very precise functional parametric results. The cost of such a system may preclude its widespread distribution, however. The logical extension of a device like this is to make a very large axial field of view tomograph to do the same, rather than two separate tomographs.

3. Alternatives to Full Ring BGO systems

One significant spin-off of the development of 3D has been the commercial development of a number of lower cost alternative systems suitable for general purpose usage.

The PENN-PET system from UGM (Philadelphia, PA, USA) is based on NaI(Tl) detectors and has evolved into both neurological and whole body (also known as the GE Quest) versions [27,28]. NaI(Tl) has much greater light output than BGO and thus these devices have very good energy resolution. The stopping power of NaI(Tl) for 0.511 MeV photons will always mean that sensitivity is compromised compared to BGO, though. However, the system does compensate by being fully 3D.

A different approach to reducing the cost without impacting greatly on performance has been the development of a partial-ring, rotating 3D PET camera (Siemens/CTI ECAT ART) with less than 50% of the components of an equivalent full ring system [29,30,31], and therefore at a potentially reduced cost, a concept originally suggested for a BGO system in the early 1980's [32]. The use of full-time 3D acquisition means that this has been achieved with virtually no compromise in terms of performance compared to an equivalent 2D PET system. This has prompted one group to investigate the feasibility of including a CT scanner in the same device in the space vacated by the detectors, to produce a combined structural/functional camera [33] of particular interest in oncological applications.

Figure 3. The ECAT ART is the first commercial partial ring BGO block detector rotating tomograph. The figure shows the two banks of detectors. On the far left the lead end shield for the detectors is seen, while opposite the detectors themselves are visible between the opposing lead shields. All electrical signals are communicated to the gantry either via slip rings or bidirectional optical transducers.

Some interest remains in alternative systems which use gas proportional counting principles, or hybrid gas/scintillator systems, for photon detection [34,35]. These are almost invariably 3D devices because of the limited intrinsic sensitivity of the detectors.

More recently, there has been considerable enthusiasm for modifying conventional dual-headed gamma cameras for coincidence detection [36]. As with most developments like this, it must be viewed as a compromise, but may prove useful for non-quantitative clinical studies or as an introduction to PET for a Nuclear Medicine department.

4. Animal Tomographs

In parallel with these developments for human scanners there has been burgeoning interest in tomographs designed specifically for small animal studies [37,38,39,40]. Small animal PET systems are useful in pharmaceutical and radiotracer development [41] as well as in animal models of disease[42], for which so many exist today, especially in the area of molecular biology and genetic studies. As they are of lower cost than full human systems, they also are capable of providing a small, fully operational prototype of future larger human systems using new detectors as an important step in the development path to a new scanner. Some recent animal systems have used conventional block detectors developed for human scanners [43,44,45] while others have explored alternative photon detection and/or light collection schema [46,47]. The problem of making smaller detectors capable of resolving structures of interest in small animals (e.g., mice, rats) is the penetration of the crystals by the annihilation photons and accurate localisation to a single detector. This is dependent on the energy of the photons (0.511 MeV), the amount of light produced by the scintillator, and the density of the scintillator. Detectors could be made smaller, but the photon would still deposit its energy over the same range in the scintillator causing light emission from a number of detectors, so that the effective intrinsic resolution of the detectors would be would be no better than for a larger detector. Also, the smaller diameter of the detector ring of an animal scanner usually suffers from a severe loss in radial resolution away from the central axis of the scanner due to oblique penetration of the crystals. One method of overcoming this problem is by measuring the depth of interaction of the photon in the detector. If this can be achieved to within 5 mm FWHM (in a 30 mm crystal) the correct line of response can be assigned to the event thus improving the positional localisation in the block and overall resolution.

5. New Scintillators

New scintillators have appeared and will feature in future PET cameras. Cerium doped lutetium oxyorthosilicate ($Lu_2(SiO_4)O:Ce$ - known as "LSO") has similar stopping power to BGO, but with 5-7 times the light output and much faster response time [48]. This promises to address the deficiencies of BGO for 3D systems. Faster scintillators and greater light output can be used to correct for the parallax error in smaller crystals, and this allows the ring diameter to be reduced. This increases sensitivity and decreases costs by using less detector to cover 2π angular sampling. Going even further, the extra light allows alternative light collection devices such as photodiodes or light pipes (optical fibres) with PMTs at a distance to be used, which can be exploited to make smaller detectors for high resolution cameras [49], or allow their use in high magnetic fields such as in an MRI scanner [40]. LSO is already being made in block detectors with various light collection strategies to test performance [49]. Another promising scintillator is $LuAlO_3$:Ce, known as "LuAP". It is a fast scintillator, similar to LSO, however, the light emitted is 2-3 times less.

6. Future Systems

The preceding discussions permit us to make an educated guess as to some of the design features of the next generation of PET cameras. As the issue of scatter correction appears to be soluble and improved scintillation detector materials are appearing, the next generation of positron tomographs will most likely have the following characteristics:

- New fast, dense detectors in a smaller block than is used today with high light output (most likely LSO);
- Fully 3D operation in a septa-less tomograph;
- Time-of-flight coincidence detection, as the new detectors will be fast enough for the coincidence timing window to be reduced and this will introduce large differences in photon arrival times in a wide transaxial field of view (whole body) tomograph from annihilations arising at the edges of the field of view;
- Individual detector readouts, rather than Anger logic, which would decrease system deadtime;
- Time of flight mode, plus the continuing development of dynamic analytical techniques, would suggest that list mode acquisition will become a more efficient and flexible method for recording the data;
- New strategies for acquiring transmission scans for attenuation correction will be required, and point source measurements with long-lived single photon emitters such as ^{137}Cs offer great promise, but will require processing to remove scatter, scale the attenuation coefficients to the appropriate energy, and possibly to segment the data into different tissue types;
- To improve spatial resolution the ring diameter should be reduced, however, this leads to radial elongation error due to crystal penetration by non-orthogonal photons. This effect can be reduced if the depth of interaction of the photon in the detector is known. For depth of interaction to be measured, high speed light-to-electrical signal converters are required. Photomultipliers are efficient for this purpose generally, but their bulk precludes them from being attached to the front and rear faces of the detectors. Photodiodes may be the solution to this as they can be manufactured to be quite small, but at present their efficiency is low (which would compromise energy resolution) and a conventional

photodiode may be too slow and therefore restrict the count rate. A compromise containing both photodiodes and photomultiplier tubes may be an alternative [50].

Full 3D operation, a smaller block (therefore less deadtime), list mode and a ^{137}Cs point source have been implemented in the latest tomograph developed by CTI in collaboration with the MRC Cyclotron Unit at Hammersmith Hospital, the ECAT EXACT3D, delivered in February 1996. This device is still BGO-based, and therefore count rate performance is anticipated to be the limiting factor in effective sensitivity. This machine will be extremely useful in studying ^{11}C compounds for longer durations, or for radiotracers for which the dosimetry restricts the amount that can be administered to the subject. The experience with these innovations will help to guide the development of the next generation of PET tomographs.

7. References

1.　Townsend D.W., Frey P., Jeavons A., Reich G., Tochon-Danguy H.J., Donath A., Christin A., and Scaller G. (1987) High Density Avalanche Chamber (HIDAC) Positron Camera, *J Nucl Med* **28**, 1554-1562.

2.　Vanstein B.K. and Orlov S.S. (1974) General Theory of Direct 3D Reconstruction. in Marr (ed), *Proceedings of the 1974 International Workshop on 3D Reconstruction Techniques*, Brookhaven National Laboratory.

3.　Orlov S. (1976) Theory of three-dimensional reconstruction. 1. Conditions of a complete set of projections, *Sov Phys Crystallogr* **20**, 312-314.

4.　Townsend D.W., Spinks T.J., Jones T., Geissbühler A., Defrise M., Gilardi M.-C., and Heather J.D. (1989) Three dimensional reconstruction of PET data from a multi-ring camera, *IEEE Trans Nucl Sci* **36**, 1056-1065.

5.　Dahlbom M., Eriksson L., Rosenqvist G., and Bohm C. (1989) A study of the possibility of using multi-slice PET systems for 3D imaging, *IEEE Trans Nucl Sci* **36**, 1066-1071.

6.　Cherry S.R., Dahlbom M., and Hoffman E.J. (1991) 3D PET using a Conventional Multislice Tomograph without Septa, *J Comput Assist Tomogr* **15**, 655-668.

7.　Bailey D.L., Lee K.-S., Stocks G., Meikle S.R., and Dobko T. (1993) Clinical 3D PET For Improved Patient Throughput, *J Nucl Med* **34**, 184P (abstract).

8.　Cutler P.D. and Xu M. (1995) Strategies to Improve 3D Whole Body PET Image Reconstruction, *J Nucl Med* **36**, 93P (abstract).

9.　Kinahan P.E., Jadali F., Sahin D., Brown M.L., Mintun M.A., Baron R.L., and Townsend D.W. (1995) A Comparison of 2D and 3D Abdominal PET Imaging, *J Nucl Med* **36**, 7P (abstract).

10.　Bailey D.L., Jones T., Watson J.D.G., Schnorr L., and Frackowiak R.S.J. (1993) Activation studies in 3D PET: evaluation of true signal gain, in Uemera

K, Lassen N, Jones T and Kanno I (eds.), *Quantification of Brain Function: Tracer Kinetics and Image Analysis in Brain PET*, Excerpta Medica, pp. 341-350.

11. Cherry S.R., Woods R.P., Hoffman E.J., and Mazziotta J.C. (1993) Improved Detection of Focal Cerebral Blood Flow Changes Using Three-Dimensional Positron Emission Tomography, *J Cereb Blood Flow Metab* **13**, 630-638.

12. Silbersweig D.A., Stern E., Frith C.D., *et al.* (1993) Detection of Thirty-Second Cognitive Activations in Single Subjects with Positron Emission Tomography: A New Low-Dose $H_2{}^{15}O$ Regional Cerebral Blood Flow Three-Dimensional Imaging Technique, *J Cereb Blood Flow Metab* **13**, 617-629.

13. Watson J.G.D., Myers R., Frackowiak R.S.J., Hajnal V., Woods R.P., Mazziotta J.C., Shipp S., and Zeki S. (1993) Area V5 of the Human Brain: Evidence from a Combined Study Using Positron Emission Tomography and Magnetic Resonance Imaging, *Cereb Cortex* **3**, 79-94.

14. Tadokoro M., Jones A.K.P., Cunningham V.J., Sashin D., Grootoonk S., Ashburner J., and Jones T. (1993) Parametric images of ^{11}C-diprenorphine binding using spectral analysis of dynamic PET images acquired in 3D, in Uemera K, Lassen NA, Jones T and Kanno I (eds.), *Quantification of Brain Function - Tracer Kinetics and Image Analysis in Brain PET*, Excerpta Medica, pp. 289-294.

15. Carson R.E., Endres C.J., and Daube-Witherspoon M.E. (1995) Quantitative Accuracy of 3D PET for Brain Receptor Imaging, *J Nucl Med* **36**, 81P (abstract).

16. Weeks R.A., Cunningham V., Walters S., Harding A.E., and Brooks D.J. (1995) A Comparison of Region of Interest and Statistical Parametric Mapping Analysis in PET Ligand Work: ^{11}C-Diprenorphine in Huntington's Disease and Tourette's Syndrome, *J Cereb Blood Flow Metab* **15**, S41(abstract).

17. Rakshi J., Bailey D.L., Morrish P.K., and Brooks D.J. (1996) Implementation of 3D Acquisition, Reconstruction and Analysis of Dynamic Fluorodopa Studies, in Myers R, Cunningham VJ, Bailey DL and Jones T (eds.), *Quantification of Brain Function Using PET*, Academic Press, San Diego, pp. 82-87.

18. Townsend D.W., Price J.C., Mintun M.A., Kinahan P.E., Jadali F., Sashin D., Simpson N., and Mathis C.A. (1996) Scatter Correction for Brain Receptor Quantitation in 3D PET, in Myers R, Cunningham VJ, Bailey DL and Jones T (eds.), *Quantification of Brain Function Using PET*, Academic Press, San Diego, pp. 76-81.

19. Jones T., Bailey D.L., Bloomfield P.M., *et al.* (1996) Performance Characteristics And Novel Design Aspects Of The Most Sensitive PET Camera Built For High Temporal And Spatial Resolution, *J Nucl Med* **37**, 85P(abstract).

20. Derenzo S.E., Budinger T.F., Cahoon J.L., Huesman R.H., and Jackson H.G. (1977) High resolution computed tomography of Positron Emitters, *IEEE*

54

Trans Nucl Sci **NS-24**, 544-558.

21. deKemp R.A. and Nahmias C. (1994) Attenuation correction in PET using single photon transmission measurement, Med Phys **21**, 771-778.

22. Karp J.S., Muehllehner G., Qu H., and Yan X.-H. (1995) Singles transmission in volume-imaging PET with a ^{137}Cs source, Phys Med Biol **40**, 929-944.

23. Yu S.K. and Nahmias C. (1995) Single-photon transmission measurements in positron emission tography using ^{137}Cs, Phys Med Biol **40**, 1255-1266.

24. Jones W.F., Vaigneur K., Young J., Moyers C., and Nahmias C. (1995) The Architectural Impact of Single Photon Transmission Measurements on Full Ring 3D Positron Tomography. in Moonier PA (ed.), Proceedings of the 1995 IEEE Nuclear Science Symposium and Medical Imaging Conference, San Francisco. Vol 2:1026-1030.

25. Xu E.Z., Mullani N.A., Gould K.L., and Anderson W.L. (1991) A segmented attenuation correction for PET, J Nucl Med **32**, 161-165.

26. Iida H., Miura S., Kanno I., Ogawa T., and Uemera K. (1996) A New PET Camera for Noninvasive Quantitation of Physiological Functional Parametric Images: Headtome-V-Dual, in Myers R, Cunningham V, Bailey DL and Jones T (eds.), Quantification of Brain Function Using PET, Academic Press, San Diego, pp. 57-61.

27. Muehllehner G., Karp J.S., Mankoff D.A., Beerbohm D., and Ordonez C.E. (1988) Design and performance of a new positron emission tomograph, IEEE Trans Nucl Sci **35**, 670-674.

28. Karp J.S., Kinahan P.E., and Mankoff D.A. (1991) Positron Emission Tomography with a Large Axial Acceptance Angle: Signal-to-Noise Considerations, IEEE Trans Med Imag **MI-10**, 249-255.

29. Townsend D.W., Wensveen M., Byars L.G., et al. (1993) A Rotating PET Scanner Using BGO Block Detectors: Design, Performance and Applications, J Nucl Med **34**, 1367-1376.

30. Townsend D.W., Bishop H., Mintun M.A., Byars L.G., Geissbühler A., and Nutt R. (1994) Physical and Clinical Performance of a Rotating Positron Tomograph, J Nucl Med **35**, 41P (abstract).

31. Bailey D.L., Young H.E., Bloomfield P.M., et al. (1996) ECAT ART - A Continuously Rotating Pet Camera: Performance Characteristics, Comparison With A Full Ring System, Initial Clinical Studies, And Installation Considerations In A Nuclear Medicine Department, Eur J Nucl Med (In Press).

32. Takami K., Ishimatsu K., Hayashi T., et al. (1982) Design Considerations for a Continuously Rotating Positron Computed Tomograph, IEEE Trans Nucl Sci **NS-29**, 534-538.

33. Townsend D.W., Kinahan P.E., and Beyer T. (1996) Attenuation Correction for a Combined 3D PET/CT Scanner, Physica Medica **XII**, 43-48.

34. Ott R.J. (1993) Wire Chambers Revisited, Eur J Nucl Med **20**, 348-358.

35. Tavernier S., Bruyndonckx P., Debruyne J., et al. (1994) First Results from a Prototype PET Scanner Using BaF_2 Scintillator and Photosensitive Wire

Chambers. in Trendler RC (ed.), *Proceedings of the 1994 IEEE Nuclear Science Symposium and Medical Imaging Conference*, Norfolk, VA, USA. Vol 4:1885-1887.

36. Lewellen T.K., Miyaoka R.S., Kaplan M.S., Kohlmyer S.K., Costa W., and Jansen F. (1995) Preliminary Investigation of Coincidence Imaging with a Standard Dual-Headed SPECT System, *J Nucl Med* **36**, 175P (abstract).

37. Rajeswaran S., Hume S.P., Cremer J.E., Young J., and Bailey D.L. (1991) Dynamic monitoring of [11C]diprenorphine in rat brain using a prototype positron imaging device, *J Neurosci Meth* **40**, 223-232.

38. Rajeswaran S., Bailey D.L., Hume S.P., Townsend D.W., Geissbühler A., Young J., and Jones T. (1992) 2-D and 3-D Imaging of Small Animals and the Human Radial Artery with a High-Resolution Detector for PET, *Trans Med Imag* **MI-11**, 386-391.

39. Watanabe M., Uchida H., Okada H., Shimizu K., Satoh N., Yoshikawa E., Ohmura T., Yamashita T., and Tanaka E. (1992) A High Resolution PET for Animal Studies, *IEEE Trans Med Imag* **MI-11**, 577-580.

40. Cherry S.R., Shao Y., Silverman R.W., *et al.* (1996) microPET: A Dedicated PET Scanner for Small Animal Imaging, *J Nucl Med* **37**, 86P (abstract).

41. Hume S.P., Luthra S.K., Brown D.J., *et al.* (1996) Evaluation of [^{11}C]RTI-121 as a selective radioligand fro PET studies of the dopamine transporter, *Nucl Med Biol* **23**, 377-384.

42. Hume S.P., Lammertsma A.A., Myers R., Rajeswaran S., Bloomfield P.M., Ashworth S., Torres E.M., Watson I., and Jones T. (1996) The potential of high resolution positron emission tomography to monitor dopaminergic function in rat models of disease, *Neurosci Meth* **67**, 103-112.

43. Cutler P.D., Cherry S.R., Hoffman E.J., Digby W.M., and Phelps M.E. (1992) Design Features and Performance of a PET System for Animal Research, *J Nucl Med* **33**, 595-604.

44. Bloomfield P.M., Rajeswaren S., Spinks T.J., *et al.* (1995) Design and Physical Characteristics of a Small Animal Positron Emission Tomograph, *Phys Med Biol* **40**, 1105-1126.

45. Hutchins G.D., Simon A.J., Winkle W., and Carlson K. (1996) Performance Evaluation of a Small Field of View (FOV) High Sensitivity/High Spatial Resolution PET Scanner, *J Nucl Med* **37**, 86P (abstract).

46. Tavernier S., Bruyndockx P., and Shuping Z. (1992) A Fully 3D Small PET System, *Phys Med Biol* **37**, 635-643.

47. Marriott C.J., Cadorette J.E., Lecomte R., Scasnar V., Rousseau J., and van Lier J.E. (1994) High-Resolution PET Imaging and Quantitation of Pharmaceutical Biodistributions in a Small Animal Using Avalanche Photodiode Detectors, *J Nucl Med* **35**, 1390-1397.

48. Melcher C.L. and Schweitzer J.S. (1991) Cerium-doped Lutetium Oxyorthorthsilicate: A Fast, Efficient New Scintillator. in Baldwin GT (ed.), *Proceedings of the 1991 IEEE Nuclear Science Symposium and Medical*

Imaging Conference, Santa Fe. Vol 1:228-231.

49. Moses W.W., Derenzo S.E., Ho M.H., Andreaco M.S., Paulus M.J., and Nutt R. (1996) Performance of PET Detector Module with LSO Scintillator Crystals and Photodiode Readout, *J Nucl Med* **37**, 87P (abstract).

50. Moses W.W., Derenzo S.E., and Budinger T.F. (1994) PET detector modules based on novel detector technologies, *Nucl Instr Meth Phys Res* **A353**, 189-194.

STRATEGIES FOR RADIOLIGAND DEVELOPMENT
PEPTIDES FOR TUMOR TARGETING

G. STÖCKLIN and H.J.WESTER
Nuklearmedizinische Klinik und Poliklinik
Klinikum rechts der Isar
der Technischen Universität München
D - 81675 München

1.   INTRODUCTION

Tumors express numerous specific receptors for peptide ligands (Eckelman 1994, Reubi 1995). Thus, the development and application of peptides and proteins, including monoclonal antibodies and their fragments for receptor mapping, tumor targeting and therapy control is one of the most active areas of radiopharmaceutical research (Miller 1993, Fishman et al. 1993).

The continuously growing number of promising radiolabeled bioactive peptides depends on several 'growth factors', such as the optimization of these biospecific agents with respect to affinity, specificity, selectivity, in vivo stability, and size. Thus, over the past three decades, agents have evolved, ranging from large proteins (i.e., intact polyclonal antibodies with low specificity) to antibody fragments (i.e., Fab of F(ab')$_2$) to smaller molecular recognition units such as antigen binding domain fragments to small biologically active synthetic peptides (Corstens and van der Meer 1991, Serafini 1993, Fishman et al. 1993, Goldenberg 1993, Goodwin 1987, Hnatowitch 1990). Moreover, the increasing knowledge of the spatial structure of the corresponding binding sites (Kaupmann et al. 1995), the cloning and sequencing of the peptide receptors and the type of receptor-ligand interaction allow the design of increasingly potent ligands (Fraser 1995, Reynor and Reisine 1992, Savarese and Fraser 1992).

In clinical application octreotide (SMS 201-995) (Pless et al. 1986), which represents an in vivo stable analog of the neuropeptide somatostatin, have played the major role during the last few years. Among a variety of labeled derivatives it is mainly ^{111}In-DTPA-D-Phe[1]-octreotide (Bakker et al. 1991), which is routinely used to visualize somatostatin receptor (SSTR) positive tumors. Probably due to the excellent results, which were achieved with this tracer, a number of emerging new peptide radiotracers have been labeled during the last years. However, in most cases the assessment of their clinical potential is still under evaluation. In table 1 we have listed

B. Gulyás and H.W. Müller-Gärtner (eds.),
Positron Emission Tomography: A Critical Assessment of Recent Trends, 57–90.
© 1998 *Kluwer Academic Publishers.*

Table 1:   Some bioactive peptides and proteins used as radiolabeled receptor ligands

Peptide	radiolabeled analog (example)	medical application or potential medical application	References
Atrial natriuretic peptide	ANP^{99-126}	diabetic nephropathy	Lambert et al. 1994, Clerico et al. 1995, Sagnella and MacGregor 1994,
Angiotensin	$[Sar^1, Ile^8]AT\ II$	receptor mapping	Gibson et al. 1994
Bombesin	$[1\text{-}B^6,Des\text{-}Met^{14}]BBN(7\text{-}13)NH_2$	small cell lung carcinomas (SCLC)	Hoffman et al. 1995
Chemotactic peptides	N-Formyl-Nle-Leu-Phe-Nle-Tyr-Lys	focal sites of bacterial infection	Babich et al. 1993, 1995; Fishman et al. 1991, 1993, Vaidyanathan et al. 1994a,
Insulin	direct labeled	hepatocellular carcinomas (HCC)	Awashti et al. 1993, Eastman et al. 1990, Kurtaran et al. 1995, Shai et al. 1989,
Low density lipoprotein	direct labeled	quantitation of hepatic LDL receptor acitivity in the case of arteriosclerosis or hyperlipo-proteinaemia	Benyai et al. 1994, Leitha et al. 1993,
Luteinizing hormone releasing hormone	SDZ 218-041 cetrorelix	LHRH+ tumors	Stolz et al. 1995 Locher et al. 1994
Neurotoxins	apamin charybdotoxin	Alzheimer disease, receptor mapping,	Wester et al. 1995b Fixmann et al. 1995
RGD - Peptides	CYT 379, PAC-2, PAC 15, PAC-26, P-280	deep venous thrombosis (DVT) angiogenesism, metastasis	Ben-Haim et al. 1993, Knight et al. 1993, 1994
Somatostatin	SMS 201-995 (octreotide)	tumor imaging	Bakker at al. 1990, 1991; Guhlke et al. 1994c, Brockmann et al. 1995, Lamberts et al. 1991, Maina et al. 1994, Smith-Jones et al. 1994, Stolz et al. 1995,
Substance P	$Arg^1\text{-}SP$	inflammatory bowel diseases, arthritis, carcinoids	Breeman et al. 1995
Transferrin	direct labeled	tumor imaging	Aloj et al. 1996, Wester et al. 1996
Vasointestinal peptide	VIP(1-28)	tumor imaging	Virgolini et al. 1994, 1995, Kurtaran et al. 1996

some peptide groups, which have been used or are being evaluated for application as radiolabeled ligands for imaging.

Most of these peptides have been labeled with SPET radionuclides, such as iodine-123, technetium-99m, and indium-111, or with therapeutic radionuclides, such as iodine-125,131, yttrium-90, and rhenium-186. PET radionuclides, such as fluorine-18 did not play a significant role sofar. This situation will most likely change in the near future, since 'the new generation' of biospecific peptides exhibits high affinity to their binding sites, reduced unspecific binding in normal tissue, and pharmacokinetics compatible with the half life of fluorine-18. Positron emitting radionuclides are, therefore, now entering the scene, which may provide essential advantages compared to SPET tracers:

- Quantitative evaluation of the pharnacokinetics and of the tumor receptor status
- Higher sensitivity and resolution
- Higher in vivo stability of fluorine-18 compared to radioiodinated products
- Small molecular changes of the original peptide of fluorinated products

Endoradiotherapy optimization requires the quantitative measurement of many parameters. Among these parameters the in vivo uptake kinetics in individual organs or regions of interest. If positron emitters of the same element as the therapeutic isotope are available, quantitative uptake kinetics and dosimetry can be obtained by PET, as has been shown for [^{86,90}Y]citrate and [^{86,90}Y]EDTMP (Herzog et al. 1996, Rösch et al. 1996) in patients and for various labeleled octreotides in rats (Brockmann et al. 1995a , Wester et al. in press). Other positron emitters, which can be used for the validation of therapeutic radionuclides, are iodine-120,124 and samarium-142. High specific activities are generally needed since the receptor density is only in the order of 1,000 to 10,000 per cell.

2. LABELING STRATEGIES
2.1. Fluorine-18

Due to the high oxidation potential of fluoride, a direct oxidative ^{18}F-fluorination starting with ^{18}F-fluoride similar to the labeling of peptides and proteins with iodine and bromine is not possible. Furthermore, ^{18}F-fluorination via electrophilic ^{18}F-fluorine leads to a low specific activity of the radiolabeled product; hence, is not suitable for the syntheses of ligands for receptor mapping. Therefore, prosthetic group labeling via two different strategies is the only suitable way:

i. ^{18}F-fluorination via homo- and heterobifunctional prosthetic groups with electrophilic nature: both functionalities, the locus for the introduction of

^{18}F-fluoride and the linker for the subsequent protein conjugation are of electrophilic nature. The electrophilic linker unit is directed to the nucleophilic side chain functionalities of the peptide and protein targets (amino-, thiol- and hydroxyl groups). Typical reactions of this conjugation approach are: *^{18}F-fluoroacylation, ^{18}F-fluoroalkylation, ^{18}F-fluoroamidination* and *^{18}F-fluorination by photochemical conjugation.*

ii. ^{18}F-Fluorination via heterobifunctional groups with electrophilic and nucleophilic nature and preconjugation of the peptide or protein: this strategy enables the introduction of the label by a nucleophilic attack of a preformed ^{18}F-labeled hapten to a protein preconjugate. Although the preparation of the ^{18}F-labeled haptens are fast and can be performed with high yields, this advantage is compensated in most cases by the laborious and sophisticated protective group chemistry on the peptide to avoid inter- and intramolecular cross-linking. Thus, this strategy seems to be a suitable alternative only for small peptides, especially when only a few nucleophilic side chains are present. Since the final prosthetic group consists of two subgroups, the steric impairment of the target peptide is generally higher. The *^{18}F-fluoroamidation* is an example for this approach.

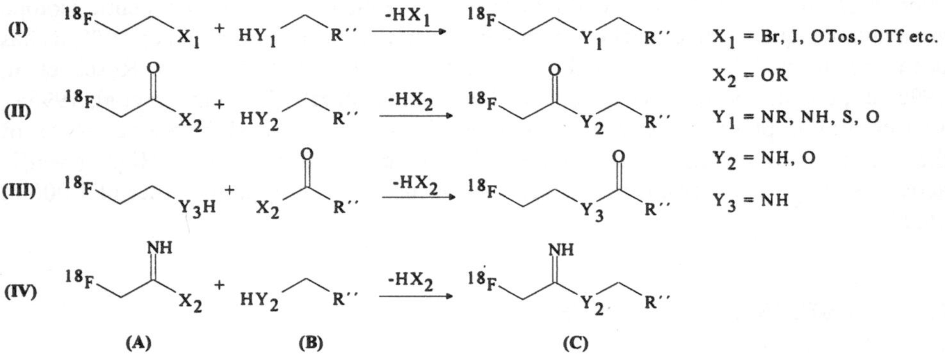

Scheme 1:   (I) Fluoroalkylation, (II) Fluoroacylation, (III) Fluoroamidation and (IV) Fluoroamidination

2.1.1. ^{18}F-Fluoroacylation

The ^{18}F-fluoroacylation is the most often used ^{18}F-labeling methodology for peptides and proteins. Thus, a variety of different reagents have been synthesized and evaluated on numerous bioactive targets. Scheme 1 illustrates the reactions of the four major radiofluorination procedures and table 2 the approaches which have been used sofar.

Table 2:   Approaches to ^{18}F-labeling of peptides and proteins

Method	^{18}F-labeling agent	preparation steps/ time[min] / RCY[%]	Ref.
acylation	2-[^{18}F]Fluoroacetic acid / DCC	2 / 80min / 50%	Müller-Platz et al. 1992
	Methyl 2-[^{18}F]fluoropropionate	1 / 15 min / 90%	Block et al. 1988
	[^{18}F]-2,3,5,6-Tetrafluorophenyl pentafluorobenzoate (*)	1 / 30 min / 32%	Herman et al. 1994
	N-Succinimidyl-8-[(4'-[^{18}F]fluoro-benzyl)amino]suberate	3 / 55-60min / 25-40%	Garg et al. 1991
	N-Succinimidyl-[^{18}F]fluorobenzoate	3 /100min / 25%	Vaidyanathan et al. 1992, 1994b, Guhlke et al. 1994, Wester et al. 1996b
	4-Nitrophenyl 2-[^{18}F]fluoro-propionate	3 / 90 min / 60%	Guhlke et al. 1994
	N-Succinimidyl 4-([^{18}F]fluoro-methyl)benzoate	1 / 30-35 min / 18%	Lang and Eckelman 1994a, 1995
amidation	1-[4-[^{18}F]Fluoromethyl)benzoyl]-aminobutane-4-amine	2 / 80min /52%	Shai et al. 1989
amidination	3-[^{18}F]Fluoro-5-nitrobenzimidate	3 / 45min /20-33%	Kilbourn et al. 1987
alkylation	N-(p-[^{18}F]Fluoro-phenyl)maleimide	4 /100min / 15%	Shiue et al. 1988
	m-Maleimido-N-(p-[^{18}F]fluoro-benzyl)benzamide	3 / 70min / 10%	Shiue et al. 1988
	[^{18}F]-Pentafluorobenzaldehyde (**)	1 / 30 min / 60 %	Herman et al. 1994
	4-[^{18}F]Fluoro-phenacylbromide	3 / 75min / 28-40%	Kilbourn et al. 1987
photochem. Conjugation	4-azido phenacyl-[^{18}F]fluoride	1 / 15min / 70%	Wester et al. 1996a,b

(*) by nucleophilic fluorination of 2,3,5,6-tetrafluorophenyl pentafluorobenzoate
(**) by nucleophilic fluorination of pentafluorobenzaldehyde

To minimize the steric impairment of the target molecule, the first trials have been performed with the smallest aliphatic acyl groups. However, despite the sucessful labeling of urokinase using for example [^{18}F]fluoroacetat activated with dicyclohexylcarbodiimide (Müller-Platz et al. 1982) and the labeling of some model compounds with methyl 2-[¹⁸F]fluoropropionate (Block et al. 1988), more potent acylation reagents were necessary to reach high conjugation yields in short conjugation times. This was obtained by using more electron withdrawing alcoholic components of the esters (i.e. 4-nitrophenyl 2-[¹⁸F]fluoropropionate instead of methyl 2-[¹⁸F]fluoropropionate (Guhlke et al. 1994)). Unfortunately, this resulted in a more complicated and time consuming preparation of the ¹⁸F-acylation reagents. Since the activation of the carbonyl carbon was significantely increased, direct nucleophilic ¹⁸F-fluorination of the corresponding precursors by bromine-for-fluorine exchange has failed and led to almost quantitative formation of the corresponding acid [¹⁸F]fluorides. Thus, protection was needed. The preparation of the commonly used ¹⁸F-acylation agents consists of three steps: a) nucleophilic fluorination of a precursor (e.g. nonactivated esters); b) formation of a suitable intermediate (e.g. hydrolysis of the esters) and c) formation of a highly reactive activated ester.

4-[¹⁸F]fluorobenzylamine succinimidylester ([¹⁸F]SFBS) (Garg et al. 1991) was utilized for the labeling of antibody fragments (Page et al. 1994b, Zalutsky et al. 1992). The reagent was prepared by nucleophilic aromatic fluorination of 4-nitrobenzonitrile, subsequent reduction with LiAlH₄ and final activation of the obtained 4-[¹⁸F]fluorobenzylamine with disuccinimidylsuberate. Immunoreactivity of a conjugated antimyosin Fab fragment prepared with HPLC purified [¹⁸F]SFBS was 75±9%, compared with about 50% when crude SFBS was used (Garg et al. 1991). 89±5% binding was found for the corresponding F(ab')₂ fragment. It is unclear, but seems possible that the nearly twofold decrease in immunoreactivity is caused by a significant amount of SFBS carrier. This carrier would lead to a loss of immunoreactivity by a factor of 0.5 in the case of F(ab')₂, since it consists of two Fab - units. This may also be reflected by two other [¹⁸F]SFBS conjugates, TP-3 with 50-61% immunoreactivity (Page et al. 1994b) and Mel-14 F(ab')₂ with 65% immunoreactivity (Page et al. 1994b).

The use of this prosthetic group seems to be limited to larger peptides or proteins, since this bulky prosthetic group can significantly affect the affinity of smaller receptor ligands.

Succinimidyl- 4-[¹⁸F]fluorobenzoate [¹⁸F]SFB represents a sterical less constraining prosthetic group. The original synthesis (Vaidyanathan and Zalutsky 1992b) consists of i) fluorine for trimethylamine exchange on the triflate of 4-trimethylammonium benzaldehyde, ii) oxidation of the 4-[¹⁸F]fluorbenzaldehyde to the substituted benzoic acid and iii) activation with DCC / N-hydroxysuccinimide. This method was improved by Guhlke et al. (1994a,b), who used the triflate of ethyl 4-trimethylammonium benzoate as a fluorination precursor. Hydrolysis led to the corresponding acid in high yields. Final activation was performed using N-succinimidyl dicarbonate. Thus, the synthesis time was shortened and the yields improved. Further improvement was reached by using tetramethyluronium tetrafluoroborate (TSTU) in the last step (scheme 2), which allows the preparation of [¹⁸F]SFB whithout HPLC purification steps (Wester et al. 1996b). [¹⁸F]SFB is extensively used for peptide and protein labeling: ¹⁸F-labeled Mel-14 F(ab')₂(Vaidyanathan et al. 1992a,b,), human serum albumin and

IgG (Wester et al.1996a, 1996b), ^{18}F-labeled apamin (Wester et al. 1995b, Fixmann et al. 1995) and ^{18}F-labeled chemotactic peptide (Vaidyanathan et al. 1994a, 1995).

For the labeling of small peptides, such as octreotide, 4-nitrophenyl 2-[^{18}F]fluoropropionate (Guhlke et al. 1994a,b,c) is preferred, since it is sterically much smaller, and the larger fluorobenzoate will increase the lipophilicity to a higher extent. Its three step preparation via fluorine for bromine exchange on ethyl

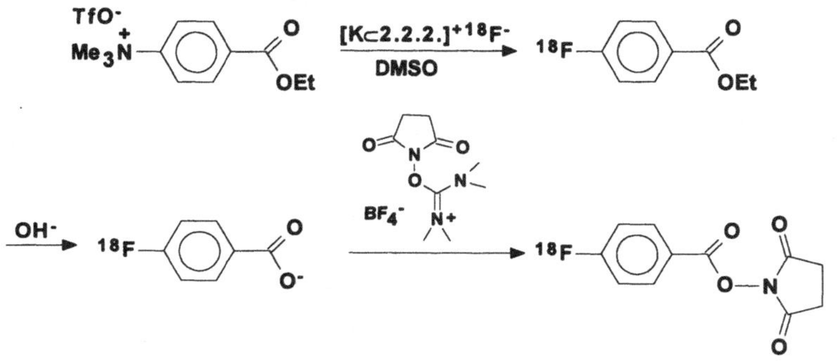

Scheme 2:   Improved synthesis of n.c.a. N-succinimidyl 4-[^{18}F]fluorobenzoate

2-bromopropionate, hydrolysis and activation with di-4-nitrophenyl carbonate is facilitated by educt trapping with dimethylaminopyridine (Guhlke et al.1994b). Using this reagent, (2-[^{18}F]fluoropropionyl-(D)Phe[1])-octreotide was synthesized by acylation of ε-Boc-Lys[5]-octreotide and subsequent deprotection of the product with TFA in 65% RCY based on the fluoroacylation reagent (Guhlke et al. 1994c). Preservation of ₁eceptor affinity of ^{18}F-octreotide was shown by in vitro (Guhlke et al. 1994c) and in vivo (Wester et al. in press) displacement experiments. Almost quantitative acylation yields were observed when 1-hydroxybenzotriazole (HOBt) was added to the reaction mixture (Fig.1). The dramatic acceleration of the acylation with 4-nitrophenyl 2-[^{18}F]fluoropropionate in the presence of different catalysts was previously shown on model peptides (Wester et al. 1994b). In the presence of HOBt, the radiochemical yield of N-terminal 2-[^{18}F]fluoropropionylated phe-gly increased from 15 to 70%. Moreover, regioselective ^{18}F-fluoroacylation of N^{α} - or N^{ε} - amino groups is possible. So, the N^{α} / N^{ε} -product ratio of the [18]F- fluoroacylation of val-lys was changed to the reciprocal value with addition of the catalyst, leading to 69% N^{α} -labeled product (without HOBt 19%). Similar results were obtained with proteins (^{18}F-avidin: 10±2 % RCY without HOBt, 61±6 % RCY with HOBt).

In a comparative evaluation (Wester et al. 1994a, Wester et al. 1996) it was recently demonstrated that conjugation yields using 4-nitrophenyl 2-[^{18}F]fluoropropionate depends on the lysine, tyrosine and histidine content of the peptides and proteins used, whereas conjugation with [^{18}F]SFB (see above) and [^{18}F]APF (see photochemical conjugation) predominantly depends on the lysine content. Owing to competing O-acylation in presence of tyrosine residues (Paul 1962, Ramachandran and Li 1963,

Martiney et al. 1979) using the HOBt/ 4-nitrophenyl 2-[^{18}F]fluoropropionate system, partially unstable conjugates are formed. Moreover, histidine is known to catalyse active ester hydrolysis (Yamashiro et al. 1972, Barany and Merrifield 1980, Imman 1981), thus limiting the maximum obtainable conjugation yield.

Recently, N-succinimidyl 4-([^{18}F]fluoromethyl)benzoate was prepared in a one step synthesis (Lang and Eckelman 1994) by fluorine for 4-nitrosulfonyloxy exchange with yields of about 18% EOS (30-35 min). In a further study (Lang and Eckelman 1995a), the direct prosthetic labeling approach was extended to a series of different active esters.

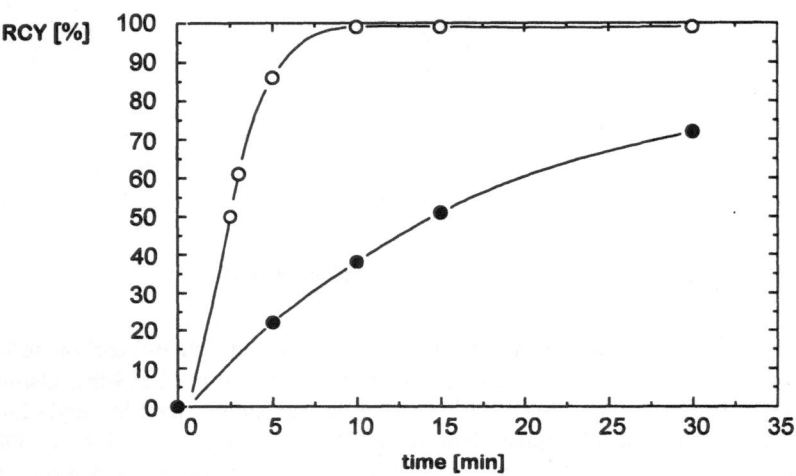

Figure 1: Radiochemical yield of the ^{18}F-fluoropropionylation of (ε-BOC-Lys^5)-(SMS 201-995) as a function of time with (o) and (•) without HOBt-catalysis.
Reaction conditions: 0.5 mg (ε-BOC-Lys^5)-(SMS 201-995), n.c.a. 4-nitrophenyl 2-[^{18}F]fluoropropionate, 2.5 µl 2,6-di-tert.-butylpyridin, 70°C in 50µl acetonitrile in the presence of (o) 1 mg hydroxybenzotriazole (HOBt), or (•) without HOBt (after Guhlke et al. 1994c)

The experiments indicated that only N-succinimidylesters without an α-hydrogen can be radiolabeled with ^{18}F-fluorine in one step. N-succinimidyl 4-([^{18}F]fluoromethyl)-benzoate was utilized for the quantification of transferrin uptake in vitro.

2.1.2. ^{18}F-Fluoroalkylation

Although alkylation is probably the most often applied reaction for the syntheses of radiopharmaceuticals, its application for peptide and protein labeling is rare.

Shiue et al. (1988) suggested the use of N-(p-[^{18}F]fluorophenyl)-maleimide and m-maleimido-N-(p-[^{18}F]fluorobenzyl)benzamide as thiol selective reagents. The reagents were prepared in 100 and 70 min with about 15 and 10% radiochemical yield. A 50% conjugation yield of Fab' from rabbit IgG was observed using the latter compound. Considering the statistically low relative amount of cystine and cysteine in peptides and proteins, the low stability of thiol groups towards oxidation and the importance of he integrity of disufide bridges for the tertiary structure of peptides, thiol groups appear not to be favorable targets for peptide modification.

α-Haloketones are known to alkylate various heteroatoms (N, O, S); thus, 4-[^{18}F]fluorophenacylbromide (Kilbourn et al. 1987) is a much more indicriminate alkylation agent. Starting with fluorine for nitro exchange on 4-nitrobenzonitrile, it can be prepared in three steps (radiochemical yield 28-40%, 75 min). Reaction of fibrinogen with 4-[^{18}F]fluorophenacylbromide gave 25-30% radiochemical yield. However, only 60-65% of the purified product retained the ability to clot when treated with thrombin. Similar to the preparation of methyl-[^{18}F]fluoro-5-nitrobenzimidate (see below, ^{18}F-amidination), the high amount of pseudocarrier (about 100μg of the precursor 4-nitro phenacylbromide) seems to be responsible for this considerable loss of biological activity. Synthesis of ^{18}F-labeled octreotide using 4-[^{18}F]fluorophenacylbromide was not successful (Downer et al. 1995). A reactivity study with different nucleophiles revealed that this reagent not only has a preference for thiols, but also alkylates amines and carboxylic acid groups, while simple alcohols (Thr^6 and Thr-ol^8 of octreotide) catalyze the decomposition of the phenacyl bromide (Downer et al. 1995). ^{18}F-fluoroalkylations with small aliphatic groups (Block et al. 1987, 1988, Chi et al. 1987) were not used for peptide or protein labeling sofar.

2.1.3. ^{18}F-Fluorination by photochemical conjugation

Photoaffinity labeling (PAL) is a commonly used tool in biochemistry (i.e. for receptor, enzyme or topological labeling) (Chowdry and Westheimer 1979, Bayley and Staros 1984, v. Iddon 1979). It allows the covalent binding of a receptor ligand or substrate to its binding sites via UV irradiation Ruoho et al. 1973). For this purpose, the ligand is modified with a photolabil group. After incubation of the modified ligand to a tissue homogenate, the formed ligand-receptor complex is irradiated with UV-light. This leads to a covalent bridging of the receptor ligand complex and allows subsequent identification of the ligand specific binding site from the receptor mixture.

In order to overcome the time consuming preparation of ^{18}F-labeled prosthetic groups, Wester et al. (1994c, 1996b) investigated the suitability of peptide and protein labeling via 4-azidophenacyl-[^{18}F]fluoride ([^{18}F]APF). [^{18}F]APF contains a masked electrophilic group (Smith et al. 1990, Hixson and Hixson 1975) (see Fig. 2), an arylnitren, that can be generated by UV irradiation. It is assumed that labeling of peptides and proteins occurs within the short lifetime of the nitren, even when no preformed complex is formed (i.e. no affinity between prosthetic group and biomolecule exists); thus, one can call this technique photochemical conjugation (PCC). This approach was also recently investigated for the iodination of antibodies using 4-azido-2-hydroxybenzoic acid (Pandey et al.1991). N.c.a. [^{18}F]APF was prepared in one step by fluorine for bromine exchange on 4-azidophenacyl bromide (RCY 70%, 15 min). In a comparative study it was shown that reactive intermediates formed during photolysis (365 nm) of [^{18}F]APF in the presence of aqueous protein solutions (human serum albumin, transferrin, avidin, IgG) mainly react with ε-amino groups of lysine residues. Radiochemical yields of up to 30% were obtained after 5-10 min irradiation time with protein concentrations of 1-2 mg / 100µl (Wester et al. 1996a,1996b) (Fig.2).

However, octreotide labeling at the N-terminal amino group (D-Phe^1) failed, presumably due to the low nucleophilicity and pronounced sterical hindrance of this amino function. This assumption is in accordance with the general proposed reaction pathways of arylnitrens: phenylnitrens mainly react by nucleophilic addition, while polyfluorinated arynitrens mainly react by insertion (Bayley and Staros 1984, Chowdry 1979, Doering and Odum 1966).

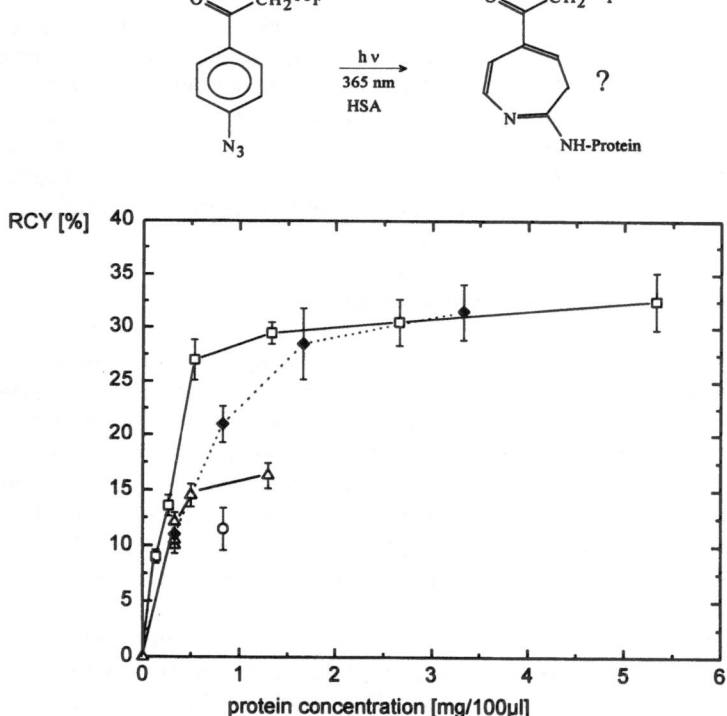

Figure 2:   Radiochemical yields of protein labeling via photolysis of 4-azidophenacyl-[^{18}F]fluoride as a function of protein concentration. (□) HSA,(◆) transferrin, (Δ) avidin and (O) IgG (Wester et al. 1996a,b).

Indeed, Panduranghi et al. (1995a,b) recently obtained a shift to more unspecific insertion by using 4-azido-2-([^{14}C]-methylamino) trifluorobenzonitrile and 4-azido-2,3,5,6-tetrafluoro-^{14}C-methylbenzoate. Using these reagents (Panduranghi et al., in press), the conjugation yields were dramatically improved (human serum albumin, 80-90%). Thus, despite the fact that polyfluorinated compounds are not ideal ^{18}F-labeling precursors (potential lowering of the specific activity), these investigations indicate the great potential of PCC for peptide and protein labeling. After further improvements, it may become an important tool in ^{18}F-labeling via prosthetic groups.

2.1.4. ^{18}F-Fluoroamidination

The reaction of imidates with proteins has been shown to be specific for ε-amino groups of lysine, forming a stable amidine derivative (Hunter and Ludwig 1962). Due to protonation of the amidine under physiological conditions, this conjugation methodology shows no disturbance of the net charge of the target molecule (Wofsy and Singer, 1963). In a comparative study Kilbourn et al. (1987) utilized methyl-[^{18}F]fluoro-5-nitrobenzimidate for the labeling of human serum albumin and fibrinogen. The reagent was prepared via a two step synthesis in about 1h with a radiochemical yield of 30-65%. In a more extensive study (Welch and Kilbourn, 1988) human transferrin, human IgG, hemoglobin, red blood cells, human platelets and HSA microspheres were labeled with the same reagent. The conjugation yields ranged form 16 to 60 % after 1h (based on benzimidate). However, due to the incomplete separation of the ^{18}F-labeling precursor 3,5-dinitrobenzonitrile in the first reaction step, the final product contained about 200-300 μg of methyl 3,5-dinitrobenzimidate. The pseudo carrier eluted close to or coincident with the radioactive peak and could not be separated by HPLC.

This common observed phenomenon -i.e. the elution of an aromatic nitro precursor close to the corresponding ^{18}F-labeled product- may generally be overcome by using a nitrophenyl RP-HPLC phase, as recently successfully utilized for the purification of butyrophenone neuroleptics (Hamacher et al. 1995). Furthermore, it should be expected that purification of methyl-[^{18}F]fluoro-5-nitrobenzimidate from its pseudo carrier not only improves the biological activity of the labeled bioactive compounds, but also may increase the conjugation yields.

2.1.5. ^{18}F-Fluoroamidation

The inverse ^{18}F-Fluoroacylation -i.e. labeling of an amine and coupling to an activated peptide- is only suitable for peptides with a small number of nucleophilic side chain functionalities. Human insulin, for instance, which consists of 51 amino acid residues exhibits only three amine groups (Geiger 1976): the N-termini of the two subchains (A-Gly[1] and B-Phe[1]) and Lys[29] in the B-chain. ^{18}F-labeling of insulin via amidation at B-Phe[1] was performed in four steps (Shai et al. 1989): a) syntheses of N-

[4-([^{18}F]fluoromethyl) benzoyl]-N'-(tert-butyloxycarbonyl) ethylen-diamine via fluorine-for-bromine exchange; b) deprotection of the product; c) amidation of A^1,B^{29}-di-Boc-B^1-[(N-hydroxysuccinimidyl)suberoyl]-insulin with the formed amine and d) final deprotection of the labeled peptide. The peptide precursor was synthesized by a) selective protection of A-Gly^1 and B-Lys^{29} with (BOC)$_2$O and b) formation of a preconjugate via acylation of A^1,B^{29}-di-Boc-insulin with an excess of disuccinimidyl suberate (DSS) at B-Phe^1. The whole synthesis was completed in 4 hours, specific activity was greater than 1,000 Ci/mmol, radiochemical yields were not reported. ^{18}F-labeled insulin retained the essential biological properties of native insulin, as measured in vitro by binding to insulin receptors on human cells and stimulating of glucose metabolism in rat adipocytes. Administration of ^{18}F-insulin in rhesus monkeys (S.A. 4,000-11,000 Ci/mmol) revealed a rapid decrease of activity in blood ($t_{1/2}$= 2-3 min). Specific receptor uptake was suggested since displacement experiments 5 min after tracer administration abruptly increased blood radioactivity and slowed clearance ($t_{1/2}$= 10-11 min) (Eastman et al. 1990).

This synthesis of ^{18}F-insulin reflects the main advantages and disadvantages of the amidation methodology: ease and straight foward preparation of the ^{18}F-labeled conjugation precursor, the time consuming and sophisticated preparation of the peptide precursor as well as the necessity of a final deprotection step on the conjugate. Thus, this reaction sequence is obviously limited to a few special peptides.

2.1.6. In vivo stability

The in vivo stability of the ^{18}F-labeled conjugates depends on three different parameters: i) the stability against in vivo defluorination, ii) the metabolic stability of the prosthetic group - peptide (protein) linkage and iii) the stability of the bioactive compound itself.

Obviously the last one is a constant for each target and is independent of the labeling strategy. Nevertherless, especially metabolic degradation of smaller peptides is often decreased by using D-amino acids instead of their L-forms at the N-terminal end or by reduction of the C-terminal end to an alcohol. However, this optimization often also affects the receptor affinity.

The stability of the prosthetic group towards defluorination is often the most important one. Undoubtedly, the aromatic bound fluorine species exhibits the highest in vivo stability. However, there are numerous examples of aliphatic ^{18}F-labeled radiopharmaceuticals (including ^{18}F-FDG (Hamacher et al. 1986, 1990)) which demonstrate sufficient stability. In several investigations it was shown that e.g. 2-[^{18}F]fluoropropionyl-labeled peptides and proteins exhibit a high in vivo stability towards defluorination, which is always indicated by the lack of activity accumulation in the bones (in the above mentioned case only about 2-3% ID/g 60 min p.i.; mice) (Guhlke et al. 1994c, Wester et al. 1996, Wester et al. in press). Thus, prosthetic groups with aromatic bound fluorine do not generally exhibit advantages with respect to in vivo defluorination.

Since acylations and amidations lead to the formation of amide bonds, the stability of this linkage is nearly equal to the high stability of the peptide bonds of the target (see e.g. Lang et al. 1995b). The stability of the other linkages is comparable to the obove mentioned results. Whereas, the activity accumulation in bones of mice after injection of ^{18}F-HSA (labeled with 4-azido phenacyl-[^{18}F]fluoride (PCC) or succinimidyl 4-[^{18}F]fluorbenzaote) shows comparable values (2.0% and 0.9% ID/g, respectively, 120 min p.i.), the faster blood clearance revealed a somewhat lower stability of the azepine linkage formed by the azide (Wester et al. 1996). Furthermore, the blood clearance of HSA (up to 4h) labeled with methyl 3-[^{18}F]fluoro-5-nitrobenzimidate (amidination) and 4-[^{18}F]fluoro-phenacylbromide (alkylation) was comparable to that obtained with iodinated HSA (via direct iodination), thus demonstrating the in vivo stability of these linkages (Kilbourne et al. 1987, Welch and Kilbourn 1988).

2.1.7. Specific activity

There are some important differences concerning the specific activity (SA) of low (lmw) and high molecular weight (hmw) radiopharmaceuticals prepared via prosthetic group labeling.

Due to the final purification of lmw tracers, undesired (potentially bioactive) side products as well as the excess of the (potentially bioactive) precursor is separated. In these cases the SA of the product is only a function of the absolute activity prepared and the dilution with isotopically labeled, non-radioactive product. The maximum obtainable SA is thus determined by the SA of the prosthetic group used .

However, these rules are not valid for prosthetic group labeling of hmw compounds (e.g. insulin, transferrin...). Except for high performance capillary electrophoresis (HPCE), which is presently limited to analytical applications (Westerberg et al. 1994), no purification method is available which allows the separation of hmw educts from their corresponding labeled products in a time compatible with the half-life of the short lived positron emitters. Thus, the SA is substituted in these cases by the apparent (or effective) specific activity (Meyer et al. 1993). This number is a function of the activity prepared, as well as the dilution with isotopically labeled, non radioactive product and unlabeled educt (both with nearly identical biologic activity), but do not depend on the SA of the prosthetic group used. However, since the use of a labeling reagent with low specific activity gives rise to multiple prosthetic groups per biomolecule (e.g. coupling to amino groups of small receptor binding proteins), which generally affects the biological integrity, the apparent specific activity is not a suitable parameter for quality assessment. Thus, two different parameters are needed for this purpose: i) the apparent specific activity of the labeled hmw ligand, and ii) the number of prosthetic groups per bioactive compound (conjugation number k). Using these parameters for quality control, the first one defines the activity per mass of peptide or protein; whereas, a low conjugation number guarantees minimal impairment of the target molecule, and thus reflects maximum preservation of the biological activity (Wester et al. 1996a).

Therfore, high specific activity is not only a prerequisite for a successfull synthesis of commonly used lmw receptor ligands, but is also a necessity for the preparation of prosthetic groups for subsequent labeling of hmw receptor ligands, in cases where final educt separation is not possible.

2.2. IODINE-120,124

There are two useful positron emitting iodine radioisotopes, iodine-120 ($T_{1/2} = 1.35$h) with a positron emission rate of 81%, and iodine-124 ($T_{1/2} = 4.2$d) with a positron emission rate of only 22%. Thus far, only iodine-124 has been applied. Due to its poor nuclear properties the use has been limited to labeling of some tumor agents such as antibodies (Bakir et al. 1992, Pentlow et al. 1991), metaiodobenzylguanidine, MIBG (Ott et al. 1992), and iodo-α-methyltyrosine, IMT (Langen et al.1990). Iodine-120 has only recently been prepared and used for labeling (Zweit et al. 1995). Its production via the Te-122(p,3n)-reaction requires a medium energy cyclotron with protons of at least 38 MeV. Furthermore, the high positron energy of up to 4.6 MeV is also a disadvantage for diagnostic imaging. However, in specific cases these radioiodines can be used in conjunction with PET, particularly in tumor research for receptor mapping, pharmacokinetics and dosimetry.

Radioiodination of peptides and proteins is well established. (for a comprehensive review see Wilbur et al. 1992). The most common procedure is the in situ production of electrophilic radioiodine from n.c.a. radioiodide by suitable oxidants such as chloramin-T, or-B, Iodo-gen or enzymatically by lactoperoxidase. Iodo-gen is particularly well suited for mild radioiodination of peptides and proteins, since it is insoluble in water, and thus provides a 2-phase reaction mixture, from which iodogen can easily be separated.

An alternative approach is prosthetic group labeling. It allows radioiodination of compounds which do not lend themselves to direct labeling. A proper choice of the prosthetic group also allows improvement of the stability against in vivo enzymatic deiodination, since direct radioiodination of a tyrosine residue gives rise to high in vivo deiodination. The iodogen method is also well suited for labeling of prosthetic groups (Krummeich et al. 1996), since the radioiodination of phenolic derivatives exhibits a maximum at pH 3-4 and 7-8 (Krummeich et al. 1996). Thus, the synthesis of radioiodinated Bolton-Hunter reagent (Bolton and Hunter 1973) as well as its O-methylated analog (Wester et al. 1995b) can advantageously be carried out in slightly acidic media, since the active ester is stable in this pH-range towards hydrolysis.

N-succinimidyl 4-radioiodobenzoate is also used as an active ester for protein labeling. This compound is superior to the Bolton-Hunter reagent with respect to in vivo deiodination (Garg et al. 1989a, Khwali et al. 1989, Murray et al. 1991, Wilbur et al. 1989). Labeling of this compound can be accomplished with 90% radiochemical yield via iododestannylation (Wilbur et al. 1989). The disadvantage of this method is the laborious synthesis of the tin precursor. An alternative technique is the Cu(I) assisted radioiododebromination using the method described by Moerlein (1990). The product 4-radioiodobenzoic acid can easily be separated by solid phase extraction from the precursor and then reacted with TSTU (Wester et al. 1995b). Radiochemical yields of 60% N-succinimidyl 4-radioiodobenzoate can be obtained (Krummeich et al. 1996).

2.3. METALLIC POSITRON EMITTERS

The vast majority of peptides and proteins have been labelled with gamma-emitting metals for SPET, such as Ga-67, Tc-99, In-111, or with therapeutic radionuclides. There are a few positron emitting radionuclides, with can be used in special cases for validation, quantification and pharmacokinetics. These are Ti-45, Fe-52, Cu-64, Ga-68, Y-86 and Tc-94m.

Some radiometals, such as technetium can be bound directly via coordination with functional groups of a peptide or protein. In most cases, however, the metal ions need a chelating agent, which can be attached covalently to the peptide. These bifunctional chelating agents can be covalently conjugated to the peptide before or after chelating the metalic nuclide, thus forming a stable radionuclide-chelate-peptide complex. In some cases a linker can be included. Many parameters can be varried to improve the chemical and biological properties of the ligand. This principle is depicted in Fig. 3.

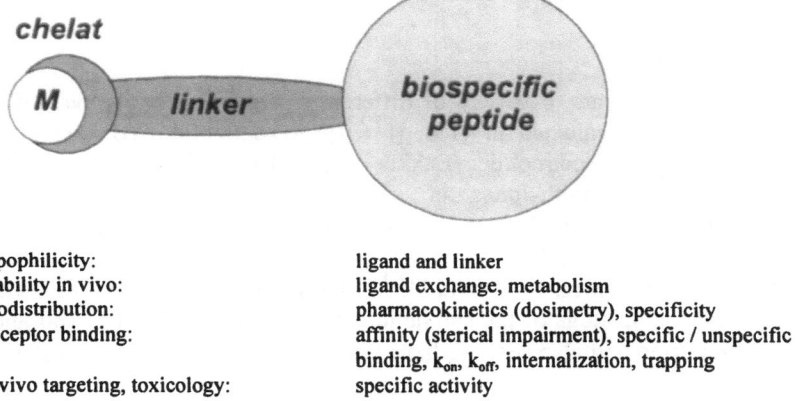

Lipophilicity:	ligand and linker
Stability in vivo:	ligand exchange, metabolism
Biodistribution:	pharmacokinetics (dosimetry), specificity
Receptor binding:	affinity (sterical impairment), specific / unspecific binding, k_{on}, k_{off}, internalization, trapping
in vivo targeting, toxicology:	specific activity

Figure 3:   Principle of peptide labeling with radiometal

Depending on their radii and coordination number of the metal, the type of ligand and linker, properties such as the in vivo stability, affinity and lipophilicity can be changed.

3. Specific examples of labeling strategies

3.1. OCTREOTIDE

Radiolabeled, synthetic derivatives of somatotropin release inhibiting factor (SRIF) are potential ligands for of the human somatostatin receptor (hSSTR) subtypes hSSTR2 to SSSTR5 (Reynor and Reisine 1992, Bruns et al. 1994). SMS 201-995 (octreotide) represents a synthetic octapeptide with improved in vivo stability (Pless et al. 1986, Kutz et al. 1986) and receptor selectivity for hSSTR2 and hSSTR5 (Raulf et al. 1994, Bruns et al. 1994).

Table 3: IC_{50}-Values [nM] of some endogenous fragments of Somatotropin Release Inhibiting Factor (SRIF) compared to SMS 201-995 (octreotide) (Bruns et al. 1994)

Peptide	SRIF (Somatostatin) Receptor Subtypes (human Subtypes)				
	hSSTR1	hSSTR2	hSSTR3	hSSTR4	hSSTR5
SRIF-14	0.40	0.10	0.25	0.79	0.59
SRIF-25	0.79	0.32	0.40	1.26	n.d.
SRIF-28	0.79	0.40	0.63	1.00	0.28
SMS 201-995	>1000	0.32	31.6	>1000	7.3

n.d.: not determined

During the last years, a variety of different radiolabeled derivatives of octreotide were prepared and evaluated in vivo: ($[^{123}I]Tyr^3$-octreotide) (Bakker et al. 1990), $[^{111}In]DTPA$-(D)Phe^1-octreotide (Bakker et al. 1991), $[^{67,68}Ga]DFO$-B-succinyl-(D)Phe^1-octreotide (Smith-Jones et al. 1994), 2-$[^{18}F]$fluoropropionyl-(D)Phe^1-octreotide (Guhlke et al. 1994), $[^{99m}Tc]PnAO$-(D)Phe^1-octreotide (Marina et al. 1994), $[^{99m}Tc]N_4Bz$-(D)Phe^1-octreotide (Stolz et al. 1994), $[^{64}Cu]TETA$-(D)Phe^1-octreotide (Anderson et al. 1994) and $[^{161}Tb]DTPA$-(D)Phe^1-octreotide. Additionally, some radiolabeled analogs of the synthetic octapeptides RC-121, RC-160 and RC-161 have been reported and also seem to have potential as tracers for somatostatin receptor imaging.

The amino acid sequence of octreotide is shown in figure 4. Starting from the hypothetical conformation of natural somatostatin and a knowledge of the minimal fragment needed for biological activity, a process of rational design has led to this optimized receptor ligand (Pless et al. 1986). The design started from the essential tetrapeptide sequence *Phe-Trp-Lys-Thr*, cyclized by cystine. Incorporation of Phe at the

NH_2-terminal end and COOH-terminal elongation of the cyclic lead by Thr further improved its activity. The substitution of Phe[1] and Trp[4] by their D-isomers improves the biological activity; whereas, the D-isomer of Phe[1] and the reduction of the C-terminal Thr to the corresponding amino alcohol further increases the in vivo stability towards enzymatic degradation.

Since derivatization in the four amino acid sequence *Phe-(D)Trp-Lys-Thr* would destroy the high affinity of octreotide to its receptor (Pless et al. 1986), the introduction of a radiolabeled moiety (labeled prosthetic group or ligand for radiometal complexation) can only be performed at the N-terminal (D)Phe[1]. Therefore, all known radiolabeled derivatives of octreotide exploit this amino acid as linkage unit. Surprisingly, the only exception, the iodinated analoge [^{123}I]Tyr[3]-octreotide, shows a higher affinity ($pK_i = 9.9\pm0.3$) (Bakker et al. 1990, 1991) to SSTR than octreotide itself ($pK_i = 9.33\pm0.06$) (Bruns et al. 1993), which suggests that this probe more accurately has access to a hydrophobic region of the receptor.

As mentioned above, experiments for the direct labeling of octreotide in aqueous solution using [^{18}F]SFB were unsuccessful. However, ^{18}F-fluorinated octreotide with high affinity to SSTR was synthesized via a two step procedure (Guhlke et al. 1994), which started with the acylation of ε-Boc-Lys[5]-octreotide using the 4-nitrophenyl 2-[^{18}F]fluoropropionate / HOBt system. Subsequent deprotection yielded 2-[^{18}F]fluoropropionyl-(D)Phe[1]-octreotide in about 65% based on the acylation agent (scheme 3). Competetive binding studies on rat cortex membranes revealed that the binding affinity of this compound is about one order of magnitude lower than that of the iodinated SRIF anaologue, but is not significantly different from that of [[nat]In]DTPA-(D)Phe[1]-octreotide (Bakker et al. 1991). The IC_{50} values obtained are in the nanomolar range (3×10^{-9} M), thus demonstrating a high affinity to SSTRs.

Influence of the lipophilicity and the net charge

The lipophilicity (Beschiavelli and Seelig 1990, 1991, 1992) and the net charge of the labeled octreotide derivatives (Seelig et al. 1993) significantly affect their biodistribution and receptor binding properties.

Table 4 shows the kidney to liver ratios and the kidney to intestine ratios of different labeled octreotide derivatives. These ratios reflect the extent of renal excretion to hepatobiliary excretion. It can be summarized that the amount of hepatobiliary excretion parallels the lipophilicity. Thus, the very lipophilic [^{99m}Tc]PnAO-octreotide (Maina et al. 1994) exhibits the highest hepatobiliary uptake among the metal - ion labeled analogs, while [^{111}In]-DTPA-octreotide (Bruns et al. 1993) is predominantly excreted by the kidneys. It should be mentioned that the interpretation of images of abdominal located tumors may be complicated by increasing hepatobiliary excretion, as also noted in previously described SPET studies using [^{123}I]Tyr[3]-octreotide (de Jong et al. 1993).

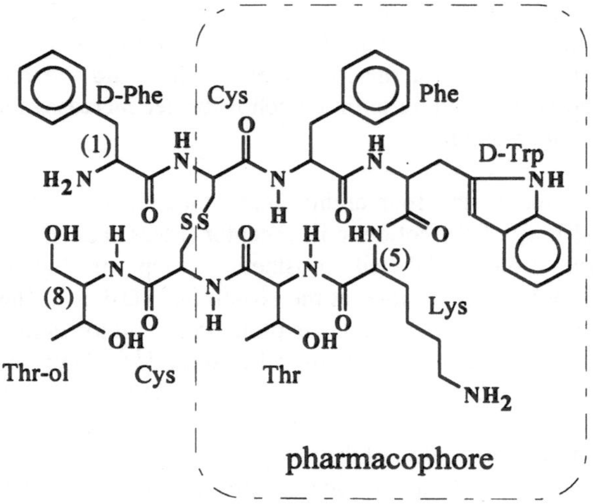

(D)Phe^1—Cys—Phe—(D)Trp—Lys^5—Thr—Cys—Thr-ol   (SMS 201-995)

Figure 4 :   SMS 201-995 (octreotide), a synthetic analog of the Somatotropin Release
Inhibiting Factor (SRIF) with pharmacophore

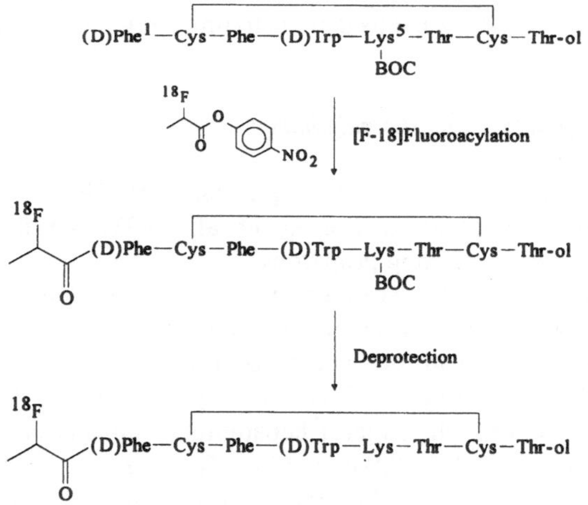

Scheme 3:   Synthesis of (2-[^{18}F]fluoropropionyl-(D)phe1)-(SMS 201-995) via ^{18}F-fluoroacylation of
(ε-BOC-Lys^5)-(SMS 201-995) with 4-nitrophenyl 2-[^{18}F]fluoropropionate and subsequent
deprotection (Guhlke et al.1994)

Table 4

Hepatobiliary excretion of different radiolabeled octreotide derivatives as a function of their lipophilicity: kidney to liver and kidney to intestines ratios.

←——————————————————— lipophilicity ——————————————————→

organ ratio	$[^{111}In]$-DTPA-octr. (1 h p.i.)	$[^{67}Ga]$-DFO-octr. (1 h p.i.)	$[^{86}Y]$-DTPA-octr. (1 h p.i.)	2-$[^{18}F]$-fluoro-prop.-octr. (1 h p.i.)	$[^{123}I]Tyr^3$-octr. (0.5 p.i.)	$[^{99m}Tc]$-PnAO-octr. (1.5 h p.i.)
kidney / liver	6.38 (Bruns et al. 1993)	5.11	1.69	1.06	1.15 (Bakker et al. 1990)	0.34 (Maina et al. 1994)
kidney / intestine	15.16 (Bruns et al. 1993)	27.60	6.47	0.68	0.56 (Bakker et al. 1990	1.05 (Maina et al. 1994

However, with the development of less lipophilic somatostatin derivatives, such as the glycated SDZ CO 611(Albert et al. 1993, Fricker et al. 1994), which exhibits predominantly renal excretion, radiohalogenation may lead to more suitable PET and SPET tracers for mapping of SSTRs.

The influence of the net charge of SMS 201-995 derivatives on the receptor binding properties is mainly attributed to an electrostatic attraction between the radiolabeled ligand (with almost positive net charge) and the negatively charged cell membrane (surface). Thus, the electrostatic field on the surface of the cells leads to an increasing concentration of positively charged ligands in the interfacial layer directly in front of the cell surface. It was found that the concentration of different SMS 201-995 derivatives with net charges of z = +0.4, +1.4 and +2.4 directly in front of the cell surface is dramatically increased. Thus, compared to a concentration of 1 nM in the bulk solution, the interfacial concentration was 3.1, 9.2 and 121 nM, respectively (Seelig et al. 1993). These increasing calculated concentrations parallel the binding affinity of the corresponding ligands.

Structure activity relation

In the case of octreotide, different labels causing various changes of the structure show little or no significant influence on the affinity of the SRIF receptor (Fig 5).

76

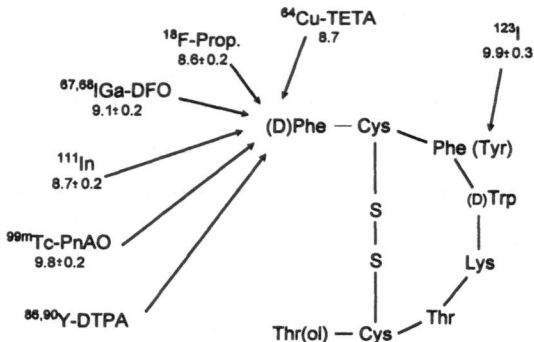

Figure 5:   Somatostatin receptor affinities (binding constants for rat cortex membranes -pK$_i$-)
(Bakker et al. 1990, 1991, Anderson et al. 1994, Guhlke et al. 1994, Smith-Jones et al.
1994, Maina et al. 1994, Brockmann et al. 1995a, Wester et al. in press)

Comparison of the tumor uptake kinetics of (2-[^{18}F]fluoropropionyl-(D)Phe^1)-octreotide (I) with that of [^{67}Ga]-DFO-B-succinyl-(D)Phe^1-octreotide (II) and [^{86}Y]DTPA-(D)Phe^1-octreotide (III) (Table 5) in tumor rats indicates that the ^{18}F-labeled tracer has the shortest residence time in tumor, k$_{off}$ being a factor of three higher than that of the radiometal complexes (Brockmann et al. 1995, Wester et al.1995a, Wester et al. in press) Due to its fast clearance from the blood, the tumor to blood ratio is significantly higher than that of the metal labeled analoges. Its behaviour resembles that of the original C-14 labeled molecule (Lemaire et al. 1989), to which its structure is also closest. It also exhibits the highest in vivo stability: even 120 min after injection only the intact tracer can be found in all specimen, while the metal complexes particularly the [^{67}Ga]-DFO-B-succinyl-(D)Phe^1-octreotide was rapidly metabolized. The low fluorine-18 activity in the bone, even after two hours, indicates high in vivo stability against defluorination.

Blocking and displacement studies on various SRIF receptor bearing organs have been carried out with the fluorine labeled compound (Wester et al., in press). The highest effects are observed for the pancreas and adrenals, but only small changes for the pituitary (Fig.6). It is interesting to note that the extent of displacement decreases when octreotide is administered at later times after the injection of the radiotracer, thus indicating internalization (Amherdt et al. 1989, Drazin et al. 1985, Morel et al. 1994, 1986) which was also been observed for ^{111}In-DTPA-(D)phe^1-octreotide by Breeman et al. (1995). A similar time dependence of displacement was observed previously for the D2 receptor ligand [^{18}F]FESP (Stöcklin 1991).

Table 5: Tumor uptake kinetics and residence time of (2-[^{18}F]fluoropropionyl-(D)phe^1)-octreotide [I], [^{67}Ga]-DFO-(D)phe^1-octreotide [II] and [^{86}Y]-DTPA-(D)phe^1-octreotide [III] in tumor bearing rats as derived from PET measurements and ex-vivo biodistribution data (Wester et al. in press)

Compd.	time post injection	tumor uptake		k_{off}
	[min]	PET [%iD/ml]	ex-vivo biodistribution [%iD/g]	[$*10^{-5}s^{-1}$]
[I]	5	0.39 ± 0.17	0.43 ± 0.28	
	60	0.40 ± 0.05	0.52 ± 0.24	10 ± 2
	120	0.20 ± 0.05	0.17 ± 0.04	
[II]	5		0.34 ± 0.03	
	60		0.82 ± 0.11	3.0 ± 0.5
	120		0.57 ± 0.18	
[III]	60	0.31± 0.05	0.44 ± 0.17	3.1 ± 1.3
	120	0.26 ± 0.01	0.39 ± 0.05	

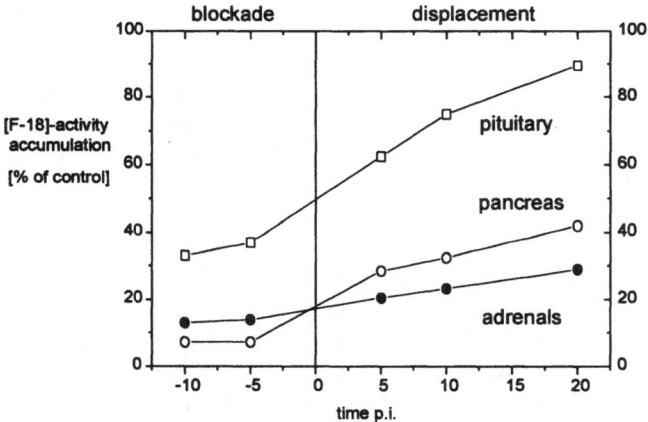

Figure 6: Pretreatment and displacement of 2-[^{18}F]fluoropropionyl-(D)phe^1)-octreotide with octreotide (1 mg / kg) on non tumor bearing male Lewis rats (Wester et al. in press)

In conclusion it can be summarized that the ^{18}F-labeled octreotide is an ideal tracer as far as stability and receptor mapping is concerned, probably due to the small changes of the structure. The tumor accumulation and the affinity, however, is very similar or even slightly poorer than that of the metal complexes studied.

3.2   VASOACTIVE INTESTINAL PEPTIDE

Vasoactive intestinal peptide (VIP) is a 28-amino acid peptide of the glucagon-secretin receptor family. Due to the high receptor density in several tumor tissues (Reubi et al. 1995), VIP was recently utilized for in vivo receptor scintigraphy after radioiodination of its two tyrosine residues (Tyr^{10} and Tyr^{22}). Iodination was performed both by the Iodo-gen and the lactoperoxidase (LPO) method. After LPO labeling, the two major ^{123}I-peaks were identified as (mono[^{123}I]iodo-Tyr^{10})VIP and (mono[^{123}I]iodo-Tyr^{22})VIP, whereas the Iodo-gen method led to the formation of (mono[^{123}I]iodo-Tyr^{10}, $MetO^{17}$)VIP and (mono[^{123}I]iodo-Tyr^{22}, $MetO^{17}$)VIP, in which the methionine was found to be oxidized (Angelberger et al. 1995). The same results were observed with chloramine -T (CAT) as oxidizing agent (Hassan et al. 1994). However, using the oxidized monoiodinated products as mixture, Virgolini et al. found in vivo a significantly elevated number of VIP receptors both in intestinal adenocarcinomas and in endocrine tumors (Virgolini et al. 1994, 1995). ^{123}I-VIP receptor scan was not helpful in the localization of primary or reccurent medullary thyroid carcinomas (Kurtaran et al. 1996). Recently, Reubi et al. (1995) investigated the VIP-R content of 339 human tumors and found that generally VIP-Rs are found much more frequently than SSTRs in all tumors tested and that the VIP-R were of high affinity and specificity for VIP.

3.3.   INSULIN

Insulin (51 amino acid residues) contains two intercatenary and one intracatenary disulfide bridge, four tyrosines (2 in the acidic (A) chain and 2 in the basic (B) chain) and three amino functions (N-terminal A-Gly^1, N-terminal B-Phe^1 and B-Lys^{26}) (Geiger 1976).

In contrast to most other peptides, the biological activity of insulin is not coupled to a discrete amino acid sequence of the peptide but rather to its tertiary structure. Thus, about 30% of all the amino acids can be substitiuted by alanine (Geiger 1976). In addition, acylation at the B-chain (both at Phe^1 and Lys^{26}) does not alter significantly the activity of insulin, whereas acylation at A-Gly^1 was found to have a more pronounced effect (Geiger 1976). In agreement with these observations, ^{18}F-amidation of insulin (see above) at B-Phe^1 yielded ^{18}F-insulin (Shai et al. 1989) with unaffected bioactivity (Eastman et al. 1990). However, the same results may also be obtained with high conjugation yields without sophisticated protection chemistry by direct ^{18}F-acylation of unprotected insulin at B-Lys^{26}.

The elimination of the disulfide bridge between A6-A11 as well as the opening of the intercatenary A7-B7 of insulin lead to peptides with reduced but still significant activity, while a complete loss of activity was obtained after enlargement of the whole structure by substitution of the Cys^7 and Cys^{20} by homocysteines (Geiger 1976). Thus, in comparison to the ^{18}F-fluorinated compound, ^{99m}Tc-labeling of insulin by direct tin reduction (Awasthi et al. 1993) should alter the structure to a higher extent. Unfortunately, no data concerning the biological activity or affinity to the binding sites of insuline were reported for the ^{99m}Tc-labeled compound.

Interestingly, the iodination of insulin using lactoperoxidase predominantly occurs at A-Tyr^{14}. The sum of side products -free iodide, [^{123}I]iodo-Tyr^{A19}-insulin, [^{123}I]iodo-Tyr^{B26}-insulin, [^{123}I]iodo-Tyr^{B16}-insulin and ^{123}I$_2$Tyr-insulin- is smaller than 50% (Kurtaran et al. 1995).

Eastman et al. showed that ^{18}F-insulin binds in vitro to insulin receptors on human cells with high affinity (Shai et al 1989), and successfull displacement and competition experiments with this tracer in vivo (rhesus monkey) were accounted to a substantial uptake to specific receptors in the liver (Eastman et al.1990). Recently, a double tracer methodology using ^{99m}Tc-neoglycoalbumin combined with ^{123}I-Tyr^{A14}-insulin was reported (Kurtaran et al. 1995). Due to a 1000-fold higher number of of specific receptors for the latter compound on human hepatocellular carcinoma (HCC) (1216± 290 pmol/mg protein on HCC to 2.4±0.8 pmol/mg protein on normal liver tissue) they obtained HCC / normal liver ratios of 1.6±0.1, whereas all HCC lesions were identified as cold spots after injection of ^{99m}Tc-NGA.

3.4. APAMIN

Eventhough this peptide is no tumor seeking agent, it should be mentioned in view of labeling strategy aspects. This octadecapeptide is found in the venom of the honey bee (Gauldie et al. 1976), is known to be a selective blocker of Ca^{2+}-dependend potassium channels ($K^+_{Ca}{}^{2+}$) (Hugues et al. 1982) and exhibits high affinity to its binding sites (K_D 15-60 pM). As recently shown by post mortem autoradiography, an anatomically discrete loss of $K^+_{Ca}{}^{2+}$-channels within the hippocampus occurs in Alzheimer disease (Ikeda et al. 1991). Since apamine is different from most peptide neurotoxins in its unusual ability to cross the blood brain barrier (Habermann 1984) radiolabeled apamin is of potential interest for in vivo receptor mapping of $K^+_{Ca}{}^{2+}$-channels with SPET and PET.

The identification of the pharmacophoric sequence of apamin was carried out by selective derivatization of Cys^1, Lys^4, Arg^{13,14}, His^{18} and Glu^7 as well as by segment substitution. (Greiner et al. 1978). These data revealed that single derivatization of Cys^1, Lys^4 and His^{18} does not affect the affinity of apamin to $K^+_{Ca}{}^{2+}$-channels and that the Ala^{12} to His^{18} sequence has to be unchanged.

Thus, apamin was labeled with ^{123}I and ^{18}F by selective acylation of Lys^{14} by [^{18}F]SFB and [^{123}I]O-methylated Bolton-Hunter reagent [^{123}I]OMeBH (Wester et al. 1995b, Fixmann et al.1995) with radiochemical yields of up to 60%. To determine the extent of competing acylation at Cys^1 and His^{18} and to check the assumption that protection is not necessary at Arg^{13,14}, initial acylations were carried out on suitable amino acid mixtures as a function of pH. In accordance with reported data on the distribution of binding sites for potassium channel ligands in rat brain (Gehlert and Gackenheimer 1993), the activity uptake of both compounds shows a significantly higher uptake in the hippocampal area (Wester et al. 1995b).

3.5. TRANSFERRIN

Considering the high molecular weight of transferrin (approx. 79 kD, 678 amino acids) (MacGillivary et al. 1983), conjugation with a small prosthetic group should not alter the biological activity. Thus, the strategy for the radiolabeling of a high molecular weight peptide receptor ligands is limited to the use of a prosthetic group with sufficient high specific activity (see *Specific activity*).

The physiologic role of transferrin is to shuttle iron insight cells by i) binding of (iron containing) holo-transferrin to the transferrin receptor (TFR), ii) internalization via receptor mediated endocytosis, iii) release of the iron, and iv) rapid recycling to the extracellular compartement via coated vesicles. In every stage of this process, the TF-TFR complex remains intact (Aloij et al. 1996). The density of the TF-receptor on tumor cells was found to be in the range of 10^5 per cell with K_D in the low nanomolar range (Eckelman et al. 1994 and references herein).

Due to the increased need of iron in tumors, ^{67}Ga, which mimics the transferrin mediated iron transport, has been used as tumor seeking agent. To obtain a tracer for a better quantitation of transferrin kinetics, ^{18}F-labeled transferrin (Welch and Kilbourn 1988, Wester et al. 1994c, 1996b, Aloij et al. 1996) was recently evaluated in vitro (Aloij et al. 1996). In contrast to a ^{67}Ga-transferrin accumulation in the cells, which linearly increases with time (liberation of the ^{67}Ga after endocytosis), the ^{18}F-labeled transferrin rapidly reached a steady state equilibrium between the intra- and the extracellular compartment, thus demonstrating the recycling of intact transferrin to the extracellular site.

3.6. CHEMOTACTIC PEPTIDES

Chemotactic peptides again cannot be used for tumor targeting, but initiates leukozyte chemotaxis by binding to high affinity receptors a the cell surface of neutrophils and phagocytes (Williams et al. 1977, Schiffmann et al. 1975). The parent compound contains the amino acid sequence N-formyl-Met-Leu-Phe. Both changes of the N-terminal formyl residue as well as substitutions of the Leu-Phe dipeptide significantly decrease the affinity to the binding sites. However, more potent analogs are obtained from a hexapetide N-formyl-Nle-Leu-Phe-Nle-Tyr-Lys. Since derivatization of the C-terminal Lys amino group does not affect the affinity, this peptide was labeled with ^{111}In (Babich 1995) and ^{99}Tc via linkage of a chelator

(Fishman 1991, Babich et al. 1993) and with ^{18}F via acylation with [^{18}F]SFB to ε-Lys^6 (Vaidyanathan et al. 1994a, 1995).

Except for the ^{18}F-fluorinated compound, which exhibited a high hepatobiliary excretion and a somewhat decreased affinity (which may be improved by using a specific activity of [^{18}F]SFB higher than 200Ci/mmol) the tracers showed a good localization at the site of infection.

3.7. 'RGD' - PEPTIDES

Integrins are heterodimeric transmembrane glycoproteins, which are involved in cell-cell and cell-matrix adhesion. These cell surface receptors consist of one α-subunit (about 1100 amino acids) and one smaller β-subunit (Haubner et al. 1996a,b and literature therein). The extracellulary part of the α-subunit is characterized by three to four divalent cation-binding (Ca^{2+}, Mg^{2+}) amino acid sequences (about 12 to 15 amino acids). The specificity for ligand binding is determined by a particular combination of α- and β-subunits.

Among the presently known integrin subfamilies, nine receptor subtypes (including the three β_3 -integrins) bind their corresponding ligands via the RGD-sequence (Arg-Gly-Asp). Two of them are the members of the β_3 - subfamily: $\alpha_v \beta_3$ and $\alpha_{IIb} \beta_3$ (often referred as GPIIb/IIIa).

The $\alpha_{IIb} \beta_3$ receptor plays an important role in hämostasis and thrombosis, since it mediates the aggregation of activated platelets. Fibrinogen, fibronectin or vitronectin are endogenous ligands for this receptor subtype. Inhibitors of these receptors are used as antithrombotic agents.

Several cyclic and linear $\alpha_{IIb} \beta_3$ receptor antagonists were labeled with Tc-99m sofar (Muto et al. 1995, Liu et al. 1996, Barret et al. 1996, Harris et al. 1994, Lister James et al. 1996) for thrombosis imaging. Muto et al. (1995) have reported success with P-280, a Tc-99m labeled dimeric peptide (Knight et al. 1993), in eight of nine patients with deep vein thrombosis (Diatech has entered phase three clinical trials with P-280). The patient, who did not have P-280 visualization of clot was imaged 42 days after the acute event.

Other peptides with high affinity to the $\alpha_{IIb} \beta_3$ are PAC-xx peptides (Knight et al. 1994) or CYT 379 (Ben-Haim et al. 1193). These peptide sequences have been developed on the base of the active binding region of the monoclonal antibody PAC1, which binds to activated platelets (Shattil et al. 1985).

The $\alpha_v \beta_3$ -integrin, a vitronectin receptor, is of great interest concerning metastasis of tumor cells. Metastasis of several tumor cell lines as well as tumor induced angiogenesis can be inhibited by small synthetic peptides acting as ligands for this receptor. Among these ligands, especially small cyclic pentapeptides seem to be promising candidates for targeting the interactions of integrin during angiogenesis and metastasis (Haubner et al. 1996a, 1996b, 1996c).

Due to the fast blood clearance of these peptides, suitable PET tracers could be obtained by ^{18}F-labeling. Furthermore, also the $\alpha_v \beta_3$ -integrin seems to be a promising target, both in view of tumor imaging and endoradiotherapy.

4. Conclusion and Outlook

While numerous ligands have been developed for central receptors, particularly for the dompaninergic and serotonergic system, peripheral receptors have been somewhat neglected. Even though they may have a larger impact on diagnostic and therapeutic nuclear medicine than the central ones. This is particularly true for tumor targeting. The rapid progress in molecular biology provide new concepts for the design of high affinity ligands.

For the radiopharmaceutical chemist, labeling and evaluating of peptides provide a challenging multidisciplinary research area which is somewhat different from the labeling chemistry of small molecules.

Among the new generation of peptides for tumor targeting, ligands for the somatostatin receptor, the vasoactive intestinal peptide receptor and RGD-containing peptides for the integrins presently seem to have the greatest potential.

For preliminary screening, radioiodination is generally sufficient for the in vivo binding studies and biodistribution in rodents. If the results are promising, labeling with PET tracers can be pursued. For quantitative measurements of the receptor status or other diagnostic purposes, fluorine-18 is the most useful tracer with respect to half-life and stability. Numerous relatively simple labeling procedures have recently been developed, but further research in peptide labeling with positron emitters is necessary. This includes metallic radionuclides as positron emitting substitutes for therapeutic radiometals which do not lend themselves to quantitative pharmacokinetics and dosimetry.

5. References

Albert R., Marbach P., Bauer W., Briner U., Fricker G., Bruns C., Pless J. (1993) SDZ CO 611: A highly potent glycated analog of somatostatin with improved oral activity. Life Sci. 53, 517-525

Aloj L., Lang L., Jagoda E., Neumann R.D., and Eckelman W.C. (1996) Evaluation of human transferrin radiolabeled with N-succinimidyl 4-[fluorine-18](fluoromethyl) benzoate. J Nucl. Med. 37, 1408-1412

Amherdt M., Patel Y.C., Orci L. (1989) Binding and internalization of somatostatine, insulin and glucagon by cultured rat islet cells. J.Clin.Invest. 84, 412-417

Anderson C.J., Edwards W.B., Pajeaun T.S., Zinn K.R., Welch M.J. (1994) Cu-64-TETA-D-Phe-1-octreotide: A positron emitting somatostatin receptor imaging agent with receptor binding and renal clearance similar to In-111-DTPA-D-Phe-1-Octreotide. J.Nucl.Med. 35, 106p (abstr)

Angelberger P., Virgolini I., Buchheit O., Egger M., Portner R., Kvaternik H. (1995) Radiopharmaceutical development of ^{123}I-vasoactive intestinal peptide (VIP) for receptor scintigraphy in oncology. J.Label. Compds. Radiopharm. 37, 502-503.

Awashti V., Gambhir S., Sewatkar A.B. (1993) ^{99m}Tc-Insulin: Labeling, biodistribution and Scintiimaging in Animals. Nucl.Med.Biol. 21, 251-254

Babich J.W. Solomon H., Pike M.C., Kroon D., Graham W., Abrams M.J., Tompkins R.G., Rubin R.H: Fischmann A.J. (1993) Technetium-99m-Labeled Hydrazino Nictotinamide Derivatized chemotactic Peptide Analogs for Imaging Focal Sites of Baceterial Infection. *J.Nucl.Med.* **34**, 1964-1974

Babich J.W., Graham W., Barrow S.A., Fischman A.J. (1995) Comparison of the infection imaging properties of a ^{99m}Tc labeled chemotactic peptide with ^{111}In-IgG. *Nucl. Med. Biol.* **22**, 643-648

Bakir M.A., Eccles S.A., Babich J.W., Aftab N., Styles J.M., Dean C.J., Ott R.J. (1992) c-erbB2 Protein overexpression in breast cancer as a target for PET using iodine-124-labeled monoclonal antibodies. *J.Nucl.Med* **33**, 2154-2160

Bakker W.H., Krenning E.P., Breeman W.H., Hofland L.J., Marbach P., Pless J., Pralet P., Stolz B., Koper J.W., Lamberts S.W.J., Visser T.J., Krenning E.P. (1991) [^{111}In-DTPA-D-phe[1]]octreotide, a potential radiopharmaceutical for imaging of somatostatin receptor positiv tumors: synthesis, radiolabeling and in vitro validation. *Life Sci.* **49**, 1583-1591

Bakker W.H., Krenning E.P., Breeman W.H., Koper J.W., Kooij P.P., Reubi J.C., Klijn J.G., Visser T.J., Docter R., Lamberts S.W. (1990) Receptor scintigraphy with radioiodinated somatostatine analogue: radiolabeling, purification and in-vivo application in animals. *J.Nucl.Med.* **31**, 1501-1509

Barany G. and Merrifield R.B. (1980) Solid phase peptide synthesis. In: The Peptides: Analysis, Synthesis, Biology. Vol.2.: Special methods in peptide synthesis (Part 1). Chap. 1, 185. Eds: E.Gross und J.Meienhofer. Academic Press, New York

Barrett J.A. Damphousse D.J., Heminway S.J., Liu S., Edwards D.S., Looby R.J., and Caroll T.R. (1996) Biological ecaluation of Tc-99m-labeled cyclic glycoprotein IIb/IIIa receptor antagonists in the canine arteriovenous shunt and deep vein thrombosis models: Effects of chelators on biological properties of [^{99m}Tc]chelator-peptide conjugates. *Bioconjugate Chem.* **7**, 196-202

Bayley H., and Staros J.V. (1984) Photoaffinity labeling and related techniques. In: Azides and Nitrenes. Ed.: E.F.V.Scriven. Chap. 9, 433-490. Academic Press, Orlando

Ben-Haim S., Kahn D., Weiner G.J., Madsen M.T., Waxman A.D., Williams C.M., Clarke-Pearson D.L., Coleman R.E., Maguire R.T. (1993) The Safety and Pharmacokinetics in Adult Subjects of the Intravenously Administered ^{99m}Tc-labeled 17 Amino Acid Peptide (CYT-379). *Nucl.Med.Biol.* **21**, 131-142

Benyai I., Lupattelli G., Li S.R., Pomgratz S., Yang Q., Böck P., Angelberger P., Virgolini I. (1994) Comparison of iodine-123 low density lipoprotein (LDL) and indium-111 LDL to mononuclear cells of healthy normolipaemic controls and patients with heterozygous familial hypercolesterolaemia. *Eur. J. Nucl. Med.* **21**, 634-639

Beschiaschvili G. and Seelig J. (1990) Peptides Binding to lipid bilayers. Binding isotherms and ζ-potential of a cyclic somatostatin analogue. *Biochemistry* **29**, 10995-11000

Beschiaschvili G. and Seelig J. (1991) Peptide binding to lipid membranes. Spectroscopic studies on the insertionof a cyclic somatostatin analog into phospholipid bilayers. *Biochimica et Biophysica Acta* **1061**, 78-84

Beschiaschvili G. and Seelig J. (1992) Peptides Binding to lipid bilayers. Nonclassical hydrophobic effect and membrane-induced pK shifts. *Biochemistry* **31**, 10044-10053

Block D., Coenen H.H., and Stöcklin G. (1988) N.c.a ^{18}F-fluoroacylation via fluorocarboxylic acid esters. *J.Label.Compds.Radiopharm.* **25**, 185-200

Block D., Coenen H.H., Laufer P., Stöcklin G.. (1986) N.c.a [^{18}F]-Fluoroalkylation via nucleophilic fluorination of disubstituted alkanes and application to the preparation of N-[^{18}F]-fluoroethylspiperone. *J.Label.Comp.Radiopharm.* **23**, 1042-1044

Block D., Coenen H.H., Stöcklin G. (1987) N.c.a ^{18}F-fluoroalkylation of H-acidic compounds. *J. Label. Comp.Radiopharm.* **25**, 201-216

Bolton A.E., and Hunter W.M. (1973) The labeling of proteins to high specific radioctivities by conjugation to a ^{125}I-containing acylation agent. *Biochem.J.* **133**, 529-539

Breeman W., Kweekeboom D.J., Kooij, P.M., Bakker W.H., Hofland L.J., Visser T.J., Ensing G.J., Lamberts S.W.J., Krenning E.P. (1995) Effect of dose and specific activity on tissue distribution of In-111-pentetreotide in rats. *J.Nucl.Med.* **36**, 623-627

Breeman W.A.P., Van Hagen M.P., Visser-Wisselaar H.E., van der Pluijm M.E., Koper J.W., Setyono-Han B., Bakker W.H., Kwekkeboom D.J., Hazenberg M.P., Lamberts S.W.J., Visser T.J., Krenning E.P. (1996) In Vitro and In Vivo Studies of Substance P Receptor Expression in Rats with the New Analog [Indium-111-DTPA-Arg[1]]Substance P. *J.Nucl.Med.* **37**, 108-117

84

Brockmann J. and Rösch F. (1995b) Labeling of small peptides with yttrium and gallium via a new DOTA-derivative of DFO using different disuccinimidyl esters as linkers. *J.Label.Compds.Radiopharm.* **37**, 516-518

Brockmann J., Rösch F., Herzog H., Stolz B., Bruns C., Stöcklin G. (1995a) Un vivo uptake kinetics and dosimetric calculations of ^{86}Y-DTPA-octreotide with PET as a model for potential endotherapeutic octreotides labeled with ^{90}Y. *J. Lab. Compds. Radiopharm.* **37**, 519-521

Bruns C., Stolz B., Albert R, Marbach P. Pless J. (1993) OctreoScan 111 for imaging of a somatostatin receptor positive islet cell tumor in rat. *Hormone and Metabolic Research* Suppl. Ser. **27**, 4-11

Bruns C., Weckbecker G., Raulf F., Kaupmann K., Schoeffter P., Hoyer D., Lübbert H. (1994) Molecular pharmacology of somatostatin receptor subtypes. *Ann.N.Y.Acad.Sci.* **733**, 138-146

Chi D.Y., Kilbourn M.R., Katzenellenbogen J.A., and Welch M.J. (1987) A Rapid and Efficient Method for the Fluoroalkylation of Amines and Amides. Development of a Method Suitable for Incorporation of the Short-Lived Positron Emitting Radionuclide Fluorine-18. *J.Org.Chem.* **52**, 658-664

Chowdry V., and Westheimer F.H. (1979) Photoaffinity labeling of biological systems. *Ann.Rev. Biochem.* **48**, 293-325

Clerico A., Iervasi G., Mafredi C., Salvadori S., Marastoni M., Del Chicca G.M., Giannessi D., Del Ry S., Andreassi M.G., Sabatino L., Iascone M.R., Biagini A., Donato L. (1995) *Eur. J. Nucl.Med.* **22**, 997-1004

Corstens F.H.M., van der Meer J.W.M. (1991) Chemotactic peptides: New locomotion for imaging of infection. *J. Nucl. Med.* **31**, 491-494

Doering W. and Odum R.A. (1966) Ring enlargement in the photolysis of phenyl azide. *Tetrahedron* **22**, 81-93

Downer J.B., McCarthy T.J., Edwards W.B., Anderson C.J., Welch M.J. (1995) Reactivity of radiolabeling agent p-[^{18}F]fluorophenacyl bromide. *J. Label. Compds. Radiopharm.* **37**, 127-129

Draznin B., Sherman N., Sussman K.E., Dahl R., Vatter R. (1985) Internalization and cellular processing of somatostatin in primary cultures of rat anterior pituitary cells. Endorinology. **177**, 980

Eastman R.C., Carson R.E., Shai Y., Channing M.A., Dunn B.B., Lesniak M.A., Baas E.J., Jones E.,Bacher J.D., Kirk K.L., Jackobson K.A., Herscovitch P., Roth J.(1990) Imaging of insulin receptors in vivo using ^{18}f-Labeled insulin. *J.Nucl.Med.* **31**, 791 (abstr.)

Eckelman W.C. (1994) The application of Receptor theory to Receptor-binding and Enzyme-binding Oncologic Radiopharmaceuticals. *Nucl.Med.Biol.* **21**, 759-769

Fishman A.J., Babich J.W., Strauss H.W. (1993) A ticket to ride: Peptide Radiopharmaceuticals. *J.Nucl.Med.* **34**, 2253-2263)

Fishman A.J., Pike M.C., Kroon D., Fucello A.J., Rexinger D., Kate C., Wilkinson R., Rubin R.H., Strauss W.H. (1991) Imaging focal sites of bacterial infection in rats with In-111-labeled chemotactic peptide analogs. *J. Nucl.Med.* **32**, 483-491

Fixmann T, Wester H.J., Krummeich C., Förmer W., Holschbach H., Hamacher K., Müller-Gärtner H.W., Stöcklin G. (1995) F-18- and I-123-labeled apamin: radiolabeling, in vitro and in vivo evaluation with mice. *Eur. J. Nucl.Med.* **22**, 770 (abstr)

Fraser C.M. (1995) Structure and functional analysis of G-protein coupled receptors and potential diagnostic ligands *J.Nucl.Med.* **36**, 17s-21s

Fricker G., Dubost V., Schwab D., Bruns C., Thile C. (1994) Heterogenity in hepatic transport of somatostatin analog octapeptides. *Hepatology* **20**, 191-200

Garg P.K, Archer G.E., Bigner D.D., Zalutsky M.R. (1989a) Synthesis of radiolabeled N-succinimidyl iodobenzoate: Optimization for use in antibody labeling. *Appl.Radiat.Isot.* **40**, 485-490,

Garg P.K., Garg S., Zalutsky M.R.. (1991) Fluorine-18 Labeling of Monoclonal Antibodies and Fragments with Preservation of Immunoreactivity. *Bioconjugate Chem.* **2**, 44-49

Garg P.K., Slade S.K., Harrison C.L., Zalutsky M. (1989b) Labeling proteins using aryl iodide acylating agents: Influence of meta vs para substitution on in vivo stability. *Nucl.Med.Biol.* **16**, 727-733

Gauldie J., Hanson J.M., Rumlanek F.D., Shipolini R., Vernon C.A. (1976) The peptide components of bee venom. *Eur.J.Biochem.* **61**, 369-376

Gehlert D.R. and Gackenheimer S.L. (1993) Comparison of the biodistribution of binding sites for the potassium channel ligands [^{125}I]Apamin, [^{125}I]charybdotoxin and [^{125}I]iodoglyburide in the rat brain. *Neuroscience* **52**, 191-205

Geiger R. (1976) Chemie des Insulins. *Chemiker-Zeitung* **100**, 111-129

Gibson E.R., Beauchamp H.T., Fioravanti C., Brenner N., Burns H.D. (1994) Receptor binding radiotracers for the angiotensin II receptor: Radioiodinated [Sar^1,Ile^8]Angiotensin II. *Nucl. Med. Biol.* **21**, 593-600

Goldenberg D.M.(1992) Cancer Imaging and therapy with radiolabeled antibodies. In Imunobiology of Proteins and Peptides, Vol 6 (Edited by E.Atassi) Plenum Press, New York

Goodwin D.A. (1987) Pharmacokinetics and antibodies. *J.Nucl.Med.* **28**, 1358-1362

Grainer C., Muller E.P., van Rietschoten J. (1978) Use of synthetic analogs for a study on the structure-activity relationship of apamin. *Eur. J. Biochem.* **82**, 193-299

Guhlke S. Wester H.J., Biersack H.J., Stöcklin G. (1994a) Simplified synthesis of n.c.a. [^{18}F]fluoroacylation agents via 2-[^{18}F]fluoropropionic acid chloride in view of remote controlled labeling of peptides and proteins. *J.Label.Compds. Radiopharm.* **35**, 194-196

Guhlke S., Coenen H.H., and Stöcklin G. (1994b) Fluoroacylation Agents Based on Small n.c.a.[^{18}F]Fluorocarboxylic Acids. *Appl. Radiat. Isot.* **45**, 715-727

Guhlke S., Wester H.-J., Bruns Ch., Stöcklin G. (1994c) (2-[^{18}F]Fluoropropionyl-(D)phe^1)-octreotide, a Potential Radiopharmaceutical for Quantitative Somatostatin Receptor Imaging with PET: Synthesis, Radiolabeling, In Vitro Validation and Biodistribution in Mice. *Nucl. Med. Biol.* **21**, 819-825

Habermann E. (1984) Apamin. *Pharmac. Ther.* **25**, 255-270

Hamacher K., and Hamkens W. (1995) Remote Controlled One-step Production of ^{18}F Labeled Butyrophenone neuroleptics Exemplified by the Synthesis of n.c.a. [^{18}F]N-Methylspiperone. *Appl.Radiat.Isot.* **46**, 911-912

Hamacher K., Blessing G., Nebeling B. (1990) Computer aided synthesis (CAS) of no-carrier-added 2-[^{18}F]fluoro-2-deoxy-D-glucose: an efficient automated system for the aminopolyether-supported nucleophilic fluorination. *Appl. Radiat.Isot.* **41**, 49-55

Hamacher K., Coenen H.H., and Stöcklin G. (1986) Efficient stereospecific synthesis of no-carrier-added 2-[^{18}F]fluoro-2-deoxy-D-glucose using aminopolyether supported nucleophilic substitution. *J.Nucl.Med.* **27**, 235-238

Harris T.D., Barret J.A., Bourque J.P., Caroll T.R., Damphousse P.R., Edwards D.S., Glowacka D., Liu S., Looby R.J., Poirier M.J., Rajopadhye M, and Yu. K. (1994) Design and synthesis of radiolabeled GPIIb/IIIa receptor antagonists as potential thrombus imaging agents. *J.Nucl.Med.* **35**, 245 (abstract)

Hassan M., Refai E., Andersson M., Schnell P.O., Jacobsson H. (1994) In Vivo Dynamical Distribution of ^{131}I-VIP in the Rat Studied by Gamma-cammera. *Nucl.Med.Biol.* **21**, 865-872

Haubner R., Gratias R., Goodman S.L., Kessler H. (1996) RGD plus X: Structure/ Activity investigations on cyclic RGD-peptides. In: Peptides: Chemistry, Structure and Biology, Kaumaya P.T.P. and Hodges R.S. (eds.) Mayflower Scientific Ltd

Haubner R., Gratias R., Diefenbach B., Goodman S.L., Jonczyk A. and Kessler H. (1996) Structural and functional aspects of RGD containing cyclic pentapeptides as highly potent and selective integrin a$_v$ß$_3$ antagonists. *J.Am. Chem.Soc.* **118**, 7461-7472

Haubner R., Schmitt W., Hölzemann G., Goodman S.L., Jonczyk A., Kessler H. (1996) Cyclic RGD peptides containing ß-turn mimetics. *J.Org.Chem.* **118**, 7881-7891

Hermann L.W., Fischmann A.J., Tompkins R.G., Hanson R.N., Boyn C., Strauss H.W., Elmaleh L.W. (1990) The use of pentafluorphenyl derivatives for ^{18}F-labeling of proteins *Nucl.Med.Biol.* **21**, 1005-1010

Herzog H., Rösch F, Stöcklin G., Lueders C., Quaim S.M, Feinendegen L.E (1993) Measurement of pharmacokinetics of yttrium-86 radiopharmaceuticals with PET and radiation dose calculation of analogous yttrium-90 endoradiotherapeutics. *J. Nucl. Med.* **34**, 2222-2226

Hixson S.H.und Hixson S.S. (1975) p-Azidophenacyl bromide, a versatile photolabile bifunctional reagent. Reaction with Glyceraldehyde-3-phosphate dehydrogenase. *Biochemistry.* **14**, 4251

Hnatowitch D.J. (1990) Antibody radioplabeling, problems and promises. *Nucl. Med. Biol.* **17**, 49-55

Hoffman T.J., Sieckamn G.L. Volkert W.A. (1995) Targeting Smal Cell Lung Cancer Using Iodinated Peptide Analogs. *J.Label. Compds. Radiopharm.* **37**, 321-323

Hugues H., Duval D., Kitabgi P., Ladzdunski M., Vincent J.P. (1982a) Preparation of a pure moniododo derivative of the bee venom neurotoxin apamin and ist binding properties to rat brain synaptosomes. *J.Biol. Chem.* **257**, 2762-2769

Hugues M., Romey G., Duval D., Vincent J.P., Lazdunski M. (1982b) Apamin as a selctive blocker of calcium-dependent potassium channels in neuroblastoma cells: Voltage-clamp and biochemical characterization of the toxin receptor. *Proc.Natl.Acad.Soc. USA* **79**, 108-1312

86

Hunter M.J. and Ludwig M.L. (1962) The reaction of imidoesters with proteins and related molecules. *J. Am.Chem. Soc.* **84**, 3491-3504

Ikeda M., Dewar D., McCulloch J. (1991) Selective reduction of [^{125}I]apamin binding sites in alzheimer hippocampus: a quantitative autoradiographic study. *Brain Research* **567**, 51-56

Imman K. (1981) Peptide synthesis with minimal protection of side-chain functions.In: The Peptides: analysis, Synthesis, Biology. Vol.3.: Protection of functional groups in peptide synthesis. Chap. 6, 254-293. Eds: E.Gross und J.Meienhofer. Academic Press, New York

Jacobson K.A., Furlano D.C., Kirk K.L. (1988) A prosthetic group for the rapid introduction into peptides and functionalized drugs. *J. Fluorine Chem.* **39**, 339-347

Jong de M., Bakker W.H., Breeeman W.A.P, van der Pluijm M.E., Kooij P.P.M., Visser T.J., Docter, R.Kreening E.P. (1993. Hepatobiliary handling of iodine-125-Tyr^3-octreotide and indium-111-DTPA-D-Phe^1-octreotide by isolated perfused rat liver. *J.Nucl.Med.* **34**, 2025-2030

Kaupmann K., Bruns C., Raulf F., Weber H.P., Mattes H., Lübbert H. (1995) Two amino acids, located in the transmembrane domain VI and VII, determine the selectivity of the peptide agonist SMS 201-995 for the SSTR2 somatostatin receptor. *The EMBO Journal*, **14**, 727-735

Kilbourn M.R., Dence C.S., Welch M.J., Mathias C.J. (1987) Fluorine-18 Labeling of Proteins. *J. Nucl.Med.* **28**, 462-470

Knight L.C., Lister-James J., Dean R.T., Maurer A.-H. (1993) Evaluation of Tc-99m Labeled Ccylic Peptides for Thrombus Imaging. *JNucl Med* **14**, 17p (abstr)

Knight L.C., Radcliffe R., Maurer A.H., Rodwell J.D., Alvarez V.L. (1994) Thrombus imaging with technetium -99m synthetic peptides based upon the binding domain of a monocloanal antibody to activated platelets. *J.Nucl.Med.* **35**, 282-288

Kragh-Hansen U. (1981) Molecular aspects of igand binding to serum albumin. *Am. Soc. Pharm. Exper.Therap.* **33**, 17-53

Krummeich C., Holschbach M., Stöcklin G. (1995) Convenient high yield direct electrophilic n.c.a. radioiodination of poorly activated anisole derivatives using the Iodo-gen™ trifluoroacetic acid system. *J.Label.Compd. Radiopharm.* **37**, 628-630

Krummeich C., Holschbach M., Stöcklin G. (1996) Convenient direct N.c.a. electrophilic radioiodination or arenes using iodo-gen. *Appl. Radiat. Isot.* **47**, 489-495

Kurtaran A., Leimer M, Kasere K, Yang Q., Angelberger P., Niederle B., Virgolini I. (1996) Combined use of ^{111}In-DTPA-D-Phe-1-Octreotide (OCT) and ^{123}I-Vasoactive Intestinal Peptide in the Loacalization Diagnosis of Medullary Thyroid Carcinoma (MTC). *Nucl. Med. Biol.* **23**, 503-507

Kurtaran A., Li S.R., Raderer M, Leimer M., Müller C., Pidlich J., Neuhold N., Hübsch P., Angelberger P., Scheidhauer W., Virgolini I. (1995) Technetium-99m-galactosyl-neoglycoalbumin combined with iodine-^{123}I-Tyr-(A14)-insulin visualizes human hepatocellular carcinomas. *J.Nucl.Med.* **36**, 1875-1881

Kutz K., Nüesch E., Rosenthaler L. (1986) Pharmacokinetics of SMS 201-995 in healthy subjects. *Scand. J. Gastroenterol.* **21**, 65-72

Kwawli L.A. and Kassis A.I. (1989) Synthesis of ^{125}I-labeled N-succinimidyl p-iodobenzoate for use in radiolabeling antibodies. *Nucl.Med.Biol.* **16**, 727-733

Lambert R., Willenbrock R., Tremblay J., Bavaria G., Langlois Y., Hpgan K., Tartaglia D., Flanagan R.J., Hamet P. (1994) Receptor imaging with atrial natriuretic peptide Part 1: High specific activity iodine-123-atrial natriuretic peptide. *J.Nucl.Med.* **35**, 628-637

Lamberts S.W.J., Krenning E.P., Reubi J.C. (1991) The role of somatostatin and its analogues in the diagnosis and treatment of tumors. *Endocrine Reviews.* **12**, 450-482

Lanathan M.V., Bass L.A., Welch M.J., Anderson C.J. (1995) Metabolism of octreotide conjugates of ^{64}Cu-TETA, ^{64}Cu-CPTA and ^{111}In-DTPA. *J. Label. Compds. Radiopharm* **37**, 510-512

Lang G., Choi C.W., Lee T.J., Webber K.O. Yoo T.M., Chang N., Le N., Jagoda E., Paik C.H., Pastan I., Carrasquillo J.A., Eckelman W.C. (1995b) Identification of metabolites of F-18 labeled TAC disulfide stabilized monoclonal antibody variable-region fragments in mice. *J. Label. Compds. Radiopharm.* **37**, 551-553

Lang G., Eckelman W.C. (1995a) Improved one-step synthesis of N-succinimidyl 4-[^{18}F]fluoromethyl-benzoate for protein labeling. *J. Label. Compds. Radiopharm.* **37**, 551-553

Lang L., Eckelmann W.C. (1994) One-Step Synthesis of ^{18}F labeled [^{18}F]-N-succinimidyl 4-(fluoromethyl) benzoate for Protein Labeling. *Appl. Radiat. Isot.* **45**, 1155-1163

Langen K.J. Coenen H.H. Roosen N., Kling P., Muzik O., Herzog H., Kuwert T., Stöcklin G., Feinendegen L.E. (1990) SPECT studies with L-3[I-123]iodo-alpha-methyltyrosine: Comparison with PET [I-124]IMT and first clinical results. *J.Nucl.Med.* **31**, 281-286

Larson S.M. (1994) Cancer or Inflammation? The holy Grail for Nuclear Medicine. *J.Nucl.Med.* **35** 1653-1655

Lathe R. (1985) Synthetic oligonucleotide probes deduced from amino acid sequence data. Theoretical and practical considerations. *J.Mol.Biol.* **183**, 1-12

Leitha T., Staudenherz A., Gmeiner B., Hermann M., Hüttinger M., Dudczak R. (1993) Technetium-99m labelled LDL as a tracer for quantitative LDL scintigraphy. *Eur.J. Nucl. Med.* **20**, 674-679

Lemaire M., Azira M., Dannecker R, Marbach P., Schweitzer A., Maurer G. (1989) Disposition of sandostatin, a new synthetic somatostatin analogue in rats. *Drug Metabol. and Dispos.* **17**, 699-703

Lister-James J., Knight L.C., Maurer A.H., Bush L.R., Moyer B.R., Dean R.T. (1996) Imaging with Tc-99m-Labeled Activated Platelet Receptor Binding Peptide. *J.Nucl.Med* **37**, 775-781

Liu S., Edwards D.S., Looby R.J., Poirier M.J., Rajopadhye M, Bourque J.P., and Caroll T.R. (1996) Labeling cyclic glycoprotein IIb/IIIa receptor antagonists with Tc-99m by the preformed chelate approach: Effects of chelators on properties of [^{99m}Tc]chelator-peptide conjugates. *Bioconjugate Chem.* **7**, 196-202

Locher M., Johnston J., Müller T., Borbe H.O., Kutscher B., Engel J. (1994) Synthesis of [U-^{14}C]Arg labelled decapeptide cetrorelix, a novel lutenizing hormone- releasing hormone antagonist. *J. Lab. Compds. Radiopharm.* **34**, 1091-1098

MacGillivary R.T.A, Mendez E., Shewale J.G., Shina S.K., Lineback-Zins J., Brew K. (1983) The primary structure of human serum transferrin. *J.Biol.Chem.* **258**, 3543-3553

Maina T., Stolz B., Albert R., Bruns C., Koch P., Mäcke H. (1994) Synthesis, radiochemistry and biological evaluation of a new somatostatin analogue (SDZ 219-387) labelled with technetium-99m. *Eur. J. Nucl. Med.* **21**, 437-444

Martinez J., Tolle J.C., Bodanszky M. (1979) Side reactions in peptide chemistry. *Int.J.Peptide Protein Res.* **13**, 22-27

Meyer G.J., Coenen H.H., Waters S., Långström B., Cantineau R., Strijckmans K., Vaalburg W., Halldin C., Crouzel C., Mazière B., Luxen A. (1993) Quality assurance and quality control of short lived radiopharmaceuticals for PET. In: Radiopharmaceuticals for Positron Emission Tomography, Methodological aspects. Eds.: G.Stöcklin and V.W.Pike. Developments in Nuclear Medicine, Vol.24, Kap.3, 91-150, Kluwer Academic Publishers

Miller L. (1993) Synthetic peptides come of age. *J. Nucl. Med.* **34**, 15N-30N

Moerlein S.M. (1990) Regiospecific aromatic radioiodination via no-carrier-added copper(I)-assisted iododebromination. *Radiochemica Acta* **50**, 55-61

Morel G. (1994) Internalization and nuclear localization of peptide hormones. *Biochemical Pharmacology.* **47**, 63-76

Morel G., Pelletier G., Heisler S. (1986) Internalization and subcellular distribution of radiolabelled somatostatin-28 in mouse anterior pituitary tumor cells. *Endocrinology* **119**, 1972-1979

Müller-Platz C.M., Kloster G., Legler G., and Stöcklin G. (1982) ^{18}F-Fluoroacetate: An Agent for Introducing No-Carrier-Added Fluorine-18 into Urokinase without loss of Biological Activity. *J Label Compds Radiopharm.* **19**, 1645

Murray J.L. Mujoo K., Williams C., Mansfield P., Wilbur D.S., Rosenblum M.G. (1991) Variables influencing tumor uptake of anti-melanoma monoclonal antibodies radioiodinated using para-iodobenzoyl (PIP)conjugate. *J.Nucl.Med.* **32**, 279-287

Muto P., Lastoria S., Varrella P., Vergara E., Salvatore M., Morgano G., Lister-James J., Bernardy J.D., Dean R.T., Wencker D., Borer J. (1995) Detecting deep venous thrombosis with Tc-99m-labeled synthetic peptide P280. *J.Nucl.Med.* **36**, 1384-1391

Nagren K, Ragnarsson U., and Langström B. (1986) The synthesis of the neuropeptide Met-Enkephaline and the two metabolic fragments labeled with ^{11}C in the methionine group. *Appl. Radiat.Isot.* **37**, 537-539.

Ott R.J., Tait D., Fowler M.A., Babich J.W. (1992) *Br.J.Radiol.* **65**, 787

Page R.L., Garg P.K., Garg S., Archer G.E., Bruland Ø.S., Zalutsky M.R. (1994a) PET Imaging of osteosarcoma in dogs using a fluorine-18-labeled monoclonal antibody Fab fragment. *J. Nucl.Med.***35**, 1506-1513

88

Page R.L., Garg P.K., Garg S., Archer G.E., Brunland O.S., Zalutsky M.R. (1994) PET imaging of osteosarcoma in dogs using fluorine-18 labeled monoclonal antibody Fab fragment. *J Nucl Med.* 34, 533

Pandey R.N., Wilson J.D., Zhao X.G., and Schlom J. (1991) Photolabeling: A new approach for radioiodination. *Antibody, Immunoconj. and Radiopharm.* 4, 399-407

Pandurangi R.S., Kuntz R.R., Volkert W.A. (1995a) Photoaffinity labeling of B72.3 antibodies using C-14-methyltetrafluoro azido benzoate. *J.Nucl.Med.* 36, 156p abstr.

Pandurangi R.S., Kuntz R.R., Volkert W.A. (1995b) Photolabeling of human serum albumin by 4-Azido-2-([^{14}C]-methylamino)trifluorobenzonitrile. A high efficiency, long wavelength photolabel. *Appl. Radiat. Isot.* 46, 233-239

Pandurangi R.S., Kuntz R.R., Volkert W.A. (in press) High efficiency photolabeling of human serum albumin and γ-globulin with 4-azido-2,3,5,6-tetrafluoro-^{14}C-methyl benzoate. *Bioconjugate Chem.*

Paul R. (1962) O-Acylation of tyrosine during peptides synthesis. *J.Org. Chem.* 28, 236

Pentlow K.S., Lambrecht R.M. Cheung N.K.V., Larson S.M. (1991) *Med. Phys.* 18, 357 (1991)

Pless J., Bauer W., Briner U., Doepfner W., Marbach P., Maurer R., Petcher T., Reubi J.C. and Vanderscher J. (1986) Chemistry and pharmacology of SMS 201-995, a long acting octapeptide analogue of somatostatin. *Int.Congr. Ser.-Excerpta Med.* 683, 319

Ramachandran J. and Li C.H. (1963) The synthesis of L-valyl-L-lysyl-L-valyl-L-tyrosyl-L-proline. *J.Org.Chem.* 28, 173-177

Raulf R., Perez J., Hoyer D., Bruns C. Differential expression of five somatostatin receptor subtypes, SSTR1-5, in the CNS and peripheral tissue. (1994) *Digestion.* 55 (suppl.3), 46-53

Reubi J.C. (1995) In vitro Identification of Vasoactive Intestinal Peptide Receptors in Human Tumors: Implications for Tumor Imaging. *J.Nucl.Med.* 36, 1846-1853

Reubi J.C. (1995) Neuropeptide Receptors in Health and Disease: The Molecular Basis for In Vivo Imaging. *J.Nucl.Med.* 36, 1825-1835

Reynor K., Reisine T.(1992) Somatostatin Receptors. *Critical. Rev. Neurobiol.* 6, 273-289

Romey G., Hugues M., Schmid-Antomarchi H., Lazdunski M. (1984) Apamin: a specific toxin to study a class of Ca^{2+} - dependend K^{+} channels. *J. Physiol.* 79, 259-264

Rösch F., Herzog H., Plag C., Neumaier B., Braun U., Müller-Gärtner H.W., Stöcklin G. (1996) Radiation doses of yttrium-90 citrate and yttrium-90 EDTMP as determined via analogous yttrium-86 complexes and positron emission tomography. *Eur.J.Nucl.Med.* 23, 958-966

Ruoho A.E., Kiefer H., Roeder P.E., Singer S.J.. The mechanism of photoaffinity labeling. Proc.Nat.Acad.Sci.USA. 70, 2567-2571 (1973)

Sagnella G.A. and MacGregor G.A. (1984) Cardiac peptides and the control of sodium excretion. *Nature* 309, 666-667

Savarese T.M. Fraser C.M. (1992) In vitro mutagenesis and the search for structure-function relationship among G-protein coupled receptors. *Biochem J.* 283, 1-19

Schiffmann E., Corcoran B.A. and Wahl S.M. (1975) N-Formylmethioyl Peptides as Chemoattractants for Leucocytes. *Proc. Nat. Acad. Sci. USA* 72, 1059-1062

Seelig J., Nebel S., Ganz P., Bruns C. (1993) Electrostatic and nonpolar peptide-membrane interactins. Lipid binding and functional properties of somatostatin analogues of charge z = +1 to z = +3. *Biochemistry* 32, 9714-9721

Serafini A.N.. (1993) From monoclonal antibodies to peptides and molecular recognition units: An overview. *J.Nucl.Med.* 34, 533-536

Shai Y., Kirk K.L., Channing M.A., Dunn B.B., Lesniak M.A., Eastman R.C., Finn R.D., Roth J., and Jacobson K.A. (1989) ^{18}F-Labeled Insulin: A Prosthetic Group Methodology for Incorporation of a Positron Emitter into Peptides and Proteins. *Biochemistry* 28, 4801-4806

Shattil S.J., Hoxie J.A., Cunningham M., Brass L.F., (1985) Changes in the platelet membrane glycoprotein IIb, IIIa complex during platelet activation. *J.Biol.Chem.* 260, 11107-11114

Shiue C.Y., Wolf A.P., Hainfeld J.F. (1988) Synthesis of ^{18}F-labeled N-(p-[^{18}F]fluorophenyl]maleimide and its derivative for labeling monoclonal antibodies with ^{18}F. *J.Label. Compds. Radiopharm.* 26, 278

Smith B.D., Nakanishi K., Watanabe K., Ito S. (1990) 4-Azido[3,5-^{3}H]phenacyl bromide, a versatile bifunctional reagent for photoaffinity radiolabeling. Synthesis of prostaglandin 4-azido [3,5-^{3}H]phenacyl-esters. *Bioconjugate Chem.* 1, 363-364

Smith-Jones P.M., Stolz B., Bruns C., Albert R., Reist H.W., Friedrich R., Mäcke H.R (1994). Gallium-

67/Gallium-68-DFO-octreotide - A potential radiopharmaceutical for PET imaging of somatostatin receptor-postive tumors: Synthesis and radiolabeling, in-vitro and preliminary in-vivo studies. *J.Nucl.Med.* **35**, 317-325

Stöcklin G. (1991) Radioligands for PET studies of D2-receptors: Butyrophenone and ergot derivatives. In: Brain dopaminergic systems: Imaging with positron emission tomography (J.C.Baron et al. Editors), Kluwer Academic Publishers.

Stolz B., Albert R. and Bruns C. (1995) LHRH Rezeptor targeting mit [^{111}In] markiertem DTPA-LHRH Analog (SDZ 218-041) - Präklinische Charakterisierung -

Stolz B., Maina T., Albert R., Briner U., Nock B., Fridrich R., Mäcke H., Bruns C. (1994) Somatostatin Rezeptor Szintigraphie mit [99mTc]N4(D)Phe1-Octreotid (SDZ 220-778): Präklinische Charakterisierung. Dt. Gesellschaft für Nuklearmedizin, 32. Int. Jahrestagung, 27-30 April 1994, Kiel; *Nucl.Med.* **33**:A27 (abstr.)

Turton D.R., Brady F., Pike V.W., Selwyn A.P., Shea M.J., Wilson R.A., DeLandsheere C.M. (1984) preparation of human serum [Methyl-^{11}C]methylalbumin microspheres and human serum [Methyl-[^{11}C]methylalbumin for clinical use. *J. Appl. Radiat. Isot.* **35**,337-344

v.Iddon B., Meth-Cohn O., Scriven E.F.V., Suschitzky H., Gallagher P.T. (1979) Entwicklungen in der Arylnitren-Chemie: Synthesen und Mechanismen. Angew.Chem. **91**, 965-981

Vaidyanathan G. and Zalutsky M.R. (1995) Fluorine-18 labeled chemotactic peptide: A potential approach for the PET imaging of bacterial infection. *Nucl.Med.Biol.* **22**, 759-764

Vaidyanathan G., Affleck D.A., Welsh P., Zalutsky M.R. (1994a) Fluorine-18 labeled chemotactic peptide: A otential agent for the imaging of focal infection. *J.Label.Compds. Radiopharm.* **35**, 365-367

Vaidyanathan G., and Zalutsky M.R. (1994b) Improved synthesis of N-succinimidyl 4- [^{18}F]fluorobenzoate and its application to the labeling of a monoclanal antibody fragment. *Bioconjugate Chem.* **5**, 352-356 122.

Vaidyanathan G., Bigner D.D., Zalutsky M.R. (1992a) Fluorine-18-Labeled Monoclonal Antibody Fragments: A Potential Approach for Combining Radioimmunoscintigraphy and Positron Emission Tomography. *J.Nucl.Med.* **33**, 1535-1541

Vaidyanathan G., Zalutsky M.R. (1992b) Labeling Proteins with Fluorine-18 using N-Succinimidyl 4-[^{18}F]Fluorobenzoate. *Nucl. Med. Biol.* **19**, 275-281

Van der Laken C.J., Boerman O.C., Oyen W.J.C., van der Ven M.T.P., Claessens R.A.M.J., van der Meer J.W.M., Corstens F.H.M. (1995) Specific targeting of infectious foci with radioiodinated human recombinant interleukin-1 in an experimental model. *Eur.J. Nucl. Med.* **22**, 1249-1255

Vincent J.P., Schweitz H., Lazdunski M. (1975) Structure - function relationship and site of action of apamin, a neurotoxic polypeptide of the bee venom with an action on the central nervous system *Biochemistry* **14**, 2521-2525

Virgolini I., Kurtaran A., Raderer M., Leimer M., Angelberger P., Havlik E., Li. S., Scheithauer W., Niederle B., Valent P. and Eichler H.G. (1995) Vasoactive intestinal peptide-receptor scintigraphy *J.Nucl.Med.* **36**, 1732-173

Virgolini I., Raderer M., Kurtaran A., Angelberger P., Banyi S., Yang Q., Li. S., Banyai M., Pidlich J., Niederle B., Scheithauer W., Valent P. (1994) Vasoactive intestinal peptide-receptor imaging for the localization of intestinal adenocarcinomas and endocrine tumors. *N.Engl. J. Med.* **331**, 1116-1121

Welch J.M. and Kilbourn M.R. (1988) Potential labeling of monoclonal antibodies with positron emitters. In: Radiolabeled Monoclonal Antibodies for Imaging and Therapy (Srivastava S.C. Ed.) p.261. Plenum Press New York

Wester H.J. (1996a) Zur praktisch trägerfreien F-18 Fluorierung von Proteinen., Peptiden und Tyrosin. Berichte des Forschungszentrums Jülich, Jül- 3206, Forschungszentrum Jülich GmbH, Jülich, Germany

Wester H.J. Hamacher K., Stöcklin G. (1996b) A comparative study of n.c.a. fluorine-18 labeling of proteins via acylation and photochemical conjugation. *Nucl.Med.Biol.* **23**, 354-372

Wester H.J., Brockmann J., Rösch F., Stolz B., Wutz W., Herzog H., Smith-Jones P.M., Bruns C., Stöcklin G. (1995a) Quantitative in-vivo uptake kinetics of ^{18}F-octreotide in tumor bearing rats as measured by PET: A comparison with ^{86}Y-DTPA-octreotide and ^{68}Ga-DFO-octreotide. *J.Nucl.Med.* **36**, 113p-114p abstr.

Wester H.J., Brockmann J., Rösch F., Wutz W., Herzog H., Schmith-Jones P., Stolz Barabara, Bruns C., and Stöcklin G. (in press) PET-pharmaconkinetics of ^{18}F-octreotide: A comparison with ^{67}Ga-DFO- and ^{86}Y-DTPA-octreotide. *Nucl.Med.Biol.*

90

Wester H.J., Guhlke S., Stöcklin G. (1994a) Comparative evaluation of n.c.a. ^{18}F-fluorination agents for peptides and proteins. *Eur.J.Nucl.Med.* **21**, 794 abstr.

Wester H.J., Guhlke S., Stöcklin G. (1994b) Regioselective ^{18}F]fluoropropionylation of peptides and proteins in aqueous solution. *J.Label.Compds.Radiopharm.* **35**, 297-299

Wester H.J., Hamacher K., Stöcklin G. (1994c) Simple and fast ^{18}F-labeling of proteins by coupling with photogenerated ^{18}F]arylnitrene. *J.Nucl.Med.* **35**, 73p abstr.

Wester H.J., Krummeich C., Fixmann A., Förmer W., Müller-Gärtner H.W., Stöcklin G. (1995b) [18]F- and ^{131}I-Labeling of the octadecapeptide Apamin: A selective blocker of the Ca^{2+} dependent K^+ channels. Synthesis and in vivo evaluation in NMRI mice. *J.Label.Compds.Radiopharm.* **37**, 513-515

Westerberg G., and Langström B. (1994) Labeling of proteins with ^{11}C in high specific radioactivity: [11C]albumin and [11C]transferrin. *Appl.Radiat.Isot.* **45**, 773-782

Wilbur D.S. (1992) Radiohalogenation of proteins: An overview of radionuclides. Labeling methods and reagents for conjugate labeling. *Bioconjugate Chem.* **3**, 433

Wilbur D.S. Hadley S.W., Hylarides M.D., Abrams P.G., Beaumier P.A., Morgan A.C., Reno J.M., Fritzberg A.R. (1989) Development of a stable radioiodinating reagent to label monoclonal antibodies for therapy of cancer. *J.Nucl.Med.* **30**, 216-226

Williams L.T., Snyderman R., Pike M.C., Lefkowitz R.J. (1977) Specific receptor sites for chemotactic peptides on human polymorphonuclear leukocytes. *Proc. Natl. Acad.Sci.USA* **74**, 1204-1208

Wofsy L.and Singer S.J. (1963) Effect of the amidination reaction on antibody activity and on the physical properties of some proteins. *Biochemistry* **2**, 104-116

Yamashiro D., Li C.H. (1972) *Int.J.Peptide Protein Res.* **4**, 181-185

Zalutsky M.R., Garg P.K., Johnson S.H., Utsunomiya H., Colemann R.E. (1992) Fluorine-18-Antimyosin monoclonal antibody fragments: Preliminary investigations in a canine myocardial infarct *model.J. Nucl. Med.* **33**, 575-580

Zweit J., Luthra S.K., Brady F., Carnochan P., Ott R.T., Jones T. (1995) Iodine-120, a new positron emitting radionuclide for PET radiopharmaceuticals. *J.Label.Compds. Radiopharm.* **37**, 823-82

DEVELOPMENT OF PET RADIOLIGANDS FOR THE QUANTITATION OF SEROTONIN RECEPTORS IN THE HUMAN BRAIN

Christer Halldin, Lars Farde, Camilla Lundkvist, Hiroshi Ito, Nathalie Ginovart, Håkan Hall, Svante Nyberg, Carl-Gunnar Swahn, Göran Sedvall Julie A. McCarron* and Victor W. Pike*.
Karolinska Institutet, Department of Clinical Neuroscience, Psychiatry Section, Karolinska Hospital, S-171 76 Stockholm, Sweden and *Cyclotron Unit, MRC Clinical Sciences Centre, Hammersmith Hospital, London, U.K.

Introduction

Pharmacological studies have shown that there are multiple serotonin receptor subtypes, which are classified in at least seven classes. [1]. There has been a lack of subtype selective serotonin receptor radioligands suitable for PET imaging of the human brain. The 5-HT1A receptor is of particular interest as it may be involved in the pathophysiology of several neuropsychiatric disorders, like anxiety, depression and schizophrenia. Recently, a potent and selective 5-HT1A receptor antagonist, WAY-100635 (N-(2-(4-(2-methoxyphenyl)-1-piperazinyl)ethyl)-N-(2-pyridyl)cyclohexanecarboxamide) was developed. Labelling of WAY-100635 in the methoxy position with [11]C [2-3] provided the first radioligand for the delineation of 5-HT1A receptors in the human brain using positron emission tomography (PET) [4]. The descyclohexanecarbonyl analogue ([^{11}C]WAY-100634) was shown to be a labelled lipophilic metabolite in primates with high affinity to 5-HT1A receptors and high ability to enter the brain, so hampering the quantitation of the uptake of the radioligand *in vivo* [5]. WAY-100635 has recently been labelled with ^{11}C also in the carbonyl position and examined in the human brain with PET [6].

A new 5-HT2A receptor antagonist, MDL 100907 ((R)-(+)-4-(1-hydroxy-1-(2,3-dimethoxyphenyl)methyl)-N-2-(4-fluorophenylethyl)piperidine) has high affinity (Ki = 0.36 nM) and selectivity for 5-HT2A receptors. The compound is currently in clinical trials [7]. We report on the preparation of both [carbonyl-^{11}C]WAY-100635 and [3-methoxy-^{11}C]MDL 100907 (Figure 1), the autoradiographic localization of both 5-HT1A and 5-HT2A receptors in the human brain using either [^{11}C]- or [^{3}H]-labelled tracers [8], metabolite studies in monkey and human plasma with HPLC [9-10], a pharmacological characterization of binding by PET in Cynomolgus monkeys [10-11] and human PET studies in healthy subjects [11-12].

91

B. Gulyás and H.W. Müller-Gärtner (eds.),
Positron Emission Tomography: A Critical Assessment of Recent Trends, 91–98.
© 1998 *Kluwer Academic Publishers.*

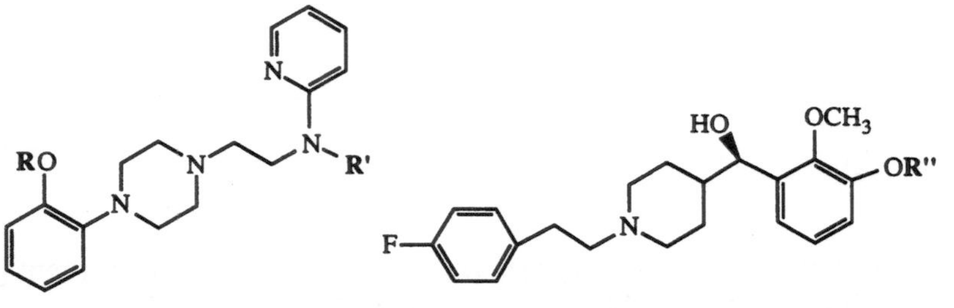

R	R'	R''	Radioligand
[^{3}H]CH$_3$	CO—⬡ 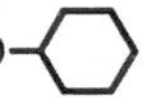		[methoxy-^{3}H]WAY-100635
CH$_3$	^{11}CO—⬡ 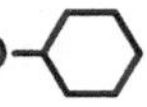		[carbonyl-^{11}C]WAY-100635
		[^{3}H]CH$_3$	[3-methoxy-^{3}H]MDL 100907
		[^{11}C]CH$_3$	[3-methoxy-^{11}C]MDL 100907

Figure 1. Structural formulas of [methoxy-^{3}H]WAY-100635, [carbonyl-^{11}C]WAY-100635, [3-methoxy-^{3}H] MDL 100907 and [3-methoxy-^{11}C]MDL 100907.

The radioligands [carbonyl-^{11}C]WAY-100635, [methoxy-^{3}H]WAY-100635, [3-methoxy-^{11}C]MDL 100907 and [3-methoxy-^{3}H]MDL 100907.

[Carbonyl-^{11}C]WAY-100635 (Figure 1) was prepared by ^{11}C-acylation of WAY-100634, essentially as described previously [2,8]. [^{11}C]CO$_2$ was reacted with cyclohexylmagnesium chloride and thionyl chloride to produce [^{11}C]cyclohexanecarbonyl chloride which was further reacted with the precursor WAY-100634 and in the presence of triethylamine to yield [carbonyl-^{11}C]WAY-100635 (Figure 2). The decay-corrected total radiochemical yield of [carbonyl-^{11}C]WAY-100635 was 15-20%. The specific radioactivity was about 1600 Ci/mmol (60 GBq/μmol) at time of administration. [Methoxy-^{3}H]WAY-100635 (Figure 1) was provided by Amersham International plc, U.K. The specific radioactivity was 80 Ci/mmol (2.2 GBq/μmol).

[3-Methoxy-^{11}C]MDL 100907 (Figure 1) was prepared as reported in detail elsewhere [10]. [^{11}C]Methyl iodide was synthesized from [^{11}C]carbon dioxide and reacted with the 3-hydroxy precursor (MDL 105725) (Figure 2). The decay-corrected total radiochemical yield after HPLC purification was 40-50% with a total synthesis time of 30 minutes. The specific radioactivity of [^{11}C]MDL 100907 obtained at time of injection was 500-1000 Ci/mmol (18-37 GBq/μmol). [3-Methoxy-^{3}H]MDL 100907 (Figure 1) was prepared in a similar way but starting from [^{3}H]methyl iodide. The specific radioactivity was 70 Ci/mmol (1.9 GBq/μmol).

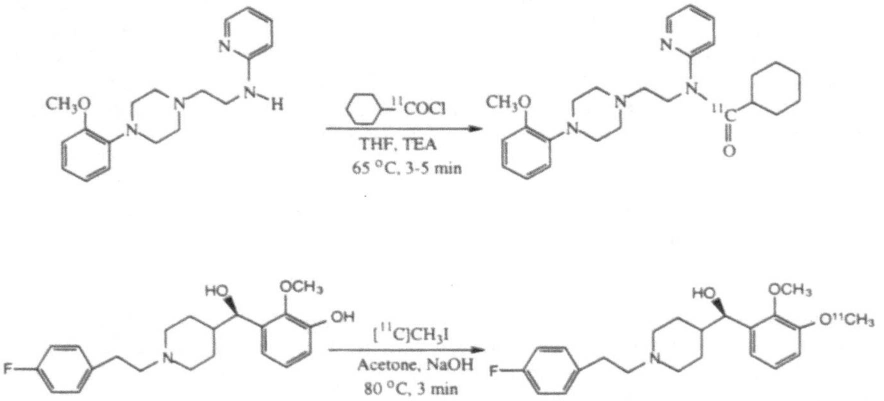

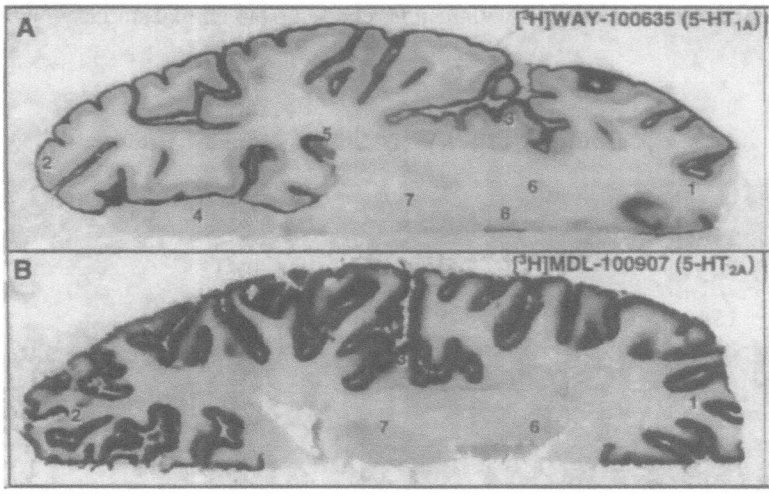

Figure 2. Labelling of [carbonyl-¹¹C]WAY-100635 and [3-methoxy-¹¹C]MDL, 100907.

In vitro autoradiography in the post-mortem human brain

Figure 3. Autoradiograms showing human whole-hemisphere autoradiography using [methoxy-³H]WAY-100635 (**A**) and [3-methoxy-³H]MDL 100907 (**B**). 1: Frontal cortex; 2: Occipital cortex; 3: Insular cortex; 4: Cerebellum; 5: Hippocampus; 6: Ventral striatum; 7: Thalamus / subthalamus; 8: Raphe nuclei.

The detailed distribution of 5-HT1A and 5-HT2A receptors was examined in the post-mortem human brain using whole hemisphere autoradiography [8]. The autoradiograms showed after incubation of [carbonyl-^{11}C]WAY-100635 very dense binding to hippocampus, raphe nuclei and neocortex, regions known to have a high density of 5-HT1A receptors (Figure 3A). [3-Methoxy-^{11}C]MDL 100907 demonstrated high binding to the 5-HT2A receptor rich neocortical regions (Figure 3B). The binding of both radioligands was inhibited by the addition of subtype selective compounds, such as 8-OH-DPAT and ketanserin, demonstrating selectivity of binding.

PET-characterization of [carbonyl-^{11}C]WAY-100635 binding to 5-HT1A receptors in the monkey and human brain

PET studies were performed on Cynomolgus monkeys injected i.v. with no-carrier-added [carbonyl-^{11}C]WAY-100635 [9,11]. There was a high accumulation of radioactivity in neocortical regions with frontal cortex to cerebellum ratios reaching about 8, at 60 min (Figure 4). There was a conspicuous accumulation of radioactivity in raphe nuclei and parahippocampus. Transient equilibrium of specific binding was obtained at about 40 min. Pretreatment of the monkeys with WAY-100635, buspirone, pindolol or 8-OH-DPAT blocked most of the specific binding. Plasma metabolite studies showed only polar labelled metabolites after injection of [carbonyl-^{11}C]WAY-100635 supporting the view that the increased signal is due to placing the radiolabel in a position that avoids the formation of the labelled lipophilic metabolite [^{11}C]WAY-100634 [9]. 5-HT1A receptor density and affinity were determined in the frontal cortex from a Scatchard analysis of two PET experiments using [carbonyl-^{11}C]WAY-100635 with high and low specific radioactivity. The Bmax and Kd in the frontal cortex were 16 pmol/ml and 1.6 nM, respectively, which is consistent with results reported from animal studies *in vitro*.

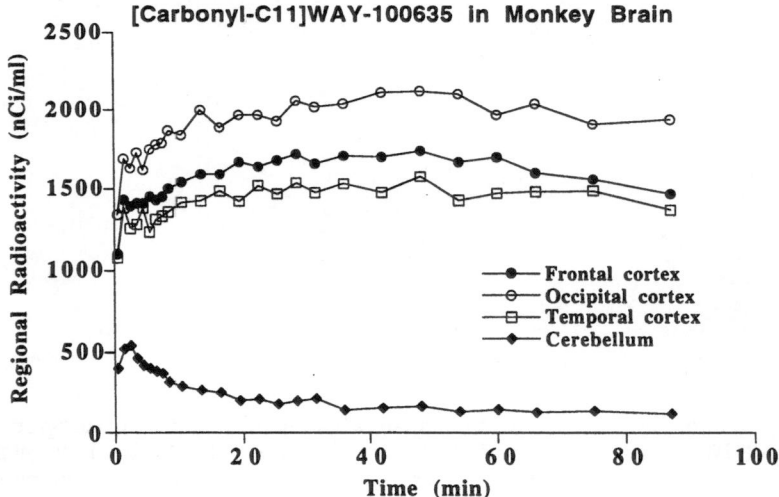

Figure 4. Time course for regional radioactivity (nCi/ml) in the brain of a Cynomolgus monkey after intravenous administration of [carbonyl-^{11}C]WAY-100635. The PET-camera system was a Siemens ECAT EXACT HR.

[Carbonyl-^{11}C]WAY-100635 was injected into healthy subjects [6,11] showing a high uptake in the neocortex and the raphe whereas cerebellum was on a low level (Figure 5). The labelling of WAY-100635 in a position that avoided lipophilic radioactive metabolites, provided a radioligand which is suitable for quantitative determination of 5-HT1A receptors in the human brain *in vivo*.

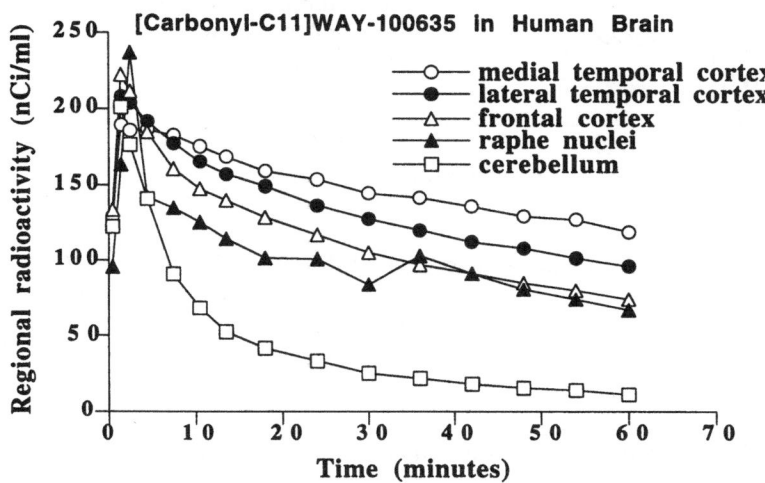

Figure 5. Time course for regional radioactivity (nCi/ml) in the brain of a healthy volunteer after intravenous administration of [carbonyl-^{11}C]WAY-100635. The PET-camera system was a Siemens ECAT EXACT HR.

PET-characterization of [3-methoxy-^{11}C]MDL 100907 binding to 5-HT2A receptors in the monkey and human brain

After i.v. injection of [3-methoxy-^{11}C]MDL 100907 in Cynomolgus monkeys there was a marked accumulation of radioactivity in neocortical regions (Figure 6) [10]. The neocortex to cerebellum ratio was about 4 after 60-80 minutes. Transient equilibrium occurred within 60 minutes. Radioactivity in the neocortex, but not in the cerebellum, was reduced after injection of ketanserin, indicating that neocortical radioactivity following injection of [3-methoxy-^{11}C]MDL 100907 represents specific binding to 5-HT2A receptors. There was no evident effect on neocortical binding after pretreatment with raclopride or SCH 23390. After injection into humans neocortex to cerebellum ratio was 2-2.5 after 75 minutes (Figure 7) [12]. The fraction of radioactivity in monkey and human plasma representing [3-methoxy-^{11}C]MDL 100907 was 15-25 and 20-40% at 60 minutes, respectively. [3-Methoxy-^{11}C]MDL 100907 has potential to become the first selective radioligand for PET-quantitation of 5-HT2A receptors in the human brain *in vivo*.

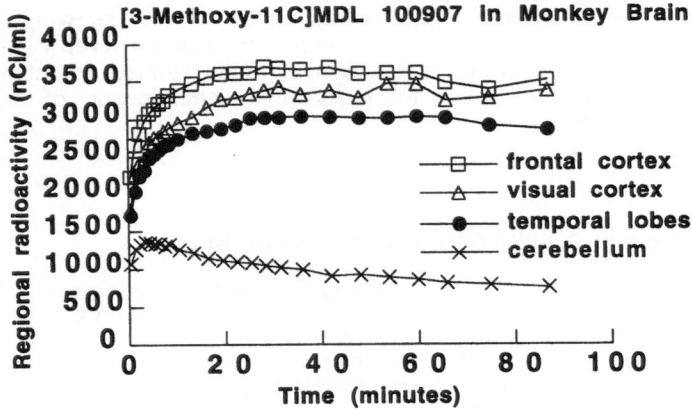

Figure 6. Time course for regional radioactivity (nCi/ml) in the brain of a Cynomolgus monkey after intravenous administration of [3-methoxy-^{11}C]MDL 100907. The PET-camera system was a Siemens ECAT EXACT HR.

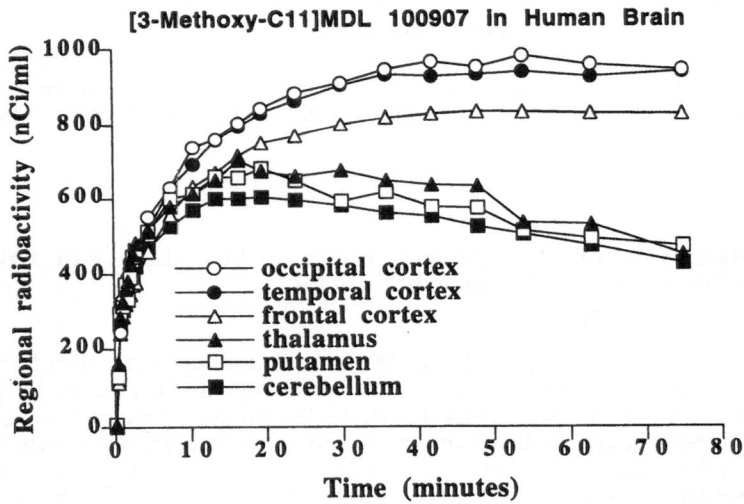

Figure 7. Time course for regional radioactivity (nCi/ml) in the brain of a healthy volunteer after intravenous administration of [3-methoxy-^{11}C]MDL 100907. The PET-camera system was a Siemens ECAT EXACT HR.

Conclusion

[Carbonyl-^{11}C]WAY-100635 binds specifically to 5-HT1A receptors. The labelling of WAY-100635 in a position that avoided lipophilic radioactive metabolites, provided a radioligand which is suitable for quantitative determination of 5-HT1A receptors in the human brain *in vivo*. The regional distribution of [3-methoxy-^{11}C]MDL 100907 in the living human brain is consistent with previous post-mortal studies, supporting its high selectivity for 5-HT2A receptors. [3-Methoxy-^{11}C]MDL 100907 has potential to become the first selective radioligand for PET-quantitation of 5-HT2A receptors in the human brain *in vivo*.

Acknowledgements

The research was supported by grants from the National Institute of Mental Health (41205 and 44814), The Swedish Medical Research Council (09114, 03560, 10360 and 11640), The Swedish Natural Research Council K-KU 9973-308 and Karolinska Institutet. The support from Hoechst Marion Roussel, Switzerland and Wyeth Research, UK is gratefully acknowledged. We are also grateful to the MRC, UK, for a studentship to Julie A. McCarron.

References

1. Peroutka, S.J. (1994) Molecular biology of serotonin (5-HT) receptors. *Synaps.* **18**, 241-260.

2. Pike, V.W., McCarron, J.A., Hume, S.P., Ashworth, S., Opacka-Juffry, J., Osman, S., Lammertsma, A.A., Poole, K.G., Fletcher, A., White, A.C., Cliffe, I.A. (1994) Preclinical development of a radioligand for studies of central 5-HT1A receptors in vivo - [^{11}C]WAY 100635. *Med. Chem. Res.* **5**, 208-227.

3. Mathis, C.A., Simpson, N.R., Mahmood, K., Kinahan, P.E., Mintun, M.A. (1994) [^{11}C]WAY 100635: A radioligand for imaging 5-HT1A receptors with positron emission tomography. *Life Sci.* **55**, PL403-PL407.

4. Pike, V.W., McCarron, J.A., Lammerstma, A.A., Hume, S.P., Poole, K., Grasby, P.M., Malizia, A., Cliffe, I.A., Fletcher, A., Bench, C. (1995) First delineation of 5-HT1A receptor in human brain with PET and [^{11}C]WAY-100635. *Eur. J. Pharmacol.* **283**, R1-R3.

5. Osman, S., Lundkvist, C., Pike, V.W., Halldin, C., McCarron, J.A., Swahn, C.G., Ginovart, N., Luthra, S.K., Bench, C.J., Grasby, P.M., Wikström, H., Barf, T., Cliffe, I.A., Fletcher, A.C., Farde, L. (1996) Characterization of the metabolites of the 5-HT1A receptor radioligand, [O-methyl-^{11}C]WAY-100635, in monkey and human plasma - A comparison of the behaviour of an identified radioactive metabolite with parent radioligand using PET. *Nucl. Med. Biol.* **23**, 627-634.

6. Pike, V.W., McCarron, J.A., Lammertsma, A.A., Osman, S., Hume, S.P., Sargent, P.A., Bench, C.J., Cliffe, I.A., Fletcher, A., Grasby, P.M. (1996) Exquisite delineation of 5-HT1A receptors in human brain with PET and [carbonyl-^{11}C]WAY-100635. *Eur. J. Pharmacol.* **301**, R5-R7.

7. Kehne, J.H., Baron, B.M., Carr, A.A., Chaney S.F., Elands, J., Feldman, D.J., Frank, R.A., Van Giersbergen, P.L., McCloskey, T.C., Johnson, M.P., McCarty, D.R., Poirot, M., Senyah, Y., Siegel, B.W., Widmaier, C. (1996) Preclinical characterization of the potential of the putative atypical antipsychotic MDL 100,907 as a 5-HT2A receptor antagonist with a favorable CNS safety profile. *J. Pharmacol Exp. Ther.* **277**, 968-981.

8. Hall, H., Lundkvist, C., Halldin, C., Farde, L., Pike, V.W., McCarron, J.A., Fletcher, A.C., Cliffe, I.A., Barf, T., Wikström, H., Sedvall, G. (1997) Autoradiographic localization of 5-HT1A receptors in the post-mortem human brain using [^{3}H]WAY-100635 and [^{11}C]WAY-100635. *Brain Res.* (in press).

9. Osman, S., Lundkvist, C., Pike, V.W., Halldin, C., McCarron, J.A., Swahn, C.G., Farde, L., Ginovart, N., Luthra, S.K., Bench, C.J., Sergeant, P., Cliffe, I.A., Fletcher, A.C., Grasby, P.M. (1997) Absence of radioactive lipophilic metabolites in monkey and human plasma from 5-HT1A receptor radioligand, [carbonyl-^{11}C]WAY-100635 - advantages in PET for signal contrast and biomathematical modelling. *Nucl. Med. Biol.* (submitted).

10. Lundkvist, C., Halldin, C., Ginovart, N., Nyberg, S., Swahn, C.G., Carr, A.A., Brunner, F, Farde, L. (1996) [^{11}C]MDL 100907, a radioligand for selective imaging of 5-HT2A receptors with positron emission tomography. *Life Sciences.* **58**, PL 187-192.

11. Farde, L., Ginovart, N., Halldin, C., Ito, H., Lundkvist, C., Swahn, C.G., McCarron, J.A. and V.W. Pike. (1997) PET-characterization of [carbonyl-^{11}C]WAY-100635 binding to 5-HT1A receptors in the monkey and human brain. *Psychopharmacology* (submitted).

12. Ito, H., Nyberg, S., Halldin, C., Lundkvist, C., Farde, L. (1997) Positron emission tomography imaging of 5-HT2A receptors with [^{11}C]MDL 100907. *J. Nucl. Med.* (submitted).

THE RENIN-ANGIOTENSIN SYSTEM

ZSOLT SZABO
The Johns Hopkins Medical Institutions
600 North Wolfe Street Baltimore, Maryland 21205, U.S.A.

1. Significance

With an overall prevalence of 25 %, hypertension is a major public health problem. A direct link between arterial hypertension and cardiovascular events such as stroke and myocardial infarction has been clearly demonstrated although the pathogenesis of arterial hypertension is still not completely understood. Angiotensin receptors play a role, but this role could not be investigated in humans for a lack of appropriate noninvasive imaging techniques. This presentation describes how specific angiotensin antagonists can be used for investigation of the receptors in order to understand their physiology, pharmacology and regulation. It also gives clinical indications for future applications of PET imaging of these receptors.

The nonpeptide angiotensin antagonist losartan has established therapeutic utility in the treatment of arterial hypertension. Losartan and other nonpeptide antagonists also proved valuable for the study of angiotensin receptor pharmacology and physiology [1-3]. The use of angiotensin antagonists in radioligand binding studies has resulted in the identification of two subtypes of angiotensin II (Ang II) receptors, termed AT_1 and AT_2 [4] with distinct pharmacological properties: biphenylimidazoles (for example DuP753 or losartan) recognize AT_1 receptors with high affinity ($K_i = 10$ nM) and AT_2 receptors with low affinity ($K_i = 10$ µM). On the other hand imidazopyridine carboxylic acids (for example PD 123177) have high affinity for the AT_2 receptors [5-7].

AT_1 receptors are located in the blood vessels, brain, liver, lungs, adrenals and kidneys. Large amounts of AT_1 receptor mRNA have been detected in these organs indicating active synthesis of the receptor protein [8-15]. In the kidney, brain and heart, co-expression of various components of the renin-angiotensin system (RAS) indicate the presence of regional molecular mechanisms through which Ang II exerts its regulatory effects [8]. Angiotensin AT_1 receptors probably mediate all physiologically important effects of Angiotensin II (Ang II) including control of aldosterone secretion, vasoconstriction, renal blood flow, glomerular filtration, sodium and water reabsorption, and drinking behavior, as well as tachycardiac responses. AT_1 receptors are involved in the regulation of blood pressure and in the pathogenesis of essential hypertension, renovascular hypertension and other types of secondary hypertension [6,16-18].

B. Gulyás and H.W. Müller-Gärtner (eds.),
Positron Emission Tomography: A Critical Assessment of Recent Trends, 99–115.
© 1998 *Kluwer Academic Publishers.*

Factors regulating the AT_1 receptor include low or high dietary sodium, ACE inhibitors, renal hypoperfusion, thyroid dysfunction and corticosteroids [12,19-26]. In the rat kidney, glomerular Ang II receptors up-regulate during sodium loading and down-regulate during sodium depletion or infusion of Ang II [27,28]. Angiotensin converting enzyme (ACE) inhibitors prevent receptor down-regulation in the glomeruli supporting the concept that down-regulation indeed is mediated by Ang II. Receptor regulation has direct functional consequences; in rats, the ability of Ang II to modulate glomerular hemodynamics is reduced during sodium depletion and enhanced during sodium loading [27]. By contrast to glomeruli, sodium depletion up-regulates tubular epithelial Ang II receptors in a manner analogous to adrenal glomerulosa cells [29], whereas tubular epithelial receptors down-regulate during sodium loading [30].

The above observations have been attributed to two recently cloned type-1 Ang II receptors in the rat which are differentially regulated by dietary sodium [31,32]. The two receptor subtypes have high structural homology but are differently expressed in different organs and tissues of the rat [33]. Sodium deprivation enhances AT_{1a} and inhibits AT_{1b} mRNA in rat brains, compared to rats on high sodium intake [12]. Conversely, rats on a high sodium diet exhibited increased levels of AT_{1a} and decreased AT_{1b} mRNA compared to those on low sodium intake. Furthermore, differential regulation of AT_{1a} and AT_{1b} was observed in bilaterally nephrectomized rats; AT_{1a} was negatively modulated in the liver whereas AT_{1b} was positively modulated in the adrenal [13].

2. Imaging Studies

A significant problem in investigating AT_1 receptors *in vivo* is the lack of noninvasive techniques. The captopril test is used in conjunction with imaging the regional GFR with Tc-99m DTPA to diagnose renovascular hypertension and to determine whether revascularization is indicated for improvement of arterial hypertension. Captopril renography examines the functional consequences of the altered renin-angiotensin system (RAS), but does not address the issue of alterations in angiotensin receptors.

The first radioligand for direct imaging of the components of the RAS 4-*cis*-[^{18}F]fluorocaptopril (^{18}FCAP). Angiotensin converting enzyme (ACE) is a zinc ion mediated peptidyldipeptide hydrolase which activates angiotensin I by reducing the decapeptide to the octopeptide angiotensin II. Captopril, one of the first clinically implemented ACE inhibitors has been labeled with fluorine-18 (^{18}FCAP) by the reaction of the triflate 2 with K18F/Kryptofix 222 in MeCN followed by hydrolysis (2N NaOH). In vivo distribution studies in rats revealed high concentration of the radioligand in lung, kidney and aorta. The dose that reduced radioligand binding by 50 % in the lung was > 5 µg/kg unlabeled captopril [34]. I human volunteers high quality images of both the lungs and the kidneys were obtained with a nearly 50 % reduction of radioligand binding after administration of 25 mg captopril orally. These experiments indicated that ^{18}FCAP could be useful to investigate the ACE in humans [34].

The drug MK-996 (L-159,282) (N-[[4'-[(2-ethyl-5,7-dimethyl - 3H - imidazo[4,5 - B]pyridin - 3 -yl) methyl] [1,1'-biphenyl]-2-yl]sulfonyl]-benzamide) is a potent, selective AT_1 receptor antagonist with an affinity for these receptors in the subnanomolar range (IC_{50} = 0.15 nM for AT_1, in comparison, IC_{50} > 2000 nM for AT_2). Another drug, L-159,884 (N-[[4'-[(2-ethyl-5,7-dimethyl-3H-imidazo[4,5-B] pyridin-3-yl)methyl][1,1'-biphenyl]-2-yl]sulfonyl]-4-methoxy-benzamide) the methoxy-substituted analog of MK-996 has been labeled with carbon-11 [35]. Labeling was achieved by [^{11}C]methylation of the corresponding precursor L-162,914. From the end of bombardment, the synthesis time was 18 minutes with a yield of 5 % and an average specific activity of 2979 mCi/µmol [35].

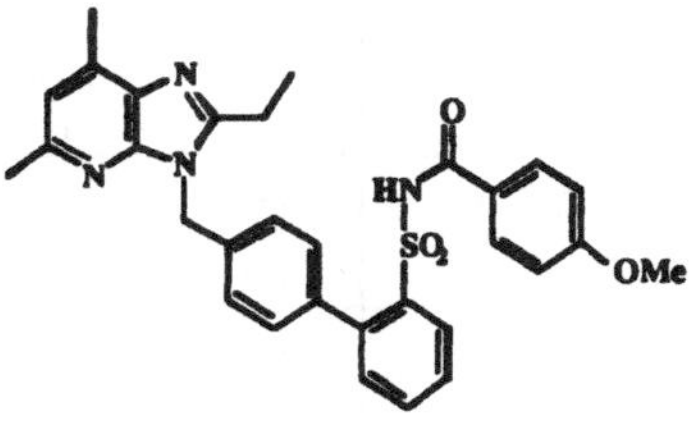

Figure 1: Chemical structure of L-159,884.

Safety studies performed on eighteen dogs included measurements of oxygen saturation of circulating blood as well as arterial blood pressure, heart rate and core temperature during PET studies with [^{11}C]L-159,884. None of these physiological parameters showed significant changes during the experiments. EKG monitoring did not show any changes in the heart rhythm and no signs of myocardial ischemia. In dogs the injected dose was less than 10 mCi and the injected amount of drug was between 0.5 and 15 µg. All dogs recovered from the anesthesia without any observable complications. Table 1 shows oxygen saturation, pulse, systolic and diastolic blood pressure and core temperature at baseline, 5, 15, 30 and 60 minutes after radiotracer injection. The parameter differences were investigated with repeat test ANOVA and paired t-test and were found to be statistically insignificant.

Table 1. Oxygen saturation, pulse, systolic and diastolic blood pressure and core temperature in the investigated animals (n=18).

time [min]	O2 [%]	Pulse [/min]	Rrsyst [mmHg]	Rrdiast [mmHg]	Temp [°F]
0	96±3	125±28	137±25	85±13	95±2
5	95±3	120±24	126±25	79±11	96±1
15	96±2	122±29	129±21	82±13	96±2
30	96±2	115±24	143±21	87±17	95±1
60	95±4	113±23	137±19	90±10	95±2

After injection of [^{11}C]L-159,884 into mice, individual organ concentrations of radioactivity were measured over 90 minutes and the radiation dose to various target organs was calculated by means of the MIRD absorbed fraction technique. The effective dose equivalent was 35 mrem/mCi, which is acceptable for PET imaging studies both in humans and in animals. The effective absorbed radiation dose equivalent to the whole body from 20 mCi [^{11}C]L-159,884 is estimated to be 0.7 rem, which is 14 % of the acceptable annual radiation exposure to radiation workers (Table 2). The calculations were based on measurements performed in mice adapted to the size of the human organs using the formulas of the Recommendations of the International Commission on Radiological Protection (1977) [36].

Table 2: Effective radiation dose equivalent for a PET study with 20 mCi [^{11}C]L-159,884

Organ	w_T	dose	Effective radiation dose	
		[mrad/mCi]	[mrad/mCi]	[mrad/20 mCi]
Gonads (testes)	0.25	38	9.5	190
Breast (*)	0.15	10	1.5	30
Red Bone Marrow	0.12	8	0.96	19.2
Lungs	0.12	8	0.96	19.2
Thyroid (*)	0.03	10	0.3	6
Bone	0.03	4	0.12	2.4
Intestine (critical organ)	0.06	260	15.6	312
Bladder wall	0.06	38	2.28	45.6
Kidney	0.06	38	2.28	45.6
Stomach	0.06	17	1.02	20.4
Spleen	0.06	15	0.9	18
Total			35.42	708.4

* Estimated from whole body dose; w_T = weighting factor

Biodistribution studies were performed in order to demonstrate that adequate target-to-nontarget ratios can be obtained and to investigate the route of excretion of the tracer. For this purpose, CD-1 mice were injected with 200 µCi (1-2 µg/kg) of [^{11}C]L-159,884. Animals were sacrificed at different time points, 5-90 min after injection. There was a rapid tracer uptake in the kidneys, lungs, and heart, followed by a slow clearance. Peak activities in kidneys, lungs, and heart were 5.6, 3.0, and 1.5 %ID/g, respectively, at 15 minutes after injection. Blood radioactivity fell rapidly to a very low level within five minutes post injection. More than 20 % of the radioactivity accumulated in the intestine over the 90 minute observation period while less than 8 % was excreted via urine, indicating that excretion of the compound occurred primarily through the hepatobiliary route. Radioligand clearance from plasma was fast. In kidneys, organs known to be rich in AT$_1$ receptors, binding was more constant.

In order to demonstrate that binding of [^{11}C]L-159,884 was specific and selective, mice were pretreated with increasing doses of drugs with high affinity for the

AT_1 receptor. These drugs were injected intravenously 10 minutes before administration of [^{11}C]L-159,884. Specific binding was defined by total blockade of the receptors with 1 mg/kg L-159,282. This resulted in 70 % inhibition of total tracer accumulation, indicating that the nonspecific binding component was 30 % in mice. L-159,282 was the most potent inhibitor of [^{11}C]L-159,884 binding, followed by cold L-159,884, then by L-159,913 and L-159,689. The median effective dose values (ED_{50}) were: L-159,282: 0.02 µmol/kg, L-159,884: 0.18 µmol/kg, L-159,689: 0.57 µmol/kg.

Selectivity of [^{11}C]L-159,884 for the AT_1 receptor was demonstrated by pretreatment of mice with a high dose (1-5 mg/kg) of the AT_2 specific antagonist PD 123319. This drug did not have any effect on the *in vivo* binding of the radiotracer in the kidneys, heart and lungs. Blood levels of [^{11}C]L-159,884 also remained unaffected by PD 123319. Furthermore, adrenergic antagonists at doses of 1 mg/kg such as prazosin (α_1), yohimbine (α_2) and propranolol (β) also did not influence [^{11}C]L-159,884 binding in the kidneys [37,38].

For PET imaging purposes, dogs were anesthetized with pentobarbital. Nineteen consecutive PET images of increasing scan time (first scan 15 sec, last scan 20 min) were obtained for each study to account for radioisotope decay. Five to ten mCi [^{11}C]L-159,884 were injected intravenously. Two experiments were performed in each animal, one control PET study and another PET study performed following intravenous injection of 1 mg/kg MK-996. The effectiveness of 1 mg/kg MK-996 in blocking the AT_1 receptors was shown in three separate animals by an Ang II challenge test. For this purpose, the animals were pretreated with 0.1 and 1 mg/kg MK-996 and were subsequently injected with 0.1 µg/kg of Ang II. The arterial blood pressure was registered continuously for 15 minutes. Peak pressure occurred 2-4 minutes after the Ang II injection; the difference from the baseline was calculated as the maximal pressure response. The usual response to ANG II was 37±4/34±6 mmHg (systolic/diastolic pressure changes respectively). After premedication with MK-996, no measurable effect on arterial blood pressure occurred with any of the applied doses of MK-996, suggesting complete blockade of AT_1 receptors.

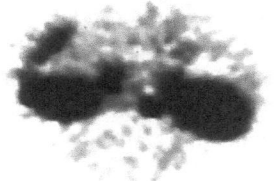

Figure 2: Image of dog kidneys obtained 55-115 minutes post injection of [^{11}C]L-159,884: baseline study.

The PET scans clearly showed accumulation of the radiotracer within the kidneys (Figure 2). In addition, there was a strong reduction of radioactivity after pretreatment with the specific receptor antagonist MK-996 [39].

3. Potential Applications

PET imaging of the AT1 receptors may gain significant applications in several distinct areas: 1) *Physiology:* It may allow investigation of AT_1 receptor expression and the effects of nutrition, stress and other environmental factors. 2) *Pharmacology:* It may be used to investigate the efficacy of AT_1 receptor antagonists at the receptor level. 3) *Pathology:* It may allow investigation of the regulatory changes of the AT_1 receptors in diseases such as chronic fatigue syndrome, aldosterone producing adrenal adenomas, renovascular hypertension, essential arterial hypertension, or hypertension of renal transplantation. 4) *Diagnosis:* It may help to recognize early changes of the AT_1 receptors which could lead to identification of patients at risk for arterial hypertension, particularly in the presence of genetic markers, risk factors or contributing diseases, such as hyperthyroidism or diabetes mellitus. 5) *Treatment:* It may aid treatment decisions in that the appropriate form of antihypertensive therapy could be based on the status of AT_1 receptors.

Essential Hypertension: As many as 50 million Americans have elevated blood pressure with a systolic blood pressure of 140 mmHg or greater and/or diastolic blood pressure of 90 mmHg or greater [41,42]. In the recent years there has been a decrease in the number of hypertensive adults attributable to healthier living habits including weight reduction, sodium restriction, alcohol moderation, cigarette smoking cessation and regular physical exercise [43]. Despite of these improvements and the advances in the pharmacology of the RAS, arterial hypertension is still a major public health problem. The percentage of persons over age 65 is increasing more rapidly than other age groups, and two thirds of people of this age have blood pressure higher than 140/90 [44].

Antihypertensive drugs clearly reduce pressure related complications such as cerebrovascular accidents, congestive heart failure and, in some cases, chronic renal failure, but atherosclerotic complications including coronary artery disease, and sudden cardiac death have not convincingly been improved [45,46]. A new approach to the diagnosis of hypertension has been recommended which should provide differentiation of patients based on underlying biochemical changes rather than symptoms [47]. In addition to pathophysiology, hemodynamic changes, end organ damage, concomitant medical diseases and problems should be investigated, and effects of demographic factors, drug therapy, patient's compliance, and health care costs should be assessed [45]. PET imaging would provide an estimate of the regional angiotensin receptors.

Hereditary/anthropological factors have a strong impact on hypertension [48]. In the African American population, prevalence of hypertension is 32 %, which is higher than 23 % found in Caucasians [49] although these differences are less pronounced with older age. African American hypertensives tend to have lower renin levels and to be more salt sensitive. They also suffer more often from cardiac hypertrophy and stroke [50].

In addition to ethnic origin, other factors, such as sex, age and body weight may contribute to hypertension. Antihypertensive therapy is less effective and its side effects are more frequent in women [51,52]. Higher incidence of hypertension with age may be related to the changes of the RAS. For example, intrarenal RAS demonstrates down-regulation of ACE mRNA and renin mRNA with aging [53]. Increased body weight is also clearly coupled with increased incidence of hypertension [54].

Tendency to develop high blood pressure with increased salt intake (salt sensitivity) exists and may include genetic or acquired mechanisms. Older men are usually more salt sensitive, and salt sensitivity is more frequent in women than men [55,56]. A direct linear correlation between salt intake and prevalence of hypertension has been described [48], and salt sensitivity is more prominent in patients with moderate to severe hypertension than in subjects with mild hypertension [56]. There is also a component of familial disposition to hypertension and salt sensitivity [55]. Although restriction of dietary sodium helps prevent or postpone the development of arterial blood pressure and its complications including coronary heart disease and stroke [57], its effects are not yet completely understood. Some authors even describe a potentially adverse impact of salt restriction on the risk factor profile for cardiovascular diseaese (increased renin, increased norepinephrine, increased cholesterol, increased blood viscosity) [58].

PET may permit investigation of the effects of dietary sodium on the AT_1 receptors and the mutual relationship between salt and RAS. Dietary sodium is a strong regulator of the RAS [5,59-61] and adaptation of kidney function to dietary sodium is based on adaptational changes of the RAS [62]. In rats, low sodium diet increases the renal mRNA levels for the AT_{1a} receptors. AT_{1a} receptor subtypes are dominant in rats and mice and are responsible for pressor effects of Ang II [63,64]. The effect of sodium on rat brain receptors is different: increase of expression of AT_{1a} receptors and decrease of expression of AT_{1b} receptors has been observed with high sodium diet [12]. Restriction of salt intake in the diet lowers the blood pressure in many subjects with high blood pressure, and this effect is additive to other blood pressure lowering drugs, particularly those that inhibit the RAS [65].

Stress is another important factor of hypertension, and inappropriate renal sodium retention has been linked with chronic stress hypertension [66]. The hypertensive effect of stress can be related to the effect of stress hormones on the RAS: corticosteroids cause up-regulation of vascular AT_1 receptors in the rat [67,68], and there is a mutual interaction between RAS and aldosteron secretion. Human adrenocortical cells are known to express AT_1 receptor subtypes and this expression is decreased by Ang II both at the mRNA and receptor protein level [69].

Essential hypertension has been linked to the RAS, particularly to the angiotensinogen gene on chromosome 1q42-43 but regulatory changes of the AT_1 receptors in human essential hypertension are largely unknown. Transplantation of kidneys from normotensive controls into hypertensive animals can normalize blood pressure of the recipient animals which indicates direct involvement of the kidneys in in the pathogenesis of essential hypertension [54]. This involvement is based on the ability of the kidneys to handle sodium and water and the high concentration of AT_1 receptors which are mediators of sodium and water reuptake. Chronic hypertension occurs if there is an abnormality of kidney function that shifts pressure natriuresis so that sodium balance is maintained at elevated blood pressure.

Investigation of the pathology and careful choice of therapy are important in hypertensive patients who have to take medicine every day for several decades. Imaging the AT_1 receptors in hypertension may prove important, and hypertension may prove to be a rather heterogeneous group of various pathologies with or without involvement of the RAS and the AT_1 receptors [70].

PET may also prove helpful for assessing the efficacy of angiotensin antagonists since measurement of blood pressure alone is not an accurate technique to assess the effect of antihypertensive drugs [70]. Compared to peptide antagonists (e.g. saralasin), losartan has significant advantages including longer duration of action, effective oral absorption and no angiotensin II agonist activity [71]. Losartan, the drug approved for human use is advantageous over the ACE inhibitors since it does not affect the metabolism of bradykinin [71,72]. Antihypertensive effects of angiotensin antagonists are probably peripheral; intracerebroventricular administration of the AT_1 antagonist EXP 3174 in hypertensive rats did not reduce blood pressure, while peripheral venous administration did [73].

Complications of Hypertension: Not only the primary mechanism of hypertension but also the consequences of hypertension involve the RAS and the AT_1 receptors. Ang II in concentrations that cause moderate hypertension induces vascular, glomerular and tubulointerstitial injury with cell proliferation, leukocyte recruitment, upregulation of proteins normally associated with smooth muscle cells, and interstitial fibrosis [74]. In hypertension induced renal injury, AT_1 receptor antagonists have a potent protective effect by inhibiting the gene expresion of renal TGF-beta 1 and extracellular matrix components [75]. ACE inhibitors, in addition to their immediate effect of reducing angiotensin concentration, also down-regulate the AT_1 receptors which adds to their long-term antihypertensive effects [23]. Protective effects of angiotensin antagonists on the development of glomerulosclerosis has been demonstrated in rats with renovascular hypertension [76], an effect which may be independent of the antihypertensive action [77].

Renovascular Hypertension: In all hypertensives, the prevalence of renovascular hypertension is less than 5 %, however, in malignant hypertension this prevalence may be as high as 43 % [78]. Anatomical stenosis of the renal artery in a hypertensive patient is not conclusive for diagnosis. Rather, the hemodynamic significance of a stenosis is to be ascertained by diagnostic tests which indicate activation of the RAS such as peripheral renin activity, captopril test, and captopril radionuclide renography [78].

The RAS, including the AT_1 receptors, plays a central role in the development of renovascular hypertension. AT_1 receptors modulate essentially all of the known intrarenal effects of Ang II [79] including modulation of renal blood flow, glomerular filtration rate, tubular epithelial transport, renin release and cellular growth. Renal hypoperfusion is a strong stimulus of the RAS and Ang II has a homeostatic role in the maintenance of GFR during renal artery narrowing [80]. Administration of ACE inhibitors will reduce the compensatory effects of Ang II on the GFR [80].

Chronic activation of the RAS in renal arterial stenosis increases the release of Ang II. In rats with 2K1C Goldblatt hypertension, increased levels of systemic Ang II enhance tubuloglomerular responsiveness and augment sodium retention and water retention thus ultimately causing arterial hypertension. This effect seems to depend on the AT_1 receptors since losartan not only normalizes hypertension in these rats but also reduces tubuloglomerular responsiveness [81]. In experimental renovascular hypertension the density of angiotensin receptors is increased [24,26,82] and the gene of the AT_1 receptors is up-regulated. A nearly sevenfold increase in AT_1 expression in the ischemic kidney and an eightfold increase in the contralateral kidney has been observed in 2K1C rats [83]. The AT_1 receptors are probably up-regulated in patients with renovascular hypertension, and this regulation can strongly be influenced by drugs affecting the RAS. AT_1 receptors in the kidney, liver and adrenal cortex are reduced by losartan and by ACE inhibitors [84] which adds to their direct pharmacological effects as antihypertensive drugs in this situation. It is still debated whether this reduction is caused by downregulation of AT_1 mRNA or enhanced receptor internalization of AT_1 receptors [85].

Renoparenchymal Hypertension: Arterial hypertension can develop in renal diseases other than renal artery stenosis. The AT_1 receptors are either up or down-regulated in such diseases. This hypothesis is based on results of animal experiments. In rats with unilateral ureteral obstruction treatment with losartan increases RBF and GFR 5-6 fold [86]. In such rats, elevated levels of renin and Ang II mRNA in the obstructed kidneys have been observed [87]. Increased level of Ang II has a major role in the development of tubulointerstitial fibrosis that follows complete unilateral ureteral obstruction in the rat. Both an ACE inhibitor and angiotensin II antagonist ameliorate the increased production of extracellular matrix protein in the tubulointerstitium of the obstructed kidney [88].

The RAS has also been linked to glomerulonephritis. The expression of glomerular angiotensin II receptors has been shown to be acutely down-regulated but chronically up-regulated in experimental anti-glomerular basement membrane nephritis in rats [89]. Biopsy materials from patients with glomerulonephritis have shown decreased renin mRNA and have made evident that the expression of this mRNA is improved if the patients are treated with ACE inhibitors [90,91].

The RAS is affected both by renal graft rejection and cyclosporin treatment [90]. In rats transplanted kidneys show up-regulation of glomerular Ang II receptor density probably as an adaptation of the intrarenal RAS system, and cyclosporin treatment causes further up-regulation of Ang II receptors [92]. When patients with renal grafts are administered ACE inhibitors, those who develop rejection show no change in renal function while in those who do not develop rejection the effective renal plasma flow increases by 23 % [93]. This is an indication that regulatory mechanisms are inefficient in the rejection group.

Cardiomyopathy: The RAS receptors have been linked to hypertensive and idiopathic cardiomyopathy. There is strong evidence for a local RAS in the human heart. Ang II has inotropic and chronotropic effect on myocardial cells and increases DNA turnover and protein synthesis in isolated atria and ventricles [6]. Ang II receptors are localized in the myocardium, cardiac adrenergic nerves and coronary vessels [94]. Ang II has both autocrine and paracrine effects that promote fibrous tissue formation which is beneficial during wound-healing after tissue injury [95,96]. Less appropriate is the activation of the local RAS during chronic hypertension and idiopathic cardiomyopathy [97,98]. Chronic hypertension causes up-regulation of the AT_1 receptors both at the protein level and at the level of mRNA [99-101], and chronic increased secretion of Ang II coupled with receptor up-regulation results in cellular proliferation, particularly of the fibrous tissue. This involves second messengers and proto-oncogenes of cellular differentiation and proliferation [102]. Ang II by means of up-regulating its AT_1 receptors causes up-regulation of the transforming growth factor-β_1 gene [103] leading to cellular hypertrophy and hyperplasia and to deposition of extracellular matrix [104]. In arterial hypertension, Ang II also increases macromolecular permeability in the coronary microvasculature [105,106].

In human end-stage heart failure, Ang II binding is significantly reduced coupled with reduction for AT_1 cDNA amplification signals [107]. ACE inhibitors and Ang II antagonists are useful in the treatment of congestive cardiac failure, ventricular remodeling, and left ventricular hypertrophy and fibrosis [108,109]. They reduce preload and afterload and suppress the local RAS. Ang II antagonists are probably more effective since recently enzymes other than ACE have been described in the heart which are capable of producing Ang II by metabolic pathways independent of the RAS route [110]. Biopsy studies have shown a 5-10 fold increase in Ang II receptor density in idiopathic cardiomyopathy [94].

Endocrine Disorders: Endocrine disorders in which regulatory changes of the RAS have been postulated include thyroid disfunction, diabetes, acromegaly, and hyperaldosteronism [22,104,111-113]. Thyroid dysfunction results in increased cardiac, liver and kidney Ang II receptors but decreased adrenal Ang II receptors [22]. In diabetes, Ang II receptors are increased in the heart, liver and adrenal gland and decreased in the kidneys [111]. Experimental proximal tubular reabsorption which is strongly regulated by Ang II is reduced in diabetic kidneys and can be improved both with ACE inhibitors and losartan [114]. Treatment with ACE inhibitors proved useful in averting diabetic glomerulosclerosis [115].

Hyperaldosteronism is a special form of secondary hypertension in which imaging the AT_1 receptors has a potential usefulness. In vitro binding and receptor cloning studies indicate that aldosteron producing adenomas express and sometimes overexpress the AT_1 receptor both at the genetic and protein level [112,113]. Ang II may have a direct involvement in the pathophysiology of these tumors both because of its stimulating effect on aldosterone production and because of its proliferative effects on the zona glomerulosa tissue [112].

Portal Hypertension: In rabbits experimental portal hypertension results in reduction of the binding capacity of the Ang II receptors in the mesenteric artery and in the adrenal cortex [116] without change a in receptor/ligand affinity.

4. References

1. Duncia, J.V., Chiu, A.T., Carini, D.J., Gregory, G.B., Johnson, A.L., Price, W.A., Wells, G.J., Wong, P.C., Calabrese, J.C., Timmermans, P.B. The discovery of potent nonpeptide angiotensin II receptor antagonists: a new class of potent antihypertensives. *J. Med. Chem.* **33**, 1312-1329, **1990**

2. Chiu, A.T., McCall, D.E., Price, W.A., Wong, P.C., Carini, D.J., Duncia, J.V., Wexler, R.R., Yoo, S.E., Johnson, A.L., Timmermans, P.B. Nonpeptide angiotensin II receptor antagonists. VII. Cellular and biochemical pharmacology of DuP 753, an orally active antihypertensive agent. *J. Pharmacol. Exp. Ther.* **252**, 711-718, **1990**

3. Blankley, C.J., Hodges, J.C., Klutchko, S.R., Himmelsbach, R.J., Chucholowski, A., Connolly, C.J., Neergaard, S.J., Van Nieuwenhze, M.S., Sebastian, A., Quin, J. Synthesis and structure-activity relationships of a novel series of non-peptide angiotensin II receptor binding inhibitors specific for the AT2 subtype. *J. Med. Chem.* **34**, 3248-3260, **1991**

4. DeGasparo, M., Husain, A., Alexander, W., Catt, K.J., Chiu, A.T., Drew, M., Goodfriend, T., Harding, J.W., Inagami, T., Timmermans, P.B.M.W.M. Proposed update of angiotensin receptor nomenclature. *Hypertension.* **25**, 924-927X, **1995**

5. Sandberg, K. Structural analysis and regulation of angiotensin II receptors. *TEM.* **5**, 28-35, **1994**

6. Griendling, K.K., Murphy, T.J., Alexander, R.W. Molecular biology of the renin-angiotensin system. *Circulation.* **87**, 1816-1828, **1993**

7. Timmermans, P.B., Wong, P.C., Chiu, A.T., Herblin, W.F., Benfield, P., Carini, D.J., Lee, R.J., Wexler, R.R., Saye, J.A., Smith, R.D. Angiotensin II receptors and angiotensin II receptor antagonists. *Pharmacol. Rev.* **45**, 205-251, **1993**

8. Gasc, J.M., Monnot, C., Clauser, E., Corvol, P. Co-expression of type 1 angiotensin II receptor (AT1R) and renin mRNAs in juxtaglomerular cells of the rat kidney. *Endocrinology.* **132**, 2723-2725, **1993**

9. Bunnemann, B., Iwai, N., Metzger, R., Fuxe, K., Inagami, T., Ganten, D. The distribution of angiotensin II AT1 receptor subtype mRNA in the rat brain. *Neurosci. Lett.* **142**, 155-158, **1992**

10. Kakinuma, Y., Fogo, A., Inagami, T., Ichikawa, I. Intrarenal localization of angiotensin II type 1 receptor mRNA in the rat. *Kidney Int.* **43**, 1229-1235, **1993**

11. Burns, K.D., Homma, T., Harris, R.C. The intrarenal renin-angiotensin system. *Semin. Nephrol.* **13**, 13-30, **1993**

12. Sandberg, K., Ji, H., Catt, K.J. Regulation of angiotensin II receptors in rat brain during dietary sodium changes. *Hypertension.* **23**, I 137-I 141, **1994**

13. Iwai, N., Inagami, T., Ohmichi, N., Nakamura, Y., Saeki, Y., Kinoshita, M. Differential regulation of rat AT$_{1a}$ and AT$_{1b}$ receptor mRNA. *Biochem. Biophys. Res. Commun.* **188**, 298-303, **1992**

14. Grone, H.J., Simon, M., Fuchs, E. Autoradiographic characterization of angiotensin receptor subtypes in fetal and adult human kidney. *Am. J. Physiol.* **262**, F326-F331, **1992**

15. Burns, L., Clark, K.L., Bradley, J., Robertson, M.J., Clark, A.J.L. Molecular cloning of the canine angiotensin II receptor - an AT$_1$-like receptor with reduced affinity for DuP753. *FEBS Letters.* **343**, 146-150, **1994**

16. Bernstein, K.E., Alexander, R.W. Counterpoint: Molecular analysis of the angiotensin II receptor. *Endoc. Rev.* **13**, 381-386, **1992**

17. Gibson, R.E., Thorpe, H.H., Cartwright, M.E., Frank, J.D., Schorn, T.W., Bunting, P.B., Siegl, P.K. Angiotensin II receptor subtypes in renal cortex of rats and rhesus monkeys. *Am. J. Physiol.* **261**, F512-F518, **1991**

18. Clark, K.L., Robertson, M.J., Drew, G.M. Role of angiotensin AT$_1$ and AT$_2$ receptors in mediating the renal effects of angiotensin II in the anaesthetized dog. *Br. J. Pharmacol.* **109**, 148-156, **1993**

19. Lewis, N.P., Ferguson, D.R. [3H]angiotensin II binding to basolateral membranes from rat proximal renal tubule: effect of sodium intake and captopril. *J. Endocrinol.* **122**, 499-507, **1989**

20. Lehoux, J.G., Bird, I.M., Rainey, W.E., Tremblay, A., Ducharme, L. Both low sodium and high potassium intake increase the level of adrenal angiotensin-II receptor type 1, but not that of adrenocorticotropin receptor. *Endocrinology.* **134**, 776-782, **1994**

21. Kitamura, E., Kikkawa, R., Fujiwara, Y., Imai, T., Shigeta, Y. Effect of angiotensin II infusion on glomerular angiotensin II receptors in rats. *Biochim. Biophys. Acta.* **885**, 309-316, **1986**

22. Sernia, C., Marchant, C., Brown, L., Hoey, A. Cardiac angiotensin receptors in experimental hyperthyroidism in dogs. *Cardiovasc. Res.* **27**, 423-428, **1993**

23. Wu, J.N., Edwards, D., Berecek, K.H. Changes in renal angiotensin II receptors in spontaneously hypertensive rats by early treatment with the angiotensin-converting enzyme inhibitor captopril. *Hypertension.* **23[part II]**, 819-822, **1994**

24. Wilkes, B.M., Pion, I., Sollott, S., Michaels, S., Kiesel, G. Intrarenal renin-angiotensin system modulates glomerular angiotensin receptors in the rat. *Am. J. Physiol.* **254**, F345-F350, **1988**

25. Tufro-McReddie, A., Chevalier, R.L., Everett, A.D., Gomez, R.A. Decreased perfusion pressure modulates renin and ANG II type 1 receptor gene expression in the rat kidney. *Am. J. Physiol.* **264**, R696-R702, **1993**

26. Sahlgren, B., Eklof, A.C., Aperia, A. Regulation of glomerular angiotensin II receptor densities in renovascular hypertension: response to reduced sympathetic and vasopressin influence. *Acta Physiol. Scand.* **146**, 467-471, **1992**

27. Ballermann, B.J., Skorecki, K.L., Brenner, B.M. Reduced glomerular angiotensin II receptor density in early and untreated diabetes mellitus in the rat. *Am. J. Physiol.* **247**, F110-F116, **1984**

28. Skorecki, K.L., Ballermann, B.J., Rennke, H.J., Brenner, B.M. Angiotensin receptor regulation in isolated renal glomeruli. *Fed. Proc.* **42**, 3064-3070, **1983**

29. Douglas, J., Catt, K.J. Regulation of angiotensin II receptors in the rat adrenal cortex by dietary electrolytes. *J. Clin. Invest.* **58**, 834-843, **1976**

30. Freedlender, A.E., Goodfriend, T.L. Angiotensin receptors and sodium transport in renal tubules. *Fed. Proc.* **36**, 4811977

31. Sandberg, K., Ji, H., Clark, A.J., Shapira, H., Catt, K.J. Cloning and expression of a novel rat angiotensin II receptor subtype. *J. Biol. Chem.* **267**, 945-948, **1992**

32. Murphy, T.J., Alexander, R.W., Griendling, K.K., Runge, M.S., Bernstein, K.E. Isolation of a cDNA encoding the vascular type-1 angiotensin II receptor. *Nature.* **351**, 233-236, **1991**

33. Burson, J.M., Aguilera, G., Gross, K.W., Sigmund, C.D. Differential expression of angiotensin receptor 1A and 1B in mouse. *American. Journal. of. Physiology.* **267**, E 260-E 267, **1994**

34. Hwang, D.R., Eckelman, W.C., Mathias, C.J., Petrillo, E.W., Jr., Lloyd, J., Welch, M.J. Positron-labeled angiotensin-converting enzyme (ACE) inhibitor: fluorine-18-fluorocaptopril. Probing the ACE activity in vivo by positron emission tomography. *J. Nucl. Med.* **32**, 1730-1737, **1991**

35. Mathews, W.B., Burns, H.D., Dannals, R.F., Hamill, T.G., Naylor, E.M., Ravert, H.T. Carbon-11 labeling of nonpeptide angiotensin-II antagonists: MK-996 and analogs. *J. Nucl. Med.* **35**, 6P, **1994**

36. Snyder, W.S., Ford, M.R., Warner, G.G., Watson, S.B. "S" Absorbed dose per unit cumulated activity for selected activity radionuclides and organs. In: *MIRD Pamphlet No 11*, New York, Society of Nuclear Medicine, **1975**,

37. Kim, S.E., Scheffel, U., Szabo, Z., Burns, H.D., Gibson, R.E., Ravert, H.T., Mathews, W.B., Hamill, T.G., Dannals, R.F. In vivo labeling of angiotensin II receptors with a carbon-11 labeled selective nonpeptide antagonist. *J. Nucl. Med.* **37**, 307-311, **1996**

38. Kim, S.E., Scheffel, U., Szabo, Z., Burns, H.D., Gibson, R.E., Ravert, H.T., Mathews, W.B., Hamill, T.G., Dannals, R.F. In vivo labeling of angiotensin II receptors with a carbon-11-labeled selective nonpeptide antagonist. *J. Nucl. Med.* **37**, 307-311, **1996**

39. Szabo, Z., Burns, H.D., Gibson, R.E., Hamill, T.G., Dannals, R.F., Ravert, H.T., Kim, S.E., Mathews, W.B., Musachio, J.L., Wagner, H.N.J. Positron emission tomography imaging of the angiotensin II/AT1 receptors. *J. Nucl. Med.* **35**, 124**1994**(Abstract)

40. Szabo, Z., Mathews, W.B., Burns, H.D., Gibson, R.E., Kivlighn, S.D., Hamill, T.G., Ravert, H.T., Dannals, R.F. Comparison of the pharmacological potency and AT1 receptor occupancy of the nonpeptide angiotensin antagonist E-3174. *J. Nucl. Med.* **37**, 288P**1996**(Abstract)

41. Joint National Committee: The fifth report of the Joint National Committee on detection, evaluation, and treatment of high blood pressure. *Arch. Intern. Med.* **153**, 154-183, **1993**

42. National High Blood Pressure Education Program Working Group: National high blood pressure education program working group report on hypertension in the elderly. *Hypertension.* **23**, 275-285, **1994**

43. Frohlich, E.D. There's good news and not so good news. *Hypertension.* **25**, 303-304, **1995**

44. Schoenberger, J.A. Epidemiology and evaluation: steps toward hypertension treatment in the 1990s. *Am. J. Med.* **90**, 3S-7SX, **1991**

45. Houston, M.C. Hypertension and coronary heart disease risk factor management. *Clin. Auton. Res.* **3**, 357-61X, **1993**

46. Lindholm, L.H. Cardiovascular risk factors and their interactions in hypertensives. *J. Hypertens. Suppl.* **9**, S3-6X, **1991**

47. Brown, M.J. Angiotensin receptor blockers in essential hypertension. *Lancet.* **342**, 1374-1375, **1993**

48. Leong, G.M., Kainer, G. Diet, salt, anthropological and hereditary factors in hypertension. *Child. Nephrl. Urol.* **12**, 96-105, **1992**

49. Burt, V.L., Whelton, P., Roccella, E.J., Brown, C., Cutler, J.A., Higgins, M., Horan, M.J., Labarthe, D. Prevalence of hypertension in the US adult population. Results from the Third National Health and Nutrition Examination Survey, 1988-1991. *Hypertension.* **25**, 305-313, **1995**

50. Saunders, E. Hypertension in blacks. *Prim. Care.* **18**, 607-22X, **1991**

51. Kitler, M.E. Differences in men and women in coronary artery disease, systemic hypertension and their treatment [editorial]. *Am. J. Cardiol.* **70**, 1077-180X, **1992**

52. Anastos, K., Charney, P., Charon, R.A., Cohen, E., Jones, C.Y., Marte, C., Swiderski, D.M., Wheat, M.E., Williams, S. Hypertension in women: what is really known? The Women's Caucus, Working Group on Women's Health of the Society of General Internal Medicine. *Ann. Intern. Med.* **115**, 287-93X, **1991**

53. Jung, F.F., Kennefick, T.M., Ingelfinger, J.R., Vora, J.P., Anderson, S. Down-regulation of the intrarenal renin-angiotensin system in the aging rat. *Journal. of. the. American. Society. of. Nephrology.* **5**, 1573-1580X, **1995**

54. Hall, J.E. Renal and cardiovascular mechanisms of hypertension in obesity. *Hypertension.* **23**, 381-394, **1994**

112

55. Murakami, K., Kojima, S., Kimura, G., Sanai, T., Yoshida, K., Abe, H., Kawamura, M., Ashida, T. The assotiation between salt sensitivity of blood pressure and family history of hypertension. *Clin. Exp. Pharmacol. Physiol.* **20 (suppl.)**, 61-63, **1992**

56. Watt, G.C.M. Does salt sensitivity exsist? *Klin. Wochenschr.* **69 (suppl. 25)**, 30-35, **1991**

57. Elliott, P. Sodium and blood pressure:A review of the evidence from controlled trials of sodium reduction and epidemiological studies. *Klin. Wochenschr.* **69 (suppl. 25)**, 3-10, **1991**

58. Weder, A.B., Egan, B.M. Potential deleterious impact of salt restriction on cardiovascular risc factors. *Klin. Wochenschr.* **69 (suppl. 25)**, 45-50, **1991**

59. Beaufils, M., Sraer, J., Lepreux, C., Ardaillou, R. Angiotensin II binding to renal glomeruli from sodium-loaded and sodium-depleted rats. *Am. J. Physiol.* **230**, 1187-1193, **1976**

60. Spangler, W.L. Pathophysiologic response of the juxtaglomerular apparatus to dietary sodium restriction in the dog. *Am. J. Vet. Res.* **40**, 809, **1979**

61. Zhuo, J., Alcorn, D., McCausland, J., Casley, D., Mendelsohn, F.A.O. In vivo occupancy of angiotensin II subtype 1 receptors in rat renal medullary interstitial cells. *Hypertension.* **23[part II]**, 838-843, **1994**

62. Jover, B., Saladini, D., Nafrialdi, N., Dupont, M., Mimran, A. Effect of losartan and enalapril on renal adaptation to sodium restriction in rat. *Am. J. Physiol.* **267**, F281-F288, **1994**

63. Du, Y., Yao, A., Guo, D.F., Inagami, T., Wang, D.H. Differential regulation of angiotensin II receptor subtypes in rat kidney by low dietary sodium. *Hypertension.* **25**, 872-877, **1995**

64. Ito, M., Oliverio, M.I., Mannon, P.J., Best, C.F., Maeda, N., Smithies, O., Coffman, T.M. Regulation of blood pressure by the type 1a angiotensin II receptor gene. *Proceedings. of. the. National. Academy. of. Sciences. of. the. United. States. ofX. America.* **92**, 3521-3525X, **1995**

65. Cappucio, F.P., Markandu, N.D., MacGregor, G.A. Dietary salt intace and Hypertension. *Klin. Wochenschr.* **69 (suppl.25)**, 17-25, **1991**

66. Koepke, J.P. Effect of environmental stress on neural control of renal function. *Miner. Electrolyte Metab.* **15**, 83-87, **1989**

67. Sato, A., Suzuki, H., Murakami, M., Nakazato, Y., Iwaita, Y., Saruta, T. Glucocorticoid increases angiotensin II type 1 receptor and its gene expression. *Hypertension.* **23**, 25-30, **1994**

68. Sato, A., Suzuki, H., Nakazato, Y., Shibata, H., Inagami, T., Saruta, T. Increased expression of vascular angiotensin II type 1A receptor gene in glucocorticoid-induced hypertension. *Journal. of. Hypertension.* **12**, 511-516, **1994**

69. Bird, I.M., Mason, J.I., Rainey, W.E. Hormonal regulation of angiotensin II type 1 receptor expression and AT(1)-r mRNA levels in human adrenocortical cells. *Endocrine. Research.* **21**, 169-182X, **1995**

70. Menard, J. Improving hypertension treatment. Where should we put our efforts: new drugs, new concepts, or new management?. *Am. J. Hypertens.* **5**, 252S-258S, **1992**

71. Siegl, P.K. Discovery of losartan, the first specific non-peptide angiotensin II receptor antagonist. *J. Hypertens. Suppl.* **11**, S19-S22, **1993**

72. Gavras, H. Angiotensin-converting enzyme inhibition and the heart. *Hypertension.* **23[part 2]**, 813-817, **1994**

73. Basso, N., Kurnjek, M.L., Ruiz, P., Cannata, M.A. Effect of EXP 3174 on blood pressure of normoreninemic renal hypertensive rats. *Hypertension.* **25**, 283-287X, **1995**

74. Johnson, R.J., Alpers, C.E., Yoshimura, A., Lombardi, D., Pritzl, P., Floege, J., Schwartz, S.M. Renal injury from angiotensin II-mediated hypertension. *Hypertension.* **19**, 464-474, **1992**

75. Kim, S., Ohta, K., Hamaguchi, A., Omura, T., Yukimura, T., Miura, K., Inada, Y., Wada, T., Ishimura, Y., Chatani, F., Iwao, H. Contribution of renal angiotensin II type 1 receptor gene expressions in hypertension-induced renal injury. *Kidney International.* **46**, 1346-1358X, **1994**

76. Imamura, A., Mackenzie, H.S., Lacy, E.R., Hutchison, F.N., Fitzgibbon, W.R., Ploth, D.W. Effects of chronic treatment with angiotensin converting enzyme Inhibitor or an angiotensin receptor antagonist in two-kidney, one-clip hypertensive rats. *Kidney International.* **47**, 1394-1402X, **1995**

77. Kakinuma, Y., Kawamura, T., Bills, T., Yoshioka, T., Ichikawa, I., Fogo, A. Blood pressure-independent effect of angiotensin inhibition on vascular lesions of chronic renal failure. *Kidney Int.* **42**, 46-55, **1992**

78. Pickering, T.G. The role of laboratory testing in the diagnosis of renovascular hypertension. *Clin. Chem.* **37**, 1831-1837, **1991**

79. DeGasparo, M., Levens, N.R. Pharmacology of angiotensin II receptors in the kidney. *Kidney International.* **46**, 1486-1491, **1994**

80. Anderson, W.P., Denton, K.M., Woods, R.L., Alcorn, D. Angiotensin II and the maintenance of GFR and renal blood flow during renal artery narrowing. *Kidney Int. Suppl.* **30**, S109-S113, **1990**

81. Braam, B., Navar, L.G., Mitchell, K.D. Modulation of tubuloglomerular feedback by angiotensin II type 1 receptors during the development of Goldblatt hypertension. *Hypertension.* **25**, 1232-1237X, **1995**

82. Cheng, H.F., Becker, B.N., Burns, K.D., Harris, R.C. Angiotensin II upregulates type-1 angiotensin II receptors in renal proximal tubule. *J. Clin. Invest.* **95**, 2012-2019, **1995**

83. Modrall, J.G., Quinones, M.J., Frankhouse, J.H., Hsueh, W.A., Weaver, F.A., Kedes, L. Upregulation of angiotensin II type 1 receptor gene expression in. *Journal. of. Surgical. Research.* **59**, 135-140X, **1995**

84. Regitzzagrosek, V., Auchschwelk, W., Hess, B., Klein, U., Duske, E., Steffen, C., Hildebrandt, A.G., Fleck, E. Tissue- and subtype-specific modulation of angiotensin II receptors by chronic treatment with cyclosporin A, angiotensin-converting enzyme inhibitors and AT1 antagonists. *Journal. of. Cardiovascular. Pharmacology.* **26**, 66-72X, **1995**

85. Chansel, D., Bizet, T., Vandermeersch, S., Pham, P., Levy, B., Ardaillou, R. Differential regulation of angiotensin II and losartan binding sites in glomeruli and mesangial cells. *Am. J. Physiol.* **266**, F384-F393, **1994**

86. Pimentel, J.L., Jr., Wang, S., Martinez-Maldonado, M. Regulation of the renal angiotensin II receptor gene in acute unilateral ureteral obstruction. *Kidney Int.* **45**, 1614-1621, **1994**

87. El-Dahr, S.S., Gee, J., Dipp, S., Hanss, B.G., Vari, R.C., Chao, J. Upregulation of renin-angiotensin system and downregulation of kallikrein in obstructive nephropathy. *Am. J. Physiol.* **264**, F874-F881, **1993**

88. Klahr, S., Ishidoya, S., Morrissey, J. Role of angiotensin II in the tubulointerstitial fibrosis of obstructive nephropathy. *American. Journal. of. Kidney Diseases.* **26**, 141-146X, **1995**

89. Timmermans, V., Peake, P.W., Charlesworth, J.A., Macdonald, G.J., Pawlak, M.A. Angiotensin II receptor regulation in anti-glomerular basement membrane nephritis. *Kidney Int.* **38**, 518-524, **1990**

114

90. Wagner, J., Volk, S., Haufe, C.C., Ciechanowicz, A., Paul, M., Ritz, E. Renin gene expression in human kidney biopsies from patients with glomerulonephritis or graft rejection. *J. Amer. Soc. Nephrology.* **5**, 1469-1475X, **1995**

91. Wagner, J., Drab, M., Gehlen, F., Langheinrich, M., Volk, S., Ganten, D., Ritz, E. PCR analysis of human renal biopsies - renin gene regulation in glomerulonephritis. *Kidney International.* **46**, 1542-1545X, **1994**

92. Rettig, R., Buch, M., Gerstberger, R., Schnatterbeck, P., Paul, M. Effects of kidney transplantation on the renin-angiotensin systems of the recipients. *Kidney International.* **46**, 1536-1538X, **1994**

93. Mimran, A., Mourad, G., Ribstein, J. The renin-angiotensin system and renal function in kidney transplantation. [Review]. *Kidney Int. Suppl.* **30**, S114-S117, **1990**

94. Urata, H., Healy, B., Stewart, R.W., Bumpus, F.M., Husain, A. Angiotensin II receptors in normal and failing human hearts. *J. Clin. Endocrinol. Metab.* **69**, 54-66, **1989**

95. Weber, K.T., Sun, Y., Guarda, E. Structural remodeling in hypertensive heart disease and the role of hormones. *Hypertension.* **23[part 2]**, 869-877, **1994**

96. Weber, K.T., Brilla, C.G. Myocardial fibrosis and the renin-angiotensin-aldosterone system. [Review]. *J. Cardiovasc. Pharmacol.* **20 Suppl 1**, S48-S54, **1992**

97. Ruzicka, M., Yuan, B., Harmsen, E., Leenen, F.H. The renin-angiotensin system and volume overload-induced cardiac hypertrophy in rats. Effects of angiotensin converting enzyme inhibitor versus angiotensin II receptor blocker. *Circulation.* **87**, 921-930, **1993**

98. Dzau, V.J. Local expression and pathophysiological role of renin-angiotensin in the blood vessels and heart. *Basic. Res. Cardiol.* **88 Suppl 1**, 1-14, **1993**

99. Fujii, N., Tanaka, M., Ohnishi, J., Yukawa, K., Takimoto, E., Shimada, S., Naruse, M., Sugiyama, F., Yagami, K., Murakami, K., Miyazaki, H. Alterations of angiotensin II receptor contents in hypertrophied hearts. *Biochemical. &. Biophysical. Research. Communications.* **212**, 326-333X, **1995**

100. Meggs, L.G., Coupet, J., Huang, H., Cheng, W., Li, P., Capasso, J.M., Homcy, C.J., Anversa, P. Regulation of angiotensin II receptors on ventricular myocytes after myocardial infarction in rats. *Circ. Res.* **72**, 1149-1162, **1993**

101. Suzuki, J., Matsubara, H., Urakami, M., Inada, M. Rat angiotensin II (type 1A) receptor mRNA regulation and subtype expression in myocardial growth and hypertrophy. *Circ. Res.* **73**, 439-447, **1993**

102. Fernandez-Alfonso, M.S., Ganten, D., Paul, M. Mechanisms of cardiac growth. The role of the renin-angiotensin system. [Review]. *Basic. Res. Cardiol.* **87 Suppl 2**, 173-181, **1992**

103. Everett, A.D., Tufro-McReddie, A., Fisher, A., Gomez, R.A. Angiotensin receptor regulates cardiac hypertrophy and transforming growth factor-b_1 expression. *Hypertension.* **23**, 587-592, **1994**

104. Brown, L., Sernia, C. Angiotensin receptors in cardiovascular diseases. *Clin. Exp. Pharmacol. Physiol.* **21**, 811-818, **1994**

105. Reddy, H.K., Campbell, S.E., Janicki, J.S., Zhou, G., Weber, K.T. Coronary microvascular fluid flux and permeability: influence of angiotensin II, aldosterone, and acute arterial hypertension. *J. Lab. Clin. Med.* **121**, 510-521, **1993**

106. Weber, K.T., Sun, Y., Guarda, E., Katwa, L.C., Ratajska, A., Cleutjens, J.P.M., Zhou, G. Myocardial fibrosis in hypertensive heart disease - an overview of potential regulatory mechanisms. *Eur. Heart J.* **16**, 24-28X, **1995**

107. Regitzzagrosek, V., Friedel, N., Heymann, A., Bauer, P., Neuss, M., Rolfs, A., Steffen, C., Hildebrandt, A., Hetzer, R., Fleck, E. Regulation, chamber localization, and subtype distribution of angiotensin II receptors in human hearts. *Circulation*. **91**, 1461-1471, **1995**

108. Johnston, C.I., Fabris, B., Yoshida, K. The cardiac renin-angiotensin system in heart failure. *Am. Heart J.* **126**, 756-760, **1993**

109. Weber, K.T., Brilla, C.G., Campbell, S.E., Guarda, E., Zhou, G., Sriram, K. Myocardial fibrosis: role of angiotensin II and aldosterone. [Review]. *Basic. Res. Cardiol.* **88 Suppl 1**, 107-124, **1993**

110. Rush, J.E., Rajfer, S.I. Theoretical basis for the use of angiotensin II antagonists in the treatment of heart failure. *J. Hypertension*. **11(suppl 3)**, S69-S71, **1993**

111. Cheng, H.F., Burns, K.D., Harris, R.C. Reduced proximal tubule angiotensin II receptor expression in streptozotocin-induced diabetes mellitus. *Kidney International*. **46**, 1603-1610, **1994**

112. Sarzani, R., Opocher, G., Dessifulgheri, P., Paci, V., Cola, G., Rocco, S., Vianello, B., Mantero, F., Rappelli, A. Expression of type 1 angiotensin II receptors in human aldosteronomas. *Endocrine. Research*. **21**, 189-195, **1995**

113. Nawata, H., Takayanagi, R., Ohnaka, K., Sakai, Y., Imasaki, K., Yanase, T., Ikuyama, S., Tanaka, S., Ohe, K. Type 1 angiotensin II receptors of adrenal tumors. *Steroids*. **60**, 28-34, **1995**

114. De Nicola, L., Blantz, R.C., Gabbai, F.B. Renal functional reserve in the early stage of experimental diabetes. *Diabetes*. **41**, 267-273, **1992**

115. Tolins, J.P., Raij, L. Angiotensin converting enzyme inhibitors and progression of chronic renal failure. *Kidney Int. Suppl*. **30**, S118-S122, **1990**

116. Sitzmann, J.V., Wu, Y.P., Aguilera, G., Cahill, P.A., Burns, R.C. Loss of angiotensin-II receptors in portal hypertensive rabbits. *Hepatology*. **22**, 559-564X, **1995**

MEMBRANE TRANSPORTERS

ZSOLT SZABO
The Johns Hopkins Medical Institutions
600 North Wolfe Street Baltimore, Maryland 21205, U.S.A.

1. Transport Processes

The science of nuclear medicine is based on three fundamental biological processes: molecular transport, bioenergetics and cellular communication [1,2]. These three processes are not independent but rather very closely related. On the one hand membrane transport provides substrates needed for energy production, on the other it also uses the energy needed to move molecules against their concentration gradients. Biological transport is also involved in cellular communication: hormones are transported by the circulation to their target tissues and even into the interior of the cells to activate intracellular/nuclear recognition sites. Termination of neurochemical signal transmission involves membrane transporters specialized in removing neurotransmitters from the synaptic cleft.

Biological transport occurs at the macroscopic, microscopic and molecular levels. At the macroscopic level, blood circulation permits movement of metabolic substrates and hormones to many organs and tissues at a time. At the microscopic level, molecules are transported in the hepatobiliary system and the glomerular-tubular system of the kidneys. Prior nuclear medicine techniques, contrast angiography, MRI and CT have focused on these mechanisms. Recently, nuclear medicine techniques have been developed to investigate transport mechanisms at the molecular level.

The cell is surrounded by a complex chemical envelope called the cellular membrane. Inside this envelope is the cytoplasm, a rather complex environment which is further divided into many subcompartments (nucleus, Golgi apparatus, endoplasmic reticulum, mitochondria). Biological membranes consist of two sheets of phospholipid a so called phospholipid bilayer. Integral proteins span the entire thickness of the membrane while peripheral proteins are attached to its external or internal surface only. Water, gases (CO_2, N_2, O_2), and small uncharged polar molecules (urea, ethanol) easily pass through the lipoprotein bilayer. Large, uncharged, polar molecules (e.g. glucose), ions (e.g. K^+, Ca^{2+}, Cl-, HCO3-) as well as charged polar molecules (e.g. amino acids and ATP) can not pass the membrane directly because the membrane is impermeable to them.

There are three basic forms of membrane transport: simple diffusion, passive transport (also called facilitated diffusion), and active transport. Facilitated diffusion can be channel mediated or carrier mediated. Channel mediated transport is accomplished

117

B. Gulyás and H.W. Müller-Gärtner (eds.),
Positron Emission Tomography: A Critical Assessment of Recent Trends, 117–131.
© 1998 *Kluwer Academic Publishers.*

by integral membrane proteins which can be in a closed or open state. Some channels are almost always open permitting free leakage of ions along the concentration/electrical gradient. Stereoconfiguration of the channel protein provides channel selectivity for specific ions.

Carrier mediated transport involves activation of a membrane transporter protein. This protein binds its substrate and undergoes a conformational change which moves the substrate and releases it on the contralateral side of the membrane. Both channels and transporters can be substrate specific and saturable (because of their limited number), however, only transporters possess the ability to move substrates against their concentration gradients [3]. Transporters are usually unidirectional (at least under physiological conditions), while channels are bidirectional. Channel mediated transport occurs at much higher speed (10^8 molecules/s) than carrier mediated transport (10^3 molecules/s).

In contrast to the curved response of saturable facilitated diffusion, the response of simple diffusion is linear, apparently nonsaturable. Transporters can be responsible for transport of a single molecule (*uniport*), two molecules in the same direction (*symport*) or two or more molecules in opposite directions (*antiport*). Coupled transport of substrates may necessitate the presence of multiple transporters. For example, recycling of 5-HT (5-hydroxytryptamine, serotonin) at the 5-HT-ergic axon terminals within the brain involves not only the 5-HT transporter but also $Na^+/K^+/ATPase$. This ion transporter accumulates K^+ inside the cell and Na^+ outside the cell (antiport). Cl^- follows Na^+. When the concentration gradient for NaCl is established and serotonin is present in the synaptic cleft, the 5-HT transporter (5-HTT) becomes activated and moves 5-HT back into the neuron together with a NaCl molecule (symport). Certain biological events, such as the activation of cardiac myocytes, involves an entire cascade of ion transport and the concerted action of multiple channels, symporters and antiporters.

The resting membrane potential of the myocardium (as well as of nerve cells and tumor cells) is established by the K^+ gradient as a result of $Na^+/K^+/ATPase$ activity [4]. This resting membrane potential is described by the Nernst equation (Nakazato et al. 1988): $\Delta \psi$ (mV) = -59 log (K^+_{in}/K^+_{out}). During action potential sodium is transported into the cell and calcium is transported out of the cell by the Na^+/Ca^{++} exchanger. Furthermore, calcium is released from the endoplasmic reticulum and interacts with the contractile elements of the myocardial tissue. Calcium channels permit influx of calcium into the myocardium. Various steps of this *ion cascade* have been objects of nuclear medicine developments: 1) Potassium analogs thallium-201 and rubidium-82 enter cells by the $Na^+/K^+/ATPase$ mechanism; 2) Lipophilic cations enter cells by direct permeation of the cellular membrane driven by the resting membrane potential; 3) Calcium channel blockers [5] have been labeled with carbon-11 to investigate the distribution and density of these channels with PET.

Many normal and abnormal cells express the $Na^+/K^+/ATPase$ in their membranes. Tumor cells often show overexpression of this protein and consequently increased uptake of specific tracers, such as Tl-201. A two-fold increase in uptake of Tl-201 has been shown in lung cancer cells with positive immunohistochemistry staining for $Na^+/K^+/ATPase$ compared to lung cancer with negative staining for the protein [6].

Lipophilic cations used for imaging purposes such as the PET radiotracer [^{11}C]MTP (methyltriphenylphosphonium) or the SPECT radiotracer [^{99m}Tc]sestamibi, move accross the membrane freely because of the driving force of the resting membrane potential. These lipophilic cations accumulate inside the cell [7] and may advance further into the mitochondria due to another potential difference between the interior of these organelles and the cytoplasm [8].

2. Multidrug Resistance

The lipophilic cation [^{11}C]MTP can be used to image tumors with increased resting membrane potential. From some tumors, however, the tracer may be actively removed by the multidrug resistance transporter (MDRT). MTP has been used as a probe to measure the function of the MDRT in bacteria [9] and mammalian tissues *in vitro* [10]. Transfected hamster cells overexpressing the multidrug resistance genes mdr1 and mdr3 demonstrate reduced intracellular accumulation of MTP, an effect which is reversed by verapamil, a competitive inhibitor of the MDRT. Also, some bacterial MDRTs such as the staphylococcal multidrug resistance (smr) protein are capable of removing MTP from the cytoplasm based on an electrogenic drug/proton antiport mechanism similar to the VMAT. Substrates of the MDRT which have been labeled for PET imaging include [^{11}C]verapamil and [^{11}C]daunorubicine [11].

Multidrug resistance of tumors involves at least three possible mechanisms: drug efflux by the MDRT, alteration of the nuclear enzyme topoisomerase II (also called atypical MDR) and activation of the glutathione detoxification pathway [12].

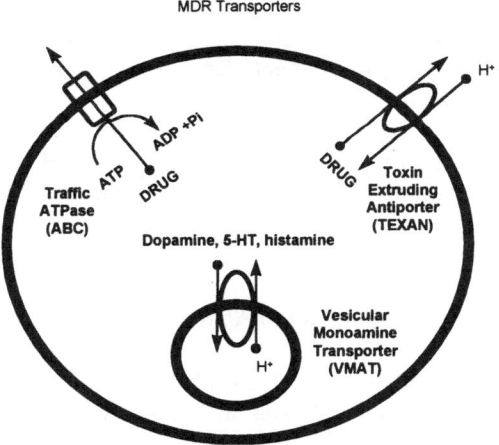

Figure 1: Multidrug resistance transporters of humans (ABC), microorganisms (TEXAN) and the related vesicular monoamine transporter (VMAT).

The MDRT (particularly the mdr1 gene) is expressed in tumors derived from tissues which, under physiological conditions express this P-glycoprotein (P-gp), such as renal, colon, liver and adrenocortical carcinomas but also in some other tumors dervied from tissues which under normal conditions do not express P-gp. In this second group are cancer of bladder, germ cell tumors, sarcomas, hematologic malignancies, breast cancer, esophageal and ovarian cancer, small cell lung cancer, sarcomas and neuroblastoma. In general, P-gp expression is an indicator of poor prognosis [12]. At present three classes of drugs are investigated for reversing drug resistance: calcium channel blockers, cyclosporine derivatives, and the antiestrogen tamoxifen. Probable mechanism of these chemosensitizers may include a competitive inhibition of the P-gp, a feature which could be utilized for positive imaging of these tumors.

A large variatey of molecules is transported by the MDRT, a phenomenon referred to as substrate promiscuity [13]. The probe of MDRT presently applied in clinical settings is the lipophilic cation [^{99m}Tc]sestamibi, a SPECT radiotracer initially developed to display regional myocardial blood flow and later used to image tumors with increased membrane potassium gradients. Accumulation of sestamibi is decrased in malignant breast cancer cell lines expressing the mdr gene [14], an effect reversed by the MDRT antagonist verapamil. Drugs with potential for imaging non-P-gp expressing multidrug resistance tumors, such as prostate cancer, are under development [15].

Although thallium-201 is a substrate of the Na$^+$/K$^+$/ATPase, and Tc-99m Sestamibi is a lipophilic cation, they both show excellent accumulation in tumor cells and are particularly useful for imaging cancers of thyroid, breast and lung [16-18]. Both radiotracers can be used to differentiate CNS toxoplasmosis from CNS lymphoma in patients with AIDS.

The human MDRT is an ATP-dependent exporter, while in other species MDRTs may depend on the pH (H$^+$-gradient). The human MDRT is also called traffic ATPase or ABC (ATP-binding-casette) transporter [19]. The human MDRT is a protein consisting of 1280 amino acids with one drug binding site, and two intracellular ATP binding sites. It has twelve transmembrane domains. It is related to the vesicular monoamine transporter (VMAT) of monoamine neurons. The VMAT enables sequestration of norepinephrine, serotonin and dopamine into the synaptic vesicles and requires a pH gradient. One proton is removed for each neurotransmitter molecule entering the vesicle.

Resistance of *E. coli* against tetracyclines and resistance of *Staph. aureus* against quinolones and antiseptics is based on the effects of MDRT. Toxine extruding antiporters (TEXANs) are MDRTs found in nonhuman species (Figure 1).

3. Glucose and Amino Acid Transport

Glucose and its analog [^{18}F]FDG (fluoro-deoxyglucose used for imaging glucose metabolic rates in brain, myocardium and tumors) are taken up by cells by means of a carrier mediated facilitated diffusion. This process involves a specific glucose transporter (GLUT) protein also called glucose permease. Maximum transport rate (V_{max}) is limited by the maximum number of glucose transporters available in the cell membrane. Another characteristic parameter of facilitated diffusion is K_m, the concentration of glucose at ½ of V_{max}. According to *the Michaelis-Menten equation*, the transport rate equals $V = V_{max}/(1 + K_m/C)$, where C is the concentration of glucose outside of the cell. The graph of V as a function of C rises sharply at first and then more slowly; it finally asymptotically approaches V_{max} at high levels of C. This type of a response permits enhanced uptake of substrate when its extracellular concentration is low while preventing overloading the cell at high substrate concentrations.

The kinetics of [^{18}F]FDG is described by compartmental rate constants k_1 (brain uptake), k_2 (brain release), k_3 (phosphorylation) and k_4 (dephosphorylation, considered to be zero in brain tissues). The parameter k_1 depends on regional cerebral blood flow, uptake of glucose through the blood brain barrier and the regional expression of the GLUT. The parameter k_3 on the other hand reflects the regional expression and activity of the enzyme hexokinase. While GLUTs are overexpressed in some CNS tumors, these transporters may be underexpressed in degenerative diseases of the CNS, such as Alzheimer's disease. This downregulation, in addition to the decrease of the regional blood flow, results in very low FDG deposition in the parietal and temporal lobes [20], a characteristic PET image of Alzheimer's disease.

Seven genes encoding glucose transporters of mammalian tissues have been identified: GLUT1-GLUT7 [21]. GLUT1 and GLUT3 are expressed in the brain and are overexpressed in various malignant tissues. GLUT1 is the transporter of the blood-brain-barrier while GLUT3 is the transporter found in the membrane of neurons. They are not regulated by insulin. On the other hand, GLUT4, abundant in adipocytes, heart and skeletal muscles, is regulated by insulin. The GLUT6 gene designates a transcript homologous to GLUT3; but as a result of multiple stop codons, the GLUT6 gene does not encode a functional glucose transporter [21]. The GLUT represent transporters of facilitated diffusion rather than active transport and are not related to the active glucose transporter of the intestinal mucosa. Glucose transporters, particularly GLUT1 are regulated by many extrinsic and intrinsic factors, hypoxia being a potent one [21]. Expression of the glucose transporters is also affected by growth factors and oncogenes. Hypoxia, growth factor overexpression and activity of oncogenes may be involved in tumor expression of the GLUTs.

Glucose, an important source of energy in tumor tissue, and its analog [^{18}F]FDG are taken up by cells by means of their glucose transporters. Nelson [22] demonstrated increased expression of the glucose transporter in human lung, human breast and human colon cancer xenografts. Such increased expression of glucose transporter constitutes the ground for increased uptake of [^{18}F]FDG in malignant tumors. Although [^{18}F]FDG is a nonspecific radioligand for glucose uptake, a significant linear correlation exists between expression of GLUT1 and [^{3}H]FDG uptake in rat mammary tumor cell lines [23]. The disadvantage of using transporter substrates for imaging purposes is due to their properties to leak out of the cell (Tl-201) or incorporate into metabolic products ([^{18}F]FDG). Ligands could be developed for more specific binding to these transporters. Such ligands of the glucose transporter are 3-radioiodofloretin, [^{125}I]-iodine-hpp-forscoline and [^{3}H]cytochalasin B. Synthesis of a radioligand for positron emission tomography (PET) or single photon emission tomography (SPECT) has not yet been successful [22].

Uptake of amino acids into neuronal and nonneuronal cellular components of the brain is facilitated by amino acid transporters. Most important in this transporter family are the cationic amino acid transporter (CAT) family, the glutamate transporter family for uptake of anionic amino acids and neutral amino acids, the proline transporter, the glycine transporter (GLYT) family and the rBAT family of transporters [24]. Cystinuria is linked to mutations of the rBAT. Glial cells also possess amino acid transporters, which can be overexpressed in gliomas and other brain tumors derived from the glial cells. Radiolabeled amino acids may become important for imaging postradiation reccurrence of brain tumors since they, in contrast to [^{18}F]FDG do not demonstrate nonspecific uptake into inflammatory cells [25]. Tumor to nontumor ratio also appears to be higher with radiolabeled amino acids than with [^{18}F]FDG.

4. Neurotransmitter Transporters, Neurodegeneration and Neurotoxicity

Neurotransmitter transporters play multiple roles in the CNS. By removing the transmitter from the synaptic cleft they terminate the signaling process and both acutely and chronically regulate the synaptic concentration of the neurotransmitter; by "recycling" they also help salvage and reuse the neurotransmitter [26].

The Na^+/K^+ - dependent glutamate transporter is an antiporter. With each molecule of glutamic acid two sodium ions are cotransported into the cell while one potassium ion and one OH$^-$ are antiported out. This process is reversed in ischemia, anoxia, hypoglycemia and other pathological situations, when glutamate accumulates in the synaptic cleft leading to overexcitation of neurons and ultimately to accumulation of Ca^{2+} and neuronal cell death [27].

Environmental neurotoxins that affect glutamate excitatoxicity include trimethylin, methylmercury, aluminum, manganese, toluene, lead, diazinon, ethanol and cyanide [28]. Abnormal excitatory amino acid metabolism has also been implicated in many neurodegenerative diseases including amyotrophic lateral sclerosis, Huntington's disease, Alzheimer's disease, Parkinson's disease, temporal lobe epilepsy, AIDS-related dementia and stroke [28].

Four molecular mechanisms responsible for neuronal cell death include oxidative stress, excitotoxic amino acid accumulation, disturbance of mitochondrial energy metabolism, and disturbance of intracellular calcium homeostasis [29]. Again, as often seen in biological systems, these four mechanisms can occur simultaneously and trigger each other.

The vesicular transport system involves two transporters: an H^+/ATPase which moves H^+ into the organelle and a monoamine transporter which exchanges the protons for less acidic amines: two H^+ for one amine molecule. This "tandem" transport system can be found not only in synaptic vesicles but also in adrenal chromaffin granules, platelet dense granules, mast cell and basophil secretory granules [30]. Monoamine transporters symport (cotransport) Na^+Cl^- with each transmitter molecule [30]. Their function depends on the sodium gradient and thus on the presence of another transporter, the Na^+/K^+/ATPase.

Environmental toxins are postulated in the development of neurodegenerative diseases, although genetic mechanisms may play an equally important role. An experimental model of Parkinson's disease is achieved with MPTP (*N*-methyl-1,2,3,6-

tetrahydro-pyridine). This toxin enters the brain by nonspecific acid transport, it is converted by means of the enzyme MAO B (monoaminooxydase B) to MPP⁺ (*N*-methyl-4-phenylpyridinium; the toxicologically active metabolite of MPTP). MAO B does not occur in dopaminergic neurons, consequently, MPP⁺ produced by glia is transported into dopamine neurons by the dopamine transporter (DAT). Inside neurons, MPP⁺ inhibits the electron transport of mitochondria, specifically complex I [27]. The toxic effects of MPP⁺ are enhanced by aging. Aging reduces the function of the VMAT, which otherwise sequesters MPP⁺ in the secretory vesicles [31]. In addition to Parkinson's disease, the DAT has also been implicated in schizophrenia and Gilles de la Tourette syndrome [32].

The 5-HT specific neurotoxine MDMA (3,4,-methylene-dioxy-methamphetamine) is a drug of abuse ("Ecstasy"). It is taken up by serotonin neurons by the means of the 5-HTT. This uptake can be prevented by specific serotonin reuptake inhibitors (SSRIs) and tricyclic antidepressants. MDMA and other amphetamines are weak bases. They enter monoamine neurons and disrupt the pH gradients between the cytoplasm and secretory vesicles. This leads to release of neurotransmitters into the cytoplasm which ultimatively leads to release into the synapse and overstimulation of the specific receptors, the mechanism of immediate effects of amphetamines. In contrast to the NMDA receptors, excitotoxicity is not responsible for cellular damage. Instead, the neurotransmitters 5-HT and DA accumulate in the cytoplasm and undergo oxidation to free radicals that injure cell organelles [27]. MDMA and the weight reducing drug fenfluramine not only dissipate ΔpH but also compete with monoamine binding at the vesicular membranes. In contrast, another serotonin releaser, PCA, only affects ΔpH but not substrate binding [33].

Endogenous neurotransmitters may play a role not only in neurotoxicology but also in neurodegeneration of Parkinson's disease. In such patients aging of the VMAT (presumably combined with environmental factors analogous to amphetamines) may contribute to cytoplasmic accumulation of DA and cytotoxic free radicals [27].

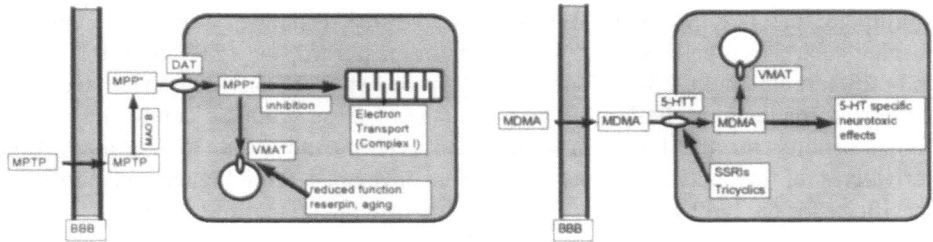

Figure 2: Neurotransmitter transporters and neurotoxicity. Legend: MPTP = *N*-methyl-1,2,3,6-tetrahydro-pyridine; MPP⁺ = *N*-methyl-4-phenylpyridinium; DAT = dopamine transporter, VMAT = vesicular monoamine transporter; MDMA = ; 5-HTT = serotonin transporter; SSRIs = selective serotonin reuptake inhibitors; BBB = blood brain barrier.

The monoamines (DA, NE, 5-HT) and the diamine histamine are formed and released by a relatively small number of neurons originating in the brainstem and projecting into widespread areas of the brain. These neurotransmitters are probably involved in physiological processes which require an overall modulation of brain function such as sleep, vigilance, hunger, brain tone and mood. They also play a role in the pathogenesis of global brain dysfunction observed in mood disorders, spinal shock and drug addiction [34]. Transporters have been identified for the monoamines but not for histamine. Radioligands for PET and SPECT have been developed for the dopamine transporter (DAT) and serotonin transporter (5-HTT), however, development of a norepinephrine (NET) specific in vivo radioligand has as far been unsuccessful.

A series of ligands have been radiolabeled for in vitro investigation of the DAT. Some of them, such as [³H]cocaine, [³H]mazindol, [³H]nomifensine, and [³H]methylphenidate are less selective for the DAT. More selective inhibitors include aryl-1,4-dialk(en)ylpiperazines such as GBR-12909, GBR-12935 and GBR-12783 [35]. Some of these have been successfully labeled with [³H].

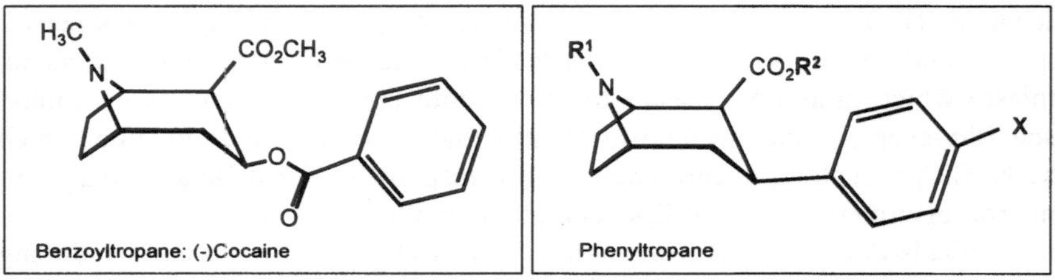

Compound	R¹	R²	X	DAT	Affinity [K_i, nM] 5-HTT	NET
β-CIT (RTI-55)	CH₃	CH₃	I	1.40	0.46	2.80
CFT (WIN 35,428)	CH₃	CH₃	F	14.7	181	635

Figure 3: Synthetic Phenyltropane Analogs of Cocaine.

The active isomer of benzoyltropane is (-)cocaine, a natural product with affinity for the DA, NE and 5-HT transporters [36]. Halophenyl-tropane analogs of cocaine such as 2β-carbomethoxy-3β-(4'-fluorophenyl)tropane (CFT or WIN 35,428) and 2β-carbomethoxy-3β-(4'-iodophenyl)tropane (b-CIT or RTI-55) have been developed for imaging the DAT. The advantage of these synthetic compounds is their higher affinity for the DAT. In addition to RTI-55 and WIN 35,428 many other derivatives of halophenyl-tropane have been developed with higher sensitivity and specificity for the DAT and much slower metabolism than (-)cocaine [37].

RTI-55, a potent cocaine analogue, has been used for in vitro imaging of the cocaine recognition sites on monoaminergic neurons [38]. The pharmacological profile of this radioligand in the striatum, globus pallidus and amygdala is consistent with binding to the DAT, while the pharmacological profile in cortical structures is consistent with binding to 5-HTT. The rank order of potency of monoamine transport blockers in human putamen membranes is mazindol > GBR12909 > GBR12935 > paroxetine >nisoxetine>desipramine>fluoxetine>citalopram. [¹¹C]WIN 35,428 has been used for PET [39] while the radioligand [¹²³I]RTI-55 has been used for SPECT imaging [40], the former being more selective for the DAT compared to other monoamine transporters (e.g. NET and 5-HTT) [37].

Assessment of 5-HT neuron damage in neurodegenerative diseases until now has been achieved by indirect tests only. Tests of serotonin function have involved lumbar puncture for measurement of CSF 5HIAA concentration, a metabolite of 5-HT [41], and endocrine challenge tests using intravenously administered L-tryptophan [42]. These tests provide only indirect information about the integrity of the serotonin neurons. Measurements of CSF 5HIAA can detect alterations in 5HT metabolism, but the nature of the alteration can not necessarily be correlated directly with neuronal degeneration. In neurodegenerative diseases distinct neuropathological lesions are related to some major clinical features. The most characteristic feature of Parkinson's disease is motor disorder (akinesia and rigidity) due to the loss of dopaminergic neurons. This is well documented both by in vitro binding studies using [^{3}H]GBR-12935 [43] and by in vivo PET studies using [^{11}C]WIN35,428 [39]. Serotonergic innervation is also damaged in Parkinson's disease, and loss of serotonergic neurons in Parkinson's disease appears to correlate with the presence of depression [44].

Serotonin insufficiency has been linked to depression, suicidality, alcoholism, aggression and obsessive compulsive disorder. Little is known about the regulation of the 5-HTT and its role in these disease although its possible involvement is indicated by the successful treatment by SSRIs [45].

Imaging the 5-HTT (5-HT transporter) is of special interest since it has been shown that the density of these proteins can be used to measure serotonergic neuronal density under various conditions of hypo-,norm-,or hyper-5-HT innervation [46]. Antidepressants which specifically bind to the 5-HT transporter have been of particular interest for radiolabeling and have been used for autoradiography. Several 5-HT transporter ligands have also been studied *in vivo* with the goal of labeling such ligands with positron or single photon emitting radionulides. Unfortunately, these radioligands are not suited for PET imaging of the 5-HT transporter because of the high nonspecific binding and low target-to-non-target ratios *in vivo*. Preliminary studies in nonhuman primates [47] demonstrate good target to background ratios for [^{123}I]iodonitroquipazine; however, the radioligand has not yet been tested in humans.

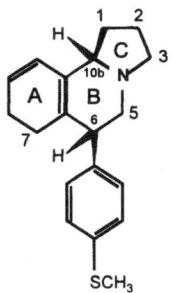

Figure 4: Structure and stereoselectivity of (1,2,3,5,6,10b-hexahydro-6-[4-(methylthio)phenyl]pyrrolo [2,1-a]isoquinoline [48]. Ring B permits boat/chair conformation. N is a labile nitrogen at B,C ring juncture responsible for cis/trans diastereisomerism. 6S,10bR is the + enantiomer, 6R,10bS is the - enantiomer.

Shank et al. [49] reported that the optical enantiomers of McN5652-Z (1,2,3,5,6,10b-hexahydro-6-[4-(methylthio)phenyl]pyrrolo [2,1-a]isoquinoline) display greatly different potencies towards the 5-HT transporter (Ki = 0.4 and 58.4 nM for the (+) and (-) isomers, respectively). In experimental animals the (+) isomer evoked behavioral effects characteristic for 5-HT but the (-) isomer did not [48]. Accordingly, [^{11}C](+) and (-)McN5652 were prepared in our laboratory and their *in vivo* kinetics were studied in mice.

Using [^{11}C](+)McN5652 in rodents the hypothalamus to cerebellum ratio, an index of specific binding of 5-HT reuptake sites, is as high as 6.0, and that ligand binding is blocked with paroxetine, a selective 5-HT reuptake inhibitor, but not by desipramine, an inhibitor of the norepinephrine transporter or by GBR-12909, an inhibitor of the dopamine transporter [52]. These results confirm the use of [^{11}C](+)McN5652 as a radioligand for investigations of central 5-HT reuptake sites.

The effective absorbed radiation dose equivalent received to the whole body from a PET study with [^{11}C](+)McN-5652 is estimated to be 1.1 rad, which is 22 % of the acceptable annual radiation exposure to radiation workers.

Organ	Weighting factor (W$_i$)	Dose (mrem)	Dose x W$_i$ (mrem)
Gonads	0.25	1,064	266
Female Breast	0.15	238	36
Red Bone Marrow	0.12	326	39
Lungs	0.12	462	55
Thyroid	0.03	238	7
Bone Surfaces	0.03	152	5
Critical Organ (intestines)*	0.06	6,380	383
Bladder	0.06	2,160	130
Kidney	0.06	1,606	96
Liver	0.06	806	48
Stomach	0.06	1,002	60
Effective Dose **	**1.00**		**1,125**

Table 1: Radiation dose of individual organs and effective radiation dose equivalent from injection of 20 mCi [^{11}C](+)McN5652. * The organ with the highest radiation dose, other than listed organs. ** Sum products of doses and weighting factors to obtain Effective Dose.

Dynamic PET studies have been performed in baboons (*Papio anubis*) anesthetized with the steroid anesthetic Saffan. Each animal was injected intravenously first with [^{11}C](+)McN5652, then with the pharmacologically inactive enantiomer [^{11}C](-)McN5652; two animals received a third study with [^{11}C](+)McN5652 after pretreatment with the specific 5-HT reuptake site inhibitor fluoxetine (5 mg/kg). Initial distribution in the brain was similar for both [^{11}C](+)McN5652 and [^{11}C](-)McN5652, although the initial uptake of the (-) enantiomer was slightly higher. At later times (55-115 min after injection) only [^{11}C](+)McN5652 showed a distribution characteristic for

5-HT reuptake sites. In contrast, in studies with [^{11}C](-)McN5652 and with [^{11}C](+)McN5652 after 5-HT reuptake site blockade with fluoxetine, radioactivity concentrations were significantly lower and the distribution pattern was relatively even.

Baboons were also investigated before and after selective lesioning of the 5-HT neurons with 3,4-methylenedioxymethamphetamine (MDMA) [53]. There was a significant reduction of specific binding expressed by the difference between the [^{11}C](+)McN5652 and [^{11}C](-)McN5652 enantiomers (100 % reduction in occipital cortex, caudate and cingulate gyrus; 80-100 % reduction in frontal cortex and parietal cortex; and 60-80 % reduction in hypothalamus and putamen).

The 5-HTT has been imaged in the human brain using [^{11}C]McN5652. Figure 5 shows the distribution of [^{11}C](+)McN5652 in the brain of a healthy individual. There is high concentration of the tracer within the hypothalamus, midbrain and pons (the hypothalamus is abundant in serotonin terminals; the midbrain and pons contain the 5-HT neuron bodies of raphe). Less activity is observed within the frontal, parietal and occipital cortex. The characteristic structures visualized with [^{11}C](+)McN5652, such as hypothalamus and midbrain, do not show enhanced binding of the pharmacologically inactive enantiomer [^{11}C](-)McN5652.

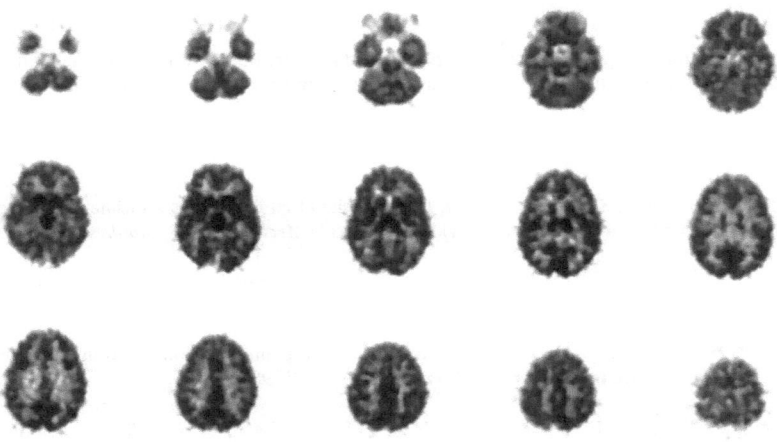

Figure 5: Distribution of the 5-HTT in the brain of a healthy subject imaged with PET and [^{11}C](+)McN5652.

In the human brain the rank order of specific binding of [^{11}C](+)McN5652 is as follows: midbrain, hypothalamus, putamen, caudate nucleus, cingulate gyrus, occipital cortex, thalamus, parietal cortex, frontal cortex, temporal cortex, cerebellum, and white matter. This rank order shows a significant correlation with known densities of 5-HT reuptake sites [54,55] as measured with [^{3}H]paroxetine in postmortem brain specimen (Spearman's correlation coefficient r = 0.91, p<0.001; linear correlation coefficient r =

0.83, p<0.001). Thus, these PET studies in healthy human volunteers, like those performed in nonhuman primates, proved the suitability of $[^{11}C](+)McN5652$ for imaging the 5-HT reuptake sites. Preliminary results of PET studies indicate that the binding of $[^{11}C](+)McN5652$ is decreased in subjects using methcathinone, a drug of abuse, and in patients with Parkinson's disease [56,57].

5. References

1. Szabo, Z. Transport systems. In: Wagner,H.N.J., Szabo,Z., Buchanan,W.J. eds., *Principles of nuclear medicine,* Philadelphia, Saunders, **1995**, p. 40

2. Wagner, H.N.J. Nuclear medicine: what it is what it does. In: Wagner,H.N.J., Szabo,Z., Buchanan,W.J. eds., *Principles of nuclear medicine*, Philadelphia, Saunders, **1995**, p. 1

3. Kakuda, D.K., MacLeod, C.L. Na⁺-independent transport (uniport) of amino acids and glucose in mammalian cells. *J. Exp. Biol. (Transporters).* **196**, 93-108, **1994**

4. Reeves, J.P., Condrescu, M., Chernaya, G., Gardner, J.P. Na^+/Ca^{2+} antiport in the mammalian heart. *J. Exp. Biol. (Transporters).* **196**, 375-388, **1994**

5. Wilson, A.A., Dannals, R.F., Ravert, H.T., Burns, H.D., Lever, S.Z., Wagner, H.N.J. Radiosynthesis of $[^{11}C]$Nifedipine and $[^{11}C]$Nicardipine. *J. Lab. Comp. Radiopharm.* **27**, 589-598, **1988**

6. Takekawa, H., Itoh, K., Abe, S., Ogura, S., Isobe, H., Furudate, M., Kawakami, Y. Thallium-201 uptake, histopathological differentiation and Na-K ATPase in lung adenocarcinoma. *J. Nucl. Med.* **37**, 955-958, **1996**

7. Seligmann, B.E., Gallin, J.I. Comparison of indirect probes of membrane potential utilized in studies of human neutrophils. *J. Cell Physiol.* **115**, 105-115, **1983**

8. Masaki, N., Kyle, M.E., Serroni, A., Farber, J.L. Mitochondrial damage as a mechanism of cell injury in the killing of cultured hepatocytes by tert-butyl hydroperoxide. *Arch. Biochem. Biophys.* **270**, 672-680, **1989**

9. Grinius, L.L., Goldberg, E.B. Bacterial multidrug resistance is due to a single membrane protein which functions as a drug pump. *J. Biol. Chem.* **269**, 29998-30004, **1994**

10. Gros, P., Talbot, F., Tang-Wai, D., Bibi, E., Kaback, H.R. Lipophilic cations: a group of model substrates for the multidrug-resistance transporter. *Biochemistry.* **31**, 1992-1998, **1992**

11. Elsinga, P.H., Franssen, E.J.F., Hendriksen, N.H., Fluks, L., Weemaes, A.M.A., van der Graaf, W.T.A., de Vries, E.G.E., Visser, G.M., Vaalburg, W. Carbon-11-labeled daunorubicin and verapamil for probing P-glycoprotein in tumors with PET. *J. Nucl. Med.* **37**, 1571-1575, **1996**

12. Theyer, G., Hamilton, G. Role of multidrug resistance in tumors of the genitourinary tract. *Urology.* **44**, 942-950, **1994**

13. Schuldiner, S., Shirvan, A., Stern-Bach, Y., Steiner-Mordoch, S., Yelin, R., Laskar, O. From bacterial antibiotic resistance to neurotransmitter uptake. *J. Exp. Biol. (Transporters).* **196**, 174-184, **1994**

14. Cordobes, M.D., Starzec, A., Delmon-Moingeon, L., Blanchot, C., Kouyoumdjian, J.C., Prevost, G., Caglar, M., Moretti, J.C. Technetium-99m-sestamibi uptake by human benign and malignant brest tumor cells: correlation with mdr gene expression. *J. Nucl. Med.* **37**, 286-289, **1996**

15. Tasaki, Y., Nakagawa, M., Ogata, J., Kiue, A., Tanimura, H., Kuwano, M., Nomura, Y. Reversal by a dihydropyridine derivative of non-P-glycoprotein-mediated multidrug resistance in etoposide-resistant human prostatic cancer cell line. *J. Urol.* **154**, 1210-1216, **1995**

16. Nakahara, H., Noguchi, S., Murakami, N., Hoshi, H., Jinnouchi, S., Nagamachi, S., Ohnishi, T., Futami, S., Flores, L.G., Watanabe, K. Technetium-99m-sestamibi scintigraphy compared with thallium-201 in evaluation of thyroid tumors. *J. Nucl. Med.* **37**, 901-904, **1996**

17. Maublant, J., Latour, M., Mestas, D., Clemenson, A., Charrier, S., Feillel, V., Le Bouedec, G., Kaufmann, P., Dauplat, J., Veyre, A. Technetium-99m-sestamibi uptake in breast tumor and associated lymph nodes. *J. Nucl. Med.* **37**, 922-925, **1996**

18. Chiti, A., Maffioli, L.S., Infante, M., Grasselli, G., Incarbone, M., Gasparini, M.D., Savelli, G., Bombardieri, E. Assessment of mediastinal involvement in lung cancer with technetium-99m-sestamibi SPECT. *J. Nucl. Med.* **37**, 938-942, **1996**

19. Higgins, C.F., Gottesman, M.M. Is the multidrug transporter a flippase? *T. I. B. S.* **17**,**1992**

20. Piert, M., Koeppe, R.A., Giordani, B., Berent, S., Kuhl, D.E. Diminished glucose transport and phosphorylation in Alzheimer's disease determined by dynamic FDG-PET. *J. Nucl. Med.* **37**, 201-208, **1996**

21. Baldwin, S.A. Mammalian passive glucose transporter: members of an ubiquitous family of active and passive transport proteins. *Biochim. Biophys. Acta.* **1154**, 17-49, **1993**

22. Nelson, C.A., Wang, Q.J., Bourque, J.P., Crane, P.D. Targeting of glucose transport proteins for tumor imaging: is it feasible. *J. Nucl. Med.* **37**, 1031-1037, **1996**

23. Hausler, M., Eilles, C., reiners, C., Moll, E., Borner, W. Zur Auswertung von funktionsszintigraphischen Magenentleerungsuntersuchungen mit der Hauptkomponentenmethode. *Nuklearmedizin.* **19**, 213-220, **1980**

24. Malandro, M.S., Kilberg, M.S. Molecular biology of mammalian amino acid transporters. *Annu. Rev. Biochem.* **65**, 305-336, **1996**

25. Kubota, K., Ishiwata, K., Kubota, R., Yamada, S., Takahashi, J., Abe, Y., Fukuda, H., Ido, T. Feasibility of fluorine-18-fluorophenylalanine for tumor imaging compared with carbon-11-L-methionine. *J. Nucl. Med.* **37**, 320-325, **1996**

26. Blakely, R.D., de Felice, L.J., Hartzell, H.C. Molecular physiology of norepinephrine and serotonin transporters. *J. Exp. Biol. (Transporters).* **196**, 263-281, **1994**

27. Edwards, R.H. Neural degeneration and the transport of neurotransmitters. *Ann. Neurol.* **34**, 638-645, **1993**

28. Dawson, R., Flint Beal, M., Bondy, S.C., Di Monte, D.A., Isom, G.E. Excitotoxins, aging, and environmental neurotoxins: implications for understanding human neurodegenerative diseases. *Toxicol. Appl. Pharmacol.* **134**, 1-17, **1995**

29. Gerlach, M., Riederer, P., Youdim, M.B. Neuroprotective therapeutic strategies. Comparison of experimental and clinical results. *Biochem. Pharmacol.* **50**, 1-16, **1995**

30. Rudnick, G., Clark, J. From synapse to vesicle: the reuptake and storage of biogenic amine neurotransmitters. [Review]. *Biochim. Biophys. Acta.* **1144**, 249-263, **1993**

31. Lesch, K.P., Heils, A., Riederer, P. The role of neurotransporters in excitotoxicity, neuronal cell death, and other neurodegenerative processes. *J. Mol. Med.* **74**, 365-378, **1996**

32. Giros, B., Caron, M.G. Molecular characterization of the dopamine transporter. *Trends. Pharmacol. Sci.* **14**, 43-49, **1993**

33. Schuldiner, S., Steinermordoch, S., Yelin, R., Wall, S.C., Rudnick, G. Amphetamine derivatvies interacti with both plasma membrane and secretory vesicle biogenic amine transporters. *Molecular. Pharmacology.* **44**, 1227-1231, **1993**

34. Bach-y-Rita, P. Nonsynaptic diffusion neurotransmission (NDN) in the brain. *Neurochem. Int.* **23**, 297-318, **1993**

35. Graham, D., Langer, S.Z. Advances in sodium-ion coupled biogenic amine transporters. *Life Sci.* **51**, 631-645, **1992**

36. Ritz, M.C., Cone, E.J., Kuhar, M.J. Cocaine inhibition of ligand binding at dopamine, norepinephrine and serotonin transporters: a structure-activity study. *Life Sci.* **46**, 635-645, **1990**

37. Neumeyer, J.L., Tamagnan, G., Wang, S.Y., Gao, Y.G., Milius, R.A., Kula, N.S., Baldessarini, R.J. N-substituted analogs of 2-beta-carbomethoxy-3-beta-(4'-iodophenyl)tropane (beta-CIT) with selective affinity to dopamine or serotonin transporters in rat forebrain. *Journal. of. Medicinal. Chemistry.* **39**, 543-548, **1996**

38. Staley, J.K., Basile, M., Flynn, D.D., Mash, D.C. Visualizing dopamine and serotonin transporters in the human brain with the potent cocaine analogue [^{125}I]RTI-55: in vitro binding and autoradiographic characterization. *J. Neurochem.* **62**, 549-556, **1994**

39. Frost, J.J., Rosier, A.J., Reich, S.G., Smith, J.S., Ehlers, M.D., Snyder, S.H., Ravert, H.T., Dannals, R.F. Positron emission tomographic imaging of the dopamine transporter with 11C-WIN 35,428 reveals marked declines in mild Parkinson's disease. *Ann. Neurol.* **34**, 423-431, **1993**

40. Shaya, E.K., Scheffel, U., Dannals, R.F., Ricaurte, G.A., Carroll, F.I., Wagner, H.N.J., Kuhar, M.J., Wong, D.F. In vivo imaging of dopamine reuptake sites in the primate brain using single photon emission computed tomography (SPECT) and iodine-123 labeled RTI-55. *Synapse.* **10**, 169-172, **1992**

41. Moir, A.T., Ashcroft, G.W., Crawford, T.B., Eccleston, D., Guldbert, H.C. Cerebral metabolites in cerebrospinal fluid as a biochemical approach to the brain. *Brain Res.* **93**, 357-368, **1986**

42. Henninger, G., Charney, D.S., Sternberg, D.E. Serotonergic function in depression: Prolactin response to intravenous tryptophan in depressed patients and healthy subjects. *Arch. Gen. Psychiatry.* **41**, 398-402, **1984**

43. Allard, P.O., Rinne, J., Marcusson, J.O. DOPAMINE UPTAKE SITES IN PARKINSONS DISEASE AND IN DEMENTIA OF THE ALZHEIMER TYPE. *Brain Research.* **637**, 262-266, **1994**

44. Paulus, W., Jellinger, K. The neuropathological basis of different clinical subgroups of Parkinson's disease. *J. Neuropathol. Exp. Neurol.* **50**, 743-755, **1991**

45. Fuller, R.W., Wong, D.T. Serotonin uptake and serotonin uptake inhibition. *Ann. NY Acad. Sci.* **600**, 68-81, **1990**

46. Descarries, L., Soucy, J.P., Lafaille, F., Mrini, A., Tanguay, R. Evaluation of three transporter ligands as quantitative markers of serotonin innervation density in rat brain. *Synapse.* **21**, 131-139, **1995**

47. Jagust, W.J., Eberling, J.L., Roberts, J.A., Brennan, K.M., Hanrahan, S.M., VanBrocklin, H., Enas, J.D., Biegon, A., Mathis, C.A. In vivo imaging of the 5-hydroxytryptamine reuptake site in primate brain using single photon emission computed tomography and [^{123}I]5-iodo-6-nitroquipazine. *Eur. J. Pharmacol.* **242**, 189-193, **1993**

48. Smith, D.F., Jensen, P.N., Poulsen, S.H., Mikkelsen, E.O., Elbaz, E., Glaser, R. Effects of pyrroloisoquinoline enantiomers ((+)- and ((-)-McN5652-Z) on behavioral and pharmacological serotonergic mechanisms in rats. *Eur. J. Pharmacol.* **196**, 85-92, **1991**

49. Shank, R.P., Vaught, J.L., Pelley, K.A., Setler, P.E., McComsey, D.F., Maryanoff, B.E. McN-5652: A highly potent inhibitor of serotonin uptake. *J. Pharmacol. Exp. Ther.* **247**, 1032-1038, **1988**

50. Suehiro, M., Ravert, H.T., Dannals, R.F., Scheffel, U., Wagner, H.N.J. Synthesis of a radiotracer for studying serotonin uptake sites with positron emission tomography: [^{11}C]McN-5652-Z. *J. Lab. Comp. Radiopharm.* **31**, 841-848, **1992**

51. Suehiro, M., Scheffel, U., Ravert, H.T., Dannals, R.F., Wagner, H.N.J. [^{11}C](+)McN5652 as a radiotracer for imaging serotonin uptake sites with PET. *Life Sci.* **53**, 883-892, **1993**

52. Suehiro, M., Scheffel, U., Dannals, R:F., Ravert, H.T., Ricaurte, G.A., Wagner, H.N.J. A PET radiotracer for studying serotonin uptake sites: Carbon-11-McN-5652Z. *J. Nucl. Med.* **34**, 120-127, **1993**

53. Scheffel, U., Szabo, Z., Mathews, W.B., Finley, P.A., Dannals, R.F., Ravert, H.T., Szabo, K., Ricaurte, G.A. Detection of MDMA neurotoxicity in vivo. PET studies in the living baboon brain. *Soc. Neurosci. Abstr.* **21**, 8811995(Abstract)

54. Laruelle, M., Maloteaux, J.M. Regional distribution of serotonergic pre- and postsynaptic markers in human brain. *Acta. Psyciatr. Scand.* **80(Supp.350)**, 56-59, **1989**

55. Laruelle, M., Vanisberg, M.A., Maloteaux, J.M. Regional and subcellular localization in human brain of [^{3}H]paroxetine binding, a marker of serotonin uptake sites. *Biol. Psychiatry.* **24**, 299-309, **1988**

56. Szabo, Z., Preziosi, T., Hoehn-Saric, R., Scheffel, U., Palmon, S., Mathews, W.B., Ravert, H.T., Dannals, R.F. PET imaging reveals reduced [C-11](+)McN5652 binding to 5-HT transporters in Parkinson's disease. *Soc. Neurosci. Abstracts.* **22**, 721,**1996**

57. Ricaurte, G., Wong, D.F., Szabo, Z., Yokoi, F., Scheffel, U., Matthews, W., Ravert, H., Dannals, R., Naidu, S. Reductions in brain dopamine and serotonin transporters detected in humans previously exposed to repeated high doses of methcathinone using PET. *Soc. Neurosci. Abstracts.* **22**,1915, **1996**

A Deformable High Resolution Anatomic Reference For Pet Activation Studies

ARTHUR W. TOGA
Laboratory of Neuro Imaging
Department of Neurology
Division of Brain Mapping
UCLA School of Medicine
710 Westwood Plaza
Los Angeles, CA 90024-1769

Abstract

Presently available anatomic atlases provide useful coordinate systems such as the ubiquitous Talairach system but are sorely lacking in both spatial resolution and completeness. An appropriately sampled anatomic specimen can provide the additional detail necessary to accurately localize activation sites as well as provide other structural perspectives such as chemoarchitecture. As part of the International Consortium for Brain Mapping (ICBM), whose goal it is to develop a probabilistic reference system for the human brain, we collected serial section postmortem anatomic data from several whole human head and brain specimens using a cryosectioning technique. Tissue imaged so that voxel resolution was 200 microns or better at full color (24 bits/pixel. The collected data sets were used in one of several ways as an anatomic reference for functional studies. First, on an individual basis as a traditional, n=1 structural atlas with unprecedented spatial resolution and complete coverage of the forebrain, midbrain and hindbrain. The dataset can be registered to a functional dataset using either anatomic landmarks or an automatic approach. Second, several of these high resolution datasets were placed within the Talairach system and used to produce a probabilistic representation. This approach represents anatomy within a coordinate system as a probability. Coordinate locations are assigned a confidence limit to describe the likelihood that a given location belongs to an anatomic structure based upon the population of specimens. These data produce an anatomic reference that is digital, high in spatial and densitometric resolution, 3D, comprehensive and, in combination, probabilistic. The superior resolution makes it possible to delineate structures impossible to visualize in other structural modalities. These data are an important and necessary part of the comprehensive structural and functional analyses that focus on the mapping of the human brain.

B. Gulyás and H.W. Müller-Gärtner (eds.),
Positron Emission Tomography: A Critical Assessment of Recent Trends, 133–142.
© 1998 *Kluwer Academic Publishers.*

1. Introduction

Accurate localization of brain structure and function in any modality is improved by correlation with higher resolution anatomic data placed within an appropriate spatial coordinate system. Recent brain mapping efforts have begun to explore this approach yet remain disadvantaged by the lack of readily available, spatially detailed 3D morphology of the normal human brain (Damasio et al., 1991). In response to this need, several computerized atlases have been developed for neurosurgical applications or for analysis of metabolic studies such as PET and SPECT (Tiede et al., 1993; Evans et al., 1991; Lehmann et al, 1991; Roland & Zilles, 1994). These atlases, based on data acquired using magnetic resonance imaging (MRI), have the advantage of intrinsic three-axis registration and spatial coordinates but have relatively low resolution and lack anatomic contrast in important subregions. High resolution MR atlases, using up to 100-150 slices, a section thickness of 2 mm, and 256^2 pixel imaging planes (Evans, et al., 1991; Lehmann et al., 1991) still result in resolutions lower than the complexity of many neuroanatomic structures.

Several digital atlases have been developed using photographic images of cryoplaned frozen specimens (Bohm et al., 1989; Greitz et al., 1991). The use of photographed material, while providing superior anatomic detail, has limitations. For accurate correlations, data must be placed in the equivalent plane as the image of interest. Digital imaging can overcome some of the limitations of conventional film photography methods. Using 1024^2, 24-bits/pixel digital color cameras, spatial resolution can be as high as 100 microns/pixel for whole human head cadaver preparations or higher for isolated brain regions (Toga et al., 1994). Cryosectioning in micron increments permits collection of data with high spatial resolution in the axis orthogonal to the sectioning plane. Acquisition of images in series directly from the consistently positioned cryoplaned blockface avoids the need for serial image registration prior to reconstruction. Serial images can be reconstructed to a 3D anatomic volume that is amenable to various resampling and positioning schemes.

The combination of cryosectioning and specimen surface photography provides both the means for acquiring anatomic image data and the potential to collect specimen tissue for histological analysis (Pech, 1987; Rauschning, 1986). Application of histochemical stains removes any remaining ambiguity in the identification of boundaries thus increasing the detail of segmentations.

We conducted experiments demonstrating the creation of spatially accurate, high resolution anatomic reference volumes from postmortem cryosectioned whole human brain. They were conducted to examine the advantages of this approach and to determine the value of the data as an anatomic reference for MR, PET and other modalities. First, we examined anatomic image data (1024^2 24-bits/pixel) to see if it contained sufficient resolution to improve the ability to delineate neuroanatomic structures. Second, we tested whether 3D reconstruction, repositioning, scaling and resampling could be performed while preserving accurate spatial relationships. Statistical morphometrics were calculated to determine the degree of precision in anatomic segmentation and placement within the Talairach coordinate system (Talairach & Tournoux, 1988). Meshes describing sulcal ribbons were used to drive deformations for higher order mappings across subjects. Finally, we evaluated tissue collected from the cryosectioned specimens for compatibility with specific histochemical staining and analysis for cytoarchitectonic delineation. The goal was to produce a series of high resolution 3D anatomic volumes with potential to serve as an anatomic reference for comprehensive human brain mapping in a variety of imaging modalities.

2. Methods

We collected high resolution image data from frozen human cadaver preparations that included fixed and unfixed whole head, fixed whole brain and fixed isolated regions of interest. Histologic sections were collected from fixed whole brain and brain regions. Image data were used for 2D anatomic segmentation, digital 3D reconstruction and visual comparison to *in vivo* MRI. Five whole head and brain data sets were reconstructed into the Talairach and Tournoux stereotactic atlas space for comparison to 3D reconstructed *in vivo* MRI from 10 normal male subjects. We calculated morphometric statistics for a small sample of neuroanatomic structures from the cryosectioned anatomy, *in vivo* MRI, and a 3D reconstruction of the Talairach atlas plates.

2.1. SPECIMEN PREPARATION AND CRYOSECTIONING

Cadavers were obtained through the Willed Body Program at the UCLA School of Medicine, optimally within 5-10 hours postmortem. All were adult or aged (54-90 yrs) with approximately equal representation of gender. Specimens were sectioned on a large sledge cryomacrotome (PMV, Stockholm Sweden) equipped with a high resolution color camera (DAGE MTI, Michigan City IN), professional flat-field macro lenses and a voltage regulated fiber optic illumination system. Whole human head specimens were sectioned in 50 micron increments with digital images acquired either every 100 microns or every 500 microns throughout the dorsal-ventral axis.

We collected histology from one fixed, decalcified head, one whole brain, cerebellum and brainstem. Tissue sections were mounted on large glass slides and stained using modified cresyl violet or von Braunmuhl silver stain protocols (19). We photographed the histology through the microscope and digitized sections using the same camera and illumination system used for blockface imaging.

2.2. MRI ACQUISITION.

MRI were obtained from 10 adult normal male subjects using a 1.5 Tesla scanner (Signa, General Electric) with TR=1500ms, TE=20ms and TI=600ms. Data for each subject (140 slices) were acquired in the horizontal plane in 1mm thickness and 256^2 resolution.

2.3. THREE DIMENSIONAL RECONSTRUCTION.

Image data from cryosectioned specimens were reconstructed to 3D anatomic volumes. Picture width (x) and height (y) in the imaging plane were determined by the camera field of view. The depth axis value (z) corresponding to the position of each serial image in the volume was determined by the distance in microns between each image. We placed anatomic data from five of the whole head and brain data sets and reconstructed MRI from all ten subjects into the spatial coordinate system described by Talairach and Tournoux (1988).

2.4. ANATOMIC SEGMENTATION.

Anatomic structures were identified in cryosectioned anatomy using visual cues provided by color, texture and contrast differences to surrounding tissue. Cerebral cortex, cerebellum, brainstem, ventricular system and selected subcortical structures including the anterior commissure, head and tail of caudate, putamen, thalamus, hippocampus and globus pallidus were outlined for segmentation. For *in vivo* MRI, structure boundaries were based on the line of highest intensity gradient visible. These boundaries were manually segmented and retained for surface model reconstruction, visualization and morphometrics.

Surface models were generated using wireframe mesh triangulation methods based in UNIX software. Shading, texture and pseudocolor were applied to enhance model visualization. Segmented structures were displayed individually or nested to emphasize spatial relationships. Texture mapping of cutplanes was used to localize structures that had been demonstrated in histochemically stained data within the greater context of the 3D surface model.

2.5. MORPHOMETRIC ANALYSIS.

We included data from five anatomic volumes of cryosectioned whole human heads, 10 MRI and the 3D reconstructed Talairach atlas model in our calculations. Morphometric statistics were computed for left and right globus pallidus, head and tail of caudate, putamen, thalamus, anterior commissure and ventricles. We compared values for surface area, volume, and center of mass within (interhemispheric) and between individual anatomic data sets, the Talairach atlas and *in vivo* MRI.

2.6. DEFORMATION STRATEGY.

Using these and other data, we have devised, implemented and tested a fast, spatially accurate technique for calculating the high-dimensional deformation field relating the brain anatomies of an arbitrary pair of subjects. The resulting 3D deformation map can be used to quantify anatomic differences between subjects or within the same subject over time, and to transfer functional information between subjects or integrate that information on a single anatomic template. This procedure is based on developmental processes responsible for variations in normal human anatomy, and is applicable to 3D brain images in general, regardless of modality. The construction of extremely complex surface deformation maps on the internal cortex is made easier by building a generic surface structure to model it. Connected systems of parametric meshes were used to model several deep sulci whose trajectories represent critical functional boundaries.

3. Results and Discussion

3.1. POSTMORTEM CRYOSECTIONED ANATOMY.

Digital images of whole head and brain possessed average spatial resolutions of 200 and 170 microns/pixel, respectively, in the imaging plane. This method produced anatomically detailed data sets for the brainstem, pons, cerebellum, cingulate cortex, and optic tract and hippocampus (figure 1). Spatial resolution in the orthogonal axis was determined by frequency of image acquisition. For whole head imaging at a frequency of 500 microns, resolution in the reconstructed axis was 200 microns/pixel.

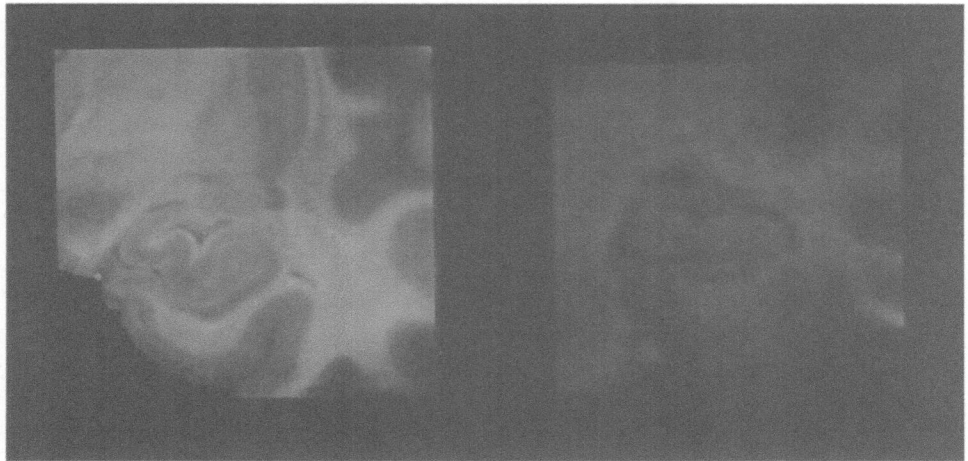

Figure 1. An optically magnified region of the hippocampus was captured with a small field of view (40 mm) to yield a spatial resolution approximately 40 microns/pixel. The optical nature of this technique permits collection of anatomic detail suitable for delineation of complex structures. At this resolution it is possible to clearly observe structures not visible in tomographic images, such as the dentate granular layer. The white matter bundle of the alveus is sharply demarcated against white subcortical matter. Also note the tail of the caudate, claustrum and putamen. Combined with histologically processed sections, these data can be used to study to any region of interest.

138

High resolution cryosectioned anatomy demonstrated gyral and sulcal anatomy as well as laminar structures and nuclear regions. We were easily able to identify subcortical structures in high resolution digital images based on color pigment differentiation and texture contrast to adjacent tissues. The densitometric gradations afforded by 24 bits provided subtle textural detail important in the segmentation of regional anatomy. This was especially apparent in deep subcortical structures such as the corpus striatum. It was possible to distinguish the internal capsule dividing the caudate and lentiform nuclei. In the thalamic region, we could observe the anterior, medial, lateral and ventral nuclei as well as the lateral geniculate bodies and pulvinar. These visual distinctions allowed us to more precisely determine the boundaries of neuroanatomic structures in cryosectioned anatomy than *in vivo* MRI (figure 2).

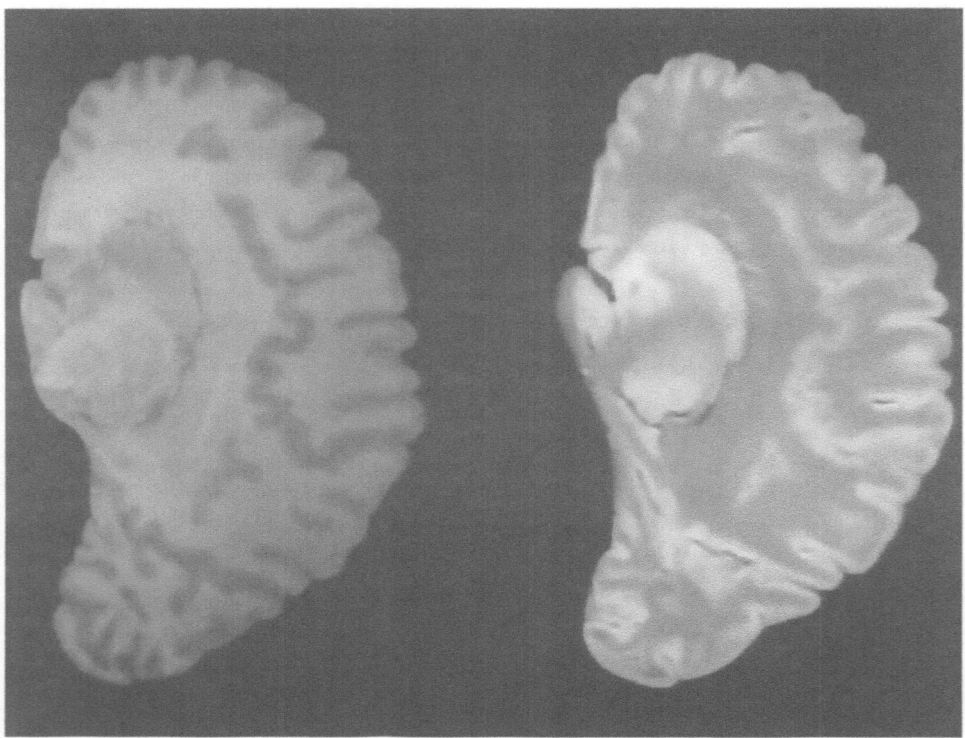

Figure 2. Nuclear and laminar boundaries of structures normally not visible in tomographic modalities can be seen in high resolution color images of cryosectioned anatomy. Whole brain in vivo MRI (right), at a resolution of 256², does not provide the anatomic detail provided by high resolution digital imaging. In addition the visual cues provided by texture and color are not available in 8-bit tomographic images. This horizontal slice through a whole human head (left) at the level of the superior colliculus shows the central tegmental tract, third nerve nucleus, and medial longitudinal fasciculus. The superior occipital region of bone was removed from this unfixed specimen to improve sectioning characteristics; however, the frozen brain has retained its in situ configuration and relevant bony landmarks such as the internal auditory meatus and infraorbital ridge are still intact.

3.2. CORRELATED HISTOLOGY.

To acquire whole-brain sections from fixed tissue it was necessary to section in larger increments producing sections up to 100 microns thick and difficult to maintain proper anatomic configuration. We were unable to obtain sections of the whole human head because the chemical decalcification process proved to be incompatible with section collection. Thinner sections (20-40 microns) were easily collected from the isolated brain region specimens such as brainstem and pons. The use of Nissl and modified silver stains enabled us to examine cell morphology under the light microscope (figure 3). For thicker whole brain sections, cytologic detail was suboptimal but we were able to appreciate the enhanced grey and white matter differentiation. By correlating each slide mounted section with the blockface image acquired during cryosectioning, we were able to record the correct spatial localization of histologic data in the gross anatomic specimen.

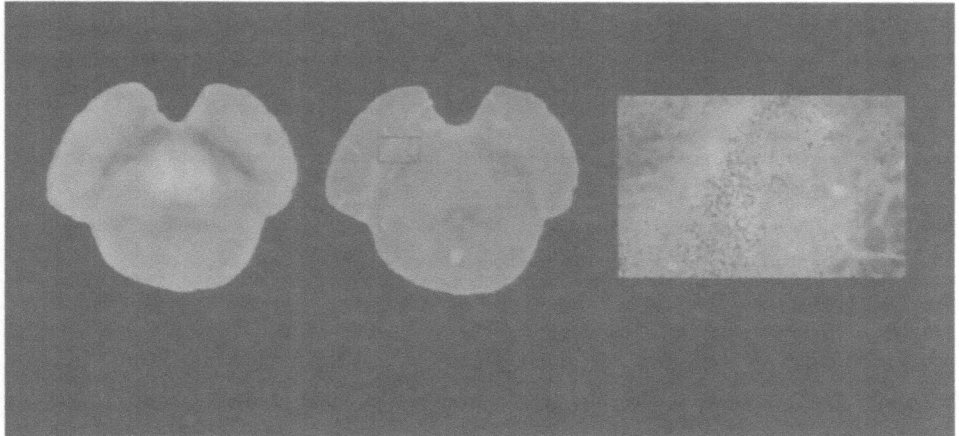

Figure 3. High resolution images from the cryoplaned blockface can be correlated with histology for three dimensional mapping of cytoarchitectonic fields. a) This axial section through the human brainstem was captured in high resolution directly from the blockface. Cryosections were stained with a modified von Braunmuhl silver stain, imaged, and digitally correlated with blockface complements. Structures possessing simple geometry, such as the brainstem, are mapped using landmark based affine transformations and require no local deformations. b) The blockface and corresponding histochemically stained section each contribute 50% of this composite image. The box indicates the light microscope view shown in C. c) Histochemically stained sections can be viewed through the light microscope to resolve ambiguities in delineation of specific nuclear regions. Here the substantia nigra is clearly seen in this 40 micron section photographed at 32X. Spatial relationships are retained for specific regions of two dimensional histologic sections within the three-dimensional anatomic structure.

3.3. MORPHOMETRY.

Morphometric statistics describing anterior commissure, caudate, putamen, globus pallidus and thalamus in image data from postmortem cryosectioned anatomy were relatively consistent across subjects and in general agreement with the Talairach atlas data. Mean surface area, volume, and center of mass demonstrated significantly less variability between hemispheres in comparison to intersubject variability. Volume was less susceptible than surface statistics to variance due to complicated geometries, for example, statistics calculated on the ventricular system were highly influenced by slight differences in the lateral extent of this structure and thus unreliable.

The goals of the present experiments were to determine if postmortem cryosectioned anatomy could be used as a spatially detailed digital anatomic reference of the normal human brain. We were able to demonstrate a spatial resolution greater than 200 microns using large fields of view capable of imaging the whole horizontal plane of human brain. Equivalent resolutions in MR have only been produced for isolated postmortem specimens using a prototype 7.0 T superconducting magnet (Boyko *et al.*, 1994). In addition, by using a 24-bit color digital camera we were able to capture additional visual cues of texture and depth not available in tomographic imaging modalities.

The ability to resolve neuroanatomic boundaries is critical for accurate structure delineation. The spatial and densitometric detail provided in high resolution images of cryosectioned anatomy significantly improved our ability to differentiate structure boundaries. When compared to MRI, we found it easier to delineate structures in the cryosectioned anatomy, for example, laminar partitions of the basal ganglia and hippocampus. Subsequent histological processing of collected tissue sections proved even more valuable for localization of additional anatomic structure. By correlating digitized histology with blockface images of it was possible to retain their respective spatial context. Histologic sections have previously been used as an anatomic reference to improve interpretation of MR and CT images (Rauschning, 1986); however, such correlations have primarily been visual. By using digital reconstruction and resampling techniques we were able to display high resolution anatomy in specified planes to precisely match image data from other modalities.

Appropriate spatial coordinate systems facilitate localization of neuroanatomic structures and serve as a prerequisite for inter-subject and between modality comparisons. We were able to accomplish accurate placement of the volumes by utilizing the same transformation that MR and PET methods employ. The spatial positioning scheme described by Talairach and Tournoux bases its registry and scaling on the bicommissural line (Talairach & Tournoux, 1988). Since this approach is so dependent on the selection of the superior and inferior margins of the AC and PC midsagittal points, the higher resolution afforded in cryomacrotomed data was advantageous.

Sulci used for deformation proved sufficiently extended inside the brain to reflect subtle and distributed variations in neuroanatomy between subjects. Integral distortion functions were successfully used to extend the deformation field required to elastically transform nested surfaces to their counterparts in the target scan. The algorithm's accuracy was tested, by warping 3D MRI volumes from normal subjects and Alzheimer's patients, full-color 1024^3 digital cryosection volumes of the human head onto MRI volumes and patients with metastatic tumors causing mechanical distortions of anatomies.

The differences between tomographic images of *in vivo* human brain and surface imaging of cryoplaned postmortem specimens must be recognized. First, frozen preparation can induce changes in the configuration of anatomic specimens. Previous studies have suggested that alterations in size, shape and attenuation values of specimens may not be significant (Ho *et al.*, 1988; Pech, 1987; Rauschning *et al.*, 1983). Our own morphometric measurements support this. However, there are certainly many changes that occur postmortem such as loss of mean arterial pressure and uneven distribution of intracranial fluids that contribute to differences between *in vivo* and *ex vivo* derived data.

These experiments demonstrated the use of postmortem anatomical volumes created from serial cryoplaned heads and brains. These data, in combination with histologically processed tissue from the specimen, provide a detailed high resolution reference for tomographically acquired images. This approach, in combination with multimodality mapping techniques, will add to the growing data base being applied to the goal of mapping the human brain.

5. Acknowledgements

This work was supported in part by the Human Brain Project funded jointly by the National Institute of Mental Health, the National Institute on Drug Abuse (P20 MH52176), the National Library of Medicine (R01 LM05639) the National Science Foundation (BIR 9322434) and the Biomedical Research Technology Program of NCRR (R01 RR05956).

142

4. References

1. Bohm C, Greitz T, Eriksson L. (1989) A computerized adjustable brain atlas. *Eur J Med*;15:687-689.

2. Boyko O, Alston S R, Fuller G, Hulette C, Johnson A, Burger P. (1994) Utility of postmortem magnetic resonance imaging in clinical neuropathology. *Arch Pathol Lab Med*;118:219-225.

3. Damasio H, Kuljis RO, Yuh W. Ehrhardt J. (1991) Magnetic Resonance Imaging of human intracortical structure in vivo *Cerebral Cortex*; 1(5):374-9.

4. Evans A, Marret S, Torrescorzo J, Ku S, Collins L. (1991) MRI-PET correlation in three dimensions using a volume of interest (VOI) atlas. *J Cereb Blood Flow Metab*;11:169-178.

5. Greitz T, Bohm C, Holte S, Eriksson L. (1991) A computerized brain atlas: construction, anatomical content, and some applications. *J Comp Assist Tomogr*;15(1):26-38.

6. Ho P, Yu S, Czervionke L, Sether L, Wagner M, Pech P, Haughton V. (1988) MR and cryomicrotomy of C1 and C2 roots. *Amer J Neuro Radiol*;9:829-831.

7. Lehmann E D, Hawkes D, Hil D, Bird C, Robinson G, Colchester A, Maisley M. (1991) Computer aided interpretation of SPECT images of the brain using an MRI derived neuroanatomic atlas. *Med Informatics*;16:151-166.

8. Mazziotta, J.C., Toga, A.W., Evans, A., Fox, P. & Lancaster J. (1995) A Probabilistic Atlas of the Human Brain: Theory and Rationale for Its Development. NeuroImage. 2:89-101.

9. Pech P (1987) Coorrelative investigations of craniospinal anatomy and pathology with computed tomography, magnetic resonance imaging, and cryomicrotomy. Uppsala University, Sweden.

10. Rauschning W. (1986) Surface cryoplaning. A technique for clinical anatomical correlations. *Uppsala J Med Sci*; 91:251-255.

11. Rauschning W, Bergstrom K, Pech P. (1983) Correlative craniospinal anatomy studies by computed tomography and cryomicrotomy. *J Comput Assist Tomogr*; 7(1):9-13.

12. Roland, P. E. and Zilles, K. (1994) Brain atlases - a new research tool. *TINS*; 17(11):458-467.

13. Talairach J, Tournoux P. (1988) *Co-planar stereotaxic atlas of the human brain*. New York: Thieme Medical Publishers, Inc,

14. Thompson, P. & Toga, A.W. (1996) A surface-based technique for warping 3-dimensional images of the brain. *IEEE Trans. on Med Imag* 15(4):402-417.

15. Thompson, P., Schwartz, C., Lin, R.T., Khan, A.A. & Toga, A.W. (1996) 3D statistical analysis of sulcal variability in the human brain. *J. Neurosci* 16(13):4261-4274.

16. Tiede U, Bomans M, Hohne K H, Pommert A, Riemer M, Schiemann T, Schubert R, Lierse W. (1993) A computerized three-dimensional atlas of the human skull and brain. *Amer J Neuro Radiol*; 14(3):551-559.

17. Toga A W, Ambach K, Quinn B, Hutchin M, Burton J S. (1994) Postmortem anatomy from cryosectioned whole human brain. *J Neurosci Meth* 54(2):239-252.

18. Toga, A.W., Goldkorn, A., Ambach, K., Chao, K., Quinn, B.C. & Yao, P. (1997) Postmortem cryosectioning as an anatomic reference for human brain mapping. (submitted).

REGISTRATION: A POWERFUL TOOL TO COMBINE INFORMATION PROVIDED BY DIFFERENT IMAGING MODALITIES

M. EMRI AND T. MÁRIÁN

PET Centre, Medical University School of Debrecen, Nagyerdei krt. 98., 4102 Debrecen, Hungary

G. KÖVÉR AND E. BERÉNYI

Diagnostic Centre, Pannon Agricultural University, Kaposvár, Hungary

AND

O. ÉSIK

Department of Radiotherapy, National Institute of Oncology, Budapest, Hungary

1. Introduction

Image registration (IR) is a fundamental topic of three-dimensional (3D) medical image processing. The need to register different imaging modalities arises for several reasons: some anatomical details, especially soft tissues, are more easily seen in MRI than CT images, but the bony structures are better visualised by CT. The visualisation of any morphological volume with a functional data set which may come from SPECT or PET is a very important method both for research purposes and for routine diagnostics.

Software engineers or researchers maintaining the software tools have to choose one or more software packages to solve the IR tasks in their PET laboratory. Registration techniques have developed very impressively during the past decade, and a number of packages are now available [1]. The development of an IR software package is a very time-consuming procedure (it probably takes a single programmer several years) and it requires considerable experience in the handling of different imaging modalities.

We have succeeded in obtaining licences for our PET Centre to install software tools which allowed us to solve our IR tasks. These packages

B. Gulyás and H.W. Müller-Gärtner (eds.),
Positron Emission Tomography: A Critical Assessment of Recent Trends, 143–151.
© 1998 *Kluwer Academic Publishers.*

helped us to start software development in order to supplement the original packages with some programs of our own.

This paper outlines the implemented packages and presents two applications of the software, i.e. PET–PET registration to check the misalignment error of the head fixation system used in our PET investigations, and PET–CT–MRI registration for 3D radiotherapy planning.

2. Image Registration Methods and Software Tools

2.1. FILE FORMAT FOR INTERNAL USE AND SOFTWARE LIBRARIES

Before adopting any software package, we chose a working file format for internal use and a C library to handle it. The Scanditronix format came with our PET scanner (GE 4096 Plus), but it was not suitable for software development. Instead, we adopted the Medical Image NetCDF (MINC) format which had been developed in the Montreal Neurological Institute (MNI) Brain Imaging Centre [2]. MINC is a self-describing, extensible, N-dimensional format with C libraries for low-level data access and volume level manipulation. This file format was chosen not only because of its good documentation and the above-mentioned properties, but also for some very useful utilities for handling the MINC format. The MINC distribution package is available from MNI via Internet.

2.2. IMAGE REGISTRATION TOOLS

The first implemented registration tool was the REGISTER program, which also originated from MNI. This program is very useful for the simultaneous display of slices of different 3D image volumes, and also for the calculation of transformations on the basis of interactively selected anatomically equivalent tag points. The MINC Automated Image Registration (MINC–AIR) package was installed in order to use Woods' AIR package, which allows the registration of intramodality volumes by the ratio image uniformity (RIU) technique without interactive processes [3–5].

Both the automatic and interactive programs read and write the MINC format image and transformation parameter files, and therefore the output files of these programs can be used as input data for the other program. The result of a registration cycle is a transformation of the source volume into the target volume's co-ordinate space. A single transformation does not usually ensure the complete alignment of two volumes. In order to increase the precision of the desired alignment, the registration cycle may be repeated, preferably by another type of registration program. From this point of view, the common file format used by the REGISTER and MINC–

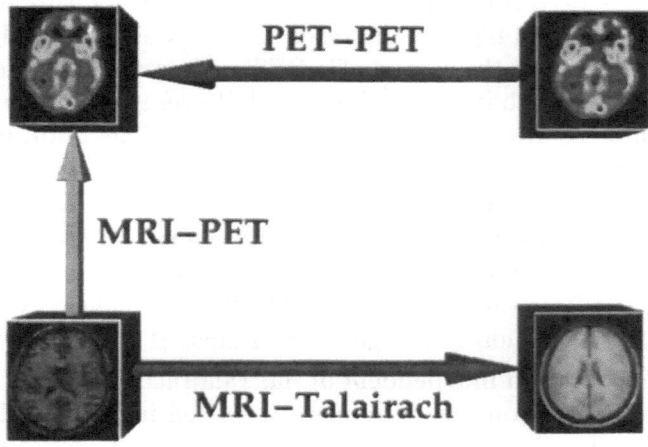

Figure 1. Brain image registration tasks

AIR programs has an advantageous feature, being instrumental in such repeated registration processes.

2.3. IMAGE REGISTRATION TASKS

Brain image registration tasks are summarised in Figure 1. The 3D boxes represent the 3D volumes and the arrows stand for the transformations of the source volume into the co-ordinate space of the target volume. The target and source volumes may originate from the same modality and same subject (intrasubject, intramodality), from the same subject but different modalities (intrasubject, intermodality) or from the same modality but different subjects (intersubject, intramodality) [6–8].

Intrasubject registration is evidently the simplest task. In these cases, the parameters of the translational and rotational transformation have to be determined by transforming one volume into the other volume's co-ordinate space. For the PET–PET registration of brain scans, this transformation can be determined by the automated RIU technique as these volumes contain only the information which needs to be registered.

To solve the intrasubject and intermodality registration task (for example in the MRI–PET or CT–PET registration), the RIU method requires removal of the non-brain (scalp, skull and meninges) voxels from the original morphological volume (segmentation), so that only brain structures are present in the MRI (or CT) volume.

The MNI registration programs offer a solution for this segmentation task only in the case of high spatial resolution T1 weighted MRI volumes, so they cannot be used in the case of CT–PET registration. Application of the MNI brain mask, which is included in the MNI average brain model data set [9], is a possible solution to strip an MRI volume. This brain mask volume determines a brain mask in the Talairach space [10]. If the transformation of a native MRI or CT volume into the Talairach space is known, its inverse can be used to transform the brain mask into the native space (the patient's proper co-ordinate system). However, to transform a native CT or MRI volume into the Talairach space is a difficult task, especially in the event of poor spatial resolution or a high level of noise [11].

A stripping method independent of the Talairach space may be useful in intrasubject registration tasks, frequently needed in routine clinical cases. The stripping of a CT volume by a simple threshold technique (i.e. selecting the threshold levels according to the grey and white matter Hounsfield values in the CT volume) gives a noisy but usable binary volume. The application of binary mathematical morphological algorithms may reduce the level of noise [12]. The native brain mask can be delineated from the low noise level binary volume by using cluster specification algorithms. We have applied this method successfully in several CT studies and, in this way, the CT-defined brain volumes were used in the automatic CT–PET registration. This brain stripping (or segmenting) method has been used in individual cases, but it has not yet been developed as a generally applicable brain stripping algorithm.

3. Intrasubject, intramodality registration

We earlier investigated in a study the effect of a single dose of Cavinton on the parameters of the cerebral glucose metabolism by the PET technique [13]. Twelve stroke patients were included in the investigation and two PET measurements were carried out on each subject. The FDG maps measured after Cavinton administration (Cavinton study, referred to as CS) were compared with those measured in the placebo study (referred to as PS). The Greitz-Bergström head fixation system was used to compare CS and PS images directly [14]. For this purpose, we had to check the reproducibility of the localisation of the patients' heads. The measure of misalignment could be determined by automatic PET–PET registration, due to the good signal-to-noise ratio of the FDG–PET images.

Automatic IR by the RIU technique was applied for these volume pairs. Using this technique, two volumes had to be defined: a target volume and a source volume. In order to reduce the calculation error of the transformation parameters, we carried out calculations to transform both volumes into the other one. With this case, the CS volume was taken as the target volume and the PS volume was regarded as the source volume, followed by the second run of calculation with the target and source volumes defined the other way round. The absolute values of the translation and rotation parameters necessary for the exact alignment of the two PET volumes were obtained for all patients. The minimal, maximal values and the average values are shown in Table 1. The displayed data are in good agreement with Greitz et al. [15].

The results argue for PET–PET registration to be performed before making any voxel level analysis of the difference of two related PET images, even if a head fixation system was used during the PET study. This is demonstrated in [13] by the calculation of difference maps from registered and non-registered pairs of FDG–PET volumes.

TABLE 1. Minimal, maximal and mean values of misalignment errors of head fixation system

	Translation [mm]			Rotation [rad]		
	dX	dY	dZ	α	β	γ
minimum	1.5	0.2	1.9	0.3	1.3	0.5
maximum	2.9	2.5	6.4	3.5	1.2	0.4
mean	1.8	2.4	4.3	0.9	1.2	0.6

4. Intrasubject, intermodality registration

4.1. METHOD

Intermodality registration very often means MRI–PET registration. This is true for research projects, but in clinical practice CT–PET, MRI–PET and CT–MRI registrations are equally important.

Intermodality registrations are of great importance in 3D radiotherapy planning. Different imaging are needed for a better definition of the different anatomical structures, planning target volume (PTV) and organ at risk (OAR). Soft tissue differences are better visualised by MRI than by CT, but CT is more informative in the case of bony structures [16]. CT matrixes contain the electron density data necessary for dose calculations relating to photon beams and charged particle radiotherapy. These facts necessitate CT–MRI registration in radiotherapy planning [17].

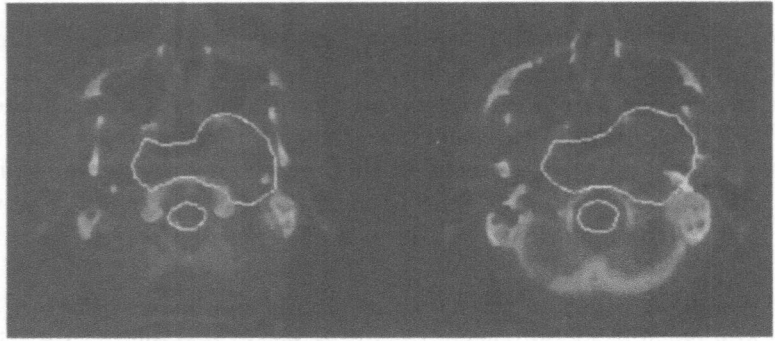

Figure 2. Contours of the original planning target volume and the organ at risk (spinal cord) in two representative slices of CT volume

To improve the sensitivity and specificity of tumour detection at the infiltrating edge, several nuclear medicine methods may be of help for anatomical studies. PET is suitable for the detection of small physiological differences between normal and diseased tissues, and SPECT with radiolabelled antibodies, tumour markers and metabolites also may enhance the accuracy of tumorous tissue localisation. These issues point to the importance of PET(SPECT)–CT registration.

Image registration allowing more adequate 3D radiotherapy planning requires solution of the following tasks:

1. the conversion of data files of all modalities into the working file format of the registration software;
2. the calculation of the transformation matrix transforming the images of the individual modalities into the co-ordinate space of the target modality (CT);
3. the transformation of the images of all modalities into the target modality space by using the evaluated matrixes and resampling all volumes to the same voxel and matrix size;
4. the conversion of the registered and resampled volumes into the file format readable by the radiotherapy planning software.

The solution of these problems depends on the modalities used, and the file formats and software package applied in the planning process.

4.2. A CASE REPORT

A 37-year-old male patient with a low-grade base of the skull tumour (adenoid cystic carcinoma with an S-phase ratio of 5%) was operated on palliatively. The postoperative CT and MRI investigations detected residual

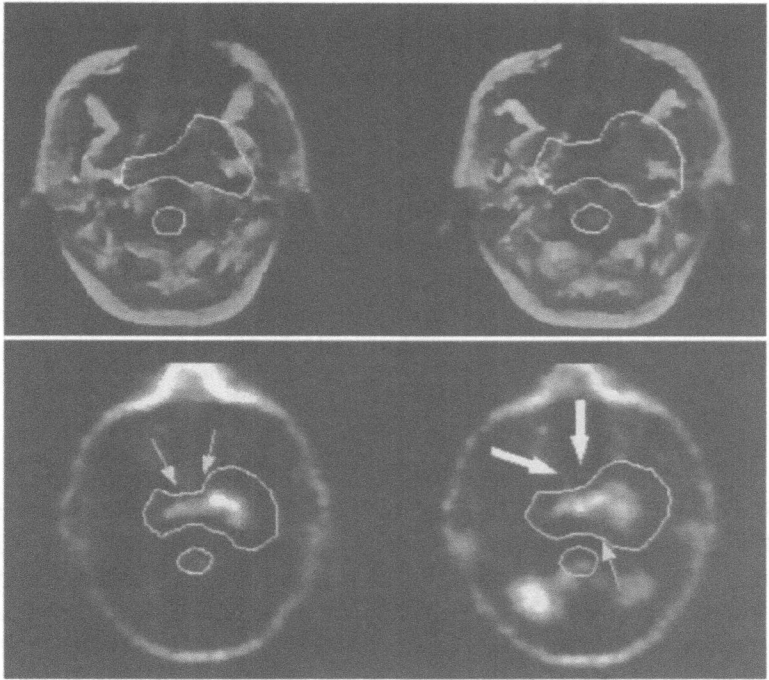

Figure 3. Contours of the original planning target volume (PTV) and the organ at risk (spinal cord) in corresponding slices of registered MRI and PET volumes. Part of the viable tumour (demonstrated by high FDG uptake) is outside the original PTV (large arrows); moreover in certain positions the safety zone around the gross tumour volume is too small (small arrows).

tumorous tissues in the left temporo-basal, parasellar and parapharyngeal regions, and in the retropharyngeal space.

For 3D radiotherapy planning involving palliative 6 MV photon irradiation, a CT investigation was performed with 4 mm slice thickness and 4 mm increments (SOMATOM DRH, Siemens). The information was transferred digitally to the 3D VIRTUOS [18] and TOMAS [19] radiotherapy planning software. The MRI volume was determined by means of 1.5 Tesla equipment (MAGNETOM, Siemens), but no CT-MRI registration was carried out in the course of the 3D therapy planning. The contours of the original PTV and OAR (spinal cord) are shown in CT slices (Figure 2).

We decided to check the accuracy of delineation of PTV and OAR by carrying out a conventional MRI–CT and PET–CT registration procedure. For this purpose, we applied the native brain-masking algorithm described in the image registration tasks section. With this method, we succeeded in stripping not only the CT volume, but also the MRI volume. The segmented MRI and CT volumes were registered by the MINC–AIR and the registration was justified by the REGISTER.

Later, a PET (GE 4096 Plus) investigation was also performed and the PET–CT registration was completed. The original PTV and OAR contours were drawn into the corresponding slices of the registered MRI and PET volumes (Figure 3). A comparison of the contours of the original PTV and the pattern of the PET and MRI images indicates that the additional information provided by these two modalities may contribute to adequate irradiation planning. The OAR (spinal cord) was basically found in the correct position in the corresponding MRI and PET slices. We plan to use the outlined procedure more systematically in the future.

5. Conclusions

Image registration is a powerful image-processing technique which should be available not only for advanced medical imaging centres, but also for a new PET centre. This technique may be useful in processing multimodality images in research projects and likewise in routine clinical practice, as it provides a tool for the combination of information from anatomic and functional imaging methods.

Application of the IR programs may quickly result in success if the implemented software tools fit perfectly into the local software development strategy.

6. Acknowledgement

We thank professor Evans and his colleagues at MNI for making their software tools available us before they were released.

References

1. Strother, S.C., Anderson, J.R., Xu, X.L., Liow, J.S., Bonar, D.C. and Rottenberg, D.A. (1994) Quantitative Comparisons of Registration Techniques Based on High-Resolution MRI of the Brain, *J. Comput. Assist. Tomogr.* **18**, 954–962.
2. Neelin, P. (1993) *MINC User Guide.* PET Imaging Lab: Montreal Neurological Institute.
3. Ward, G. (1995) *Automated Registration Using AIR and minctracc.* PET Imaging Lab: Montreal Neurological Institute.
4. Woods, R.P., Cherry, S.R. and Mazziotta, J.C. (1992) Rapid Automated Algorithm for Aligning and Reslicing PET Images, *J. Comput. Assist. Tomogr.* **16**, 620–633.
5. Woods, R.P., Mazziotta, J.C. and Cherry, S.R. (1993) MRI-PET Registration With Automated Algorithm, *J. Comput. Assist. Tomogr.* **17**, 536–546.
6. Woods, R.P., Mazziota, J.C. and Cherry, S.R. (1993) Automated Image Registration,. In *Quantification of Brain Function: Tracer Kinetics and Image Analysis In PET*, edited by Uemura, K., Lassen N., A., Jones, T. and Kanno, I., pp. 391–400. Elsevier Science, Amsterdam.

7. Evans, A.C., Marret, S., Torrescorzo, J., Ku, S. and Collins, L. (1991) MRI-PET Correlation in Three Dimensions Using a Volume-of-Interest (VOI) Atlas, *J. Cereb. Blood Flow Metab.* **11**, A69–A78.

8. Evans, A.C., Marrett, S., Neelin, P., Dai, W., Milot, S., Meyer, E., Bub, D., Collins, L. and Worsley, K. (1992) Anatomical Mapping of Functional Activation in Stereotactic Coordinate Space, *Neuroimage* **1**, 43–53.

9. Evans, A.C., Collins, D.L., Mills S.R. Brown E.D. Kelly, R.L. and Peters, T.M. (1993) 3D Statistical Neuroanatomical Models from 305 MRI Volumes, In *Proc. IEEE Nuclear Science Symposium / Medical Imaging Conference*, pp. 1813–1817.

10. Talairach, J. and Tournoux, P. (1988) *Co-planar Stereotactic Atlas of the Human Brain: 3-Dimensional Proportional System: an Approach to Cerebral Imaging.* Georg Thieme Verlag, Stuttgart.

11. Collins, D.L., Neelin, P., Peters, T.M. and Evans, A.C. (1994) Automatic 3D Intersubject Registration of MR Volumetric Data in Standardized Talairach Space, *J. Comput. Assist. Tomogr.* **18**, 192–205.

12. Haralick, R.M., Stenberg, S.R. and Zhuang, X. (1987) Image Analysis Using Mathematical Morphology, *IEEE Trans. Pattern Anal. Machine Intell.* **PAMI-9**, 532–549.

13. Trón, L., Szakáll S., Jr., Veress, G., Németh, F. and Galuska, L. (1996) Cavinton Affects the Kinetic Constants of FDG Accumulation: an Application of Registration and Kinetic Modelling, In *Positron Emission Tomography: A Critical Assessment of Recent Trends*, edited by Gulyás, B. and Müller-Gärtner, H.W., Kluwer Academic Publisher, Dordrecht.

14. Bergström, L., Boëthius, J., Ericson, L., Greitz, T., Ribbet, T. and Widén, L. (1981) Head Fixation Device for Reproducible Position Alignment in Transmission CT and Positron Emission Tomography, *J. Comput. Assist. Tomogr.* **5**, 136–141.

15. Greitz, T., Bergström, M., Boëthius, J., Kingsley, D. and Ribbe, T. (1980) Head Fixation System for Integration of Radiodiagnostic and Therapeutic Procedures, *Neuroradiology* **19**, 1–6.

16. Dobbs, H.J., Parker, R.P., Hodson, N.J., Hobday, P. and Husband, J.E. (1983) The Use of CT in Radiotherapy Treatment Planning, *Radiother. Oncol.* **1**, 133–142.

17. Webb, S. (1993) *The Physics of Three-Dimensional Radiation Therapy.* Medical Science Series. IOP Publishing Ltd., Bristol and Philadelphia.

18. Bendl, R., Pross, J., Hoess, A., Keller, M.A., Preiser, K. and Schlegel, W. (1994) VIRTUOSE - a Program for Virtual Radiotherapy Simulation and Verification, In *Proceedings of the 11th International Conference on the Use of Computers in Radiation Therapy*, edited by Hounsell, A.R., Wilkinson, J.M. and Williams, P.C., pp. 226–227.

19. Pross, J., Bendl, R. and Schlegel, W. (1994) TOMAS, a Tool for Manual Segmentation Based on Multiple Image Data Sets, In *Proceedings of the 11th International Conference on the Use of Computers in Radiation Therapy*, edited by Hounsell, A.R., Wilkinson, J.M. and Williams, P.C., pp. 192–193.

QUANTIFICATION OF FDG UPTAKE USING KINETIC MODELS

L. BALKAY, T. MOLNÁR, I. BOROS, AND SZ. LEHEL
PET Centre, University Medical School of Debrecen
Nagyerdei krt 98., 4012 Debrecen, Hungary.

AND

T. GALAMBOS
Research Group of the Hungarian Academy of Sciences

1. Introduction

It is well-known that the results of PET investigations can be quantified, i.e. biochemical parameters rendered to a whole image or to a given region of interest can be determined. For this purpose, defined mathematical transformations have to be performed on the reconstructed images. From stored primary data of PET scans images can be reconstructed in terms of radioactivity concentration. True activity concentration values can also be derived from the data, as all the necessary corrections can be carried out with high precision. (distortions are produced, among others, by tissue attenuation, scattered photons, and random coincidence). Modeling the kinetics of the biochemical processes of a supposed mechanism, allows us to calculate the interrelationship between measured activity concentration and the biochemical parameters of the tracer used based on which numerical values of these parameters can be evaluated.

Various radioligands are used for the PET imaging of tumors, however, the most frequently used radiopharmaceutical is the glucose analogue, 2-[F-18]fluoro-2-deoxy-D-glucose (FDG). The transport of this molecule and on the involved metabolic processes were published previously [1], and can be summarized by the three compartment model (Fig.1). The final goal of any applied FDG quantification method is to calculate the local glucose metabolic rate (LGMR). To solve this problem for the most general case, a dynamic PET investigation is required. This method demands the follow up of the FDG concentration in the blood by taking blood samples and acquiring a dynamic set of images. These data allow us to calculate the time course of blood activity (blood curve) and tissue activity within a defined region of interest (ROI TACT) as the two input functions of the kinetic tracer model calculations (Fig. 2). The calculation results in a set of numerical data for the kinetic constants k_1, k_2, k_3, and k_4 which belong to the chosen ROI(s). To calculate the quantitative FDG uptake several methods are available. These methods differ

153

B. Gulyás and H.W. Müller-Gärtner (eds.),
Positron Emission Tomography: A Critical Assessment of Recent Trends, 153–162.
© 1998 *Kluwer Academic Publishers.*

154

from each other in the details of the compartment model used and, consequently, the mathematics involved.

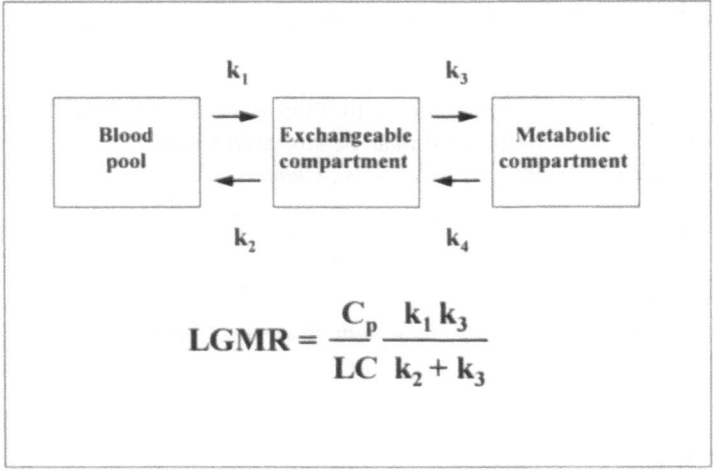

$$LGMR = \frac{C_p}{LC} \frac{k_1 k_3}{k_2 + k_3}$$

Figure 1. The three compartment model of FDG accumulation. Kinetic constants k_1, k_2, k_3, and k_4 refer to the facilitated transport from plasma to tissue(cell), the reversed transport from cells to plasma, the phosporylation of FDG, and the dephosphorylation, respectively. As FDG does not enter the Szentgyörgyi-Krebs cycle, it accumulates in the intracellular space in proportion to the rate of glycolysis. Glucose utilization can be calculated using the formula displayed, where Cp is the arterial plasma glucose concentration, and the constant LC (lump constant) takes care of the differences between the kinetics of glucose and FDG.

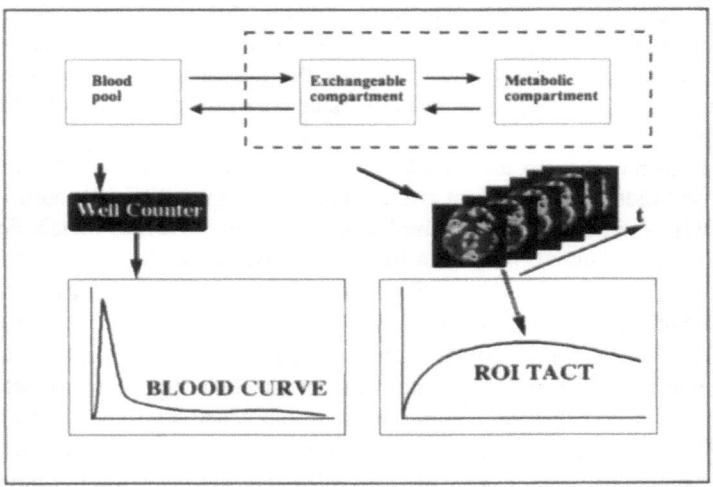

2. The basic operation equations

The first model calculations for the FDG utilization were carried out by Sokoloff and coworkers [2], and his model was adapted to PET by Phelps and his colleagues [3,4]. Sokoloff's three compartment model of FDG accumulation contains three kinetic constants: k_1, k_2, and k_3. Phelps extended this model by adding the k_4 rate constant to describe the dephosphorylation of FDG-6-PO$_4$ from the metabolic compartment. In this case the following set of differential equations has to be hold :

$$\frac{d[C_v(t)]}{dt} = -k_1 C_v(t) + k_2 C_e(t) \tag{1}$$

$$\frac{d[C_e(t)]}{dt} = k_1 C_v(t) - (k_2 + k_3) C_e(t) + k_4 C_m(t), \tag{2}$$

$$\frac{d[C_m(t)]}{dt} = k_3 C_e(t) - k_4 C_m(t), \tag{3}$$

where C_v, C_e, and C_m are the FDG concentrations in the vascular, exchangeable and metabolic compartment, respectively. The measured tissue activity concentration can be expressed as

$$C_{tis}(t) = C_e(t) + C_m(t). \tag{4}$$

Substituting the solution of the previous set of differential equations (Eq. 1-3) for the $k_4 = 0$ case (Sokoloff model) into Eq.4. we get:

$$C_{tis}(t) = k_1 \left[\frac{k_3}{k_2 + k_3} \int_0^t C_p(\tau) d\tau + \frac{k_2}{k_2 + k_3} \exp(-(k_2 + k_3)t) \int_0^t C_p(\tau) \exp((k_2 + k_3)\tau) d\tau \right], \tag{5}$$

With $k_4 \neq 0$ (Phelps model), we obtain:

$$C_{tis}(t) = \frac{k_1}{\alpha_2 - \alpha_1} \left[\int_0^t [(k_3 + k_4 - \alpha_1)e^{-\alpha_1\tau} + (\alpha_2 - k_3 - k_4)e^{-\alpha_2\tau}] C_p(t - \tau) d\tau \right], \tag{6}$$

where α_1 and α_2 are:

$$\alpha_1 = [k_2 + k_3 + k_4 - \sqrt{(k_2 + k_3 + k_4)^2 - 4k_2 k_4}]/2 \qquad (7)$$

$$\alpha_2 = [k_2 + k_3 + k_4 + \sqrt{(k_2 + k_3 + k_4)^2 - 4k_2 k_4}]/2 \qquad (8)$$

Parameter (k_1, k_2, k_3, k_4) estimation can be done by nonlinear regression by minimizing the difference between the calculated and measured $C_{tis}(t)$ data. This mathematical procedure requires only a blood curve of good statistics. With known values of k's, LGMR can be calculated on the basis of the equation displayed in Fig.1. As this optimization (minimization) is a lengthy procedure and requires a whole set of dynamic PET frames, Sokoloff and coworkers developed a simpler protocol for the calculation of LGMR :

$$LGMR = \frac{C_a}{LC} \cdot \frac{C_{tis}(T) - k_1 \left[\int_0^T e^{-(k_2+k_3)\tau} C_p(T-\tau)d\tau \right]}{\left[\int_0^T C_p(\tau)d\tau - \int_0^T e^{-(k_2+k_3)\tau} C_p(T-\tau)d\tau \right]} . \qquad (9)$$

Similarly, Phelps and coworkers also introduced a simpler procedure for the $k_4 = 0$ case:

$$LGMR = \frac{C_a}{LC} \cdot \frac{C_{tis}(T) - \dfrac{k_1}{\alpha_2 - \alpha_1} \left[\int_0^T [(k_4 - \alpha_1)e^{-\alpha_1\tau} + (\alpha_2 - k_4)e^{-\alpha_2\tau}]C_p(T-\tau)d\tau \right]}{\dfrac{k_2 + k_3}{\alpha_2 - \alpha_1} \left[\int_0^T [e^{-\alpha_1\tau} + e^{-\alpha_2\tau}]C_p(T-\tau)d\tau \right]} . \qquad (10)$$

Both formulas represent approximations and both are based on the assumption that k_1, k_2, k_3 and k_4 values do not depend on location. Although for these simplified calculation successive blood samples have to be taken the same way as in the more general case, the only required single PET scan at around T=40 min simplifies the protocol remarkable. LGMR can be calculated using average values of k from the literature.

3. Generalizations and constraints

3.1. ATTEMPTS TO LINEARIZE OPERATION EQUATIONS

3.1.1 Patlak graphical method

A fairly rapid calculation of LGMR can be applied if one is not interested in the numerical value of the rate constants and if $k_4=0$ may be assumed [5,6]. Furthermore, with the assumption that in the integrals of Eq.5 $C_p(t)$ can be regarded constant, compared to the strong time dependence of $exp((k2+k3)t)$, - it will be true for sufficiently large t - , that the equation 5 can be rewritten:

$$\frac{k_1 k_3}{k_2 + k_3} = \frac{C_{tis}(t)/C_p(t) - k_1 k_2 /(k_2 + k_3)^2}{\left[\int_0^t C_p(\tau)d\tau\right]/C_p(t)} . \tag{11}$$

By rearrangement, we get

$$\frac{C_{tis}(t)}{C_p(t)} = \frac{k_1 k_3}{k_2 + k_3} \frac{\int_0^t C_p(\tau)d\tau}{C_p(t)} + \frac{k_1 k_2}{(k_2 + k_3)^2} . \tag{12}$$

Eq. 12 describe a linear relationship between $C_{tis}(t)/C_p(t)$ and $\int_0^t C_p(\tau)d\tau /C_p(t)$, thus the slope of the linear plot $C_{tis}(t)/C_p(t)$ vs. $\int_0^t C_p(\tau)d\tau /C_p(t)$ can be determined by a simple and rapid linear regression. The product of the slope and factor C_p/LC is the local glucose metabolic rate.

3.1.2. The linearization approach by Blomqvist

The Patlak linearization is based on two assumptions:
   a/ $k_4=0$,
   b/ at sufficiently large t, the second integral term of equation 5. can be simplified.
Blomqvist proposed [7] an alternative linearization method which does not require assumption b. This method is based on a second order differential equation which can be obtained from Eq.1-4, with $k_4 = 0$:

158

$$\ddot{C}_{tis}(t) + (k_2 + k_3)\dot{C}_{tis}(t) = k_1\dot{C}_p(t) + k_1k_3C_p(t) . \tag{13}$$

Integrating this equation twice over the time interval [0 , t_i], - t_i are the time points of the PET measurement - results in a system of equations,

$$C_{tis}(t_i) = k_1 \int_0^{t_i} C_p(\tau)d\tau + k_1k_3 \int_0^{t_i}\int_0^{\tau} C_p(\omega)d\omega d\tau - (k_2 + k_3)\int_0^{t_i} C_t(\tau)d\tau \,/i = 1..n/ \tag{14}$$

containing three constants k_1, k_1k_2, $k_2 + k_3$ as variables (n is the number of sequential PET scan). As parameters $C_p(t)$, $C_{tis}(t)$ and all the integrals can be calculated, the above system of equations is represented by the following four dimensional linear equation:

$$X^1 = k_1X^2 + k_1k_3X^3 - (k_2 + k_3)X^4 \tag{15}$$

Standard least square method offers a rapid and convenient solution of this equation for k_1, k_2, and k_3, allowing the evaluation of LGMR.

3.2. CONSIDERATION OF THE VASCULAR ACTIVITY

It has been shown experimentally that the vascular compartment can give a contribution to the tissue activity curve - $C_{tis}(t)$ -, which can be described as a function of the vascular activity [8,9]. This fraction is very large for the first couple of minutes. For improving the accuracy of the calculated k's and LGMR an additional equation has to be added to the basic operation equations (5,6):

$$C_{tot}(t) = C_{tis}(t)(1 - CBV) + C_p(t)CBV \tag{16}$$

Here $C_{tot}(t)$ represents the volume weighted average of the tissue and vascular activity and CBV expresses the fractional vascular space. Substituting this formula into the more general operation (Eq.6.), we get:

$$C_{tot}(t) = \frac{k_1}{\alpha_2 - \alpha_1} \cdot \left[\int_0^t [(k_3 + k_4 - \alpha_1)e^{-\alpha_1\tau} + (\alpha_2 - k_3 - k_4)e^{-\alpha_2\tau}]C_p(t - \tau)d\tau \right] \cdot$$
$$(1 - CBV) + C_p(t)CBV \tag{17}$$

Although the solution of Eq.17 for the five parameters makes the calculation somewhat lengthier, this set of parameters assures a more perfect fit of the measured curve(s).

3.3. MICHAELIS-MENTEN CONSTRAINTS TO IMPROVE THE ACCURACY OF
PARAMETER ESTIMATION

It is well known that nonlinear regression is very sensitive to the noise level of the input
curve and the number of parameters to be determined . Decrease in the size of the region
of interest results in an increased relative noise of the measured tissue uptake curve. At the
same time, ROI(s) of bigger size may make the selected anatomical area heterogeneous.
With $LGMR$ and k_1, k_2, k_3, and k_4 not being constant within the region(s), only the
average value of these parameters can be determined. Nonlinear regression methods
require long calculation times which is an additional drawback. The calculation time
necessary for the determination of the kinetic parameters can significantly be reduced by
any (derived or assumed) interrelationship between the variables. Kuwabara and
coworkers [10] postulated Michaelis-Menten kinetics for glucose transport and
phosphorylation yielding the following constraint:

$$V_e = \frac{k'_1}{k'_2} = \frac{k_1}{k_2} = \frac{K_t V_d + M_e}{K_t + C_a} = \text{constant} , \tag{18}$$

where K_t denotes half saturation constant of glucose transport, V_d is the brain water
volume, M_e is the total brain glucose contents, C_a stands for the glucose concentration in
the arterial plasma, and comma (') refers to glucose. Based on experimental observations
and measured data they further assumed:

$$\tau = \frac{k_1}{k'_1} = \text{constant} , \tag{19}$$

$$\varphi = \frac{k_3}{k'_3} = \text{constant} , \tag{20}$$

$$\tau = 1.1 \quad \varphi = 0.3 \quad K_t = 4.8 mM \quad V_d = 0.77 \text{ml} / \text{g} . \tag{21}$$

Operation equation 17. can be rewritten using Eq.18-21 :

$$C_{tot}(t) = (k_1 - K)e^{-\beta \tau}\left[\int_0^t e^{\beta \tau} C_p(\tau)d\tau\right] + K\int_0^t C_p(\tau)d\tau + C_p(t)CBV . \tag{22}$$

Exponent β is an explicit function of k_1, K, τ, φ, K_t, V_d, C_a and $K = k_1 k_3/(k_2+k_3)$,
thus only three parameters $(k_1, K,$ and $CBV)$ are left to be determined by the fitting
procedure.

4. Matrix representation of basic operation equation

Let us consider the basic differential equations (1-3). Another equation can be added which describes the integrated activity of the tissue directly measured by the scanner:

$$C_i(t) = \int_0^t [C_e(\tau) + C_m(\tau)]d\tau . \tag{23}$$

Equations (1-3, 21) can be written in matrix representation :

$$\begin{bmatrix} \dot{C}_e(t) \\ \dot{C}_m(t) \\ \dot{C}_i(t) \end{bmatrix} = \begin{bmatrix} -(k_2 + k_3) & k_4 & 0 \\ k_3 & -k_4 & 0 \\ 1 & 1 & 0 \end{bmatrix} \begin{bmatrix} C_e(t) \\ C_m(t) \\ C_i(t) \end{bmatrix} + \begin{bmatrix} k_1 \\ 0 \\ 0 \end{bmatrix} C_p(t) , \tag{24}$$

or in a more simplified form substituting the X activity vector, the $\hat{\mathbf{A}}$ rate matrix and the $\mathbf{B}$ rate vector,

$$\dot{\mathbf{X}} = \hat{\mathbf{A}}\mathbf{X} + \mathbf{B}C_p(t) . \tag{25}$$

This representation is one of the simplest forms used in state-space methods for analyzing the behavior of a dynamic system model [11]. The state-space method is a standard method in system control engineering. By dividing the total acquisition time (t_{tot}) into N equal time intervals and assuming that $C_p(t)$ remains constant over the interval kT < t < (k+1)T (where T = t_{tot}/N, and k=1..N), the discrete solution of Eq.25 has the form:

$$\mathbf{X}((k+1)T) = e^{\hat{\mathbf{A}}T}\mathbf{X}(kT) + \left[\int_0^T e^{\hat{\mathbf{A}}\tau}d\tau\right]\mathbf{B}C_p(t) . \tag{26}$$

By using notations:

$$\hat{\phi}(T) = e^{\hat{\mathbf{A}}T}, \quad \mathbf{G}(T) = \left[\int_0^T e^{\hat{\mathbf{A}}\tau}d\tau\right]\mathbf{B} , \tag{27}$$

we get

$$\mathbf{X}((k+1)T) = \hat{\phi}(T)\mathbf{X}(kT) + \mathbf{G}(T)C_p(kT) . \tag{28}$$

The most important mathematical function in this simple recursive equation is the matrix exponential, which can be evaluated faster than the calculation based on the convolution equation (Eq 6). By using the MATLAB [12] software package and some imported own C source code program, the iteration time could be reduced to 1 sec for a given tissue uptake curve. The generalization and restriction of compartment models is very easy using the matrix representation. To include the vascular compartment, we only have to rewrite the rate matrixes:

$$\hat{A} = \begin{bmatrix} -(k_2 + k_3) & k_4 & 0 \\ k_3 & -k_4 & 0 \\ 1 - CBV & 1 - CBV & 0 \end{bmatrix}, \quad B = \begin{bmatrix} k_1 \\ 0 \\ CBV \end{bmatrix}. \tag{29}$$

For the case of restriction let us suppose that k's do not depend on location and only the observed organ accumulates the tracer from the blood. Thus Eq.24 can be rewritten:

$$\begin{bmatrix} \dot{C}_e(t) \\ \dot{C}_m(t) \\ \dot{C}_i(t) \\ \dot{C}_p(t) \end{bmatrix} = \begin{bmatrix} -(k_2 + k_3) & k_4 & 0 & k_1 \\ k_3 & -k_4 & 0 & 0 \\ 1 - CBV & 1 - CBV & 0 & CBV \\ k_2 & 0 & 0 & k_1 \end{bmatrix} \begin{bmatrix} C_e(t) \\ C_m(t) \\ C_i(t) \\ C_p(t) \end{bmatrix}, \tag{30}$$

or using the vector representation, we got

$$\dot{X} = \hat{A}X. \tag{31}$$

This is a simple homogeneous vector differential equation, and the solutions have to satisfy the general solution:

$$X(t) = e^{\hat{A}t} X(t_0). \tag{32}$$

An advantageous feature of this calculation is that it does not require any integration.

4. References

1. Horton, R.W., Meldrum, B.S., Bechelard, H.S. (1973) Enzymatic and cerebral metabolic effect of 2-deoxy-D-glucose. *J. Neurochem*, **21**,506-520.
2. Sokoloff, L., Reivich, M., Kennedy, C., et al (1977)The (C^{14})-deoxyglucose method for the measurement of local cerebral glucose utilization: theory, procedure, and normal values in the conscious and anesthetized albino rat. *J. Neurochem* **28**, 897-916
3. Phelps, M.E., Huang, S.C., Hoffman, E.J., et al (1979) Tomographic measurement of local cerebral glucose metabolic rate in humans with 2-[F-18]fluoro-2-deoxy-D-glucose: validation of method. *Ann Neurol* **6**, 371-388
4. Huang, S.C., Phelps, M.E., Hoffman, E.J., et al (1980) Noninvasive determination of local cerebral glucose metabolic rate in man. *Am. J. Physiol* **238**, E69-E82
5. Gjedde, A. (1982) Calculation of cerebral glucose phosphorylation from brain uptake of glucose analogs in vivo: a re-examination. *Brain Res Rev* **4**, 234-277

162

6. Patlak, C.S., Blasberg, R.G. (1985) Graphical evaluation of blood to brain transfer constants from multiple time uptake data. Generalizations. *J. Cereb Blood Flow Metab* **5**, 584-590

7. Blowqvist, G. (1984) On the construction of functional map in positron emission tomography.*J. Cereb Blood Flow Metab* **4**, 629-632

8. Evans, A.C., Diksic, M., Yamamoto, Y.L. et al (1986) Effect of vascular activity in the determination of rate constants for the uptake of F^{18}-labeled 2-Fluoro-2-deoxy-D-glucose: error analysis and normal values in older subjects. (1986)*J. Cereb Blood Flow Metab* **6**,724-738

9. Hawkins, R.A., Phelps, M.E., Huang, S.C. (1986) Effect of temporal sampling, glucose metabolic rates, and disruptions of the blood-brain barrier on the FDG. Model with and without a vascular compartment: studies in human brain tumors with PET. *J. Cereb Blood Flow Metab* **6**,170-183

10. Kuwabara, H., Evans, A.C., Gjedde, A. (1990) Michaelis-Menten constrains improved cerebral glucose metabolism and regional lumped constant measurement with F^{18}Fluorodeoxyglucose.*J. Cereb Blood Flow Metab* **10**,180-189

11. Hoh, C.K., Dahlbom, M., Hawkins, R.A., et al (1994) Basic principles of positron emission tomography in oncology: quantification and whole body techniques. *Wien Klin Wochenschr* **106/15**, 496-504

12. MATLAB, the Math Works, Inc. Cochituate Plance, 24 Prime Park Way, Natick, MA 01760, USA

CAVINTON AFFECTS THE KINETIC CONSTANTS OF FDG ACCUMULATION: AN APPLICATION OF REGISTRATION AND KINETIC MODELLING

L. TRÓN, SZ. SZAKÁLL JR. AND G. VERESS
PET Centre, University Medical School of Debrecen, Nagyerdei krt. 98., 4012 Debrecen, Hungary

F. NÉMETH
Research Group of the Hungarian Academy of Sciences

AND

L. GALUSKA
Department of Nuclear Medicine, University Medical School of Debrecen

1. Introduction

Cavinton (vinpocetine: ethyl apovincaminate) is frequently used in the therapy of patients with cerebrovascular disorders [1, 2]. Vinpocetine has been shown to dilate cerebral blood vessels, increase the cerebral blood flow, improve the cerebral utilization of oxygen, accelerate the cerebral turnover of noradrenaline, and increase the cerebral glucose uptake [3-6]. The pharmacokinetics of the drug in healthy volunteers and laboratory animals was studied previously [7 and references therein]. Vinpocetine was found to increase the cerebral energy metabolism in conscious mice within half an hour after oral administration or intraperitoneal injection [6]. In patients with stenosis or occlusion of the internal carotid arteries, an increase in cerebral blood flow was detected on use of the Xe inhalation method [7], with the maximum appearing 32 min after the start of the infusion.

These findings were obtained during the past twenty years by means of techniques such as thermography, Xe clearance, invasive radiochemical analysis, etc. Since these investigations begun, however, there has been considerable progress as concerns the possibilities and precision of these and related methods. We set out to supplement the existing data with the results of PET investigations. A study was therefore carried out to detect

163

B. Gulyás and H.W. Müller-Gärtner (eds.),
Positron Emission Tomography: A Critical Assessment of Recent Trends, 163–172.
© 1998 *Kluwer Academic Publishers.*

any changes in the glucose metabolism of the brain following a single dose of 20 mg of Cavinton, administered IV in 500 ml of physiological saline solution.

2. Materials and methods

2.1. PATIENTS

Twelve stroke patients (4 women, 8 men, aged 62.3 ± 4.9 years, range 55–70 years) participated in the study. They had had their last stroke 13.4 ± 11.9 months (range 5.5–41 months) before PET scanning. The criteria for exclusion included diabetes and the administration of any drug affecting the circulation. In one case, the majority of the stroke-affected region was in the white matter; in another case, the blood vessels were sclerotized to such an extent that the blood sampling necessary for the kinetic analysis was not possible; in a third case, the cerebral glucose metabolic rates (CMRGl) were extremely low, which is typical in diabetic cerebral hypoglycaemia. For these reasons, only the data on 9 patients were subjected to statistical analysis. The subjects had been informed about the objective, details and risks of the investigation, and all had given written consent in agreement with the Helsinki Declaration and the Code of Federal Regulation: Protection of Human Subjects [8, 9]. The study was approved by the Ethical Committee of the University Medical School, Debrecen.

MRI investigations documented the stroke in all cases and localized the affected area in the region of the arteria cerebri media. Naturally, neurological investigations, SPECT scans, transcranial Doppler and the necessary laboratory investigations were also carried out.

Two FDG–PET (FDG: ^{18}F-labelled deoxyglucose-6-phosphate) measurements were performed on each subject. The reference state of the cerebral glucose metabolism of the patients was defined by a dynamic PET scan started at the end of a 45–min infusion of 500 ml of physiological saline solution. PET measurement was repeated the next day after the IV administration of 20 mg of vinpocetine dissolved in 500 ml of physiological saline solution. The cubital veins were cannulated in both arms. FDG (0.15 $\pm$ 0.02 mCi/kg in 5 ml of physiological saline) was given as a bolus injection (in 10 sec). Data acquisition and blood sampling for the purpose of kinetic analysis were started immediately after the FDG injection. Blood samples were taken from the vein of the other arm as often as possible during the first 2 min, followed by 6 samples at 30-sec, 5 samples at 1-min, 4 samples at 5-min and 3 samples at 10-min intervals, with the arm warmed to 40 °C to arterialize the venous blood.

2.2. BRAIN SCANNING

MR imaging was performed with a SHIMADZU SM magnetic resonance tomograph (1 tesla) resulting in 2 × 18 transaxial slices (18 proton-enhanced and 18 T1-weighted images), 14 parasagittal (T1-weighted images) and 14 coronal slices (T1-weighted images).

PET measurements were made with a GE 4096 PLUS whole body camera with 6.0 mm in-plane resolution and 6.86 mm inter-slice distance. The camera produced 15 transaxial slices. Head fixation [10, 11] was used to ensure identical positioning in both scans on each of the patients. All PET investigations were of dynamic type with the number of scans x the length of the exposure as follows: 12 × 30 sec, 6 × 1 min, 6 × 3 min, 6 × 5 min, resulting in a total of 30 frames with a total length of data acquisition of 60 min.

2.3. DATA EVALUATION

Quantitative data were obtained by using the three-compartment model of FDG accumulation [12, 13]. Kinetic constants k_1, k_2, k_3 and k_4 and the regional cerebral metabolic rate of glucose (rCMRGl) were determined for three regions of interest (ROIs): the whole brain, intact hemisphere and stroke area (the latter area was delineated on the basis of the FDG−PET map, but the appropriate MRI images were also taken into account). Analysis was performed by using the software package MATLAB. Pixel by pixel analysis was also carried out in several cases.

To characterize the effect of vinpocetine, the difference images were evaluated by subtraction of the appropriate images measured before and after vinpocetine treatment. Subtraction was accomplished with and without registration of the PET images involved. For details of registration (both PET−PET and PET−MRI), see in [14].

3. Results and Discussion

The kinetics of the accumulation of deoxyglucose-6-phosphate was analysed according to the three-compartment Sokoloff model. These compartments comprise the vascular compartment, the combined compartment of the interstitial and free intracellular space (with exchangeable tracer molecules) and the metabolic compartment (with the phosphorylated form of FDG produced by hexokinase). The kinetic constants characterizing the transfer of FDG across the barriers separating these compartments are defined as illustrated in Fig. 1.

Parameters k_1 and k_2 refer to the forward and reverse capillary transport of FDG, with a contribution from the side of the transport through

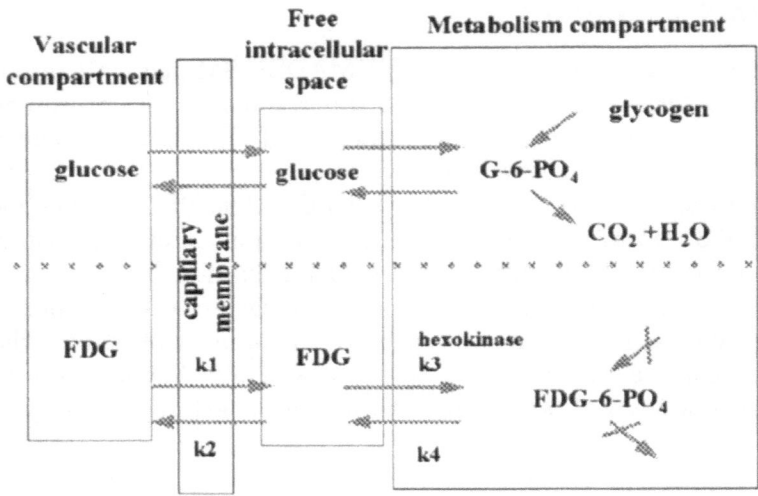

Figure 1. Three-compartment model (by Sokoloff et al. 1977) for quantification of FDG accumulation data.

the cytoplasmic membrane. Kinetic constants k_3 and k_4 refer to the phosphorylation of FDG and the dephosphorylation of FDG-6-phosphate, respectively. Calculated values of these parameters allow an evaluation of the regional cerebral metabolic rate of glucose (rCMRGl) as

$$rCMRGl = \frac{C_p}{LC} \frac{k_1 k_3}{k_2 + k_3}$$

where C_p denotes the glucose concentration in the plasma and LC (lumped constant) is an experimentally determined constant correcting for the difference between FDG and glucose, as all the kinetic parameters (as the output data of the mathematical analysis) relate to FDG.

Numerical values of k_1, k_2, k_3, k_4 and rCMRGl were evaluated separately in three ROIs for all of the patients. Figure 2 presents an illustrative example for the delineation of the stroke ROI, together with the obvious definition of the whole brain ROI and the intact hemisphere ROI. Three-dimensional ROIs were constructed from the individual two-dimensional ROIs for the different slices and the kinetic calculations were carried out on these three-dimensional ROIs in all patients. The results of the calculations are displayed in Table 1. Mean values of the kinetic constants with standard deviations were determined from data on the individual patients by using EXCEL for WINDOWS 5.0. The same software was applied for the statistical analysis; a paired two sample for means version of the t-test

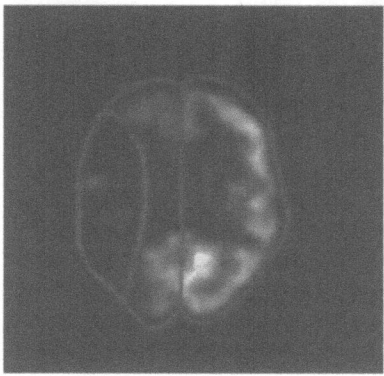

Figure 2. FDG-PET scan of a 66-year-old female stroke patient. The figure shows the delineation of the stroke-affected ROI. The whole brain, and the intact hemisphere ROIs within this slice are also indicated.

was chosen to assess the significance of the effect of Cavinton treatment on the mean values of the parameters.

TABLE 1. Numerical values of the means and standard deviations of the kinetic constants k_1, k_2, k_3, and rCMRGl relating to the different ROIs. Data are displayed separately for the post-treatment and pre-treatment cases (referred to as Cav. and pre., respectively).

		Whole brain		Intact hemisphere		Stroke-affected area	
		Cav.	pre.	Cav.	pre.	Cav.	pre.
rCMRGl	mean	6.21	6.24	6.80	6.86	4.99	5.47
[/min/100g]	SD	0.92	0.89	1.16	1.11	0.77	1.28
k_1	mean	0.12	0.10	0.13	0.11	0.09	0.08
[1/min]	SD	0.04	0.02	0.03	0.02	0.03	0.02
k_2	mean	0.26	0.20	0.26	0.21	0.22	0.18
[1/min]	SD	0.08	0.05	0.06	0.05	0.07	0.05
k_3	mean	0.11	0.11	0.11	0.11	0.10	0.10
[1/min]	SD	0.04	0.02	0.03	0.03	0.03	0.03

The tabulated data suggest an increase in k_1 for the intact hemisphere in response to the Cavinton infusion. This change proved to be statistically significant (the level of significance was chosen as 5 % throughout the experiments). k_1 for the whole brain ROI also increased significantly. This should definitely be a consequence of the Cavinton-induced increase in k_1 for the healthy part of the brain, as the increase in k_1 for the stroke-affected area

168

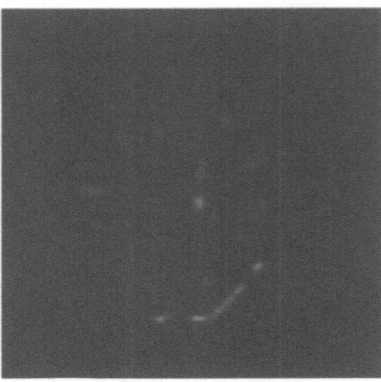

Figure 3. Difference maps of FDG accumulation after a single treatment of Cavinton infusion relative to the pre-treatment case (66-year-old female patient). Note the bands indicating increased FDG accumulation around the stroke area and in the occipital region. Slices are at the height of the caudate nuclei (left) and the top of the third ventricule (right).

was not significant. A general increase was detected in k_2 for FDG accumulation for all three ROIs analysed. The single Cavinton infusion-induced rise in k_2 also proved to be significant. In contrast, only a slight change of about 3 % (far from significant) was observed in k_3. Finally, we did not observe significant changes in rCMRGl for any area. However, rCMRGl decreased for the stroke area and its level of significance (p=6.3 %) was very close to the conventional threshold value (5 %).

For a clearer insight into the changes due to the Cavinton treatment, difference maps were obtained by subtracting the pre-treatment and post-treatment maps (Fig. 3) of FDG accumulation. It is interesting that there is a suggestion of an increased glucose accumulation in a narrow band around the stroke region. For this to be interpreted as a true increase, possible artifacts due to misalignment must be ruled out. An intense white band close to the contour of the difference map in the occipital region may indicate the incomplete overlay of the two images.

It is clearly seen in Fig. 3, that the two images might be misaligned. To investigate this point, we modified the FDG maps in such a way that, above a low threshold value, we used homogeneous black (pre-treatment map) and grey staining (post-treatment map) for subtraction of the FDG accumulation maps (Fig. 4). In case of overlay of the two images with the pre-treatment (post-treatment) map on top results in narrow grey (black) bands on the lower-left (upper-right) part of the contour of the whole slice and the region with physiological (or near physiological) glucose accumulation (Fig. 5 left). This is a clear indication of the fact that the head fixation helmet does not ensure an accurate, well-reproducible localization

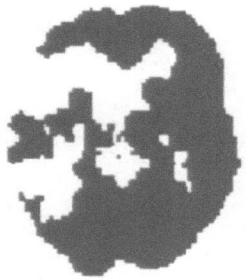 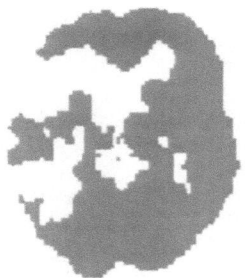

Figure 4. Schematic representation of FDG–PET maps with a special tone code. Pixels with counts above a set low threshold value are designated by black or grey (pre-treatment and post-treatment maps, respectively). Subthreshold pixels are white.

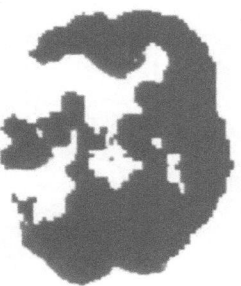 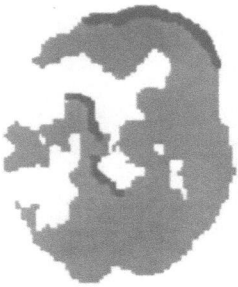

Figure 5. Overlay of the two FDG accumulation maps of Fig. 4, with the pre-treatment one on top (left) or the post-treatment one on top (right).

of the patient's head.

These black and grey bands disappeared from the overlaid pair of black and grey maps after a PET–PET registration had been made according to [14]. The registration algorithm yielded a value of 4 mm for the linear displacement of the two images in Fig. 3. Such a displacement is consistent with the published accuracy of this head fixation system. Figure 6 shows the difference maps from Fig. 3, together with the same difference maps demonstrating the subtraction after PET–PET registration. For easy comparison, the original difference maps constructed without PET–PET registration are also displayed. It seems quite obvious that the structure in the difference map disappears as a result of the PET–PET registration. Accurate alignment of the images before subtraction renders the resulting difference map of glucose accumulation very random.

In order to check for any localized effect of the Cavinton treatment in the cerebral glucose metabolism, we abandoned the ROI concept, as any

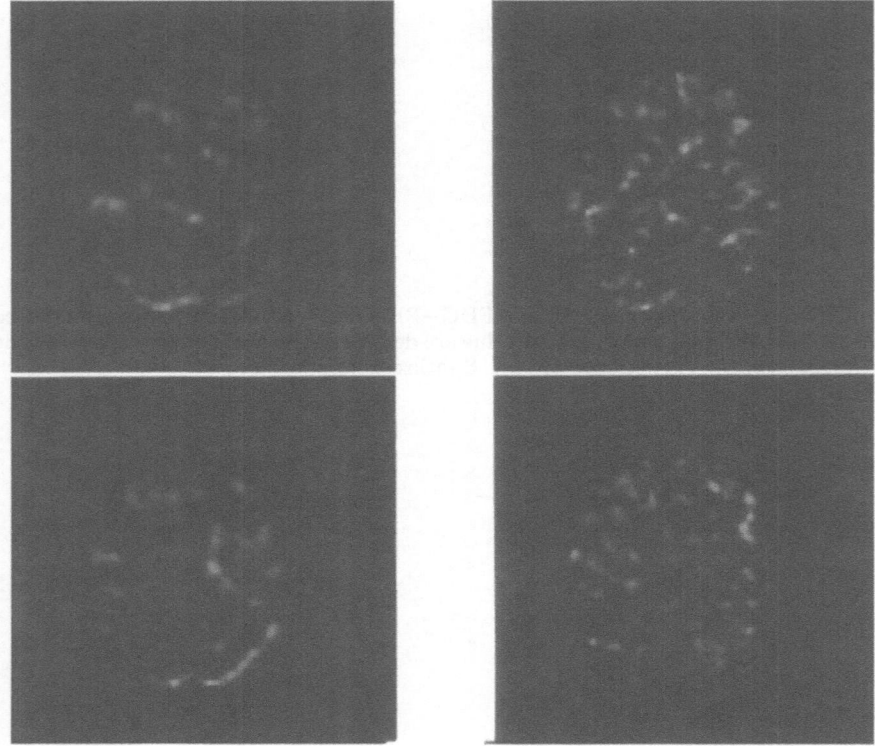

Figure 6. Difference maps of FDG accumulation from Fig. 3, (left panels) and the same difference maps evaluated with PET−PET registration (right panels). Pixels with negative differential counts are black.

local effect can be smeared away by averaging over a large enough area beyond the range of the real effect. Instead of evaluating the kinetic parameters of the Sokoloff model of FDG accumulation for predetermined ROIs we decided to make a pixel by pixel kinetic analysis of the entire dynamic PET investigation. Using the method described in [13], we evaluated the maps of k_1, k_2, k_3, k_4 and the lCMRG, on a pixel by pixel basis. To facilitate comparison, Fig. 7 displays both the FDG accumulation map and the lCMRGl map. The general features of these two images are quite similar, but the pixel by pixel map of the glucose utilization appears more structured. Accordingly, we were interested in investigating the effects of Cavinton treatment on difference maps relating to the pre-treatment and post-treatment lCMRG maps of the same patient (Fig. 8).

The difference maps of glucose accumulation and local cerebral glucose

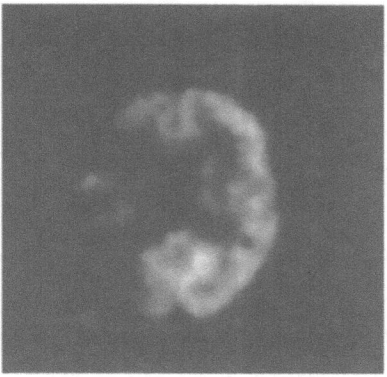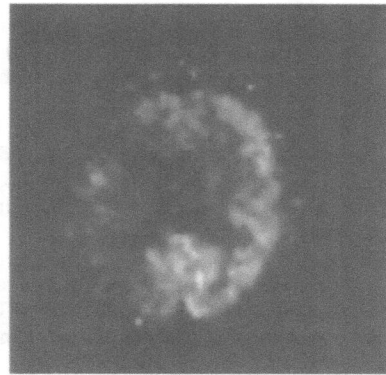

Figure 7. FDG accumulation map (left) and lCMRGl map (right) of the same slice of a 66-year-old female stroke patient (see Fig. 3, right panel) after Cavinton treatment.

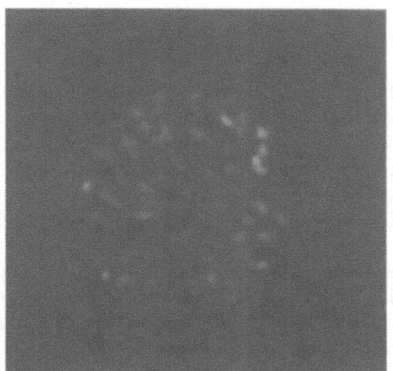

Figure 8. Difference maps of FDG accumulation (left) and lCMRGl (right) of the same slice of a 66-year-old female stroke patient (see Fig. 3 right panel) after Cavinton treatment relative to the appropriate pre-treatment maps.

metabolism display no characteristic structure. The difference map of lCM-RGl seems to be very random. This seems quite obvious since pixels with negative differential counts are also represented as black ones. The random pattern of the lCMRGl map can be regarded as an argument against any localized effect of Cavinton specific either for a given anatomical structure or related to the stroke–affected or peristroke area. It must be noted, however, that pixel by pixel analyses have so far been carried out for only a small number of slices, and this evidence may be strengthend or weakened by results of further analyses of the same type. Such investigations are clearly necessary.

172

References

1. Hadjiev, D. and Yanceva, S. (1976) Rheoencephalographic and psychological studies with ethyl apovincaminate in cerebral vascular insufficiency, *Arzneimittelforsch* **26**, 1947–1950.
2. Szobor, A. and Klein, M. (1976) Ethyl apovincaminate therapy in neurovascular disease, *Arzneimittelforsch* **26**, 1984–1989.
3. Kiss, B., Lapis, E., Pálosi, E., Groó, D. and Szporni, L (1982) In *Protection of Tissue against Hypoxia*, edited by Wanquier, A., Borgers, M. and Amery, W.K. Elsevier, Amsterdam.
4. Biro, K., Kárpáti, E. and L., Szporny (1976) Protective activity of ethyl apovincaminate on ischemic anoxia of brain, *Arzneimittelforsch* **26**, 1918–1920.
5. Bencsath, P., Debreczeni, L. and Takács, L. (1976) Effect of ethyl apovincaminate on cerebral circulation of dogs under normal conditions and in arterial hypoxia, *Arzneimittelforsch* **26**, 1920–1923.
6. Shibota, M., Kakihan, M. and Nagaoka, A. (1982) The effects of vinpocetine on the brain glucose uptake in mice, *Folla Pharmacologica Japonica* **80**, 221–224.
7. Polgár, M., Vereczkei, L. and Nyáry, I. (1985) Pharmacokinetics of vinpocetine and its metabolite, apovincaminic acid, in plasma and cerebrospinal fluid after intravenous infusion, *J. Pharmac. Biomed. Anal* **3**, 131–139.
8. World Medical Association (1964) Declaration of Helsinki, Code of Ethics, *Br. Med.J.*
9. Code of Federal Regulations (1989) *Protection of Human Subjects* Title 45, Part 46.
10. Greitz, T., Bergström, M., Boëthius, J., Kingsley, D. and Ribbe, T. (1980) Head fixation system for integration of radiodiagnostic and therapeutic procedures, *Neuroradiology* **19**, 1–6.
11. Bergström, L., Boëthius, J., Ericson, L., Greitz, T., Ribbet, T. and Widén, L. (1981) Head fixation device for reproducable position alignment in transmission CT and positron emission tomography, *Journal of Computer Assisted Tomography* **5**, 136–141.
12. Sokoloff, L., Reivich, M., Kennedy, C., Des Rosiers, M.H., Patlak, C.S., Pettigrew, K.D., Sakurada, O. and Shinohara, M. (1977) The (C-14) deoxyglucose method for the measurements of local cerebral glucose utilization: theory, procedure and normal values in the conscious and anaesthetized albino rat, *J. Neurochem.* **28**, 897–916.
13. Balkay, L., Molnár, T., Boros, I., Lehel, Sz. and Galambos, T. (1996) Quantification of FDG uptake using kinetic models, In *Positron Emission Tomography: A Critical Assessment of Recent Trends*, edited by Gulyás, B. and Müller-Gärtner, H.W., Kluwer Academic Publisher, Dordrecht.
14. Emri, M., Márián, T., Kövér, G., Berényi, E., and Ésik, O. (1996) Registration: a powerful tool to combine information by different imaging modalities, In *Positron Emission Tomography: A Critical Assessment of Recent Trends*, edited by Gulyás, B. and Müller-Gärtner, H.W., Kluwer Academic Publisher, Dordrecht.

WHOLE BODY SCANNING

Magnus Dahlbom
Division of Nuclear Medicine
Department of Molecular and Medical Pharmacology
UCLA School of Medicine
Los Angeles, CA 90095
USA

Abstract

Whole body PET scanning has become an important addition to the traditional single organ procedures in clinical PET. Although the image quality is in most cases adequate for diagnostic purposes, there are two significant limitation in this imaging technique. First, because of the relatively short acquisition time, the total number of counts is limited, thus, the images suffers from relatively high statistical noise which may reduce lesion detectability. Second, due to the limited time most patients can tolerate to be immobilized, transmission scans for attenuation correction are not routinely performed. Consequently, the reconstructed images contain image artifacts and are furthermore non-quantitative, which prevents monitoring of tumor response during or following therapy. In this paper strategies for improving S/N ratio and quantitation in whole body PET imaging are discussed.

Introduction

In conventional PET procedures, the organ to be studied (e.g., brain, heart) is positioned in the center of the field-of-view (FOV) of the scanner. Following the tracer injection, data are acquired either in a sequence of dynamic frames or the integrated activity uptake is acquired in a single static frame, all dependent on the particular physiological process to be studied. In whole body PET (WB-PET), where the goal is to image the entire or a large portion of the body, the images need to be acquired in multiple segments or axial bed positions due to the limited axial FOV [1, 2]. From these data sets one can form both planar rectilinear projection images and tomographic cross sectional images. From the volume formed by the transaxial images one can extract coronal and sagittal cross sectional images.

The major limitation in whole body imaging is the limited time available to acquire the entire study, since most patients cannot tolerate a total imaging time exceeding

B. Gulyás and H.W. Müller-Gärtner (eds.),
Positron Emission Tomography: A Critical Assessment of Recent Trends, 173–181.
© 1998 *Kluwer Academic Publishers.*

60-90 minutes. For a system with an axial FOV of 15 cm to cover a distance of 150 cm, the acquisition time per position is therefore limited to 6-9 minutes. This is in contrast to the 30-40 minute acquisition time typically used in static brain and cardiac imaging. Consequently, the WB-PET data will contain significantly less number of counts/plane than single organ studies, and will inherently contain more statistical noise. Due to the time constraints, transmission scans for attenuation correction are not routinely performed, since this would require the patient to be immobilized for another 60-90 minute in the scanner.

Another concern in WB-PET related to the overall acquisition time is the tracer re-distribution. Since the images are acquired in multiple segments, it is necessary that the tracer distribution does not significantly change between the time of the first and last bed positions. FDG, for instance, has shown to have a very slow washout following a 30-45 min uptake time. The radioactive decay, may also be of concern for tracers labeled with isotopes of shorter half life such as ^{11}C.

Signal-to-noise ratio

Due to the relatively low number of counts acquired/plane in WB-PET, the reconstructed images will have a low signal-to-noise ratio (S/N). Thus, the main goal in improving the image quality in WB-PET is to improve the S/N in the emission scan itself. One strategy is to improve the overall system sensitivity, which can be accomplished by increasing the number of acquired coincidence plane combinations. In older generation PET systems, coincidences were only acquired in direct and cross planes, where the direct planes were made up of coincidences between detectors within one detector ring, and cross planes of coincidences between detectors in two adjacent rings [3, 4]. A gain in detection efficiency in these systems can be accomplished by adding 2nd order cross plane combinations to the direct planes [5]. For the ECAT 931 [6], this method improves the overall detection efficiency by approximately 30%. The drawback is that the addition of the 2nd order cross plane degrades the axial spatial resolution at tangential offsets in the FOV, due to the geometric divergence and the septa shadowing [5, 7].

A more dramatic increase in sensitivity is achieved by operating the scanner in 3D [8-10]. In this mode the interplane septa, used in the conventional 2D mode to reduce inter-plane scatter, are removed and every possible coincidence plane combination in the system are acquired. It has been shown that the sensitivity can be increased by up to a factor of 7 by operating in 3D compared to 2D. However, since the septa are removed, there is also a significant increase in scatter, which reduces image contrast. Due to the increased acceptance angle of each detector, there is also dramatic increase in random events. Although these events are corrected for, they are a source of increased statistical noise in the net true rate and also a major source of dead time in the system [11, 12]. In order to reduce the randoms and the dead time in the system when operated in 3D mode, it is necessary to reduce the injected activity, which in turn results in a reduction in the total number of counts acquired. For WB-PET, the S/N in 3D may therefore not be significantly higher than in 2D.

Axial sampling

Another strategy to improve the S/N in WB-PET is to modify the axial sampling scheme. The conventional way of acquiring the whole body scan is to move the patient bed in steps equal to the axial FOV [1]. To assemble the set of transaxial images into a volumetric data set, it is necessary to normalize the data to ensure a axial uniformity. This normalization will not only equalize the efficiency of all planes by scaling the counts in low efficiency planes (i.e., direct planes), but will also scale up the noise in the low efficiency planes. When reorienting the data into longitudinal cross sections (i.e., coronal and sagittal cross sections), the noise distribution is not uniform but appears as a "zebra" pattern [5]. This problem can be overcome by using a continuous axial sampling scheme, where each axial data point is measured by each detector plane. Using this scheme, a uniform axial sensitivity is achieved and there is no need for axial normalization. Furthermore, there is no scaling that would introduce non-uniformities in the noise pattern, and the end result is an improvement in S/N [5].

The continuous sampling mode is mainly beneficial in systems where the variations in the axial sensitivity is large, such as in older systems where the cross plane efficiency is almost twice the efficiency of the direct planes. In the latest generation of PET systems, the axial plane efficiency is fairly uniform in the center of the axial FOV, because of the use of 2nd and higher order cross plane combinations [13-15]. The benefit of the continuos axial sampling would therefore be less for these systems. However, these systems do have a rapid roll-off at the ends of the axial FOV, where the efficiency at the edge can be as low as 1/8 compared to the center of the FOV. In order to avoid the scaling noise in these low efficient planes and to compensate for the efficiency roll-off, whole body scans on these system are acquired by overlapping the end-planes at the different axial positions [11].

Image reconstruction

The S/N in WB-PET can also be improved by optimizing the method of image reconstruction. The conventional way of reconstructing images in PET is to use the filtered backprojection (FBP) algorithm [16]. The advantages of this algorithm is that it is fast and, for a given reconstruction filter, has a unique solution. The drawback is that FBP amplifies noise and it is therefore necessary to apply a spatial smoothing filter in order to avoid excessive noise amplification. The amount of smoothing is dependent on the amount of statistical noise in the raw data and is typically a compromise between loss in resolution and improvements in contrast. In WB-PET it is not unusual to smooth the data to final image resolution of 10-12 mm FWHM [1].

An alternate reconstruction algorithm to the filtered backprojection is the maximum likelihood reconstruction method [17, 18]. This method is an iterative reconstruction method that tries to reconstruct an image that best fits the acquired data. In each iteration step, the forward projection of the estimated image is compared to the actual measurement. Adjustments are then made to the image estimate, and the iterations are repeated until a pre-defined stopping criterion is fulfilled (e.g., mean square errors are minimized). The attractive feature of EM algorithm is its noise reducing properties. The drawback is that, in comparison to the filtered backprojection method, it is slow since it may take a large number of iterations for the algorithm to converge. However, with the continued increase in CPU power and also improvements in the algorithm itself [19], the iterative image reconstruction may be a realistic alternative to filtered backprojection [20].

Attenuation correction

In conventional PET imaging, a transmission scan is typically acquired prior to the injection of the radioactive tracer. This scan is used to create an attenuation correction for the self absorption of the photons in the body. If the emission data are not corrected for attenuation, the reconstructed resulting images will contain artifacts, which are produced by the image reconstruction. For instance, structures located deep in the body will appear less active that more superficial ones. Structures with low attenuation, such as the lungs, will have an artificially high uptake than surrounding soft tissue. It is therefore important to correct for attenuation, in order to obtain images that accurately represent the true activity distribution. The application of the attenuation correction does also amplify the noise in the emission data, especially where the attenuation is the highest and the number of emission counts is the lowest [21].

MEASURED ATTENUATION CORRECTION

The transmission scan is performed by placing a number of ring sources, or more recently, rotating rod sources around the patient body [22, 23]. Taking the ratio of a reference or blank scan (i.e., a scan without the patient in the FOV) and the transmission scan, the amount of attenuation is directly measured. In order to get a good estimate of the attenuation and to avoid contamination of the emission images with excessive statistical noise, the length of the transmission scan is typically not shorter than 20 min. To apply the same technique in whole body PET, where a transmission scan is needed for each axial bed position, would translate into an unacceptably long acquisition time. For this reason, most sites perform whole body PET scans without attenuation correction. Thus, the resulting images contain attenuation artifacts and these has to be considered by the diagnosing physician while evaluating the images.

Due to the problem of the attenuation artifacts in the whole body images, it has been investigated whether short transmission scans can be acquired prior to the injection of the tracer for attenuation correction [24]. In these studies, the length of the transmission scan is kept to approximately the same length as the emission scan (i.e., less than

1 hr). Due to the short scan time, the noise levels in these transmission scans are very high and it is therefore necessary apply noise reducing processing techniques to the data.

SPATIAL SMOOTHING

A method for reducing noise in conventional transmission scan is to apply spatial smoothing filters to the data [21]. For short transmission scans the same strategy can be used where the smoothing filters are adjusted (i.e., widened) to compensate for the higher noise levels. The drawback of using "smoother" filters is the distortion that is introduced at interfaces where there is a significant change in attenuation coefficient [24, 25].

IMAGE SEGMENTATION

An alternate approach to the spatial smoothing is to apply an image segmentation technique to the transmission [26, 27]. In this technique, transmission image are reconstructed from the natural logarithm of the attenuation correction data. The resulting images are maps of the attenuation coefficient distribution. In the image segmentation step, the continuous distribution of attenuation coefficients are grouped into a few discrete values, representing different tissue types (i.e., soft tissue, lung tissue, bone, etc). From the segmented attenuation map, one can calculate the appropriate attenuation for each line of response in the system by forward projection. The segmentation of the transmission images effectively eliminates all noise in the attenuation correction. However, it makes the assumption that the attenuation coefficient is uniform within each tissue type, which may not be true in the lungs or the abdomen. Several investigators have therefore proposed refinements to the image segmentation method, where less assumptions are made about the attenuation coefficients and a the segmented images are based on a probabilities derived from the actual measurement [24, 28, 29].

ITERATIVE IMAGE RECONSTRUCTION

Another approach to reduce the noise contamination from the transmission scan is to reconstruct the transmission images using an iterative reconstruction algorithm [30]. From the reconstructed transmission images, the attenuation would then be estimated by forward projection. The advantage of this method over the image segmentation method is that it does not make any assumptions about the attenuation coefficients or the distribution.

POST-INJECTION TRANSMISSION SCAN

With the introduction of rod sources for transmission scans in newer PET system, it is possible to acquire transmission scan after the activity has been injected into the patient [23, 31]. Since photons originating from the patient and the transmission rods sources have the same energy they are indistinguishable by the detectors in the system. However, at any given position, photons originating from the rotating rods will only produce valid coincidences between certain detector pairs. By knowing the angular position of the rod source, and dynamically enabling valid detector pairs (i.e., rod windowing), the transmission events can be separated from the emission events. If the activity in the rod sources is high enough, the contamination of emission events in the transmission window is minimal and will not seriously affect the attenuation data. Like the pre-injection scan, it is necessary to perform some kind of noise suppression of the post-injection data. If an image segmentation is applied to the data, the effects of possible emission contamination can be adjusted for by using pre-determined attenuation coefficients for the different tissue types.

The main advantage of this method is that the patient only has to stay in the scanner during time of the emission and the transmission scan. However, since two scans are performed, the total scan time is in most cases impractical

SIMULTANEOUS EMISSION AND TRANSMISSION SCAN

A very attractive feature of the transmission rod sources is the possibility of actually performing simultaneous emission and transmission scans [32]. In the rod-transmission scan, described above, the events from the transmission sources are separated from the emission events by only accepting events within the rod window. If all events that fall outside the rod window are stored separately, these would then form the emission scan. Thus, it is in theory possible to simultaneous acquire emission and transmission data. In practice it is necessary to apply some corrections to the raw data in order to fully separate the emission and transmission components. First, it is necessary to correct for the spillover of transmission events into the emission window, and vice versa. Second, it is necessary to reduce the activity in the rod sources in order to reduce the contamination of random events in the emission window produced by the rod sources [32]. Although there is little spillover of true transmission events into the emission window is small, most of the random events produced by the rod sources will be detected in the emission window. If the activity in the rod sources is high enough, the random event rate may be close or even exceed the true emission event rate. Even though the randoms are subtracted from the prompt events (trues + randoms), the correction increases the noise in the net true count rate. It is therefore necessary to reduce the activity in the rod sources, in order to minimize the noise contamination from randoms. The drawback of reducing the rod sources activity is that the appropriate activity for simultaneous emission and transmission scan is to low for conventional and post-injection transmission scans. This means that two sets of rod sources are needed to accommodate both acquisition modes. Furthermore, current PET system designs do not provide a means for automatic source exchange, thus, the sources have to be manually exchanged for the different study types, which potentially may cause a problem of high radiation exposure to personnel.

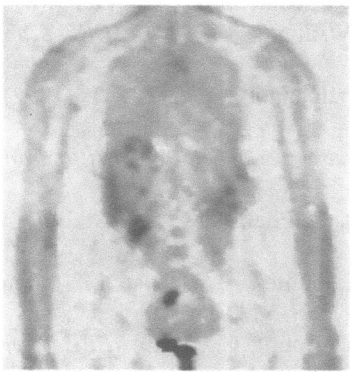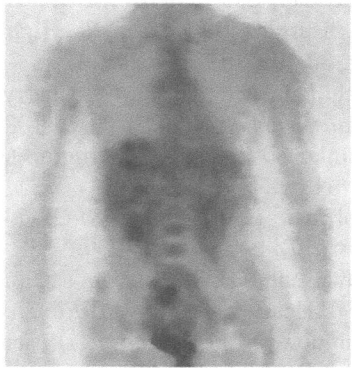

Figure 1. Comparison of a whole body PET scan without (left) and with (right) correction for attenuation. The attenuation correction was acquired simultaneously as the emission scan using the rod-windowing technique. Both images were reconstructed using an iterative algorithm. (Images courtesy of Dr. S.R. Meikle, Royal Prince Alfred Hospital, Sydney)

An example of a whole body study acquired using the simultaneous emission and transmission scan technique is shown in figure 1. The two images show a coronal cross section without (left) and with correction for attenuation (right). In the corrected image, the attenuation correction was created by segmenting the resulting transmission images to reduce noise, as discussed above. Both images were reconstructed using an iterative image reconstruction technique.

Summary

Due to the low number of counts acquired per image plane in whole body PET, the resulting images have relatively high noise levels. It is therefore important that all processing steps of the WB-PET data are optimized to avoid losses in S/N. Also the acquisition protocols of WB-PET data should be optimized to minimize losses in S/N. The lack of routine attenuation correction is a limitation in WB-PET, since it prevents quantitation. However, with careful optimization of short transmission scan protocols (pre-,post- or simultaneous) and data processing, it is possible to provide quantitative WB-PET.

References:

1. Dahlbom, M., Hoffman, E. J., Hoh, C. K., Schiepers, C., Rosenqvist, G., Hawkins, R. A. & Phelps, M. E. (1992) *J Nucl Med* **33,** 1191-1199.
2. Guerrero, T., Hoffman, E. J., Dahlbom, M., Cutler, P. D., Hawkins, R. A. & Phelps, M. E. (1990) *IEEE Trans Nucl Sci* **NS-37,** 676-680.
3. Eriksson, L., Bohm, C., Kesselberg, M., Blomqvist, G., Litton, J., Widen, L., Bergstrom, M., Ericson, K. & Greitz, T. (1982) *IEEE Trans Nucl Sci* **NS-29,** 539-543.

180

4. Hoffman, E. J., Phelps, M. E., Huang, S. C., Kuhl, D. E., Crabtree, M., Burke, M., Burgiss, S., Keyser, R., Highfill, R. & Williams, C. (1981) *IEEE Trans Nucl Sci* **28,** 99-103.

5. Dahlbom, M., Yu, D.-C., Cherry, S., Chatziioannou, A. & EJ, H. (1992) *IEEE Trans Nucl Sci* **NS-39,** 1079-1083.

6. Spinks, T. J., Jones, T., Gilardi, M. C. & Heather, J. D. (1988) *IEEE Trans Nucl Sci* **NS-35,** 721-725.

7. Hoffman, E. J., Huang, S.-C., Plummer, D. & Phelps, M. E. (1982) *J Comput Assist Tomogr* **6,** 987-999.

8. Townsend, D. W., Spinks, T., Jones, T., Geissbühler, A., Defrise, M., Gilardi, M. C. & Heather, J. (1989) *IEEE Trans Nucl Sci* **NS-36,** 1056-1065.

9. Dahlbom, M., Eriksson, L., Rosenqvist, G. & Bohm, C. (1989) *IEEE Trans Nucl Sci* **NS-36,** 1066-1071.

10. Cherry, S. R., Dahlbom, M. & Hoffman, E. J. (1991) *J Comput Assist Tomogr* **15,** 655-68.

11. Dahlbom, M., Cutler, P., Digby, W., Luk, P. & Reed, J. (1994) *IEEE Trans Nucl Sci* **NS-41,** 1571-1576.

12. Eriksson, L., Wienhard, K. & Dahlbom, M. (1994) *IEEE Trans Nucl Sci* **NS-41,** 1566-1570.

13. Wienhard, K., Eriksson, L., Grootoonk, S., Casey, M., Pietrzyk, U. & Heiss, W. D. (1992) *J Comput Assist Tomogr* **16,** 804-13.

14. Wienhard, K., Dahlbom, M., Eriksson, L., Michel, C., Bruckbauer, T., Pietrzyk, U. & Heiss, W. D. (1994) *J Comput Assist Tomogr* **18,** 110-8.

15. DeGrado, T. R., Turkington, T. G., Williams, J. J., Stearns, C. W., Hoffman, J. M. & Coleman, R. E. (1994) *J Nucl Med* **35,** 1398-406.

16. Shepp, L. A. & Logan, B. F. (1974) *IEEE Trans Nucl Sci* **NS-21,** 21-43.

17. Shepp, L. A. & Vardi, Y. (1982) **MI-1,** 113-122.

18. Lange, K. & Carson, R. (1984) *J Comput Assist Tomogr* **8,** 306-316.

19. Hudson, H. M. & Larkin, R. S. (1994) *IEEE Trans Med Imag* **MI-13,** 601-609.

20. Meikle, S. R., Hutton, B. F., Bailey, D. L., Hooper, P. K. & Fulham, M. J. (1994) *Phys Med Biol* **39,** 1689-1704.

21. Dahlbom, M. & Hoffman, E. J. (1987) *IEEE Trans Nucl Sci* **NS-34,** 288-293.

22. Huang, S. C., Hoffman, E. J., Phelps, M. E. & Kuhl, D. E. (1979) *J Comput Assist Tomogr* **3,** 804-814.

23. Ranger, N. T., Thompson, C. J. & Evans, A. C. (1989) *J Nucl Med* **30,** 1056-68.

24. Meikle, S. R., Dahlbom, M. & Cherry, S. R. (1993) *J Nucl Med* **34,** 143-50.

25. Chatziioannou, A. & Dahlbom, M. (1996) *IEEE Trans Nucl Sci* **NS-43,** 290-294.

26. Huang, S. C., Carson, R. E., Phelps, M. E., Hoffman, E. J., Schelbert, H. R. & Kuhl, D. E. (1981) *J Nucl Med* **22,** 627-637.

27. Xu, E. Z., Mullani, N., Gould, K. L. & Anderson, W. L. (1991) *J Nucl Med* **32,** 161-165.

28. Xu, M., Luk, W. K., Cutler, P. D. & Digby, W. M. (1994) *IEEE Trans. Nucl. Sci.* **NS-41,** 1532-1537.

29. Chatziioannou, A.-X. F. (1996) in *Measurements and calculations towards quantitative whole body PET imaging.* University of California, Los Angeles.
30. Mumcuoglu, E. U., Leahy, R. M. & Cherry, S. R. (1996) *Phys Med Biol* **41**, 1777-1807.
31. Carson, R. E., Daube-Witherspoon, M. E. & Green, M. V. (1988) *J Nucl Med* **29**, 1558-67.
32. Meikle, S., Bailey, D., Hooper, P., Eberl, S., Hutton, B., Jones, W., Fulton, R. & Fulham, M. (1995) *J Nucl Med* **36**, 1680-1688.

Part three:
PET in oncology

POSITRON EMISSION TOMOGRAPHY IN ONCOLOGY

In vivo measurement of protein synthesis in tumors

A.M.J. PAANS, J. PRUIM, A. VAN WAARDE,
A.T.M. WILLEMSEN, W. VAALBURG
PET-Center, Groningen University Hospital
Groningen, the Netherlands

1. Introduction

Nowadays a variety of two- and three-dimensional medical imaging modalities is available. The different modalities can be divided into two groups: one giving anatomical and one giving functional or metabolic information. Images made by the first group display tissue density (ultrasound), tissue absorption of X-rays (Computerized Tomography, CT) or physical parameters with respect to the proton density or proton relaxation times (Magnetic Resonance Imaging, MRI). Images from the second group represent functional information such as perfusion, metabolism or receptor density. Single Photon Emission Computerized Tomography (SPECT) and especially Positron Emission Tomography (PET) are generating this functional information by using a whole spectrum of radioactive compounds (radiopharmaceuticals). By measuring the 3D-spatial and temporal distributions of the radiopharmaceutical PET is able to quantify biochemistry in vivo. PET combines a high spatial resolution with a very high sensitivity and is able to quantify the measured radioactivity in absolute terms, i.e. in Becquerel per pixel. Examples of quantitative in vivo measurements are the glucose consumption, blood flow and protein synthesis rate (PSR). Glycolysis and protein synthesis is of broad interest since energy metabolism and proteins are involved in cellular homeostasis, enzyme function, growth and development, plasticity and the response to drugs. In pathological states (e.g. brain tumors, inborn errors of metabolism, neurodegeneration) the metabolism may be altered.

PET uses radioactive nuclides of the elements carbon, nitrogen and oxygen, while fluorine is often used as a replacement for a hydrogen atom in a molecule. Using the former nuclides, molecules can be synthetized which are chemically identical to endogenous organic compounds or to registered drugs. Since the employed radionuclides have short half-lives (^{11}C = 20 min, ^{13}N = 10 min, ^{15}O = 2 min and ^{18}F = 110 min), these radionuclides must be produced in house. For this

185

B. Gulyás and H.W. Müller-Gärtner (eds.),
Positron Emission Tomography: A Critical Assessment of Recent Trends, 185–198.
© 1998 *Kluwer Academic Publishers.*

purpose, a small compact cyclotron is the accelerator of choice. Since a transformation of chemical elements takes place, e.g. ^{11}C is produced by the proton irradiation of nitrogen, the required nuclide can be obtained with a very high specific activity, i.e. the ratio between the activity of the nuclide (e.g. ^{11}C) and the mass of the stable isotope (e.g. ^{12}C). If this high specific activity is maintained during the synthesis of the radiopharmaceutical, then the injected dose represents only a very small mass (1-100 nmol). Combined with the high sensitivity of PET cameras this enables the measurement of drugs at picomolar concentrations. Thus, in vivo studies of receptors, pharmacokinetics and pharmacodynamics are possible and no pharmacological effects will occur when the radiopharmaceutical is administered to human volunteers.

Since a PET camera can only measure the concentration of the radionuclide, and not the radiochemical form in which this nuclide is present in the body, it is essential that the amount of radiolabeled metabolites is limited and that they can be quantified separately. For this reason not only the choice of the radiopharmaceutical is important but also the position of the radionuclide within the molecule is of eminent importance.

In PET the detection method explicitly takes advantage of the decay process of a positron emitting radionuclide. The emitted positron "annihilates" ($E = mc^2$), after having been slowed down by collisions within the tissue, and fusion with a normal electron. Since the physical conservation laws in this annihilation process have to be obeyed, two co-linear γ-quanta of 511 keV each, are generated. A PET camera measures these two γ-quanta using a coincidence technique. Due to this coincidence technique a quantitative measurement in terms of Bq/pixel of the radionuclide concentration is possible. This quantitative nature of PET-data is unique in comparison to data generated with other nuclear imaging techniques. Because of the co-linear properties of the generated γ-radiation a PET camera is build in a cylindrical architecture. Multiple rings of detectors (nowadays up to 32) make it possible to simultaneously image 63 planes in a 3D-data acquisition and reconstruction mode. The axial field of view amounts to 15-20 cm in the present cameras. A spatial resolution of close to 4 mm FWHM is achievable with an extremely high sensitivity in the 3D-mode, as compared to more conventional nuclear imaging techniques. There is a fundamental limitation to the spatial resolution achievable with a PET camera. Since the point of annihilation is always different from the position at the moment of decay of the radionuclide due to the kinetic energy of the positron, sub-millimeter spatial resolution as in CT and MRI images is not possible with PET. The energy of a positron is specific for every radionuclide and covers a mean range of 1 mm or more.

2. PET in Oncology

Metabolism of many malignant tumors is high compared to normal tissue. This fact is used in PET to visualize tumors and to analyse and quantify their metabolism. At present methodology is available to quantify oxygen consumpties of tumortissue, glycolysis, protein synthesis and DNA synthesis. Besides these parameters PET allows the study of tumor hypoxia, estrogen and progesterone receptors. These receptor studies may be of clinical value to determine the hormonal sensitivity of e.g. breast tumors. In comparison to SPECT, CT and MRI in oncology, PET has a very high sensitivity and specificity to detect tumor tissue. This is of clinical value in staging and for the detection of residual tumor tissue or recurrence. Other interesting clinical PET applications are: to grade tumors, to access the heterogeneity, to differentiate tumors and to access the effect of chemo- and radiotherapy with the hypoxia markers and receptorligands the radiosensitivity and hormonal sensitivity can be studied.

3. Protein syntheses in tumors

Amino acids labeled with positron emitting radionuclides are worthwhile radiopharmaceuticals for the study of the amino acid metabolism in tumors in vivo. An amino acid suitable to measure protein synthesis rates has to meet several criteria. In order of importance these are:
- high incorporation and retention in proteins
- insignificant metabolic pathways to non-proteins
- supply of amino acid from plasma only
- rapid clearance from plasma to a low steady state
- a small pool of free amino acid in tissue and a rapid turnover of the tissue precursor pool
- or PSR measurements in brain tissue, high blood-brain barrier transport
- availability of a convenient procedure for labelling

4. Choice of amino acid

Quantification of the regional rate of glucose metabolism, using the tracer 2-[^{18}F]-fluoro-2-deoxy-D-glucose (FDG), has also found wide clinical application. Although FDG-PET has been very successful over the years, it is evident that it is not the optimal tracer in all cases. Especially in neuro-oncology, sensitivity and specificity of tumor detection are hampered in regions with a high basal metabolism such as the brain [1,2,3,4] while gliomas may exhibit both elevated and reduced rates of glucose metabolism. Furthermore, inflammatory cells have been reported to exhibit increased glucose metabolism [5,6,7].

To circumvent these problems, the in vivo study of other metabolic processes may be fruitful, such as the rates of protein, RNA or DNA synthesis [8,9]. The local protein synthesis rate in tissue can be quantified in vivo with positron emitting amino acids, or amino acid analogues, in combination with PET and adequate kinetic models. If infused tracers are used, only the protein synthesis for exogenous amino acids can be quantified [8]. One should realize that the incorporation of an amino acid is the result of the synthesis of many different proteins. The first step in protein synthesis is the enzymatic conversion of an L-amino acid, after being transported into the cell, into amino-acyl-tRNA. L-amino acids also serve as substrates for other metabolic pathways as transamination and decarboxylation. D-amino acids are not incorporated into proteins. The desired properties for a radiopharmaceutical to quantify the PSR in vivo comprise a whole spectrum of qualities like a high blood-brain-barrier (BBB) permeability, rapid clearance and turnover, and favourable metabolic pathways as discussed by Phelps et al [9]. If the brain uptake index of L-amino acids is considered, phenylalanine, leucine or tyrosine, with respective relative brain uptake indices of 55%, 54% and 50%, seem to be the amino acids of choice [10].

To translate the quantitative PET data into a PSR, tracer kinetic models are required. Such models have already been proposed at an early date [11] for the L-isomer of a carboxylic labeled amino acid. For the evaluation of such a model knowledge on the kinetics of non-protein metabolites is essential [12]. To determine the protein synthesis rate (PSR), various labeled amino acids have been studied. These include L-[methyl-^{11}C]methionine [13,14,15,16,17,18], L-[1-^{11}C]methionine [19,20], L-[1-^{11}C]leucine [21,22,23], L-[1-^{11}C]tyrosine [24,25,26,27,28], ^{11}C-valine [29,30], ^{18}F-fluorotyrosine [31,32] and [11]C- or ^{18}F-labeled phenylalanine [33,34]. Metabolic pathways of the tracer, kinetic modeling and the clinical applicability have all been studied to some extent.

It was shown by Ishiwata [35] that the accumulation of the labeled amino acids L-[methyl-^{11}C]methionine, and L-2-[^{18}F]fluorotyrosine represent amino acid transport whereas L-[1-^{11}C]leucine represents protein synthesis rate. This indicates that only the in vivo uptake of carboxyl labeled amino acids, such as L-[1-^{11}C]leucine and L-[1-^{11}C]tyrosine, represents the PSR. Non-carboxyl labeled amino acids may still be of clinical value in tumor detection, as they appear to be reliable markers of tumor tissue. However, for tumor staging, patient follow-up or therapy evaluation, a quantitative analysis such as the PSR, may be essential. Furthermore, although the radiochemistry of methyl-labeled amino acids [8] is considered to be relatively simple as compared to the synthesis of carboxyl labeled amino acids, our present synthesis strategy for L-[1-^{11}C]tyrosine shows that a fast and simple synthesis is feasible and facilitates the use of carboxyl labeled amino acids in clinical settings. The tissue metabolite activities of 2 to 3% for L-[1-^{11}C]tyrosine compare favorably with similar studies using L-[methyl-^{11}C]methionine or L[1-^{11}C]methionine showing metabolite activity values of 10 to 17% [20]. Apparently, this relates to the different metabolic pathways for tyrosine and methionine. Thus the application of L-[1-^{11}C]tyrosine and L-[1-^{11}C]leucine is superior to the other labeled amino acids mentioned. In our institution the potential of PSR measurement in humans and in animals, using L-[1-^{11}C]tyrosine and PET, has been investigated extensively.

5. Metabolic pathways of L-[1-^{11}C]tyrosine

On the time scale of a PET study ($\pm$ 60 min in the case of ^{11}C-labeled radiopharmaceuticals), the main metabolic pathway of L-[1-^{11}C]tyrosine in tissue is irreversible incorporation of carbon-11 into protein. Minor metabolic pathways are decarboxylation and synthesis of L-[p-hydroxy]-[1-^{11}C]-phenylpyruvic acid, L-[p-hydroxy]-[1-^{11}C]lactic acid and L-[1-^{11}C]DOPA [12]. Further metabolism results in the release of the label as ^{11}CO$_2$. Carbon-11 labeled protein may be found in plasma as it is produced in the liver and released in the circulation. These plasma metabolites can be determined from arterial blood samples. For human tissues, however, only the total radioactivity can be directly established from PET data. Therefore, to assess the metabolites in tissue, metabolic studies with L-[1-^{11}C]tyrosine as well as PSR determination in rats have been preformed [12]. The metabolic studies revealed a significant amount of ^{11}CO$_2$ in plasma although in tissue the amount was negligible. The total amount of activity in the plasma decreases very rapidly. Other labeled non-protein metabolites in tissue were less than 4% of total tissue activity. Non-protein metabolites in plasma may reach 10 % and proteins in plasma reached 85 % leaving less than 10% free tyrosine at 60 minutes. Paans et al. [36] established the PSR in rat rhabdomyosarcoma as 0.4 $\pm$ 0.1 nmol/(g.min) while the faster growing Walker 256 carcinosarcoma showed a PSR of 0.8 $\pm$ 0.2 nmol/(g.min).

6. Material and methods

6.1 SYNTHESIS OF L-[1-^{11}C] TYROSINE

L-[1-^{11}C]tyrosine is synthesized via a microwave induced Bücherer-Strecker synthesis [37]. The synthesis time is 40 minutes, including HPLC purification, with a radiochemical yield of 30-60%. Specific activity is > 40 TBq/mmol. with an average yield of 800 MBq.

6.2 DATA ACQUISITION

The data acquisition protocol for ^{11}C-tyrosine studies in humans is the following. Subjects have fasted for at least 6 hours when they are positioned in the PET camera (Siemens ECAT 951/31, in plane resolution FWHM = 6 mm). Photon attenuation is measured using a retractable ring source. ^{11}C-tyrosine (370 MBq preferably) is injected i.v. over a 1 min period. Dynamic image acquisition starts simultaneously with the injection. A controlled and reproducible injection technique is employed using a programmable infusion pump. Arterial plasma samples are obtained at an adequate frequency. All blood samples are centrifuged. The plasma samples are analyzed for ^{11}CO$_2$ and ^{11}C-proteins. Because of their low prevalence other metabolites are neglected.

6.3 METABOLITE ANALYSIS

The following radioactive metabolites may be formed: proteins, CO_2, L-DOPA, p-hydroxyphenylpyruvic acid and p-hydroxy-phenyllactic acid. Reversed-phase HPLC of arterial samples showed that the contribution of L-DOPA, p-hydroxy-phenylpyruvic acid and p-hydroxy-phenyllactic acid to total radioactivity in human plasma was negligible ($\leq$ 8% during the first 40 min, Willemsen et al 1995). Therefore, the cumbersome and time-consuming analysis of plasma samples by HPLC was skipped in further experiments.

Plasma levels of $^{11}CO_2$ are determined using a modification of the procedure described for ^{11}C-palmitate [38]. Labeled protein in plasma is determined using the method of Ishiwata et al [12]. It has been shown that incorporation of ^{11}C-TYR into serum proteins is initially negligible, but rises from 2% at 20 min to > 18% at 40 min post injection. Formation of $^{11}CO_2$ is already detectable within 0.5 min. The fraction of plasma radioactivity representing $^{11}CO_2$ rapidly rises to a maximum of ca. 25% which is maintained from 15 to 40 min post injection [39].

7. Compartment model

Based on our understanding of the biochemical behaviour of L-[1-^{11}C]tyrosine a number of animal (in-vivo and ex-vivo) and human studies were performed. These measurements were designed to assess the validity of the model assumptions. Furthermore they enabled an estimation of the effect of these model assumptions on the calculated protein synthesis rate. The model assumptions are:
- transport of labeled protein from the tissue under investigation into plasma can be neglected on the time-scale of the PET measurements.
- recycling of amino-acids by proteolysis of labeled proteins can be neglected on the time-scale of the PET measurements.
- the plasma concentration of L-[1-^{11}C]tyrosine is at tracer levels.
- labeled nonprotein metabolites including labeled carbon dioxide in the cell are negligible.

In plasma, labeled plasma metabolites can be detected almost immediately. They consist of $^{11}CO_2$ and (after 20 min) also of labeled proteins released by the liver. Therefore, all data on plasma radioactivity must be corrected for these metabolites. This also shows that the presented model cannot be applied to the liver as this organs releases labeled proteins to the circulation; therefore the first model assumption is not valid.

The resulting kinetic model describing the transport of L-[1-^{11}C]tyrosine and its labeled metabolites is shown in Fig. 1. It consists of five compartments: <1> free tyrosine in plasma <2> free tyrosine in tissue and <3> proteins in tissue. Carbon-11 is released as <4> $^{11}CO_2$ with rate constant k_4, which in turn is transported to the plasma <5> with rate constant k_5. From this model the activity

Plasma   **Tissue**

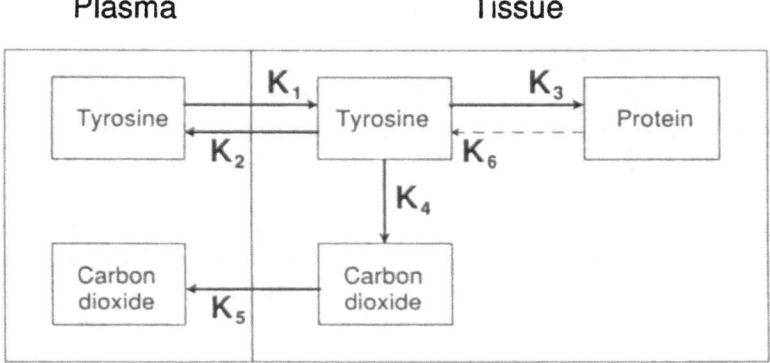

Fig. 1 Compartment model of L-[1-^{11}C]tyrosine, for details see text.

as measured by a PET camera can be expressed as a function of the plasma activity, time and the model parameters [39]. A graphical analysis can also be performed using equation 1.

$$C_i/C_p = V_bC_b/C_p + (1-V_b)k_1(1+k_4/k_5)/(k_2+k_3+k_4) +$$
$$\{(1-V_b)k_1k_3/(k_2+k_3+k_4)\}\{[\int C_p(\tau)d\tau]/C_p(t)\} \qquad \text{Eq. 1}$$

with C_i^* = total radioactivity in tissue, C_p^* = plasma radioactivity, corrected for metabolites, C_b^* = blood radioactivity, V_b = blood volume in the tissue under study, and k_n = the various rate constants. The first term on the right hand side of Eq. 1, V_bC_b/C_p, is considered a pertubation which was calculated to be in the order of 1-2%. The second term is a constant and does not influence the slope. The third term determines the slope of the resulting curve, $(1-V_b)/K$, from which the amino acid incorporation can be calculated. The PSR is given by:

$$\text{PSR} = k_1k_3/(k_2+\phi k_3+k_4) \ C_p = K\epsilon C_p \qquad \text{Eq. 2}$$
with
$$K = k_1k_3/(k_2+k_3+k_4)$$
and
$$\epsilon = (k_2+k_3+k_4)/(k_2+\phi k_3+k_4)$$

where C_p represents the endogenous level of unlabeled tyrosine in plasma.

The correction term ϵ results from the recycling of amino acids from endogeneous (unlabeled) protein. For the brain it was estimated as 1.1 assuming 50% amino acid recycling from data of the brain in volunteers. For brain studies the blood volume is less than 5% in which case the $(1-V_b)$ term can be ignored. For other studies, separate measurements may be required to assess the blood volume (e.g. PET studies using inhaled $C^{15}O$).

It is important to emphasize that the PSR should be understood as the amino acid incorporation rate. Because the transport of amino acids between plasma and precursor pool is faster than the amino acid incorporation rate the effect of recycling on the amino acid incorporation rate can be neglected. Consequently this will have little effect on the calculated amino acid incorporation rates.

The average protein synthesis rate in normal brain tissue of 0.7 nmol/ml.min as found by us compares well with previously published data on both humans and animals [13,22,36,40]. A correction of plasma activity for the presence of metabolites is essential and a graphical Patlak analysis further enhances this method as it has a robustness against metabolites [39].

In conclusion, by acquiring arterial blood samples from which the L-[1-^{11}C]-tyrosine plasma time activity data can be determined, and by acquiring a dynamic PET scan to measure the tissue time activity data, the protein synthesis rate can be determined in vivo in man using a simple graphical analysis [41,42].

8. Applications of L-[1-^{11}C]-tyrosine

Most of the applications of L-[1-^{11}C]tyrosine will be in the field of oncology. However, applications in the field of metabolic diseases, like phenylketonuria where amino acids are implied, may also be important. Oncology research in our PET-Center aims primarily at therapy evaluation, since this makes full use of the quantitative power of PET. However, a number of studies deal with tumor detection and staging making use of the sensitivity of the PET-methodology.

8.1 BRAIN TUMORS

The use of PET in the clinical diagnosis of brain tumors and in therapy evaluation is acknowledged throughout the world. Mostly, ^{18}FDG is used for this purpose. However, the use of this tracer is complicated because of the high uptake of ^{18}FDG in normal brain tissue. In certain differential diagnostic problems, therefore, the use of ^{18}FDG has its limitations. Certainly in the case of brain tumors with a hypo-metabolism of glucose, as is common in low-grade gliomas, a pitfall exists. The awareness of these drawbacks have led many groups to search for alternative tracers, mostly amino acids such as methyl-labeled methionine. As stated, we have focused on the measurement of PSR using L-[1-^{11}C]tyrosine instead. A series of patients has been studied.

The calculated PSR of brain tumors varied greatly with an average of 1.1±0.6 nmol/(ml.min). Average PSR in the contralateral (normal) cortex was 0.7±0.4, thus yielding an average tumor-to-background ratio of 1.9±0.6. No clear relation with the malignancy grade of the tumor could be established in these rather small series [43].

In a special study the response of brain gliomas to radiotherapy was monitored. If patients are referred for radiotherapy after neurosurgery, the response to radiotherapy is usually evaluated by MRI or CT. Since MRI or CT do not discriminate between tumor progression or regression from tumor necrosis, the use of PET has been proven to be a better alternative. In this study we examined patients with inoperable but biopsied brain gliomas, who were scheduled for radiotherapy. PET scanning was performed using L-[1-^{11}C]tyrosine before and after 54-60 Gy external beam radiotherapy. Average PSR values of segmented regions in the PET-images were calculated using the Patlak graphical analysis for L-[1-^{11}C]tyrosine [39]. Prior to radiotherapy all gliomas were clearly visualized with an increased PSR in the tumor region. After radiotherapy all gliomas remained visible with a PSR value significantly higher than the PSR values of normal brain tissue. These PSR results seem to indicate the persistent viabilty of gliomas after conventional external beam radiotherapy. To evaluate the effect of radiotherapy, the tumor volume should also be examined. Since gliomas do not grow as a single mass the partial volume effect, due to the finite spatial resolution of a positron camera, hampers a detailed analysis of the tumor volume and adequate tools have to be found.

8.2 HEAD AND NECK CANCER

In the treatment of patients with squamous cell carcinomas of the upper aero-digestive tract, the assessment of the presence of small cervical lymph node metastases remains difficult. However, the extent of the surgery is determined by the size of the tumor in combination with the presence of metastatic lymph node disease. Although anatomical imaging techniques have improved the staging of the neck, there is still room for further improvement. In a preliminary study in 23 patients with histologically specified primary squamous cell carcinoma a comparison was made between the sensitivity and specificity of tyrosine-PET and FDG-PET on the one hand, and anatomical imaging techniques as MRI and CT on the other hand [44]. The high specificity and predictive values make PET with tyrosine suitable for further evaluation of the already infested neck (stage N1 and higher). On the basis of their high sensitivity and high negative predictive values both PET with FDG and with tyrosine are suitable techniques for confirmation of the negative (stage N0) neck.

8.3 BREAST CANCER

In our institution PET using FDG and L-[1-^{11}C]tyrosine have been compared in patients with primary breast carcinoma. All tumors were visualized with both tracers. FDG uptake was also seen in fibrocystic disease, whereas L-[1-^{11}C]tyrosine-uptake appeared to be more tumor-specific. To investigate whether less complex protocols can be implemented the quantitative data analysis was compared with standard uptake values. The standard uptake value is defined as the ratio between the regional radioactivity concentration and the total mean radioactivity concentration. A high correlation was found between the rate of glucose consumption, the standard uptake value and plasma glucose. Also, a high correlation was observed between the protein synthesis rate and the standard uptake value. Therefore, the standard uptake value technique appears to be a valuable, and less invasive means of quantification of tumor metabolism.

8.4 SOFT TISSUE SARCOMAS AND MELANOMAS

The response to chemotherapy using Hyperthermic Isolated Limb Perfusion (HILP) in patients with primarily unresectable soft tissue sarcomas was studied using ^{18}FDG and L-[1-^{11}C]tyrosine. The chemotherapy has been a combination of Tumor Necrosis Factor-alpha (TNFα) and Melphalan. In the FDG group PET clearly showed a reduction in ^{18}FDG uptake after HILP. However, many patients also showed a 'rim' of activity around the original tumor that could histologically be related to inflammation. This finding confirms reports that the metabolic role of glucose in inflammation obscures the view on the efficacy of therapy using FDG.

In case of the ^{11}C-tyrosine study, all malignant lesions were correctly identified as 'hot spots' in all patients participating in this protocol. In that series two lipomas were visible as 'cold spots', and one Schwannoma as a 'hot spot'. The difference between malignant and benign tumors was significant (p=0.006). After therapy the PSR appeared to distinguish the patients with >60% necrosis from the patients with >90% necrosis and the patients with no viable tumor. So far, presence of a 'hot spot' after perfusion indicated viable tumor tissue in all cases. These findings indicate a potential clinical applicability of ^{11}C-tyrosine for the evaluation of the therapy-response in soft-tissue sarcomas.

A similar protocol is performed in patients with melanomas of the extremity. The data reveal similar results as in patients with soft-tissue sarcomas. Preliminary analysis indicates that PET with L-[1-^{11}C]tyrosine may predict a response to therapy very well, but the technique can probably not differentiate between complete response and microscopic residual tumor. Macroscopic residual 'hot spots' have thus far coincided with viable tumor tissue.

8.5 LUNG CANCER

In the Western world lung cancer is the leading cause of cancer related death in men, while the incidence is increasing in women. Lung cancer is an aggressive tumor with limited therapeutic options. Irrespective of the stage of the disease the overall survival rate is 10-15%. The use of (poly-)chemotherapy, whether or not in combination with external radiotherapy, to achieve curation or to improve quality of life (palliation), will become increasingly important. The potential of L-[1-

[11C]tyrosine for early evaluation of lung cancer therapy is investigated. Both patients with small cell lung cancer (SCLC) and patients with non-small cell lung cancer (NSCLC), all stage III, were studied.

L-[1-[11C]tyrosine-PET was performed before, during and after therapy. Dynamic studies were made with full quantification of the protein synthesis rate (PSR) of the tumor. A contrast enhanced thoracic-CT was performed within 1 week of the PET scan.

All tumors were clearly visible on the first PET scan. Remarkably, SCLC and NSCLC did not differ in their respective baseline PSR-values. During therapy, a significant reduction of the PSR was seen in 8 of 10 patients. Six of these 8 patients showed a further decline of the PSR or were stable in the last PET scan. Two patients showed a small increase of the PSR, but remained well below the baseline value. In all 8 patients CT and clinical course were indicative of a partial remission or stable disease. In contrast, the 2 (of the 10) other patients (1 SCLC, 1 NSCLC) showed a gradual increase of the PSR, which coincided with progressive disease on CT and clinical course. An important finding was also that the images were not affected by radiation-associated pneumonitis [45]. Again these data indicate that PET with L-[1-[11C]tyrosine can be used as an instrument for assessment of early response on cancer therapy. The clinical benefit of this finding should be adressed in forthcoming studies.

9. Conclusion

The amino acid incorporation rate, generally described as protein synthesis rate or PSR, can be assessed in vivo using carboxylic labeled amino-acids such as L-[1-[11C]tyrosine and PET. In animals, labeled tissue metabolites are below 4% of total tissue radioactivity and are therefore neglected in the model. Labeled plasma metabolites on the other hand rise continuously to 50% of total plasma radioactivity at 40 min. After correction of the total plasma radioactivity for the metabolite fraction, a Patlak analysis may be performed to calculate the PSR. A number of applications in the field of oncology were presented. It is concluded that the application of ^{11}C-tyrosine-PET in a clinical oncology is of interest for an increasing number of different tumors.

196

REFERENCES

1. Ishiwata K, Takahashi T, Iwata R, et al. (1992) Tumor diagnosis by PET: potential of seven tracers examined in five experimental tumors including an artificial metastasis model. *Nucl Med Biol* **19**, 611-618.
2. Hawkins RA, Phelps ME, Huang SC. (1986) Effects of temporal sampling, glucose metabolic rates and disruptions of the blood-brain barrier on the FDG model with and without a vascular compartment: Studies in human brain tumors with PET. *J Cereb Blood Flow Metab* **6**, 170-183.
3. Di Chiro G. (1987) Positron emission tomography using [18F]fluorodeoxyglucose in brain tumors, a powerful diagnostic and prognostic tool. *Invest Radiol* **22**, 360-371.
4. Ericson K, Lilja A, Bergstrom M, et al.(1985) Positron emission tomography with ([^{11}C]methyl)-L-methionine, [^{11}C]D-glucose, and [^{68}Ga]EDTA in supratentorial tumors. *J Comput Assist Tomogr* **9**, 683-689.
5. Sasaki M, Ichiya Y, Kuwabara Y. (1990) Ringlike uptake of [^{18}F]FDG in brain abscess: a PET study. *J Comput Assist Tomogr* **14**, 486-487.
6. Kubota R, Yamada S, Kubota K, Ishiwata K, Tamahashi N, Ido T. (1992) Intratumoral distribution of fluorine-18-fluorodeoxyglucose in vivo: High accumulation in macrophages and granulation tissues studied by microautoradiography. *J Nucl Med* **33**, 1972-1980.
7. Wahl RL, Fisher SJ. (1993) A comparison of FDG, L-methionine and thymidine accumulation into experimental infections and reactive lymph nodes. *J Nucl Med* **34**, 104P.
8. Vaalburg W, Coenen HH, Crouzel C, et al. (1992) Amino acids for the measurement of protein synthesis in vivo by PET. *Nucl Med Biol* **19**, 227-237.
9. Phelps ME, Barrio JR, Huang SC, Keen RE, Chugani H, Mazziotta JC. (1984) Criteria for the tracer kinetic measurement of cerebral protein synthesis in humans with positron emission tomography. *Ann Neurol* **15** Suppl, S192-S202.
10. Oldendorf WH and Szabo J. (1976) Brain uptake of amino acids, amines, and hexoses after arterial injection, *Am J Physiol* **221**, 1629-1639.
11. Smith CB, Davidson L, Deibler G, et al (1980). A method for the determination of local rates of protein synthesis in brain. *Trans Am Soc Neurochem* **11**, 94.
12. Ishiwata K, Vaalburg W, Elsinga PH, Paans AMJ, Woldring MG. (1988b) Metabolic studies with L-[1-^{11}C]tyrosine for the investigation of a kinetic model to measure protein synthesis rates with PET. *J Nucl Med* **29**, 524-529.
13. Ericson K, Blomquist G, Bergström M, Eriksson L, Stone-Elander S. (1987) Application of a kinetic model on the methionine accumulation in intracranial tumours studied with positron emission tomography. *Acta Radiol* **28**, 505-509.
14. Lilja A, Lundquist H, Olsson Y. (1989) Positron emission tomography and computed tomography in differential diagnosis between recurrent or residual glioma and treatment-induced brain lesions. *Acta Radiol* **30**, 121-128.
15. Fujiwara T, Matsuzawa T, Kubota K, et al. (1989) Relationship between histologic type of primary lung cancer and carbon-11-L-methionine uptake with positron emission tomography. *J Nucl Med* **30**, 33-37.
16. Ishiwata K, Ido T, Abe Y. (1988a) Tumor uptake studies of S-adenosyl-L-[methyl-^{11}C]methionine and L-[methyl-^{11}C]methionine. *Nucl Med Biol* **15**, 123-126.
17. Planas AM, Prenant C, Mazoyer BM, Comar D, Di Giamberardino L. (1992) Regional cerebral L-[^{11}C-methyl]methionine incorporation into proteins: evidence for methionine recycling in the rat brain. *J Cereb Blood Flow Metab* **12**, 603-612.
18. Lindholm P, Leskinen-Kallio S, Minn H, et al. (1993) Comparison of fluorine-18-fluorodeoxyglucose and carbon-11-methionine in head and neck cancer. *J Nucl Med* **34**, 1711-1716.
19. Bolster JM, Vaalburg W, Elsinga PH, Ishiwata K, Vissering H, Woldring MG. (1986) The preparation of ^{11}C-carboxylic labelled L-methionine for measuring protein synthesis. *J Label Comp Radiopharm* **23**, 1081-1082.

20. Ishiwata K, Vaalburg W, Elsinga PH, Paans AM, Woldring MG. (1988c) Comparison of L-[1-^{11}C]methionine and L-methyl-[^{11}C]methionine for measuring in vivo protein synthesis rates with PET. *J Nucl Med* **29**, 1419-1427.

21. Keen RE, Barrio JR, Huang SC, Hawkins RA, Phelps ME. (1989) In vivo cerebral protein synthesis rates with leucyl-transfer RNA used as a precursor pool: Determination of biochemical parameters to structure tracer kinetic models for positron emission tomography. *J Cereb Blood Flow Metab* **9**, 429-445.

22. Hawkins RA, Huang SC, Barrio JR, et al. (1989) Estimation of local cerebral protein synthesis rates with 1-[^{11}C]leucine and PET: Methods, model, and results in animal and humans. *J Cereb Blood Flow Metab* **9**, 446-460.

23. Barrio JR, Keen RE, Ropchan JR, et al. L-[1-^{11}C]leucine: routine synthesis by enzymatic resolution (1983). *J Nucl Med* **24**, 515-521.

24. Daemen BJ, Zwertbroek R, Elsinga PH, Paans AM, Doorenbos H, Vaalburg W. (1991) PET studies with L-[1-^{11}C]tyrosine, L-[methyl-^{11}C]methionine and ^{18}F-fluorodeoxyglucose in prolactinomas in relation to bromocryptine treatment. *Eur J Nucl Med* **18**, 453-460.

25. Luurtsema G, Medema J, Elsinga PH, Visser GM, Vaalburg W. (1994) Robotic synthesis of L-[1-^{11}C]tyrosine. *Appl Radiat Isot* **45**, 821-828.

26. Halldin C, Schoeps KO, Stone-Elander S, Wiesel FA. (1987) The Bücherer-Strecker synthesis of D- and L-(1-^{11}C)tyrosine and the in vivo study of L-(1-^{11}C)tyrosine in human brain using positron emission tomography. *Eur J Nucl Med* **13**, 288-291.

27. Daemen BJ, Elsinga PH, Paans AMJ, Wieringa AR, Konings ATW, Vaalburg W. (1992) Radiation-induced inhibition of tumor growth as monitored by PET using L-[1-^{11}C]tyrosine and fluorine-18-fluorodeoxyglucose. *J Nucl Med* **33**, 373-379.

28. Go KG, Prenen GH, Paans AMJ, Vaalburg W, Kamman RL, Korf J. (1990) Positron emission tomography study of ^{11}C-acetoacetate uptake in a freezing lesion in cat brain, as correlated with ^{11}C-tyrosine and 18F-fluorodeoxyglucose uptake, and with proton magnetic resonance imaging. *Adv Neurol* **52**, 525-528.

29. Mitsuka S, Diksic M, Takada A, Yamamoto YL. (1992) Influence of the tumor mass on the valine rate constants and on valine incorporation into proteins in an experimental brain tumor model. *Neurochem Int* **20**, 537-551.

30. Kirikae M, Diksic M, Yamamoto YL. (1989) Quantitative measurements of regional glucose utilization and rate of valine incorporation into proteins by double-tracer autoradiography in the rat brain tumor model. *J Cereb Blood Flow Metab* **9**, 87-95.

31. Coenen HH, Kling P, Stöcklin G. (1989) Cerebral metabolism of L-[2-^{18}F]fluorotyrosine a new PET tracer of protein synthesis. *J Nucl Med* **30**, 1367-1372.

32. Wienhard K, Herholz K, Coenen HH, et al. (1991) Increased amino acid transport into brain tumors measured by PET of L-(2-^{18}F)fluorotyrosine. *J Nucl Med* **32**, 1338-1346.

33. Bodsch W, Coenen HH, Stöcklin G, Takahashi K, Hossmann KA. (1988) Biochemical andautoradiographic study of cerebral protein synthesis with [^{18}F] and [^{11}C]fluorophenylalanine. *J Neurochem* **50**, 979-983.

34. Van Langevelde A, Van der Molen HD, Journee-de Korver JG, Paans AMJ, Pauwels EK, Vaalburg W. (1988) Potential radiopharmaceuticals for the detection of ocular melanoma. Part III. A study with ^{14}C and ^{11}C labelled tyrosine and dihydroxyphenylalanine. *Eur J Nucl Med* **14**, 382-387.

35. Ishiwata K, Kubota K, Murakami M, et al. (1993) Re-evaluation of amino acid PET studies: Can the protein synthesis rates in brain and tumor tissues be measured in vivo? *J Nucl Med* **34**, 1936-1943.

36. Paans AMJ, Elsinga PH, Vaalburg W. (1993) Carbon-11 labeled tyrosine as a probe for modelling the protein synthesis rate, in Mazoyer BM, Heiss WD, Comar D (eds): PET Studies on Amino Acid Metabolism and Protein Synthesis. Dordrecht, Kluwer Academic: 183-196.

37. Paans AMJ and Willemsen ATM eds., Annual Report PET-Center Groningen University Hospital,1993, 3.

38. Fox KAA, Abendschein DR, Ambos HD, Sobel BE, Bergmann SR. (1985) Efflux of metabolized

198

and nonmetabolized fatty acid from canine myocardium. Implications for quantifying myocardial metabolism tomographically. *Circ Res* **57**, 232-243.

39. Willemsen ATM, Van Waarde A, Paans AMJ, Pruim J, Luurtsema G, Go KG, Vaalburg W. (1995) In vivo protein synthesis rate determination in primary or recurrent brain tumors using L-[1-^{11}C]-tyrosine and PET. *J Nucl Med* **36**, 411-419.

40. O'Tuama LA, Guilarte TR, Douglass KH, et al. (1988) Assessment of [^{11}C]-L-methionine transport into the human brain. *J Cereb Blood Flow Metab* **8**, 341-345.

41. Patlak CS, Blasberg RG, Fenstermacher JD. (1983) Graphical evaluation of blood-to-brain transfer constants from multiple-time uptake data. *J Cereb Blood Flow Metab* **3**, 1-7.

42. Patlak CS, Blasberg RG. (1985) Graphical evaluation of blood-to-brain transfer constants from multiple-time uptake data. Generalizations. *J Cereb Blood Flow Metab* **5**, 584-590.

43. Pruim J, Willemsen ATM, Molenaar WM, Van Waarde A, Paans AMJ, Heesters MAAM, Go KG, Visser GM, Franssen EJF, Vaalburg W. (1995) Brain tumors: L-[1-^{11}C]-tyrosine PET for visualization and quantification of protein synthesis rate. *Radiology* **197**, 221-226.

44. Braams JW, Pruim J, Nikkels PGJ, Roodenburg JLN, Vaalburg W, Vermey A. (1996) Nodal spread of squamous cell carcinoma of the oral cavity detected with PET-tyrosine, MRI and CT. *J Nucl Med* **37**, 897-901.

45. Appel M, Pruim J, van der Mark ThW, Willemsen ATM, Smit EF, Paans AMJ, Koëter GH, Vaalburg W. (1996) Protein synthesis rate: Evaluation of cytotoxic therapy in lung cancer using L-[1-^{11}C]-tyrosine and positron emission tomography in Metabolic imaging of cancer (Paans AMJ, Pruim J, Franssen EJF, Vaalburg W eds.) PET Center Groningen, ISBN 90-802859-1-9, 22-24.

DIAGNOSIS, DIFFERENTIAL DIAGNOSIS, AND FOLLOW-UP OF TUMOURS BY MEANS OF FDG PET

O. ÉSIK[1], B. GULYÁS[2] and L. TRÓN[3]
[1]*Department of Radiotherapy, National Institute of Oncology, Budapest H-1122 Budapest, Ráth György u. 7-9, Hungary*
[2]*Laboratory for Brain Research and PET, The Nobel Institute of Neurophysiology, Karolinska Institute, Stockholm, Sweden*
[3]*PET Centre, Debrecen University of Medicine, Debrecen, Hungary*

Abstract
Ninety-two [18]fluoro-deoxy-D-glucose (FDG) PET examinations performed on oncological patients yielded conclusive results for primary diagnostics, staging, restaging and therapeutic monitoring in 39%, 85%, 82% and 93% of the cases, respectively. The 77% overall incidence of conclusive results indicates that the FDG PET technique is a powerful clinical tool. It should be emphasized that this result was attained through careful histopathological investigations, high-resolution anatomic imaging procedures and an active follow-up policy. Additional information was obtained concerning the proliferative capacity of the tumour cells and the extent of the viable tumorous tissue. The metabolically-based determination of the tumour margins by PET and CT-MRI-PET image fusion helped in delineation of the gross tumour volume and its subclinical extent for three-dimensional (3D) radiotherapy planning ensuring a more adequate dose coverage. The value of FDG PET investigations in oncology may be summarized as follows: limited in the primary diagnostics, strong in staging and restaging, very strong in therapeutic monitoring, and investigational in estimations of the proliferative capacity of tumorous tissue and 3D radiation treatment planning.

Key words: PET, FDG, tumour, oncology, diagnostics, staging, restaging, therapeutic monitoring, proliferative capacity, image fusion, 3D, radiotherapy, planning

1. Introduction
Between November 1994 and September 1996, 92 [18]fluoro-deoxy-D-glucose (FDG) PET examinations were performed on patients from the Department of Radiotherapy, National Institute of Oncology, Budapest. The clinical questions (Table 1) concerned the primary diagnostics (18 cases), initial staging (13 cases), restaging (33 cases) and therapeutic monitoring (28 cases).

B. Gulyás and H.W. Müller-Gärtner (eds.),
Positron Emission Tomography: A Critical Assessment of Recent Trends, 199–212.
© 1998 *Kluwer Academic Publishers.*

Table 1. Indications and results of FDG PET examinations

Indication	Number of cases	Conclusive results Number	%
Primary diagnostics	18	7	39
Initial staging	13	11	85
Restaging	33	27	82
Therapeutic monitoring	28	26	93
Total	92	71	77

2. Primary diagnostics

During the primary diagnostics, the majority of the questions involved a _search for unknown primary tumours_, which is a real problem in about 5% of all cancer cases. The diagnostic efficiency of PET is around 30% in these cases [12].

The most frequently discovered occult primaries are lung, breast, colorectal, ovarian and pancreatic cancers, and malignant melanoma, but several types of tumours are easily missed [7, 11, 12]. The lack of visualization of occult primary tumours might be explained by their small size or low proliferative capacity (the latter being in contrast with the clonal selection of the actively metabolizing cells of high proliferative capacity and the considerable FDG uptake in metastases primarily arousing medical interest). Physiological tracer uptake by the surrounding tissues may also prevent identification of an occult primary.

Differentiated thyroid cancer (papillary or medullary) may be missed due to the physiological uptake in the thyroid bed (about 30% of the individuals display some uptake in this region), and the typically low proliferative capacity of this type of tumour resulting in a low FDG uptake not distinct from that of the neighbouring tissues. Several per cent of the primary nasopharyngeal poorly differentiated carcinomas (lymphoepitheliomas) are undetectable by any diagnostic method, due to the small size of the cancer. The localization of occult primary tumours in the oropharyngeal region (the hypopharynx, and especially the sinus piriformis; the base of the tongue, etc.) may also be disturbed by the physiological tracer uptake of the muscles (swallowing or talking) and tonsils (inflammation). The diagnosis of prostatic and gynecological cancer is hampered by the proximity of the urinary bladder, which accumulates radioactivity of the excreted tracer.

We have investigated 15 patients for occult primary tumours following histological verification of a metastasis, with inconclusive results of immunopheno-type determination, conventional diagnostic imaging and laboratory tests (Table 2). The diagnostic yield was 27% (4 cases), 3 of which were lung cancers, with a pronounced FDG uptake accompanied in 1 case (Fig. 1) by a low proliferative capacity determined by DNA analysis. The typically high FDG uptake of lung tumours is a well-known phenomenon; it gives an easy-to-recognize contrast to the relatively low

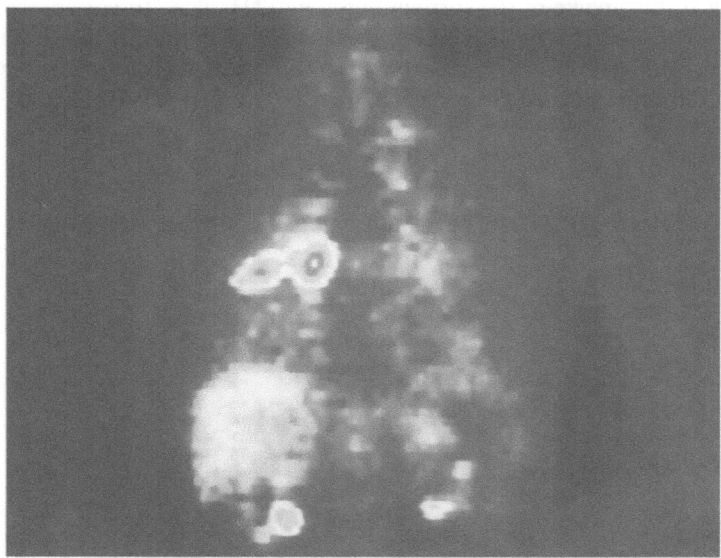

Figure 1. A FDG PET search for an unknown primary tumour resulted in the discovery of an occult primary cancer in the right lung with right-sided hilar lymph node metastases in 1995. In 1990, 1991 and 1995, solitary cerebral (twice) and thyroid metastases (once) were surgically removed. All diagnostic investigations had inconclusive results before the FDG PET examination. The poorly differentiated carcinoma metastases phenotypically expressed some squamous characteristics, and in each case a low proliferative capacity with an S phase ratio of 2%, 8% and 8%. (The primary tumour was not investigated, as the patient was lost to follow-up due to a cerebral vascular event.)

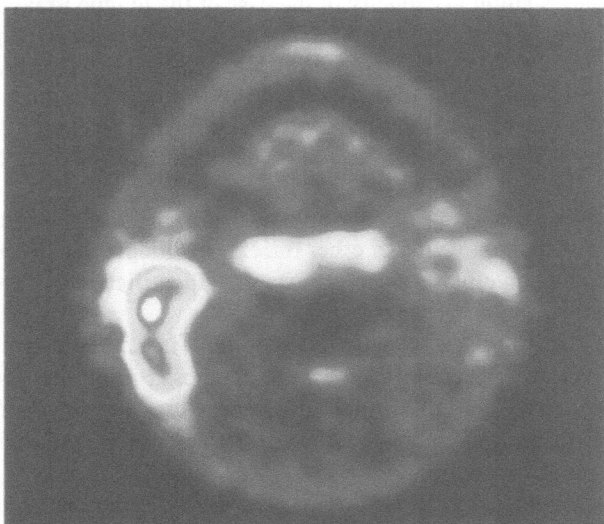

Figure 2. FDG PET examination was unable to detect the primary cancer of a patient with extensive metastatization (right cervical and axillary, mediastinal, and even intraabdominal foci), and it was later discovered at autopsy as a right-sided base of the tongue tumour. Retrospective evaluation of the PET images revealed a slight asymmetry of the base of the tongue in favour of the right side. The metastatic right-sided cervical lymph nodes are clearly visualized.

physiological tracer uptake of the surrounding lung [21]. The 4th case proved to be a colon cancer. The 27% diagnostic efficiency of FDG PET in this setting is comparable to the results of others [12], and it must be taken into account that these investigations were carried out on a selected patient population for which the efficiency of all other conventional diagnostic procedures had proved to be 0%.

Table 2. Results of FDG PET examinations for primary diagnostics

Indication	Number of cases	Conclusive results	
		Number	%
Search for unknown primary	15	4	27
Search for multiple primary/ cancer family syndrome	2	2	100
Solitary pulmonary nodule	1	1	100
Total	18	7	39

As for the indeterminate results of PET examinations, we were not able to find the primary poorly differentiated carcinoma in the nasopharyngeal region. One occult primary with extensive multiple metastases at the initial presentation was discovered at autopsy on the right side of the base of the tongue. Retrospective evaluation of the PET images revealed only a slight asymmetry of the base of the tongue in favour of the right side (Fig. 2). It is worthy of note that PET failed to detect the suspected primary breast tumours in 3 women. Axillary lymph node (LN) metastases, however, were biopsy-proven in all three cases, while mammography, ultrasonography and dynamic MRI investigation were negative, and the primary tumours have not yet manifested during a follow-up period of 6-26 months. It seems that in such situations FDG PET examination has only limited potency.

The low yield in the search for an unknown primary could be improved by an active patient follow-up, application of other tracers (methionine or thymidine), a more detailed immunophenotyping, receptor status identification, and molecular pathological investigations.

The _search for multiple primaries_ is important in cases of typical multifocal tumours, such as colorectal or lung cancers, as diagnosed multiplicity fundamentally alters the treatment strategy. This situation often emerges in _genetically determined cancer family syndromes,_ generally appearing with an autosomal dominant inheritance and expressing multiple primaries.

We have investigated Lynch syndrome (Table 2), i.e. hereditary non-polyposis colorectal cancer [16]. The genetic background of the disease involves errors in DNA repair genes. This may result in an excess of synchronous and metachronous colonic cancers (Lynch syndrome type I), or a similar colonic phenotype accompanied by a high risk of endometrial carcinoma (Lynch syndrome type II). Both types of

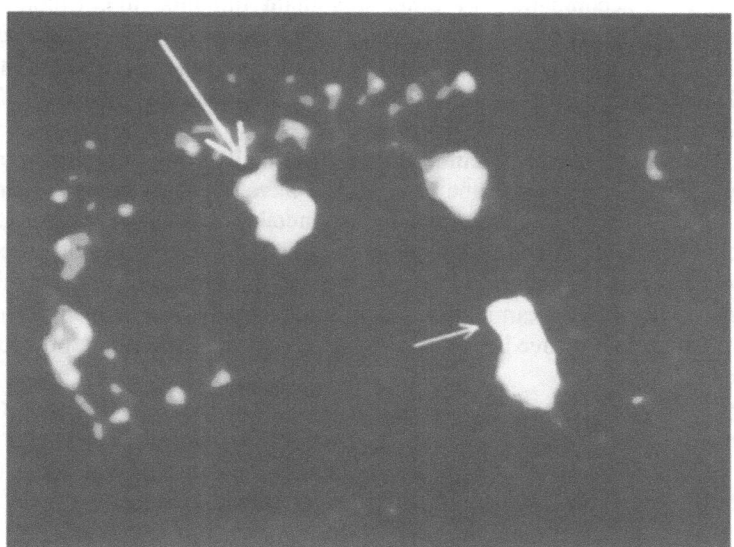

Figure 3. Lynch syndrome type II case. The patient had an endometrial adenocarcinoma in 1972 (operated and irradiated), and later a phenotypically different coecum adenocarcinoma in 1987 (operated). In March 1996, she had a local recurrence in the right pelvis, as proved by CT and operative findings (only biopsy was performed). Before palliative local irradiation, an FDG PET investigation was indicated to decide about other primaries or manifestations. In May 1996, PET confirmed not only the local recurrence, but also lymph node metastases in the hilus of the liver (big arrow), later proved by CT. The small arrow indicates the left pyelon.

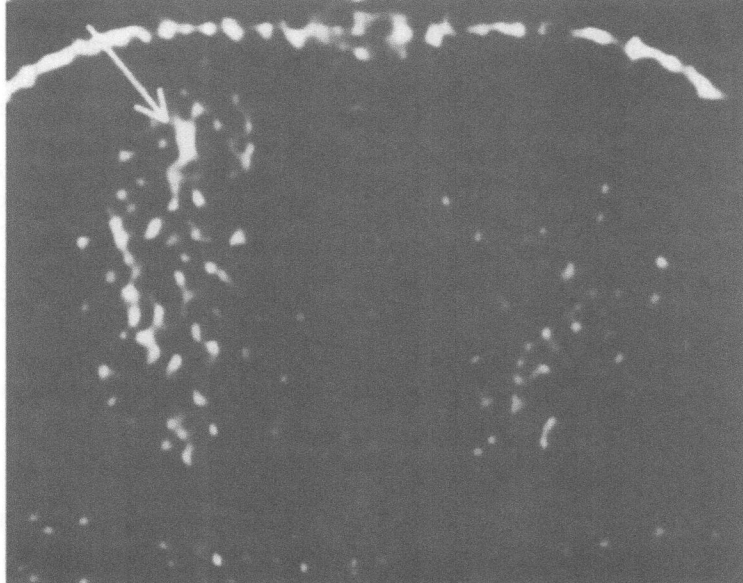

Figure 4. Pulmonary metastases (inhomogeneous FDG uptake and arrow) were proved by means of PET in a patient with a high-grade, Clark V malignant melanoma. The metastases were not detected initially by CT.

syndromes may be accompanied by other malignant tumours, too. Colon tumours generally develop before age 44 years, typically on the right side of the colon, and their natural history is relatively benign, although they are poorly differentiated p53 negative adenocarcinomas (with diploid DNA index, in contrast with the frequent aneuploidy of other colon adenocarcinomas).

Three generations of a large family living in a small Hungarian city are afflicted by histology-proven Lynch syndrome type II. We have investigated two members of this family so far. The PET scan of one family member did not reveal a pathologic accumulation of FDG following surgical intervention for his first right-sided colon tumour. The other case was investigated after a local recurrence of her second cancer was detected. The PET examination identified a previously unsuspected new involvement, which later proved to be LN metastases in the liver hilus, probably from her second cancer (Fig. 3).

Further investigations are needed to evaluate the significance of PET examination in the search for tumours in genetically determined cancer family syndromes. FDG PET scans may be of great help theoretically in identifying tumours of affected family members following the discovery of the proband.

FDG PET has an important role in the *differential diagnosis of solitary pulmonary nodules* [9]. Their estimated incidence is considerable: around 50 cases per 100 000 people. The diagnostic accuracy of conventional imaging methods (CT/MRI) is 60-70%, and that of transthoracal needle aspiration is 80-95% (with a pneumothorax incidence of 20% and a chest tube incidence of 5%). Although surgical intervention can result in a final diagnosis, it involves a mortality rate between 3 and 7%. Since 60% of the resected nodules are benign, a non-invasive method may effectively help the medical decision-making in these cases.

FDG PET examination is indicated in cases of solitary pulmonary nodules if benignity is strongly suspected, i.e. a long anamnesis, a tumour diameter less than 3 cm, and regular contours, with a calcification not characteristic (spiculated) of malignancy. The diagnostic accuracy of PET for differentiating a benign from a malignant solitary nodule is 92% [9]. Although a PET scan can help in the mediastinal nodal staging of tumour-positive cases, its real value is the avoidance of unnecessary thoracotomy for patients with benign nodules. The causes of false-negative results are small, low-grade tumours (typically adenocarcinomas). False-positive results may be caused by inflammation (e.g. histoplamosis, aspergillosis, tuberculosis, sarcoidosis, etc.).

We have indicated FDG PET for 1 case (Table 2) of a solitary pulmonary nodule, which had been seen unchanged for several years on CT, and the examination confirmed the benignity of the nodule.

3. Intitial staging

Initial staging of cancers with a propensity for early lymphogenous and haematogenous dissemination can be performed with high accuracy by PET. Its role is especially important in malignant lymphoma, melanoma, lung-, colorectal, breast, head and neck cancers, and sarcomas.

FDG PET gives valuable information concerning the _nodal staging_, for at least two reasons. Firstly, conventional diagnostic procedures judge basically LNs on the basis of the size, which has major drawbacks: LNs of several millimeters could contain microscopic tumorous foci, and LNs of 2 cm or more may be reactive. (Ultrasonography and MRI can give some structural information of help in differentiation between malignancy and benignity.) Secondly, in several troublesome localizations the assessment of LNs is hindered due to the surrounding anatomical structures (vessels, nerves, muscles, etc.), in supraclavicular, mediastinal, internal mammary, deep cervical, retrocrural, etc. lymphatic regions. Providing a metabolic map, PET can distinguish effectively between benign and malignant LNs. However, it must be borne in mind that false negativity can occur in cases of microscopic tumorous foci, and inflammatory reactions can result in false positivity.

As an illustrative example, PET would be of great importance in replacing surgical axillary staging in breast cancer by this non-invasive method. Other diagnostic imaging methods, such as lymphoscintigraphy, ultrasonography and CT have rather low sensitivity and specificity (e.g. the sensitivity and specificity of CT are only 50% and 75%, respectively [17]), and consequently they are not routinely applied procedures. The commonly performed surgical staging causes arm and breast oedema with incidences of 2-23% and 17%, respectively. The sensitivity and specificity of multicentre FDG PET studies for axillary nodal staging each proved to be 96% [1, 11, 20, 30, 31]. This means that only a small percentage of patients would be expected to be misclassified by PET as false-negative, but PET can spare all negative patients from axillary LN dissection. Furthermore, by means of PET internal mammary LN staging would also be feasible, which is otherwise practically out of question for treatment decision-making (because of the considerable difficulties in the involved diagnostics), although 20% of breast cancer patients harbour metastases in this region.

As for _extranodal staging_, PET is especially important in the following localizations: bone marrow (focal involvements of malignant lymphoma and melanoma), pulmonary (Fig. 4) and brain metastases (melanoma), suprarenal glandular involvement (lung cancer), and hepatic metastasis (colorectal cancer).

We indicated FDG PET examinations for initial staging in 13 patients, with an incidence of conclusive results of 85% (11 cases).

4. Restaging

PET is of great significance in restaging for recurrence [2], including differential diagnosis between a viable tumour (Fig. 5) and late sequelae of surgery, radiotherapy [28] or chemotherapy.

In marker-producing carcinomas (such as thyroid cancer), PET is extremely useful in cases of an elevated serum tumour marker level accompanied by negative results of traditional imaging methods, or in cases of a normal serum tumour marker level accompanied by positive findings by conventional diagnostic imaging [27].

We indicated FDG PET examinations for restaging in 33 patients, with an incidence of conclusive results of 82% (27 cases).

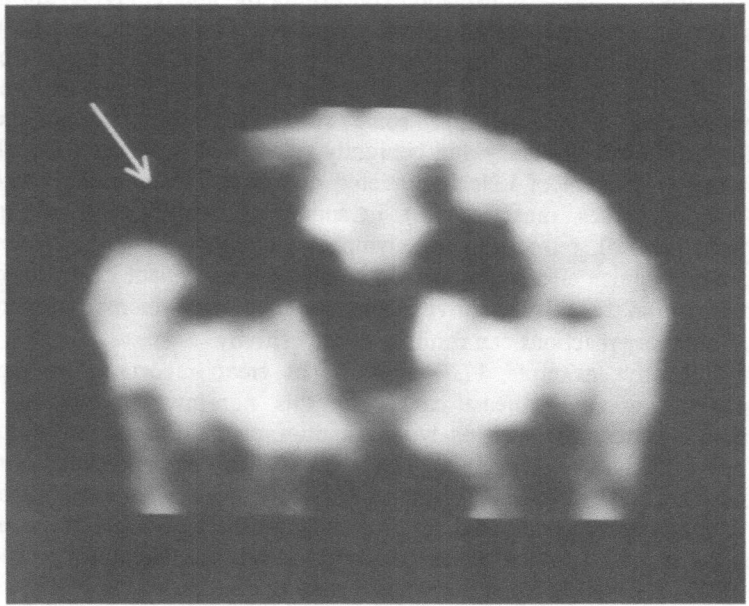

Figure 5. The patient was operated on and irradiated for an astrocytoma grade 3 (S phase ratio of 33%). One month later, an MRI investigation resulted in the questionable diagnosis of a recurrent/residual tumour or postirradiation reactions. The subsequent FDG PET study ruled out the presence of viable tumour cells (postirradiation sequelae were to be seen, arrow), and 2 years following the completion of primary therapy, the patient is disease-free.

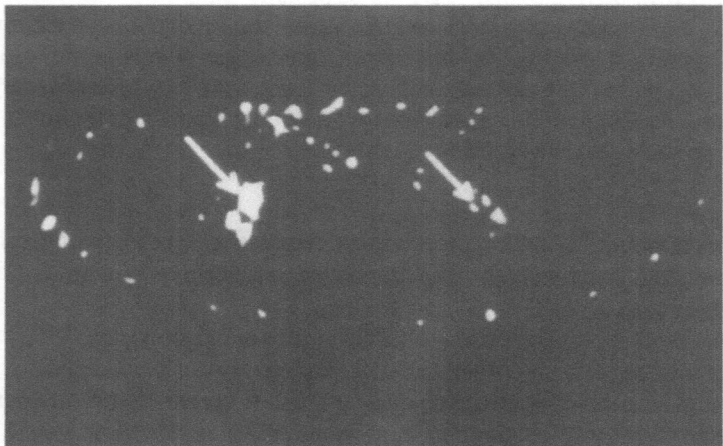

Figure 6. Two months after completion of primary radiotherapy/chemotherapy for Hodgkin disease, FDG PET examination revealed considerable FDG tracer uptake in CT-proved normal-sized supraclavicular lymph nodes (arrows), indicating only partial remission.

5. Therapeutic monitoring

PET is the best-known method for estimating the early (within 24 hours) and late (following several years) responses to radiotherapy, chemotherapy (Fig. 6) and hormone therapy, as it is able to differentiate between viable tumour tissue and treatment-related changes. This is especially important, as the assessment of treatment-altered tissues is impossible by using other imaging methods.

We indicated FDG PET examinations for therapeutic monitoring in 28 patients, (including 4 cases before/after bone marrow transplantation), with an incidence of conclusive results of 93% (26 cases).

6. Inconculsive results on staging, restaging and therapeutic monitoring

Among the causes of inconclusive results relating initially staged, restaged and monitored patients are: small tumour size, low proliferative capacity of the cancer, cystic tumour, inflammation and short time interval between the completion of radiotherapy/chemotherapy and the PET investigation (Table 3). These difficulties and problems could be partially avoided by using other tracers (e.g. methylspiperone or raclopride for pituitary adenoma, methionine or thymidine for other tumours, etc.), simultaneously applied other, mainly anatomic imaging methods (e.g. CT/MRI for cystic tumours), and an adequate waiting period between the completion of therapy and the PET investigation. A precise evaluation of the end-result of radiotherapy/chemotherapy/hormone therapy generally needs several weeks, and sometimes even months or years. Finally, it should be emphasized again that PET is not able to detect microscopic tumorous involvement.

Table 3. Causes of inconclusive FDG PET results in staging, restaging and therapeutic monitoring (n=10)

Small tumour	
- microadenoma (pituitary)	3
Tumours with low proliferative capacity	
- low-grade astrocytoma	3
- papillary thyroid cancer (solitary bony metastasis)	1
Cystic tumour	
- LN metastases (papillary thyroid cancer)	1
Inflammation	
- pelvic tumour recurrence	1
Short time interval between completion of systemic therapy and PET investigation	
- Hodgkin disease	1

7. Estimation of proliferative capacity

Besides qualitative oncological diagnostic information, FDG PET examinations provide additional data, most significant of which is the proliferative capacity of the transformed cells in tumorous tissue. The non-invasive estimation of this parameter relies on the fact that metabolically active, rapidly proliferating tumours consume more glucose, and thus FDG accumulation means a significant contribution to individual prognosis estimation.

The proliferative capacity and the genetic instability of the tumours can also be characterized by DNA parameters (S phase fraction, ploidy, number of subpopulation and polyploid fraction) using flow cytometric or digital image analysis of samples with appropriately stained DNA. Independent PET and DNA analysis data provide a reliable prognosis of the clinical course of the disease. Numerous publications report on a direct correlation between distinct DNA parameters (ploidy, S phase fraction) and the FDG/methionine uptake of tumorous tissues [4, 10, 13, 18, 19, 24].

We investigated the correlation between FDG uptake, the DNA parameters and the clinical course of the disease in medullary thyroid cancer [8] and LN metastases of different histological types [15]. The FDG metabolic rate in primary tumours and metastatic LNs was determined semiquantitatively, and DNA analysis was performed with a digital image analyser [29]. A direct correlation (Fig. 7) was found between the FDG uptake, the S phase ratio and the clinical course of the disease.

8. Three-dimensional (3D) radiation treatment planning

The goal of radiotherapy is to eradicate a tumour without causing damage to healthy tissue. About one-third of all malignancies are localized at the initial diagnosis. In these cases a carefully conducted local treatment modality, e.g. radiotherapy, could result in long-term disease control. For a precise delineation of the anatomical structures (gross tumour volume and its subclinical extent, radiosensitive critical organs, body contour, etc.), an obvious need emerged for the direct use of digitalized volumetric tomographic (CT/MRI/SPECT/PET) information in 3D radiotherapy planning software.

The margins of body contour and critical organs are well visible, and their delineation is easily handled by anatomic imaging methods (especially CT). However, defining the exact edge of the tumorous tissue and its extent (including nodal dissemination) is generally difficult with CT and MRI, as the margins of the tumours may be blurred or invisible, and the LN size affords only a rough estimation of its involvement. This may cause the worst possible error in radiotherapy planning, as improper definition of the tumorous tissue can never be corrected during the treatment course, and these errors fundamentally influence the survival chance of patients with localized tumours.

At present, the 3D radiotherapy planning is based on CT and MRI information. It means a tremendous improvement, as replacing the conventional treatment planning by CT-based planning, and the CT-based planning by CT- and MRI-based planning involved an overall change of the treatment plans in one-third and 53% of the cases, respectively, mostly because of the correction of inadequate tumour coverage [5, 26]. However, these anatomic imaging methods are not able to adequately delineate

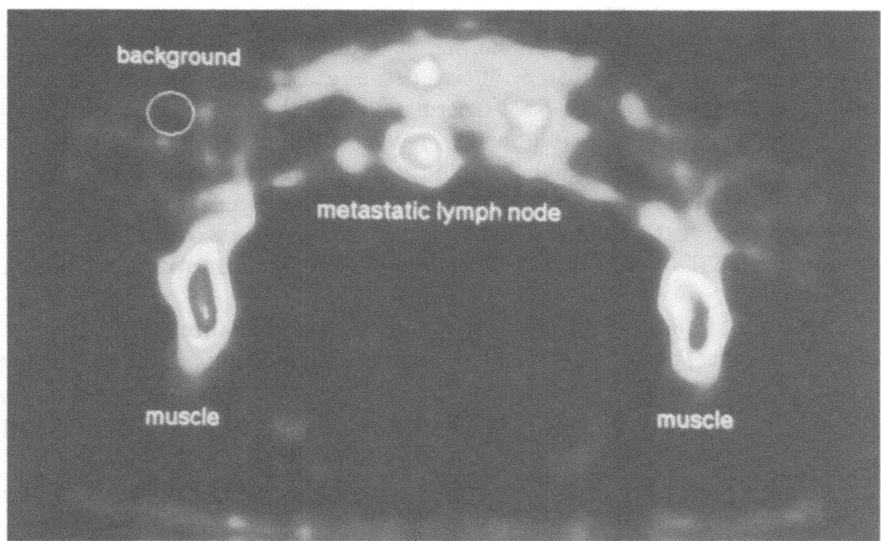

Figure 7. Before a second surgical intervention, an enlarged residual paratracheal lymph node was seen on the CT of a papillary thyroid cancer patient. FDG PET investigation indicated low tracer uptake in the node (tumour-to-background ratio of 2:1, ROI: 1.3 cm^2), and histopathological examination confirmed the low proliferative capacity of the tumour, as DNA analysis resulted in a low S phase ratio (7%).

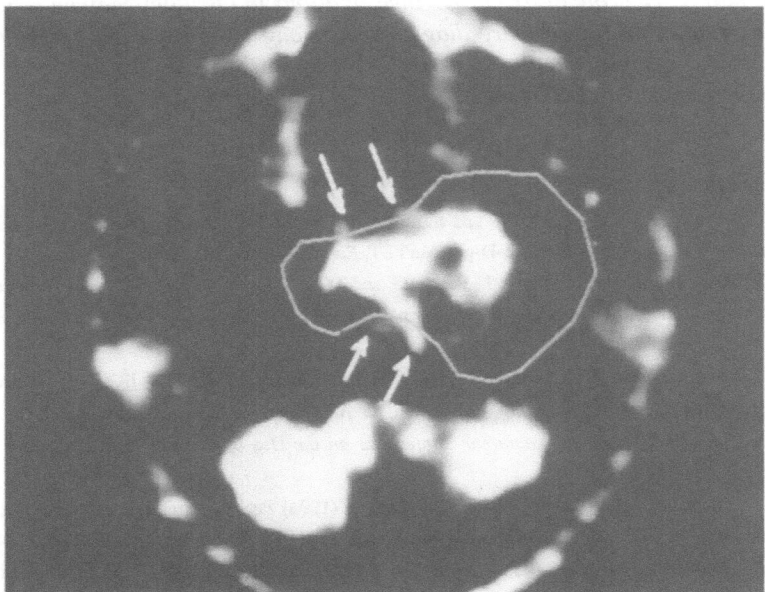

Figure 8. Following palliative surgical removal of a low-grade base of the skull tumour (adenoid cystic carcinoma with an S phase ratio of 5%), the planning target volume (PTV) for the residual cancer was determined by using digital CT information within the VIRTUOS/TOMAS 3D radiotherapy planning software. (MRI images were also inspected and "eyeball registered".) Following a CT-MRI-PET digital image fusion, the original contours of the CT-based PTV were superimposed onto the FDG PET images. A geographical miss is demonstrated: part of the residual tumorous tissue is beyond the PTV (arrows).

the tumorous tissue. As PET with its physiological tracers is very sensitive to small physiological differences between normal and diseased tissue (e.g. differentiation between viable tumorous or therapy-altered tissues can be perfomed), it is expected to contribute substantially to the exact determination of the tumour margins. Similarly SPECT, with its metabolic information gained through the use of radiolabelled antibodies, tumour markers and metabolites, thallium, etc., may also help in the definition of the tumorous tissue. For the best results, these funcional nuclear imaging studies need to be registered with the anatomic imaging studies, due to the low spatial resolution (SPECT 5-7 mm, and PET 3.5-5 mm) in comparison with the 1 mm of CT and MRI.

With appropriate transformation, we made the digitalized PET information acceptable for the TOMAS/VIRTUOS 3D radiotherapy planning software package [3, 23] developed by the Schlegel group in Heidelberg, a prerequisite of which is digital image fusion [14, 22, 25, 32, 33]. When the first fusion was performed for this purpose [6], a severe geographical miss was detected in the tumour volume coverage (Fig. 8). Thus, it can be predicted that some radiotherapy-treated patients may benefit from future anatomic/functional image fusions.

Altogether, clinically conclusive results were achieved in 77% of our cases by means of FDG PET, indicating that this is a powerful clinical tool. It must be emphasized that this result was attained through careful histopathological investigations, high-resolution anatomic imaging procedures and an active clinical follow-up policy. The role of FDG PET investigations in oncology may be summarized as follows: limited in the primary diagnostics, strong in staging and restaging, very strong in therapeutic monitoring, and investigational in estimations of the proliferative capacity of tumorous tissue and 3D radiation treatment planning.

9. References

1. Adler, L.P., Crowe, J.P., Al-Kaisi, N.K. and Sunshine, J. (1993) Evaluation of breast masses and axillary lymph nodes with [F-18] 2-deoxy-2-fluoro-D-glucose PET, *Radiology* **187**, 743–750.

2. Ágoston, P., Ésik, O., Gulyás, B., Boros, I., Forrai, G. and Trón, L. (1996) Whole-body PET in the search for unknown primary tumours and restaging of cancer patients after treatment, *Radiother. Oncol.* **40**, Suppl (1) S141.

3. Bendl, R., Pross, J., Hoess, A., Keller, M.A., Preiser, K. and Schlegel, W. (1994) VIRTUOS - A program for VIRTUal RadiOtherapy Simulation and verification, in A.R. Hounsell, J.M. Wilkinson and P.C. Williams (eds.), *Proceedings of the 11th International Conference on the Use of Computers in Radiation Therapy*, Manchester, pp. 226–227.

4. Crowe, J.P. Jr., Adler, L.P., Shenk, R.R. and Sunshine, J. (1994) Positron emission tomography and breast masses: comparison with clinical, mammographic, and pathological findings, *Ann. Surg. Oncol.* **1**, 132–140.

5. Dobbs, H.J., Parker, R.P., Hodson, N.J., Hobday, P. and Husband, J.E. (1983) The use of CT in radiotherapy treatment planning, *Radiother. Oncol.* **1**, 133–142.

6. Emri, M., Márián, T., Kövér, G., Berényi, E. and Ésik, O. (1996) Registration: a powerful tool to combine information by different imaging modalities, in B. Gulyás and H.W. Müller-Gärtner (eds.), *Positron emission tomography: a critical assessment of recent trends*, Kluwer Academic Publishers, Dordrecht, this volume.

7. Engel, H., Steinert, H., Buck, A., Berthold, T., Huch Böni, R.A. and von Schulthess, K. (1996) Whole-body PET: physiological and artifactual fluorodeoxyglucose accumutations, *J. Nucl. Med.* **37**, 441–446.

8. Ésik, O., Márián, T., Gulyás, B., Tóth, E., Lövey, J. and Trón, L. (1996) FDG PET investigation in medullary thyroid cancer, in A.M.J. Paans, J. Pruim, E.J.F. Franssen and W. Vaalburg (eds.), *Proceedings of the European Conference on Research and Application of Positron Emission Tomography in Oncology*, Groningen, p. 137.

9. Gupta, N.C., Maloof, M. and Gunel, E. (1996) Probability of malignancy in solitary pulmonary nodules using flourine-18-FDG and PET, *J. Nucl. Med.* **37**, 943–948.

10. Haberkorn, U., Strauss, L.G., Reisser, C., Haag, D., Dimitrakopoulou, A., Ziegler, S., Oberdorfer, F., Rudat, V. and van Kaick, G. (1991) Glucose uptake, perfusion, and cell proliferation in head and neck tumors: relation of positron emission tomography to flow cytometry, *J. Nucl. Med.* **32**, 1548–1555.

11. Hoh, C.K., Hawkins, R.A., Glaspy, J.A., Dahlbom, M., Tse, N.Y., Hoffman, E.J., Schiepers, C., Choi, Y., Rege, S., Nitzsche, E., Maddahi, J. and Phelps, M.E. (1993) Cancer detection with whole-body PET using 2-[18F]fluoro-2-deoxy-D-glucose, *J. Comput. Assist. Tomogr.* **17**, 582–589.

12. Kole, A.C., Nieweg, O.E., Vermey, A., Braams, J.W., Pruim, J., Hoekstra, H.J., Roodenburg, J.L.N., Schraffordt Koops, H. and Vaalburg W. (1996) Detection of unknown primary tumors using whole body PET with 18FDG, in A.M.J. Paans, J. Pruim, E.J.F. Franssen and W. Vaalburg (eds.), *Proceedings of the European Conference on Research and Application of Positron Emission Tomography in Oncology*, Groningen, p. 135–136.

13. Leskinen-Kallio, S., Någren, K., Lehikoinen, P., Routsalainen, U. and Joensuu H. (1991) Uptake of ^{11}C-methionine in breast cancer studied by PET. An association with the size of S-phase fraction, *Br. J. Cancer* **64**, 1121–1124.

14. Levin, D.N., Pelizzari, C.A., Chen, G.T.Y., Chen, C.-T. and Cooper, M.D. (1988) Retrospective geometric correlation of MR, CT, and PET images, *Radiology* **169**, 817–823.

15. Lövey, J., Ésik, O., Gulyás, B., Tóth, E., Molnár, T. and Trón, L. (1996) FDG PET for the evaluation of metastatic lymph nodes, in B. Gulyás and H.W. Müller-Gärtner (eds.), *Positron emission tomography: a critical assessment of recent trends*, Kluwer Academic Publishers, Dordrecht, this volume.

16. Lynch, H.T., Smyrk, T.C., Watson, P., Lanspa, S.J., Lynch, J.F., Lynch, P.M., Cavalieri, R.J. and Boland, C.R. (1993) Genetics, natural history, tumor spectrum, and pathology of hereditary nonpolyposis colorectal cancer: an updated review, *Gastroenterology* **104**, 1535–1549.

17. March, D.E., Wechsler, R.J., Kurtz, A.B., Rosenberg, A.L. and Needleman, L. (1991) CT-pathologic correlation of axillary lymph nodes in breast carcinoma, *J. Comput. Assist. Tomogr.* **15**, 440–444.

18. Minn, H., Joensuu, H., Ahonen, A., Klemi, P. (1988) Fluorodeoxyglucose imaging: a method to assess the proliferative activity of human cancer in vivo. Comparison with DNA flow cytometry in head and neck tumors, *Cancer* **61**, 1776–1781.

19. Miyazawa, H., Arai, T., Iio, M. and Hara, T. (1993) PET imaging of non-smal-cell lung carcinoma with carbon-11-methionine: relationship between radioactivity uptake and flow-cytometric parameters, *J. Nucl. Med.* **34**, 1886–1891.

20. Nieweg, O.E., Kim, E.E., Wong, W.-H., Broussard, W.F., Singletary, S.E., Hortobagyi, G.N. and Tilbury, R.S. (1993) Positron emission tomography with fluorine-18-deoxyglucose in the detection and staging of breast cancer, *Cancer* **71**, 3920–3925.

21. Nolop, K.B., Rhodes, C.G., Brudin, L.H., Beaney, R.P., Krausz, T., Jones, T. and Hughes, J.M.B. (1987) Glucose utilization in vivo by human pulmonary neoplasms, *Cancer* **60**, 2682–2689.

22. Pelizzari, C.A., Chen, G.T.Y., Halpern, H., Chen, C.-T. and Cooper, M.D. (1987) Three-dimensional correlation of PET, CT and MRI images, *J. Nucl. Med.* **28**, 682.

212

23. Pross, J., Bendl, R. and Schlegel, W. (1994) TOMAS, a TOol for MAnual Segmentation based on multiple image data sets, in A.R. Hounsell, J.M. Wilkinson and P.C. Williams (eds.), *Proceedings of the 11th International Conference on the Use of Computers in Radiation Therapy*, Manchester, pp.192-193.

24. Reisser, C., Haberkorn, U. and Strauss, L.G. (1993) The relevance of positron emission tomography for the diagnosis and treatment of head and neck tumors, *J. Otolaryngol. (Canada)* 22, 231–238.

25. Shad, L.R., Boesecke, R., Schlegel, W., Hartmann, G.H., Sturm, V., Strauss, L.G. and Lorenz, W.J. (1987) Three dimensional image correlation of CT, MR, and PET studies in radiotherapy treatment planning of brain tumors, *J. Comput. Assist. Tomogr.* 11, 948–954.

26. Shuman, W.P., Griffin, B.R., Haynor, D.R., Johnson, J.S., Jones, D.C., Cromwell, L.D. and Moss, A.A. (1985) MRI imaging in radiation therapy planning, *Radiology* 156, 143–147.

27. Szakáll, S. Jr., Trón, L., Gulyás, B. and Ésik, O. (1996) FDG-PET in the follow-up of patients with differentiated thyroid cancer, *Radiother. Oncol.* 40, Suppl (l) S140.

28. Székely, J., Poller, I., Gulyás, B., Balkay, L., Trón, L. and Ésik, O. (1996) PET in the follow-up of CNS tumours after radiotherapy, *Radiother. Oncol.* 40, Suppl (l) S140.

29. Szentirmay, Z., Tusnády, G. and Tóth, E. (1997) A daganatok kóros DNS tartalma [Cellular DNA content of human tumours, in Hungarian], Orvosi Hetilap 138, 000-000.

30. Tse, N.Y., Hoh, C.K., Hawkins, R.A., Zinner, M.J., Dahlbom, M., Choi, Y., Maddahi, J., Brunicardi C., Phelps M.E. and Glaspy, J.A. (1992) The application of positron emission tomographic imaging with fluorodeoxyglucose to the evaluation of breast disease, *Ann. Surg.* 216, 27–34.

31. Wahl, R.L., Cody, R.L., Hutchins, G.D. and Mudgett, E.E. (1991) Primary and metastatic breast carcinoma: initial clinical evaluation with PET and the radiolabeled glucose analog 2-[F-18]-fluoro-2-deoxy-D-glucose, *Radiology* 179, 765–770.

32. Wahl, R.L., Quint, L.E., Cieslak, R.D., Aisen, A.M., Koeppe, R.A. and Meyer, C.R. (1993) "Anatometabolic" tumor imaging: fusion of FDG PET with CT or MRI to localize foci of increased activity, *J. Nucl. Med.* 34, 1190–1197.

33. Woods, R.P., Mazziotta, J.C. and Cherry, S.R. (1993) MRI-PET registration with automated algorithm, *J. Comput. Assist. Tomogr.* 17, 536-546.

CLINICAL APPLICATION OF WHOLE-BODY [18F]-FDG-PET IN MALIGNANT MELANOMA

Hans C. Steinert

University Hospital Zürich, Division of Nuclear Medicine
Zürich, Switzerland

1. Introduction

The incidence of malignant melanoma has greatly increased in recent years (1). Proper tumor staging is a key prerequisite for choosing the appropriate treatment strategy in oncology in general and this is equally true for melanoma (2). While the cross-sectional imaging modalities sonography, computed tomography (CT) and magnetic resonance imaging (MRI) are sensitive to morphologic changes identification of tumor tissue e. g. in normal sized lymph nodes is difficult with these methods (3). Furthermore, morphological imaging procedures are better used to evaluate a given region of the body than the entire body. A single imaging modality with the potential to distinguish accurately between benign and malignant lesions, and to define the extent of disease would be extremly helpful in the evaluation and management of malignant melanoma.

One imaging modality with this potential is Positron Emission Tomography (PET) using a metabolic tracer. The metabolic properties of neoplastic tissue have been studied and used by many investigators to

B. Gulyás and H.W. Müller-Gärtner (eds.),
Positron Emission Tomography: A Critical Assessment of Recent Trends, 213–222.
© 1998 *Kluwer Academic Publishers.*

detect malignant tissue (4-6). Glucose metabolism can be studied by Deoxyglucose labeled with the positron emitter [18]Fluorine, 2-[^{18}F]-fluoro-2-deoxy-D-glucose (FDG) (7,8). Excellent FDG uptake into various tumors has been described.

Until recently the problem with PET screening for metastases has been its inability to cover large areas of the body fast. With the advent of new large field of view (15 cm) high sensitivity scanners, whole-body PET in 60 minutes has become a reality (9,10). Hence whole-body FDG PET could serve as a suitable screening modality for patients with malignant melanoma. While the treatment of metastising melanoma has not been very successful so far, this may be partly due to the fact that its spreading pattern is very erratic making identification of metastases and staging difficult. We therefore undertook a prospective study in patients with high risk melanoma to assess the value and the efficacy of whole-body FDG PET in staging of malignant melanoma.

2. Materials and Methods

Till now 130 whole-body PET scans in 89 consecutive patients with known metastatic melanoma or with newly diagnosed melanoma and a high risk of metastases (Breslow thickness > 1,5 mm) were evaluated. All patients were evaluated by clinical history, physical examination, routine clinical laboratory tests, image chest x-ray and ultrasound of the regional lymph nodes and of the abdomen. Any lesion suspicious for metastases was either confirmed by biopsy or CT and/or MRI. All imaging procedures were performed within 3 weeks of PET scanning. CT follow-up was performed after 3 months in those patients who did not undergo surgery.

For PET scanning, the patients were kept in a fasting state for at least 4 hours prior to the study in order to suppress myocardial glucose utilization. Fourty minutes before scanning they received an intravenous injection of FDG labeled with F-18. After the injection of approximately 370 MBq of FDG, the patients were asked to remain at rest for the time until scanning. PET scanning was performed on a GE Signa Advance PET Scanner (GE Medical Systems, Milwaukee, WI, USA) using the whole-body mode implemented as standard software. In this mode the scanner acquires 2D data over an axial field of view of 14.6 cm during 5 minutes. Once the measurement is accomplished, the table position is automatically incremented by 14.6 cm and the acquisition process starts anew. Up to 13 increments can be preprogrammed allowing a total axial field of view of up to 190 cm in approximately 65 minutes of acquisition time. The acquired data were reconstructed using standard back-projection techniques in transaxial slices and they were arbitrarily reformated into coronal and sagittal views. The tomographic images consisted of 455 transaxial, 79 coronal and 79 sagittal images. In accordance with other groups no attempt was made to also collect transmission data for attenuation correction, as this would have prolonged the procedure beyond patient tolerance (10). Furthermore, if not very long transmission acquisition times are used, the transmission corrected images contain inacceptable image noise. The reconstructed transaxial, coronal and sagittal scans were documented using a laser color copier with a digital interface (A & S, Münster, FRG) using a level and window setting which brought the brain into an intensity range just below the upper window level. This assured relative uniformity in scan appearance. For reading the PET scans the presence, number and location of lesions detected by other preceeding examinations were not known and correlative imaging studies were not available.

A lesion was defined as a focus of increased FDG uptake above the intensity of the surrounding activity, excluding the renal pelvis, urinary bladder, and myocardium if present. They were either classified as non-metastatic (class 0) or if lesions were found, as possibly metastatic (class I) or very likely metastatic (class II). A lesion was considered as class II lesion, if it showed intense FDG uptake comparable to that of physiological high brain FDG uptake, and if it had a nodular appearance in an area where no physiological FDG uptake should be expected. While this classification is somewhat arbitrary, it is practical because brain FDG uptake is presumably fairly constant in the brain. Class I lesions were either circumscribed focal lesions of distinct, but clearly lower FDG uptake, if compared to brain FDG uptake, or they were lesions which had a less circumscribed appearance. Again, the finding had to be in a region where no physiological FDG accumulation should be expected.

Since no attenuation correction was performed, "artifactual" enhancement of surface structures was present. Lesion classification therefore also occured in relation to the surrounding structures. As an example, a cutaneous lesion could appear more intense to be classified still as class I lesion than a lesion lying within the body. However, we considered this acceptable because artifactual enhancement is in relation to lesion location only. The relative intensities of the lesions compared to their surroundings are maintained if no attenuation correction is performed. If class I or II lesions were identified on a PET examination and had not been known from previous studies, the patients were reexamined with other imaging methods deemed appropriate to also identify these lesions, and/or fine needle aspiration or surgical removal of the lesions was performed if indicated.

After all diagnostic material was gathered, a standard of reference was generated in a consensus conference in which all available results of the 89 patients were reviewed. In this standard of reference all PET lesions were classified as positive, if they were clearly identified in at least one other imaging modality or if they were confirmed by fine needle aspiration or surgery.

3. Results

Using the histopathologic and combined imaging information (US, CT, MRI, PET) it was established that the 89 patients suffered from a total of 169 suspicious lesions. Of these, 130 proved to be metastases of melanoma, 27 corresponded to enlarged benign lymph nodes and 11 were due to an inflammatory lesion. The 130 lesions served as the standard of reference against which PET was evaluated.

PET identified a total of 123 suspicious lesions of which 30 belonged to class I and 110 to class II. 14 lesions of class I and 109 lesions of class II proved to be metastatic. PET showed 17 false postive lesions, 16 classified as class I and one classified as class II, where increased FDG uptake was due to either a recent operation with resulting tissue repair or due to local infection. There were seven false negative lesions in PET, four being skin metastases smaller than 3 mm in diameter. If all PET lesions were considered to be metastatic, independently of whether they were assigned to class I or II, a sensitivity of 94.6 %, a specifity of 56.4 % and an accuracy of 85.7 % was found. Using a highly specificity reading considering only class II lesions as positive the sensitivity was 83.8 %, the specificity was 97.4 % and the accuracy was 86.9 %.

4. Discussion

The results of this study show that whole-body PET is an excellent imaging method to screen for metastases of malignant melanoma in patients at risk. Strong FDG accumulation in the body occurs physiologically only in a few sites. Hence, the "signal to noise" of identifying a lesion is usually very high, making PET an useful screening modality very much like bone scanning. In contrast to the morphological cross sectional imaging modalities, PET permits in essence an "at a glance" diagnosis. It is therefore not surprising to find that PET with the high FDG accumulation in tumor sites has a high specifity for the identification of metastases, which was approaching 97% in our study if a high specificity reading approach was taken, with a concurrent and acceptable sensitivity of 84%.

The false positive findings (n = 17) were caused by inflammation in the form of infectious or post-operative reparation processes. The false negative findings (n = 7) in our study were four skin metastases of small diameter (<5 mm), two brain metastases and one liver metastasis.

Potential sources of false positive or negative results are lesions close or within structures taking up FDG naturally. The brain has a high glucose metabolic rate, and correspondingly a high FDG accumulation. In our study only four patient had a brain metastasis which was easily detected by PET because of its higher uptake in comparison to the surrounding. FDG is excreted via the urine. So the urinary system has a high uptake (renal collecting system, ureter, bladder). In our patient population only one metastasis close to the kidney was seen, so lesions in such problem areas are potentially underrepresented thereby increasing sensitivity. Some patients showed a significant myocardial FDG uptake even if they were imaged in a fasting condition. In fasting state the heart predominantely metabolizes fatty acids. Choi et al., however, have reported that myocardial uptake may occur in 60 % of humans even after 16 hours of fasting (12). The lung, liver, spleen, bone marrow and gastrointestinal tract consistently show relative little FDG uptake.

This study clearly has some limitations. As in all studies evaluating the effectiveness for staging, establishing a gold standard or standard of reference against which the method is compared is difficult, because it cannot be justified ethically to obtain cytological or histological proof of the diagnosis for all lesions identified. We have nevertheless used great care to establish our standard of reference, biopsying lesions identified whenever possible, or at least confirming the findings definitively by at least one imaging modality in addition to PET. A further study imperfection is that not all regions were prospectively examined by sonography and CT or MRI. As a result the data base permitting to compare sonography, CT or MRI with PET was not obtained. Again, this limitation is difficult to avoid particularly in melanoma with its erratic spreading pattern, as it is hardly possible to subject each melanoma patient to sonographic or CT examinations covering large parts of the body in a prospective search for metastases. In fact, these examinations are used more on the basis of suspicion in a given region of the body when a patient suffers from melanoma. At our institution, high risk melanoma patients will be examined with abdominal ultrasound or CT, but other body regions are not examined consistently or routinely with either of the two modalities, and MRI is performed only when there is a suspicion that brain metastases are present.

Gritters et al. (11) have published data on the use of PET for the identification of melanoma metastases, but this study was limited by the small field of view of an older scanner.We have chosen not to evaluate lesions quantitatively because without correcting for attenuation the values obtained would not be readily interpretable. However, as stated in the methods section, this limitation may not be so important because relative intensity of lesions to surroundings is maintained. In our study imaging time did not exceed 70 minutes including imaging of the lower extremities. With the introduction of 3D data acquisition in the near future, the acquisition time will decrease by a factor of 2, thus potentially reducing a whole-body PET examination to of the order of 30 minutes (13,14). Given the fact that a state of the art PET scanner is comparable in price to an MR scanner, PET scanning costs become comparable to those of MRI examinations. These considerations together with the fact that sonography, CT and MRI are usually charged on the basis of the number of body regions scanned, may make FDG PET a competitive method for staging patients with metastatic disease.

Hence, despite the limitations, we feel that the results of this study are valid, although whole-body PET is still in a early stage of use and much more clinical data will have to be accumulated to definitively establish its value in tumor staging. Furthermore, therapeutic outcome studies will be necessary. Particulary in metastatic melanoma with its poor response to any therapy except for surgical lesion removal, this outcome has to be looked at carefully.

In conclusion, this study suggests, that whole-body PET is a very effective imaging modality to screen for tumor metastases in metastatic melanoma. The procedure takes less than one hour of scan time and is virtually non-invasive. It is certainly worthwile, to examine large patient series with other oncologic diseases to establish the definitive value of whole body FDG PET for tumor staging.

References

1. Skolnick AA. (1991) Melanoma epidemic yields grim statistics. JAMA 265, 3217-3218.

2. Horgan K, Hughes LE. (1993) Review. Staging of melanoma. Clin Radiol 48, 297-300.

3. Buzaid AC, Sandler AB, Maani S, et al. (1993) Role of computed tomography in the staging of primary melanoma. J Clin Oncl 11, 638-643.

4. Warburg O. (1931) The metabolism of tumors. New York: Richard R. Smith, Inc., 167-181.

5. Hatanaka M. (1974) Transport of sugars in tumor cell membranes. Biochem Biophys Acta 355, 77-104.

6. Weber G. (1977) Enzymology of cancer cells (part 2). N Engl J Med 296, 541-555.

7. Kubota K, Kubota R, Yamada S. (1993) FDG accumulation in tumor tissue. Glycolysis and cancer cells. J Nucl Med 34, 419-421.

8. Som P, Atkins HL, Bandoypadhyay D, et al. (1980) A fluorinated glucose analog, 2-fluoro-2-deoxy-D-glucose (F-18): nontoxic tracer for rapid tumor detection. J Nucl Med 21, 670-675.

9. Dahlbom M, Hoffmann EJ, Hoh CK, et al. (1992) Whole body positron emission tomography: part I. Methods and performance characteristics. J Nucl Med 33, 1191-1199.

10. Hoh CK, Hawkins RA, Glaspy JA, et al. (1993) Cancer detection with whole-body PET using 2-[18F] fluoro-2-deoxy-D-glucose. J Comput Assist Tomogr 4, 582-589.

11. Gritters LS, Francis IR, Zasadny KR and Wahl RL. (1993) Initial assessment of positron emission tomography using 2-fluorine-18-fluoro-2-deoxy-D-glucose in the imaging of malignant melanoma. J Nucl Med 34, 1420-1427.

12. Choi Y, Brunken RC, Hawkins RA, et al. (1991) Determinats of myocardial glucose utilization assessed with dynamic FDG PET [abstract]. Circulation 84, 1690.

13. Cherry SR, Dahlbom M, Hoffman EJ. (1991) 3D PET using a conventional multislice tomograph without septa. J Comput Assist Tomogr 15, 655-668.

14. Bailey DL. (1992) 3D acqisition and reconstruction in positron emission tomography. Ann Nucl Med 6, 123-130.

EVALUATION OF METASTATIC LYMPH NODES BY MEANS OF FDG PET

J. LÖVEY[1], O. ÉSIK[1], B. GULYÁS[2], E. TÓTH[3],
T. MOLNÁR[4] AND L. TRÓN[4]

[1]Department of Radiotherapy, National Institute of Oncology, Budapest
H-1122 Budapest, Ráth György u. 7-9, Hungary
[2]Laboratory for Brain Research and PET, The Nobel Institute of
Neurophysiology, Karolinska Institute, Stockholm, Sweden
[3]Department of Molecular Pathology, National Institute of Oncology,
Budapest, Hungary
[4]PET Centre, Debrecen University of Medicine, Debrecen, Hungary

Abstract

Seventeen oncological patients with different tumours were investigated by means of [18]fluorodeoxy-D-glucose (FDG) positron emission tomography (PET) for primary staging and restaging after therapeutic interventions. Patients also underwent other staging procedures, including serum tumour marker level determination, radionuclide planar scintigraphy and radiological investigations (ultrasonography, computed tomography and magnetic resonance imaging). The results of different imaging methods in the search for metastatic lymph nodes (LNs) were reviewed. To check the applicability of FDG PET for estimation of the proliferative capacity of the tumours, a comparison was carried out of the patients' history, results of DNA analysis of the tumour cells and the FDG uptake as measured by PET. FDG PET proved to be superior to conventional imaging methods in staging tumours by the determination of metastatic LNs. Staging by PET was more effective for tumours in the case of LNs localised to the mediastinum (12 cases), the deep (paratracheal and paraoesophageal) cervical (5 cases), the axillary (2 cases), the supraclavicular (1 case) and the iliac regions (1 case). DNA analysis was performed in 14 cases. A direct correlation was found in all but one case between the tumour (LNs) to background FDG uptake ratio (TBUR) and the S-phase proportion of the tumour cells: high (>5) TBUR values were associated with a high ($>10\%$) S-phase proportion, while cases with TBUR ≤ 5 involved an S-phase fraction of $\leq 10\%$. In the 14th case the tumour was aneuploid and a high TBUR value coincided with a low (1%) S-phase proportion. Patients with LNs displaying high TBUR and high S-phase ratio values have a greater chance of developing distant metastases (3/4 cases) than those with low values of both parameters (3/9 cases). An additional advantageous feature of FDG PET is that it can provide individual patient prognoses via estimation of the proliferative capacity of metastatic LNs.

Key words: PET, FDG, tumour, oncology, staging, lymph node, proliferative capacity, DNA analysis, tracer uptake

B. Gulyás and H.W. Müller-Gärtner (eds.),
Positron Emission Tomography: A Critical Assessment of Recent Trends, 223–228.
© 1998 Kluwer Academic Publishers.

1. Introduction

The use of positron emission tomography (PET) in oncology has proved to be advantageous in the staging and restaging of tumours and in therapy monitoring, and it also has a promising role in estimation of the proliferative capacity of tumours [4,6,9].

PET differentiates between tumorous and normal tissue on the basis of their metabolic activity. While conventional imaging methods such as ultrasonography (US), computed tomography (CT) and magnetic resonance imaging (MRI) provide more anatomical than functional images, PET rather supplies functional information. Due to this feature PET can provide valuable additional information relative to that yielded by the former techniques, and thus helps in establishment of the appropriate staging.

A number of investigations have been carried out to determine the correlation between the proliferative features of tumours and the [18]fluorodeoxy-D-glucose (FDG) uptake in head and neck tumours [8,5,11] in breast cancer [2] and in medullary thyroid cancer [3] as measured by FDG PET. There have also been some similar methionine-PET studies on breast cancer [7] and non-small cell lung cancer [9].

The present report compares results of the investigation of metastatic lymph nodes (LNs) by conventional imaging methods and PET. An examination was also made of the applicability of PET for estimation of the proliferative capacity of LN metastases by comparing FDG uptake as measured by PET, the S-phase proportion of the tumour cells and the clinical course for each patient.

2. Patients and methods

Seventeen patients with thyroid cancer (10), malignant lymphoma (4) or LN metastases (3) of occult primary tumours (phenotypically carcinomas) were investigated by FDG PET.

Patients received intravenously 5-10 mCi activity of FDG 30 minutes prior to scanning by a GE 4096 PLUS whole body camera. They also underwent conventional radiological (US, CT, MRI) and nuclear imaging investigations (iodine for papillary and meta-iodobenzylguanidine (MIBG) planar scintigraphy for medullary thyroid cancer). Appropriate serum marker levels (human thyroglobulin, calcitonin and carcinoembryogenic antigen (CEA)) were routinely determined.

The results of the different imaging methods were compared. LNs appearing with a diameter of less than 1 cm in radiological (CT, MRI and US) images were regarded as negative, while those with a diameter of over 1 cm were classified as positive. LNs were considered metastatic on the basis of FDG PET scans with a tumour to background FDG uptake ratio higher than 2 (TBUR>2).

As an approximation of the metabolic activity, the tracer uptake (nCi/cm^3) of the metastatic LNs and the background was measured semiquantitatively within round regions of interest (ROIs) in the tumorous areas and in healthy reference areas. All ROIs defined for a given patient were of the same size, but there was a patient to patient variation in the diameter of the applied ROIs. In tumorous regions, the diameter was set not to extend beyond the visible maximum of the high uptake area. ROIs delineated in this way contained 20 to 95 pixels. The activity ratios of the tumour and the background ROIs were calculated.

The proliferative capacity of the tumour cells was investigated by DNA analysis, in 14 of the 17 cases carried out with a digital image analyser and Feulgen staining [12]. Extensive fibrosis did not permit the DNA analysis in 2 Hodgkin lymphoma cases, and in one case the histological specimen was damaged.

As this study was performed retrospectively, DNA analysis was carried out following surgery, in 12 of the 14 cases on LNs different from those measured by FDG PET. Only in 2 cases did the DNA and FDG PET data relate to the same LNs. For evaluation of the results, it was presumed that the principal proliferative characteristics of the LN metastases did not change with time or location [1].

The patients' history was reviewed with special attention to the relationship between the occurrence of distant metastases, the FDG uptake and the results of DNA analysis.

3. Results

In 16 of the 17 cases, the PET examinations gave additional information as compared with that provided by conventional radiological and nuclear imaging methods, as the FDG PET scans showed LN metastases in new regions relative to those diagnosed earlier (Table 1). In 16 of the cases, the PET investigations resulted in a more accurate delineation of the tumorous area within the already identified regions.

TABLE 1. Regions of metastatic LNs diagnosed by different imaging methods

Diagnosis	Regions of metastatic LNs by appropriate planar scintigraphy (iodine, MIBG)	Additional regions of metastatic LNs by conventional imaging methods	Regions of metastatic LNs diagnosed only by PET
papillary thyroid cancer	w.p.u.	neck, mediastinum	none
papillary thyroid cancer	w.p.u.	none	mediastinum
papillary thyroid cancer	neck, upper mediastinum	none	lower mediastinum
papillary thyroid cancer	neck, upper mediastinum	none	lower mediastinum
papillary thyroid cancer	w.p.u.	none	neck
medullary thyroid cancer	neck	none	upper mediastinum
medullary thyroid cancer	w.p.u.	none	neck
medullary thyroid cancer	w.p.u.	supraclavium	mediastinum
medullary thyroid cancer	w.p.u.	neck	mediastinum
medullary thyroid cancer	w.p.u.	none	neck, mediastinum, supraclavium
NHL, B-cell	n.p.	neck, axilla, abdomen	mediastinum
NHL, B-cell	n.p.	none	iliac region
Hodgkin's lymphoma, NS	n.p.	neck, axilla, supraclavium	mediastinum
Hodgkin's lymphoma, NS	n.p.	mediastinum	neck
occult primary carcinoma	n.p.	neck	axilla, mediastinum
occult primary carcinoma	n.p.	none	axilla
occult primary carcinoma	n.p.	none	mediastinum

NHL = non-Hodgkin lymphoma, NS = nodular sclerosis, MIBG = meta-iodobenzylguanidine
w.p.u. = without pathological uptake, n.p.= not performed

226

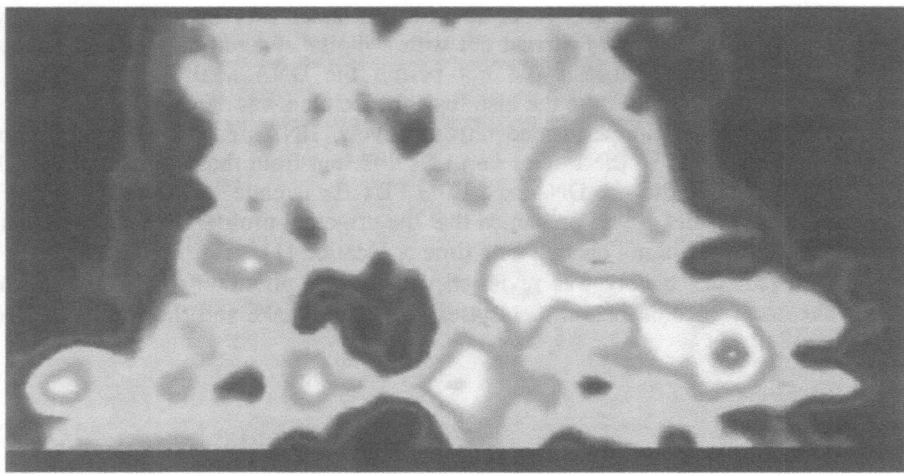

Figures 1 and 2. A 32-year old male patient with medullary thyroid cancer was investigated by FDG PET for recurrence indicated by elevated serum calcitonin and CEA levels, following surgery and external radiotherapy. US and CT failed to reveal local recurrence or enlarged LNs in the cervical region. PET demonstrated LN metastases in the left and right parajugular chain and in the left supraclavicular fossa (upper panel). The patient was referred to surgery, but only left neck dissection and left supraclavicular excision were carried out. The postsurgical PET scan revealed a still existing metastasis in the right parajugular chain (lower panel).

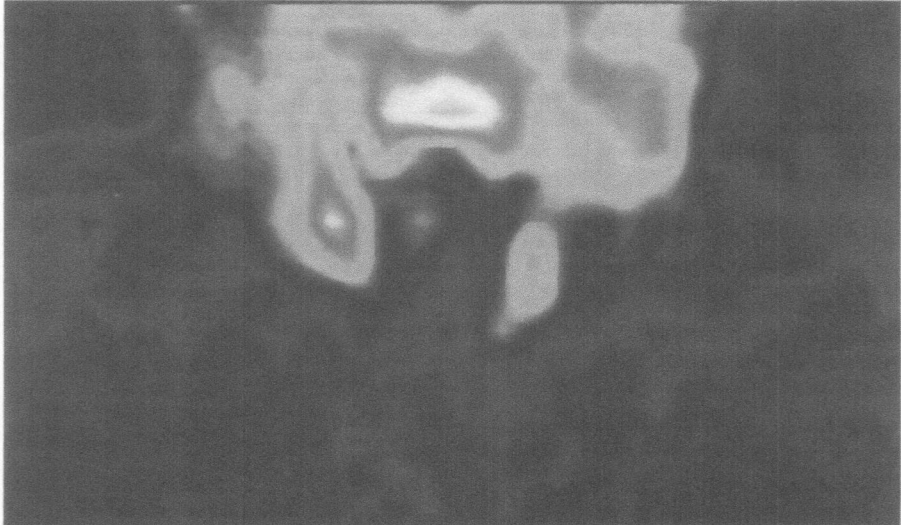

Comparison of the measured TBUR and the DNA analysis parameters shows the close interrelationship of these entities. A positive correlation was observed in 13 of the 14 DNA analysis cases: high values (>5) of TBUR were associated with a high proportion (>10%) of S-phase cells, and low (≤5) TBUR values appeared together with a low (≤ 10%) abundance of S-phase cells. There was only a single case (male, 24 y, papillary thyroid cancer) where this correlation was not satisfied, with a relatively high tracer uptake (TBUR=6.6) accompanied by a rather low (1%) S-phase cell fraction. It is interesting that DNA analysis of this case revealed extreme aneuploidity, which is a rare event within papillary thyroid cancer at a young age. Patients with tumours showing an increased tracer uptake and a high S-phase fraction were found to be at higher risk of developing distant metastases (Table 2).

Table 2. Correlation between S-phase proportions, tracer uptake ratios and distant metastases*

	Tracer uptake ratio ≤ 5 S-phase ≤ 10%	Tracer uptake ratio > 5 S-phase > 10%
n	9	4
cases with distant metastases	3 (33%)	3 (75%)

*patients without DNA analysis are excluded

A convincing correlation could not be demonstrated between the DNA parameters (the ploidity of the tumour cells, the polyploid fraction and the number of the subpopulation of the tumour), the clinical course and the FDG uptake, because of the small size of the group of patients and the heterogeneity of the investigated tumours.

4. Conclusions

FDG PET is a very effective method for the nodal staging of tumours. It helps in the adequate establishment of oncological staging and restaging, especially in troublesome anatomical localisations such as the mediastinum, deep cervical regions or the supraclavicular fossa. In these cases, PET can be the most reliable diagnostic method. The additional information afforded by PET can influence decision-making concerning the therapy, e.g. the choice of treatment modality, the extent of surgical intervention, the definition of the planning target volume for radiotherapy, and the indication of radionuclide therapy or chemotherapy. By promoting more adequate decisions, PET contributes to more effective therapy, resulting in better patient survival.

The correlation observed between the S-phase proportion of the cells, the FDG uptake and the clinical course of the tumours indicates that the metabolic rate of the tumours as defined by PET is suitable for estimation of the proliferative capacity of the tumours and can be used to predict the biological behaviour of the disease and to assess the individual prognosis.

228

5. References

1. Barlogie, B., Raber, M.N., Schumann, J., Johnson, T.S., Drewinko, B., Swartzendruber, D.E., Göhde, W., Andreeff, M. and Freireich, E.J. (1983) Flow cytometry in clinical cancer research, *Cancer Research* **43**, 3982-3997

2. Crowe, J.P. Jr., Adler, L.P., Shenk, R.R. and Sunshine, J. (1994) Positron emission tomography and breast masses: comparison with clinical, mammographic, and pathological findings, *Ann. Surg. Oncol.* **1**, 132-140.

3. Ésik, O., Márián, T., Gulyás, B., Tóth, E., Lövey, J. and Trón L. (1996) FDG PET investigations in medullary thyroid cancer, in A.M.J. Paans, J. Pruim, E.J.F. Franssen, W. Vaalburg (eds.), *Proceedings of the European Conference on Research and Application of Positron Emission Tomography in Oncology*, Groningen, p. 137.

4. Ésik, O., Gulyás, B., and Trón, L. (1996) Diagnosis, differential diagnosis, and follow-up of tumours by means of FDG PET, in B. Gulyás and H.W. Müller-Gärtner (eds.), *Positron emission tomography: a critical assessment of recent trends*, Kluwer Academic Publishers, Dordrecht, this volume

5. Haberkorn, U., Strauss, L.G., Reisser, C., Haag, D., Dimitrakopoulou, A., Ziegler, S., Oberdorfer, F., Rudat, V. and van Kaick, G. (1991) Glucose uptake, perfusion, and cell proliferation in head and neck tumours: relation of positron emission tomography to flow cytometry, *J. Nucl. Med.* **32**, 1548-1555.

6. Hoh, C.K., Hawkins, R.A., Glaspy, J.A., Dahlbom, M., Tse, N.Y., Hoffman, E.J., Schiepers, C., Choi, Y., Rege, S., Nitzsche, E., Maddahi, J. and Phelps, M.E. (1993) Cancer detection with whole body PET using 2-[18F]fluoro-2-deoxy-D-glucose, *J. Comput. Assist.Tomogr.* **17**, 582-589.

7. Leskinen-Kallio, S., Någren, K., Lehikoinen, P., Routsalainen, U., and Joensuu H. (1991) Uptake of ^{11}C-methionine in breast cancer studied by PET. An association with the size of S-phase fraction, *Br. J. Cancer* **64**, 1121-1124.

8. Minn, H., Joensuu, H., Ahonen, A. and Klemi, P. (1988) Fluorodeoxyglucose imaging: a method to assess the proliferative activity of human cancer in vivo. Comparison with DNA flow cytometry in head and neck tumours, *Cancer* **61**, 1776-1781.

9. Miyazawa, H., Arai, T., Iio, M. and Hara, T., (1993) PET imaging of non-small cell lung carcinoma with carbon-11-methionin: relationship between radioactivity uptake and flow-cytometric parameters, *J. Nucl. Med.* **34**, 1886-1891.

10. Nieweg, O.E., Kim, E.E., Wong, W.H., Broussard, W.F., Singletary S.E., Hortobagyi, G.N. and Tilbury, R.S. (1993) Positron emission tomography with fluorine-18-deoxyglucose in the detection and staging of breast cancer, *Cancer* **71**, 3920-3925.

11. Reisser C., Haberkorn U. and Strauss L.G. (1993) The relevance of positron emission tomography for the diagnosis and treatment of head and neck tumours, *J. Otolaryngol. (Canada)* **22**, 231-238.

12. Szentirmay, Z., Tusnády, G., and Tóth, E. (1997) A daganatok kóros DNS tartalma [Cellular DNA content of human tumours, in Hungarian], *Orvosi Hetilap* **138**, 000-000.

DIAGNOSIS, DIFFERENTIAL DIAGNOSIS, AND FOLLOW-UP OF INTRACRANIAL TUMORS WITH PET

K. BORBELY, *M.D., Ph.D.*
National Institute of Neurosurgery
Budapest, Amerikai út 57
1145, Hungary
Tel: (36-1) 251-2999
* (36-1) 220-5192*
Fax: (36-1) 251-5678

Abstract

Recent progress in neuroimaging has resulted in the development of functional imaging techniques, including positron emission computed tomography (PET). Whereas computed tomography (CT) and magnetic resonance imaging (MRI) display exquisitely anatomical details, PET provides us with information on pathophysiological and biochemical changes as well.

Since neither CT nor MRI prove always to be successful in grading of tumors or differentiating recurrent tumors from radiation necrosis the recent application of PET method in clinical management of patients harboring brain neoplasms has proved to be of vital importance.

PET-[18F]fluorodeoxyglucose (FDG) technique based on Warburg's hypothesis, using the Sokoloff's kinetic model clearly demonstrates the effectiveness of glucose metabolic mapping in distinguishing high-grade from low-grade lesions, and radiation necrosis from residual or recurrent tumor.   It can show malignant degeneration in tumors originally low-grade in nature, and response to therapy not achievable using morphological imaging techniques.   Also, the PET-FDG findings have a prognostic value for tumors.

The present status of PET radiotracers, particularly 18F-FDG will be discussed in determining the diagnosis, prognosis, and response of tumor tissue to therapy and in distinguishing tumors of varying grades and histology.

229

B. Gulyás and H.W. Müller-Gärtner (eds.),
Positron Emission Tomography: A Critical Assessment of Recent Trends, 229–236.
© 1998 *Kluwer Academic·Publishers.*

Discussion

Currently-used routine methods for the diagnosis and follow-up of brain tumors are CT and MRI. Both technologies demonstrate exquisitely the tissue morphology and use of contrast enhancement for differential diagnosis. Although MRI and CT can provide precise anatomical delineation of intracranial lesions, frequently reveal changes in tumor grading or volume in response to different therapeutic manipulation, they do not have the ability to assess either biological behavior or response to therapy, particularly in evaluation of changes following surgery or radiotherapy. Little or no information is revealed that might be used to predict response to therapy other than referral to the knowledge of the natural history of similar lesions after the diagnosis has been made. In addition, the categorization by routine biopsy is not a "gold standard". The pathological examination of biopsy might be inconclusive, since sampling is limited by surgical interventions.

There are two technologies that appear to provide such information: MR spectroscopy (MRS) and PET. MRS is capable of providing useful biochemical information, but it has so far found only limited clinical utility.

Almost 70 years ago, Otto Warburg first suggested that neoplasms display higher rates of anaerobic glycolysis with increasing degrees of malignancy (1). Also, a consistent observation was made that the transformation from slow-growing, well-differentiated tumor tissue to rapidly growing, poorly differentiated neoplasms is accompanied by a progressive increase in the utilization of anaerobic glycolysis (2).

The pioneering work of Di Chiro and his co-workers based on Warburg's hypothesis, using the Sokoloff's kinetic model clearly demonstrated the effectiveness of PET-FDG technique in distinguishing high-grade from low-grade lesions, and radiation necrosis from residual or recurrent tumor (3-6).

GLIOMAS

Gliomas constitute more than 50% of primary brain tumors. Primary malignant forms recur within 1/2 to 1 year after surgery, and even totally removed benign forms may recur. 50% of recurrent astrocytomas are more malignant than the original tumor. The time elapsing until recurrence strongly depends on the degree of malignancy and the surgical removal. Patients survival also depends on the therapy. Effective therapy requires a knowledge of the degree of malignancy as well as differentiation between recurrent tumor versus (vs.) radiation necrosis, thus influance the choice of conservative vs. surgical therapy or combination thereof.

Oxygen-based tracers
In studies of tumors, PET has revealed lower regional cerebral blood flow (rCBF) compared with contralateral homologous gray matter (7). The regional cerebral oxygen utilization (rCMRO2) was also lower and this low demand for oxygen was

reflected in a low regional oxygen extraction fraction (rOER). On the other hand, tumor rCBF was found to be extremely variable. For large high grade tumors, rCBF can range from virtually zero to values in excess of normal brain. This is not surprising due to the heterogenity of these tumors with areas of necrosis, hemorrhage or cyst formation. The low rCMRO2 and low rOER findings suggest that these tumors have a more than adequate blood supply to meet metabolic demands. No relationship between tumor perfusion and vascularity has been found.

RCBF and oxygen utilization were lower in regions of peritumoral edema than in contralateral white or gray matter. The rOER was never raised, and in some cases was even lower than normal, which indicates a reduced demand for oxygen in the edematous region (8). This suggests that the reduction in rCBF is not a primary phenomenon due to compression of the vascular bed by increased tissue pressure. If local ischemia existed, the rOER would be elevated.

Glucose analogs

Tyler (1987) reported that, despite the fairly large range of regional cerebral glucose utilization (rCMRGl) values, the average rate of glioma glucose metabolism was lower than that of contralateral gray matter. Although high grade gliomas represent a fairly heterogenous group of tumors, certain metabolic and hemodynamic similarities were found in this study (9). Decreased rOER and rCMRO2, and increased regional cerebral blood volume (rCBV) were common characteristics, while rCBF was quite variable within tumors. There were metabolic differences between untreated tumors and tumors recurring after therapy, in which high rCMRGl values have been observed.

Di Chiro and his team have found the FDG technique to be effective in distinguishing high-grade from low-grade lesions, and radiation necrosis from residual or recurrent tumor (3-6). Using the peak glucose utilization rate within the tumor region as a metabolic measure, they found a significantly greater metabolism in high grade tumors than in low grade tumors.

FDG images of the normal brain show a biphasic uptake, with high activity in gray-matter areas and low activity in white-matter areas. The appearance of the tumor, therefore, depends on its location. A cold tumor within a white-matter region will not stand out, but will be hypoactive or, more likely, isoactive with its surroundings. If a cold tumor invades a gray-matter area such as the cortex, it appears as a cold spot within the area.

The appearance of hot tumors also depends on their location. If found within normal white matter, the hot appearance is usually unmistakable. However, hot tumors that invade the cortex or other gray matter might easily be confused with normal structures. Also, there are numerous entities that may demonstrate high, intermediate, or low uptake of FDG that are not cancers, including seizure foci, abscess and other inflammatory lesions, necrosis, and infarct. For precise interpretation of PET-FDG findings both clinical data and morphological imaging data are required (CT or MRI).

Malignancy can be demonstrated by multiple histological patterns, variable cell types, extent of differentiation, and differences in antigenicity, immunogenecity, biochemical properties, growth behavior, and cellular susceptibility to chemotherapeutic drugs (10,11). Although tumors of the same histological type often follow a relatively predictable and reproducible pattern of clinical progression in most hosts (11), the biological behavior of one cancer can not always be inferred from that of another because of the presence of heterogenous cell populations, which may dominate as the tumor grows and so change the behavior (10).

By 1982 it was evident that the rate of glucose utilization correlated with the histological grading of brain tumor malignancy, and that this rate was a good predictor of the patient's life expectancy (3). FDG is currently the principal radiotracer used for tumor grading, differential diagnosis and prognosis in primary brain tumors (3-6,12-15), but some investigators have reported variability in the rates of glucose utilization within tumors of similar pathological grades, suggesting that a direct correlation of glucose utilization with the histological grade of tumor malignancy may not always hold (9).

Elevated or decreased FDG uptake may be seen in other conditions that are frequently included in the differential diagnosis of intracranial mass lesions or that may simultaneously occur with tumor, such as infection, inflammation, seizure activity, radiation necrosis, edema, and infarct (16,17).

The PET technique should not be used as a screening method. In order to create an in vivo metabolic profile of a tumor, biologically observable radiotracers are required that localize in the neoplasm at concentrations sufficient to detect changes in uptake. An understanding of in vivo biochemical heterogeneity may ultimately lead to a better patient management and choice of therapeutic protocols.

The PET-FDG findings have prognostic value for tumors. A high metabolic ratio represents a poor prognostic indicator. The aggressive gliomas, although all histologically malignant, do not progress or grow at the same rate. Histology does not give any indication in regard to the future biologic behavior of the high grade gliomas. PET-FDG, however, offers the capability of predicting the survival time of the patients suffering from these neoplasms. Finally, PET scans with FDG may be used to monitor the tumor responce to a given treatment.

Di Chiro, Patronas et al. (4,5) were the first to use the FDG method in the evaluation of radiation necrosis of the brain. These investigators were able to differentiate between radiation necrosis and recurrent tumors. The rate of glucose utilization in the necrotic lesions was less than that in the corresponding region of contralateral normal brain, while recurrent tumors had higher rates.

Di Chiro and his team found a regional depression of glucose utilization in peritumoral edema and in cortical regions that were adjacent or neurally connected to the tumor (6). The latter findings may be explained by deafferentation. In addition, there is evidence of supressed metabolism in the cerebellar hemisphere contralateral to the supratentorial lesion ("crossed cerebellar diaschisis").

Amino acids

Amino acid transport across tumor cell membranes as measured with PET may also serve to differentiate malignant from nonmalignant tumors (18-21). Several investigators have found 11-C-labeled methionine (11-C-Met) to be superior to 11-C-deoxy-D-glucose or FDG for delineation of tumor margins and in the differential diagnosis of tumor recurrence from radiation necrosis (22, 23).

In brain tumor imaging, 11-C-amino-cyclobutane carboxylic acid has proved effective in detecting primary tumors, with a sensitivity, specificity, and accuracy that rivals FDG (24).

11-C-thymidine, labeled either in the 5-methyl position or in the 2-position of the base ring, has been studied in clinical trials with primary brain neoplasms and peripheral lymphomas (16,17,25). Multiparametric analysis techniques of experimental and human brain tumors using radiolabeled tracers have included markers for perfusion and metabolism such as 15-O-water (perfusion), 11-C-Met (amino acid uptake) and 18-F-FDG (glucose utilization) (22,25-27). A hypometabolic lesion on a FDG scan, which could represent a low-grade tumor, infarct, or necrosis, would require a follow-up study with 11-C-Met in order to evaluate the presence of viable tissue - i.e. tumor (25). By choosing the appropiate combination of radiotracers, it may be possible to improve substantially the differential diagnosis, as entities that frequently confuse the anatomical picture may be separated on the basis of differences in metabolic profiles. The in vivo study of other metabolic processes may be fruitful, such as the rates of protein, RNA or DNA synthesis (28-30).

Recently, it was shown that the accumulation of the labeled amino acids (L-[methyl-3-H]methionine , and L-2-[18F]fluorotyrosine) represent amino acid transport whereas L-[1-14-C]leucine represents the protein synthesis rate. This indicates that only in vivo uptake of carboxyl labeled amino acids, such as L-[1-11-C]leucine and L-[1-11-C]tyrosine, is representative of protein synthesis rate (30).

Kubota has demonstrated, that carbon-14-methionine (14-C-Met) and 18F-FDG showed different distributions in tumor tissue. 14-C-Met uptake by the tumor was mostly by viable cancer cells. The uptake by macrophages and other cellular components was low. The uptake was higher in the highly proliferative tumor but was not correlated with protein synthesis. Both rapidly and slowly growing tumors had a lower uptake ratio of 14-C-Met than of 18F-FDG, reflecting de novo DNA synthesis. In other words, the 14-C-Met uptake represents the presence of viable cancer cells. *14-C-Met may be suitable for treatment evaluation of individual tumors, but does not show the growth rates of different tumors. Fluorine-18-FDG reflects tumor-host immune system reaction and is an excellent tool for pretreatment evaluation of tumors and determination of tumor proliferative activity* (31).

MENINGIOMAS

Intracranial meningiomas represent approximately 14% of primary intracranial tumors. They are usually considered benign lesions, curable with surgical removal. In some cases they may be aggressive, as indicated by invasion of brain matter and metastasis, and lead to an unfavorable, even fatal, outcome. In addition, even after complete resection, meningiomas may recur. Recurrence has been reported in 29% of histologically benign meningiomas; more than 50% of metastasizing meningiomas described in the literature were from benign primary tumors. Cortical invasion by tumor has also been given some weight as a factor facilitating recurrence, and invasion of the dural sinuses, and bone usually makes complete resection difficult and regrowth likely. Progestin and estrogen receptors of meningiomas have been evaluated, but the results do not appear to correlate with recurrence rate.

Di Chiro, Hatazawa et al (1987) were the first to demonstrate that the glucose metabolic rates of nonrecurrent meningiomas were significantly lower than those of recurrent or regrowing tumors, suggesting that the glucose utilization rate of intracranial meningiomas may be an index of recurrence. Meningiomas displaying high glycolytic rates tend to recur relatively soon after surgery, whereas a lower glucose utilization rate is associated with nonrecurrence or late recurrence. The FDG studies regarding viability and probability of recurrence may be a valuable asset in patient management, e.g., by indicating the use of radiation therapy as an alternative method of treatment for recurrent and for malignant meningiomas (12).

SCHWANNOMAS

Cranial and spinal schwannomas, including acoustic neuromas associated with neurofibromatosis type 2, can be visualized with 18F-FDG (15). These lesions, being extracerebral, are well seen, since surrounding normal structures generally demonstrate low accumulation of tracer. Preliminary studies also suggested that the lesions can be graded with regard to aggressivity and recurrence potential.

ADENOMAS

De Souza and Francavilla reported the usefulness of FDG studies in the management of patients suffering from pituitary microadenomas and macroadenomas (14,32). They showed that energy metabolism, demonstrated by PET-FDG, is more directly related to tumor growth capability than it is to hormonal production and secretion. The method may have clinical utility, particularly in the treatment of nonfunctional adenomas. As no hormone is produced, the therapeutic effect of pharmacological therapies may be difficult to monitor. The method offers the potential of monitoring tumor response to therapies with both hormone-producing and nonfunctional adenomas, and of differentiating between recurrent tumor and scarring. PET-FDG may be a useful method for predicting and defining the growth potential of pituitary adenomas.

The density of dopamine receptors is often increased in pituitary adenomas, and appropriately labeled receptor ligands can be used to image those sites in tumor cells (33).

METASTASES

PET-FDG is clearly superior to CT and MRI in the distinction between recurrent metastases and radiation effects after gamma knife radiosurgery and is of significant importance to patients in individual cases (34). With 11-C-Met an increased accumulation is seen in a majority of brain tumors regardless of differentiation and blood-brain barrier damage. Metastases are, with regard to FDG and Met accumulation, similar to malignant gliomas, but it seems that metastases of non-nervous system tumors have a higher tendency to show normal or even reduced FDG accumulation than primary CNS tumors.

PET-FDG studies should not be used as a screening method after appearance of neurological symptoms or disease. They should be last in the order of imaging techniques or diagnostic methods. First we must learn the morphological changes in the brain. Other imaging techniques beyond the clinical data are necessary to determine the basic parameters of the lesion. When the patient management is of concern, e.g., surgery vs. radiotherapy, aggressive vs. conservative medicine or combination thereof, degree of malignancy, delineation of tumor margins following surgery, differentiation between residual or recurrent tumor and necrosis, etc.), then PET imaging can be not only helpful, but even mandatory.

References

1. Warburg, O. (1930) The metabolism of tumors, *London, Arnold Constable* 75-327.
2. Dickens, F., Simer, F. (1930) The metabolism of the normal and tumor tissue, *J. Biochemistry* **24**, 1301-1326.
3. Di Chiro, G., DeLaPaz, R.L., Brooks, R.A., et al. (1982) Glucose utilization of cerebral gliomas measured by [18F]fluorodeoxyglucose and positron emission tomography, *Neurology* **32**, 1323-1329.
4. Patronas, N.J., Di Chiro, G., Brooks, R.A., et al. (1982) Progress: [18F] fluorodeoxyglucose and positron emission tomography in the evaluation of radiation necrosis of the brain, *Radiology* **144**, 885-889.
5. Di Chiro, G., Patronas, N.J., Oldfield, E.H., et al. (1985) PET, CT and NMR of cerebral necrosis following radiotherapy or intra-arterial chemotherapy for cerebral tumors, *AJNR* **6**, 473-474.
6. Di Chiro, G., Brooks, R.A., Bairamian, D., et al. (1985) Diagnostic and prognostic value of positron emission tomography using [18F]fluorodeoxyglucose in brain tumors, in: M. Reivich, A. Alavi (eds.), *Positron emission tomography*, Alan R. Liss, Inc., New York, pp. 291-309.
7. Beaney, R.P. (1984) Positron emission tomography in the study of human tumors, *Sem Nucl Med* **14**, 4.
8. Lammertsma, A.A., Wise, R.J.S., Jones, T. (1984) Regional cerebral blood flow and oxygen utilization in edema associated with cerebral tumors, in: K.G. Go, A. Baethmann (eds), *Recent progress in the study and therapy of brain edema*, Plenum, New York, pp. 331-344.
9. Tyler, J.L., Diksic, M., Villemure, J.G., et al. (1987) Metabolic and hemodynamic evaluation of gliomas using positron emission tomography. *J Nucl Med* **28**, 1123-1133.

236

10. Henson, D.E. (1982) Heterogeneity in tumors. *Arch Pathol Lab Med* **106**, 597-598.

11. Poste, G., Greig, R. (1982) On the genesis and regulation of cellular heterogeneity in malignant tumors, *Invasion Metastasis* **2**, 137-176.

12. Di Chiro, G., Hatazawa, J., Katz, D.A., et al. (1987) Glucose utilization by intracranial meningiomas as an index of tumor aggressivity and probability of recurrence: a PET study, *Radiology* **164**, 521-526

13. Alavi, J.B., Alavi, A., Chawluk J., et al. (1988) Positron emission tomography in patients with glioma: A prediction of prognosis, *Cancer* **62**, 1074-1078.

14. Francavilla, T.L., Miletich, R.S., DeMichele, D. (1991) Positron emission tomography of pituitary macroadenomas: Hormone production and effects of therapies, *Neurosurgery* **28**, 826-832.

15. Borbély, K., Fulham, M.J., Brooks, R.A., et al. (1992) PET-Fluorodeoxyglucose of cranial and spinal neuromas, *J Nucl Med* **33**, 1931-1934.

16. Conti, P.S., Camargo, E.E, Grossman, S.A., et al. (1991) Multiple radiotracers for evaluation of intracranial mass lesions using PET, *J Nucl Med* **32**, 954.

17. Schmall, B., Conti, P.S., Kleinert, E.L., et al. (1992) Tumor and organ biochemical profiles determined in vivo following uptake of a combination of radiolabeled substrates: Potential applications for PET, *Am J Physiol Imaging* **7**, 2-11.

18. Foster, D.O., Pardee, A.B. (1969) Transport of amino acids by confluent and nonconfluent 3T3 and polyoma virus-transformed 3T3 cells growing on glass cover slips, *J Biol Chem* **244**, 2675-2681.

19. Isselbacher, K.J. (1972) Sugar and amino acid transport by cells in culture: Differences between normal and malignant cells, *N Engl J Med* **286**, 929-933.

20. Johnstone, R.M., Scholefield, P.G. (1965) Amino acid transport in tumor cells, *Adv Cancer Res* **9**, 143-226.

21. Parnes, J.R., Isselbacher, K.J. (1978) Transport alterations in virus transformed cells, *Prog Exp Tumor Res.* **22**, 79-122.

22. Ericson, K., Lilja, A., Bergstrom, M., et al. (1985) Positron emission tomography with 11-C-methyl-L-methionine, 11-C-D-glucose, and 68-Ga-EDTA in supratentorial tumors, *J Comput Assist Tomogr* **9**, 683-689.

23. Lilja, A., Bergstrom, K., Hartvig, P., et al. (1985) Dynamic study of supratentorial gliomas with L-methyl-11-C-methionine and positron emission tomography, *AJNR* **6**, 505-514.

24. Hubner, K.,F. (1993) Unnatural amino acids in PET oncology. 5th Annual Meeting of the Institute for clinical PET, *Arlington, VA.* Oct, 28-31.

25. Wagner, H.N.Jr., Szabo Zs, Buchanan J.W. (1995) Principles of Nuclear Medicine. W.B. Saunders Company, Philadelphia, London, Toronto, Montreal, Sydney, Tokyo.

26. Kameyama, M., Tsurumi, Y., Shirane, R., et al. (1987) Multiparametric analysis of brain tumor with PET, *J Cereb Blood Flow Metab* **7**, S466.

27. Meyer, G.J., Schober, O., Gaab, M.R., et al. (1989) Multiparametric studies in brain tumors, in C. Beckers, A. Goffinet, A. Bol (eds), *Positron emission tomography in clinical research and clinical diagnosis.*, Kluwer Academic Publishers, Dordrecht, pp. 229.

28. Vaalburg, W., Coenen, H.H., Crouzel, C., et al. (1992) Amino acids for the measurement of protein synthesis in vivo by PET, *Nucl Med Biol* **19**, 227-237.

29. Phelps, M.E., Barrio, J.R., Huang, S.C., et al. (1984) Criteria for the tracer kinetic measurement of cerebral protein synthesis in humans with positron emission tomography, *Ann Neurol* **15**, S192-202.

30. Willemsen, A.T.M., Van Wardee, A., Paans, A.M.J., et al. (1995) In vivo protein synthesis rate determination in primary or recurrent brain tumors using L-[1-11-C]-Tyrosine and PET, *J Nucl Med* **36**, 411-419.

31. Kubota, R., Kubota, K., Yamada, S. (1995) Methionine uptake by tumor tissue: a microautoradiographic comparison with FDG, *J Nucl Med* **36**, 484-492.

32. De Souza, B., Brunetti, A., Fulham, M.J. (1990) Pituitary Microadenomas: a PET study, *Radiology* **177**, 39-44.

33. Muhr, C., Bergstrom, M., Lundberg, P.O. (1986) Dopamine receptors in pituitary adenomas: PET visualization, *J Comput Assist Tomogr* **10**, 175-180.

34. Mogard, J., Kihlström, L., Ericson, K., et al. (1994) Recurrent tumor vs radiation effects after gamma knife radiosurgery of intracerebral metastases: Diagnosis with PET-FDG. *J Comput Assist Tomogr* **2**, 177-181.

Part four:
PET in neurology and psychiatry

PET studies in Neuropharmacology.

Novel approaches

David J Brooks

MRC Cyclotron Unit, Hammersmith Hospital, UK

Abstract

Recently, novel methods for analysing ligand binding have been developed (spectral analysis, statistical parametric mapping). This chapter presents these approaches and shows examples where their use in epilpesy and various movement disorders enabled changes in ligand binding to be localised that might not have been predicted. Additionally, the potential of using these techniques for studying changes in neurotransmitter fluxes under activating conditions is discussed.

Introduction

PET provides a means of examining cerebral pharmacology *in vivo* under both resting and activating conditions. A list of some of the neurotransmitter systems now amenable to study and the more commonly employed tracers is detailed below in Table 1:

Table 1: PET tracers available for studying neuropharmcology

Dopaminergic system:	dopa decarboxylase	^{18}F-dopa, ^{18}F-metatyrosine
	MAOB	^{11}C-deprenyl
	dopamine transporter	^{11}C-CFT, ^{11}C-RTI 32, ^{11}C-FP-CIT
		^{11}C-nomifensine
	vesicle transporter	^{11}C-dihydrotetrabenazine
	dopamine D$_1$	^{11}C-SCH23390 / 39166
	dopamine D$_2$	^{11}C-raclopride
		^{11}C-methylspiperone
		^{18}F-fluoroethylspiperone
		^{76}Br-bromospiperone
Serotonergic system:	serotonin reuptake	^{11}C-β-CIT (RTI 55)
	HT$_{1a}$	^{11}C-WY100635
	HT$_2$	^{18}F-altanserin
		^{18}F-setoperone
	HT$_{2a}$	^{11}C-methylspiperone

239

B. Gulyás and H.W. Müller-Gärtner (eds.),
Positron Emission Tomography: A Critical Assessment of Recent Trends, 239–248.
© 1998 *Kluwer Academic Publishers.*

Benzodiazepine sites:	central	^{11}C-flumazenil
	peripheral	^{11}C-PK11195
Opioid system:	μ	^{11}C-carfentenil, ^{18}F-cyclofoxy
	μ,κ,δ	^{11}C-diprenorphine
Histamine:		^{11}C-dothiepin
Acetylcholine:	muscarinic M_1,M_2	^{11}C-tropanyl benzylate
	vesicle transporter	^{11}C-vesamicol
Noradrenergic:	α_2	^{11}C-methoxyidazoxan
Glutamatergic:	NMDA	^{11}C-ketamine

Modelling approaches

KINETIC MODELLING

In order to derive information about the integrity of neurotransmitter systems from the time course of regional cerebral tracer uptake and arterial plasma activity it is necessary to use a kinetic model describing the tracer distribution between different physical and metabolic compartments. PET detects 511 keV photons and so cannot distinguish native tracer activity from that of a metabolite. Fortunately, most of the ligands used as PET tracers are not metabolised significantly by brain tissue. The rate of plasma tracer metabolite formation can be determined using serial HPLC measurements enabling a metabolite-corrected input function to be generated.

Traditionally, brain uptake of PET tracers has been described using compartmental analysis. The simplest approaches assume the brain comprises one or two compartments and arterial plasma comprises one compartment - see fig 1. More complex models include brain compartments representing non-specific ligand binding, ligand metabolism, and ligand diffusion to binding sites. At tracer doses of ligands, a series of first order rate constants can be used to describe the kinetic behaviour of a ligand. Rate constants are obtained by solving the linear simultaneous differential equations describing the kinetics of tracer binding using non-linear regression analysis. The ratio of the two rate constants describing ligand passage from free to bound brain compartments, k_3/k_4, is known as the binding potential (BP) and, at tracer doses of ligands, reflects $f_2.B_{max}/K_d$ where f_2 is the tissue free tracer fraction, B_{max} the receptor availability, and K_d the receptor affinity [1]. Measurement of the BP in the presence of different levels of cold ligand along with conventional Scatchard analysis allows B_{max} and K_d

Figure 1

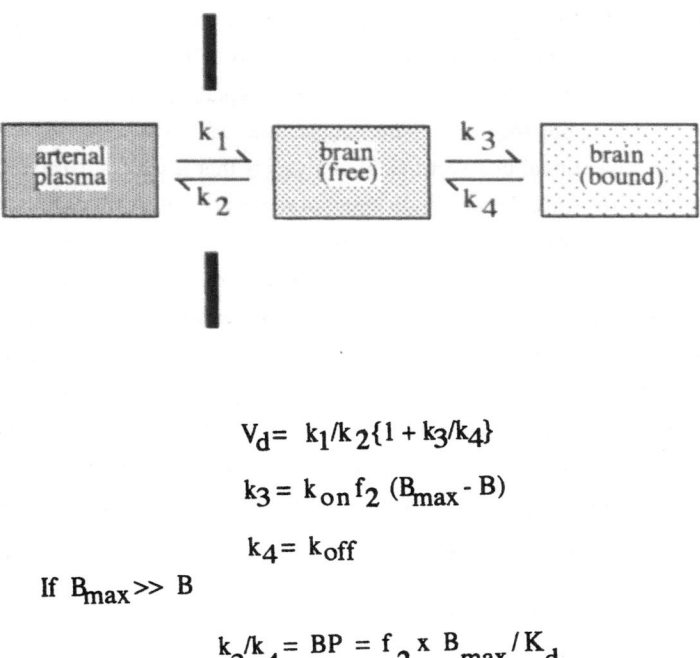

$$V_d = k_1/k_2 \{1 + k_3/k_4\}$$

$$k_3 = k_{on} f_2 (B_{max} - B)$$

$$k_4 = k_{off}$$

If $B_{max} \gg B$

$$k_3/k_4 = BP = f_2 \times B_{max}/K_d$$

to be independently estimated [2]. As in most neurodegenerative conditions K_d is unchanged, measurements of BP provide a direct reflection of changes in B_{max} without necessitating displacement studies. Compartmental modelling requires an input function and regional brain uptake curves. The input function is generally a metabolite-corrected arterial plasma curve but if a non-specific brain reference area is available (eg cerebellum for dopaminergic tracers, occipital cortex for opioid tracers) then a brain reference uptake curve can often be substituted as an input function avoiding the need for arterial cannulation [3,4].

A problem with conventional compartmental analysis is that it makes *a priori* assumptions about the number of pharmacological brain compartments present. A way around this problem is to use spectral analysis which estimates the minimum number of kinetic components required to describe the data set without making any assumptions about their physiological nature [5]. In essence, the brain tissue function is modelled as a convolution of the measured plasma input function with a sum of k exponential terms where k is the number of frequency components required to describe brain uptake. This exponential function is known as an impulse function and when integrated yields parametric images of tracer total volume of distributions (V_d) which can then be sampled using a conventional region of interest (ROI) approach or subjected to statistical parametric mapping (SPM). Specific V_d for a region can be estimated by subtracting the reference area V_d from the total V_d.

ROI ANALYSIS VERSUS STATISTICAL PARAMETRIC MAPPING

Region of interest (ROI) analysis is conventionally used to determine regional brain levels of tracer uptake and to compare these in normal and pathological situations. There are, however, a number of problems that arise when the ROI approach is employed: First, *a prioi* assumptions have to be made about the shape and size of ROI's used to sample brain activity. This can, in part, be overcome by co-aligning all functional images to MRI and using the latter to define ROI's, however, brain orientation may vary from subject to subject. Alternatively, all functional images can be transformed into standard stereotactic space though this inevitably results in some loss of functional resolution. Second, an *a prioi* selection of the locations of ROI's relevant to the analysis has to be made. It is not feasible to analyse every cortical area and subcortical nucleus with the ROI approach. Third, brain shape can vary significantly between subjects. Standardised ROI templates do not allow for this while individually tailored ROI's result in non-uniform analyses.

An alternative approach to the analysis of changes in ligand binding associated with pathological situations or the presence of competitve agents is to use statistical parametric mapping (SPM) [6]. Here, the whole 3D brain volume is considered without any requirement for *apriori* definition of ROI's. In essence, parametric images describing ligand influx (K_i) or binding (V_d) are transformed into standard stereotactic space. Images of group mean tracer uptake with associated standard deviations are then generated on a voxel-by-voxel basis. This data set can then be used to compare tracer uptake by individual or groups of patients with that of control subjects, or to examine the effects of competitve doses of cold ligands or activating paradigms on regional tracer binding. The location of volumes of significantly altered K_i or V_d can be detected at preassigned thresholds (generally $p<.001$ to avoid spurious changes engendered by the multiple comparisons performed) and the magnitude of these changes measured. The SPM approach has the clear advantage that changes in regional ligand binding can be detected on a voxel basis that might not have been otherwise predicted. Its disadvantage is the requirement for stereotactic transformation of 3D data sets.

Applications

LOCALISATION OF EPILEPTIC FOCI

Functional imaging provides a highly sensitive means of detecting the location of epileptic foci and may obviate the need for invasive depth electrode studies. What is less clear is whether PET is more sensitive than quantitative volumetric MRI in this situation. 11C-flumazenil PET enables central benzodiazepine (BDZ) site binding to be assessed *in vivo* . The central BDZ site is a component of the $GABA_A$ complex and BDZ agonists increase the frequency of chloride channel opening by GABA. With 11C-flumazenil PET, the SPM approach has revealed that patients who have focal cortical dysplasia evident on MRI have a volume of dysfunction that extends far beyond the apparent anatomical limits and often have further occult cortical dysplastic lesions [7]. The same is true for patients with cortical migration abnormalities such as nodular or band heterotopias and schizencephaly. These 11C-flumazenil PET findings may explain the poor outcome of such patients when only those lesions evident on MRI are surgically removed. In contrast, MRI and 11C-flumazenil PET demonstrate similar extents of structural and functional involvement in patients with hippocampal sclerosis [8]. An area of current research concerns whether 11C-flumazenil PET is able to demonstrate cortical dysfunction in patients with clinical and EEG evidence of focal epilepsy where volumetric MRI is normal. In a recent 11C-flumazenil PET study we have detected focal abnormalities of BDZ binding in eight out of nine such patients though only one of these eight patients had isolated unilaterally decreased hippocampal binding (Koepp MJ; unpublished observations).

MOVEMENT DISORDERS

Parkinson's disease

18F-dopa PET, a marker of dopa decarboxylase activity, has been widely used to demonstrate changes in striatal dopamine terminal function in Parkinson's disease. Additionally, subclinical changes can be detected in the ipsilateral putamen of early cases with focal limb involvement and also in around 30% of asymptomatic adult relatives of PD cases with familial disease. It is well known that there is a strong dopaminergic projection from the midbrain tegmentum to prefrontal areas and that the substantia nigra also is rich in dopamine terminals. To date, however, PET has failed to reveal abnormal 18F-dopa signals in these areas in PD.

We have recently developed a means of generating 3D parametric images of 18F-dopa influx constants (K_i) for normal subjects and PD patients enabling SPM to be applied [9]. With SPM early cases of hemiparkinsonism show bilateral striatal decreases in 18F-dopa uptake, the dorsal putamen being most affected. Additionally, increases in anterior cingulate and mesial prefrontal uptake can be detected suggesting that upregualtion of mesofrontal dopaminergic projections is occurring. In patients with established bilateral disease SPM shows both striatal and nigral decreases in 18F-dopa K_i while the initially increased prefrontal binding has now normalised.

We have also used 11C-diprenorphine PET to examine opioid binding in PD. Current theories of basal ganglia function suggest that dyskinesias may result from dysfunction of the "indirect" striatal-external pallidal pathway whose projections contain GABA and enkephalin. In early PD patients with a sustained non-dyskinetic response to levodopa no abnormalities of 11C-diprenorphine are evident [10]. In dyskinetic patients, however, SPM reveals that there are significant reductions in striatal, thalamic, and cingulate opioid binding with increases in prefrontal 11C-diprenorphine uptake [11]. This suggests that the development of involuntary movements in these patients may in part be due to alterations in opioid transmission.

Essential tremor

The aetiology of essential tremor (ET) is still unknown but it affects around 5% of the population and is familial in 50% of cases. Animal models have suggested that it may arise from overactivity of olivary cerebellar projections which in turn project to the red nucleus and ventral thalamus. $H_2^{15}O$ PET activation studies from our unit have shown cerebellar, red nuclear, and thalamic overactivity in ET, the cerebellar overactivity being evident even at rest in the absence of tremor [12,13]. When ET patients are given ethanol 70% obtain transient relief from their tremor. We have shown that the ethanol, which acts at $GABA_A$ sites, results in a depression of cerebellar activity along with tremor relief [14]. In an attempt to image $GABA_A$ site function directly in ET we have recently performed ^{11}C-flumazenil PET on a group of these patients at rest in the absence of tremor. SPM showed bilaterally increased ^{11}C-flumazenil binding in the ventral thalamus suggesting that disordered GABA transmission may be associated with this disorder (Boecker H; unpublished observations).

Huntington's disease

Affected Huntington's disease gene carriers show a severe (mean 60%) parallel loss of striatal D_1 and D_2 binding with ^{11}C-SCH23390 and ^{11}C-raclopride PET which correlates with the degree of rigidity present [15]. By symptom onset striatal D_1 and D_2 binding is reduced by around 30%. Reduced striatal D_1 and D_2 binding has also been detected in 50% of asymptomatic adult HD gene carriers [16]. Post-mortem studies on HD gene carriers with early choreic or preclinical disease have suggested that there may be a selective loss of "indirect" striatal-external pallidal enkephalin projections [17,18]. ^{11}C-diprenorphine PET shows clear cut reductions in striatal opioid binding in choreic HD patients using a conventional region of interest approach. Interestingly, when SPM is applied to parametric images not only are these striatal reductions confirmed but cingulate reductions and thalamic and prefrontal increases in opioid binding are revealed [19]. These changes are strikingly similar to those seen in dyskinetic PD patients and again suggest that disruption of opioid binding may underlie chorea in these two distinct conditions.

Tourette's syndrome

This condition is characterised by simple and complex motor and vocal tics along with obsessive and compulsive behaviour. In its most severe form coprolalia and copropraxia are features. Its aetiology is uncertain but it is often familial. It has been widely hypothesised to be a basal ganglia disorder and involve disturbance of dopamine transmission based on rather soft evidence. It is known that dopaminergic agents can exascerbate and dopamine blocking and depleting agents relieve symptoms. An autopsy study has also suggested that striatal dopamine reuptake sites may be increased. However, there is also evidence that a disturbance of opioid transmission may underlie this condition. At post-mortem a loss of pallidal dynorphine staining has been reported and patients have been reported to respond to naltrexone [20,21].

We have performed ^{11}C-diprenorphine PET in TS patients and found no change in basal ganglia opioid binding. Interestingly, however, SPM of parametric images detected clear reductions in cingulate and rises in prefrontal ^{11}C-diprenorphine binding [19]. This would suggest that TS may be a disorder of frontal rather than basal ganglia function. This makes a degree of sense as patients with cingulate seizures can make vocal utterances and perform stereotyped movements similar to the tics of TS while ictal and often report that they feel compelled to do this.

ACTIVATED MICROGLIA IN MUTIPLE SCLEROSIS

Peripheral benzodiazepine sites are not normally expressed in the central nervous system. In the presence of trauma, ischemia, neoplasia, and inflammatory disorders, microglia become activated and macrophage invasion occurs leading to expression of peripheral BDZ receptors [22]. ^{11}C-PK11195 is an antagonist at such sites and can be used to detect the presence of neuroinflammation. Our preliminary data suggest that PET provides a sensitive means of detecting not only the active multiple sclerosis plaques revealed by Gd-enhanced MRI, but also occult distant expression of peripheral BDZ binding (Banarti R, unpublished observations). It is likely that some of the microglial activation detected represents a reaction to disconnection of regions by the plaques of demyelination. In the future, ^{11}C-PK11195 PET should enable the effect of immunotherapy, such as β-interferon, on neuroinflammatory processes to be directly monitored.

Ligand activation studies

To date, PET ligand studies have been generally performed under resting conditions. In principle, however, by measuring levels of ligand binding under activating situations it should be possible to indirectly monitor endogenous neurotransmitter fluxes. That PET can detect changes in receptor availability following changes in synaptic neurotransmitter levels has already been demonstrated by pharmacological challenge studies. Significant displacement of ^{11}C-raclopride, a reversibly bound dopamine D_2 antagonist, occurred from striatal sites after adminstration of the dopamine reuptake inhibitor methylphenidate [23]. More recently, we have shown displacement of the non-selective opioid PET ligand ^{11}C-diprenorphine from all brain opioid sites following inhalation of 30% nitrous oxide (Dagher A; unpublished observations). This agent is known to have analgesic properties due to its induction of endorphin release.

We have now used ^{11}C-diprenorphine to study changes in endorphine release in patients exposed to cutaneous capsaicin. At onset of pain, 40-50' into the study, the thalamic V_d for ^{11}C-diprenorphine binding was reduced (Dagher A; unpublished observations). Patients with seizures also show displacement of ^{11}C-diprenorphine from opioid sites during ictal activity. Those with primary generalised absence seizures show displacement of ^{11}C-diprenorphine from all association cortical areas [24] while patients with reflex epilepsy induced by reading text showed reduced parietal ^{11}C-diprenorphine binding while experiencing throat and jaw contractions along with a feeling of detachment [25]. We are currently investigating the effects of changes in mood and active learning on endorphine and dopamine release.

Conclusions

Using SPM with parametric 3D images of ligand binding makes it possible to detect focal changes in brain function that could not have been predicted. This approach with ^{11}C-flumazenil PET has revealed occult cortical GABA$_A$ binding abnormalities in patients with focal epilepsy and abnormal thalamic GABA$_A$ binding in patients with essential tremor. Used with ^{18}F-dopa PET this approach has revealed abnormal midbrain and frontal dopamine storage in addition to striatal dysfunction in Parkinson's disease. Cases of Tourette's syndrome have shown abnormal frontal ^{11}C-diprenorphine binding suggesting that this syndrome may represent a frontal rather than a basal ganglia disorder. Striatal, thalamic, and frontal opioid binding can be shown to be disrupted in dyskinetic Parkinson's disease and choreic Huntington's disease. Finally, focal displacement of ^{11}C-diprenorphine binding from cingulate and thalamus can be demonstrated in subjects while experiencing painful stimuli and from association cortex in epileptic patients during their seizure activity.

References

1. Mintun, M.A., Raichle, M.E., Kilbourne, M.R., Wooten, G.F. and Welch, M.J. (1984) A quantitative model for the in vivo assessment of drug binding sites with positron emission tomography, *Ann.Neurol.* **15**, 217-227.
2. Farde, L., Eriksson, L., Blomquist, G. and Halldin, C. (1989) Kinetic analysis of central [11C]raclopride binding to D$_2$-dopamine receptors studied by PET - A comparison to the equilibrium analysis, *J.Cereb.Blood Flow Metabol.* **9**, 696-708.
3. Brooks, D.J., Salmon, E.P., Mathias, C.J., et al. (1990) The relationship between locomotor disability, autonomic dysfunction, and the integrity of the striatal dopaminergic system, in patients with multiple system atrophy, pure autonomic failure, and Parkinson's disease, studied with PET, *Brain* **113**, 1539-1552.
4. Lammertsma, A.A., Bench, C.J., Hume, S.P., et al. (1996) Comparison of methods for analysis of clinical [11C]Raclopride studies, *J.Cereb.Blood Flow Metabol.* **16**, 42-52.
5. Cunningham, V. and Jones, T. (1993) Spectral analysis of dynamic PET studies, *J.Cereb.Blood Flow Metabol.* **13**, 15-23.
6. Friston, K.J., Frith, C.D., Liddle, P.F., Dolan, R.J., Lammertsma, A.A. and Frackowiak, R.S.J. (1990) The relationship between global and local changes in PET scans, *J.Cereb.Blood Flow Metab.* **10**, 458-466.

248

7. Richardson, M.P., Koepp, M.J., Brooks, D.J., Fish, D.R. and Duncan, J.S. (1996) Benzodiazepine receptors in focal epilepsy with cortical dysgenesis: An ^{11}C-flumazenil PET study, *Ann.Neurol.* **40**, 188-198

8. Koepp, M.J., Richardson, M.P., Brooks, D.J., Poline, J.B., Van Paesschen, W. and Duncan, J.S. (1996) Cerebral benzodiazepine receptors in hippocampal sclerosis: an objective in vivo analysis, *Brain* (In Press)

9. Rakshi, J.S., Uema, T., Ito, K., et al. (1996) Statistical parametric mapping of three dimensional ^{18}F-dopa PET in early and advanced Parkinson's disease, *Movement Disorders* **11** Supp 1, 147.

10. Burn, D.J., Rinne, J.O., Quinn, N.P., Lees, A.J., Marsden, C.D. and Brooks, D.J. (1995) Striatal opioid receptor binding in Parkinson's disease, striatonigral degeneration, and Steele-Richardson-Olszewski syndrome: An 11C-diprenorphine PET study, *Brain* **118**, 951-958.

11. Piccini, P., Weeks, R.A., Burn, D.J. and Brooks, D.J. (1996) PET studies on opioid receptor binding in Parkinson's disease patients with and without levodopa-induced dyskinesias, *Neurology* **46 Supp**, A454.

12. Jenkins, I.H., Bains, P.G., Colebatch, J.G., et al. (1993) A PET study of essential tremor: evidence for overactivity of cerebellar connections, *Ann.Neurol.* **34**, 82-90.

13. Wills, A.J., Jenkins, I.H., Thompson, P.D., Frackowiak, R.S.J., Findley, L.J. and Brooks, D.J. (1994) Red nuclear and cerebellar but no olivary activation associated with essential tremor: A positron emission tomography study, *Ann.Neurol.* **36**, 636-642.

14. Boecker, H., Wills, A.J., Ceballos-Baumann, A., et al. (1996) The functional effect of ethanol on alcohol-responsive essential tremor: A positron emission tomography study, *Ann.Neurol.* **39**, 650-658.

15. Turjanski, N., Weeks, R., Dolan, R., Harding, A.E. and Brooks, D.J. (1995) Striatal D_1 and D_2 receptor binding in patients with Huntington's disease and other choreas: A PET study, *Brain* **118**, 689-696.

16. Weeks, R.A., Piccini, P., Harding, A.E. and Brooks, D.J. (1996) Striatal D_1 and D_2 dopamine receptor loss in asymptomatic mutation carriers of Huntington's disease, *Ann.Neurol.* **40**, 49-54.

17. Albin, R.L., Reiner, A., Anderson, K.D., Penney, J.B. and Young, A.B. (1990) Striatal and nigral neuron subpopulations in rigid Huntington's disease: implications for the functional anatomy of chorea and rigidity-akinesia, *Ann Neurol* **27**, 357-365.

18. Preproenkephalin messenger RNA-containing neurons in striatum of patients with symptomatic and presymptomatic Huntington's disease: An in situ hybridisation study, *Ann.Neurol.* **30**, 542-549.

19. Weeks, R.A., Cunningham, V., Waters, S., Harding, A.E. and Brooks, D.J. (1995) A comparison of region of interest and statistical parametric mapping analysis in PET ligand work: 11C-diprenorphine in Huntington's disease and Tourette's syndrome, *J.Cereb.Blood Flow Metabol.* **15 Supp** 1, S41.

20. Haber, S.N., Kowell, N.W., Vonsattel, J.P., Bird, E.D. and Richardson, E.P. (1986) Gilles de la Tourette's syndrome: a postmortem neuropathological and immunohistochemical study, *J.Neurol.Sci.* **75**, 225-241.

21. Kurlan, R. (1989) Tourette's syndrome: Current concepts, *Neurology* **39**, 1625-1630.

22. Ramsay, S.C., Weiller, C., Myers, R., et al. (1992) Monitoring by PET of macrophage accumulation in brain after ischaemic stroke, *Lancet* **339**, 1054-1055.

23. Volkow, N.D., Wang, G-J., Fowler, J.S., et al. (1994) Imaging endogenous dopamine competition with [11C]raclopride in the human brain, *Synapse* **16**, 255-262.

24. Bartenstein, P.A., Duncan, J.S., Prevett, M.C., et al. (1993) Investigation of the opioid system in absence seizures with positron emission tomography, *J.Neurol.Neurosurg.Psychiat.* **56**, 1295-1302.

25. Koepp, M.J., Richardson, M.P. , Watabe, H., et al. (1994) Endogenous opioid release in reading epilepsy, *Epilepsia* **35 Supp 7**, 83.

PET STUDIES OF THE DOPAMINE HYPOTHESES IN SCHIZOPHRENIA

A-L. NORDSTRÖM
Department of Clinical Neuroscience
Psychiatry Section
Karolinska Hospital
S-171 76 Stockholm
Sweden

1. Introduction

The *dopamine hypothesis for the pathophysiology of schizophrenia* states that the schizophrenic symptoms are related to an overactivity in central dopaminergic transmission [1]. This hypothesis was stimulated by the *hypothesis for antipsychotic drug action* which postulates that the antipsychotic effect is mediated by a blockade of dopamine receptors [2]. Further, the psychotic symptoms induced by the dopamine releasing agent amphetamine resemble those of paranoid schizophrenia [3]. Later the dopamine hypothesis for schizophrenia was supported by post-mortem studies of brains from schizophrenic patients by the finding of an increased density of D2 dopamine receptors in the basal ganglia [4]. Some studies have reported increased densities only in patients who were maintained on neuroleptic drug treatment until the time of death indicating that the increase in D2 receptor density rather is induced by neuroleptic drug treatment.

The development of positron emission tomography (PET) has provided a tool for the study of receptor binding in the living human brain [5, 6]. Applying PET and selective radioligands, several studies have been performed aiming to test the dopamine hypotheses in schizophrenia *in vivo*. The studies can be devided into those aiming to test the dopamine hypothesis for antipsychotic drug action and to those aiming to test the dopamine hypothesis for the pathophysiology of schizophrenia.

2. The dopamine hypothesis for antipsychotic drug action

Quantitative methods have been developed to calculate the degree of D2 receptor occupancy induced by antipsychotic drug treatment in psychotic patients by using PET and D2 receptor selective radioligands. A widely accepted approach is to calculate a ratio between the uptake of the

249

B. Gulyás and H.W. Müller-Gärtner (eds.),
Positron Emission Tomography: A Critical Assessment of Recent Trends, 249–257.
© 1998 *Kluwer Academic Publishers.*

radioligand in the striatum (high density of D2 receptors) and the uptake in the cerebellum (negligable density of D2 receptors). The reduction of the striatum/cerebellar ratio (S/C ratio) in drug treated patients compared to unmedicated controls is used as an indicator of the degree of D2 receptor occupancy [7].

The dopamine hypothesis for antipsychotic drug action was supported in initial open PET studies by consistent findings of a high D2 receptor occupancy in patients treated with antipsychotic drugs [7-11]. More recent studies have focused on the relationship between degree of D2 receptor occupancy and clinical effects. Other recent studies have taken the advantage of PET to explore the mechanism of action of atypical antipsychotic drugs. In the studies of D2 receptor occupancy summarized below, [11C]raclopride has been used as the radioligand.

2.1 CLASSICAL NEUROLEPTICS

D2 receptor occupancy was examined in 22 schizophrenic patients treated with conventional doses of classical neuroleptics and who were all rated as much or very much improved. The D2 dopamine receptor occupancy was significantly higher in the patients who had extrapyramidal side effects (EPS) than patients who did not [12]. The finding indicates that the neuroleptic induced EPS are related to the degree of D2 receptor occupancy in the basal ganglia.

The relationship between D2 receptor occupancy and clinical effects was further explored in a double blind controlled study of 17 patients who were treated with the antipsychotic drug raclopride [13]. The patients were randomized to treatment with 2, 6 or 12 mg raclopride daily for four weeks. Thirteen patients participated in PET examinations for determination of D2 receptor occupancy. A statistically significant relationship ($p<0.05$) was found between the degree of D2 receptor occupancy and the antipsychotic clinical effect. The finding provided principally new evidence for the dopamine hypothesis for antipsychotic drug action.

Based on these findings, a hyperbolic relationship has been suggested between the dose (or plasma concentration) of the antipsychotic drug and the degree of D2 receptor occupancy where the threshold in D2 receptor occupancy for antipsychotic effect is lower than the threshold for antipsychotic effect [14]. This suggested relationship between D2 receptor occupancy and clinical effects has recently received further support by the finding of Kapur et al (1996) who reported that the D2 receptor occupancy in 7 patients treated with a low dose of haloperidol (2 mg daily) was 53-74%. Five of the seven patients showed substantial clinical improvement whereas none showed important side effects [15].

2.2 ATYPICAL ANTIPSYCHOTIC DRUGS

Clozapine is the prototype atypical antipsychotic drug. The frequency of EPS is low [16] and clozapine is efficacious in some patients who do not respond to classical neuroleptics [17]. In a PET study of 16 clozapine treated patients [18] the D2 receptor occupancy was significantly lower (20-67%) than in patients treated with classical neuroleptics (70-89%) [12].The finding indicate that a high D2 receptor occupancy is not an absolute prerequisite for antipsychotic effect. A simplistic explanation for the lack of EPS during clozapine treatment is that the D2 receptor occupancy is too low to induce EPS.

Several receptors other than the D2 receptor subtype have been suggested to be involved in the mechanism of action of clozapine [19]. By using PET and the D1 dopamine receptor antagonist [11C]SCH23390 it has been demonstrated that clozapine induces a relatively high D1 receptor occupancy (36-59%) [18] compared to that of classical neuroleptics (0-44%) [20]. In an open study of schizophrenic patients treated with the D1 receptor antagonist SCH-39166 no evident antipsychotic effect was recorded [21]. Based on this preliminary report it is not likely that the atypical effects of clozapine are related to a selective action on D1 receptors. However, the report does not preclude the possibility that a combined D1 and D2 receptor antagonism may contribute the the atypicality of clozapine.

It has been claimed that the atypical effects of clozapine are related to a combined effect on D2 dopamine and 5-HT2 receptors [22]. By using PET and [11C]N-methylspiperone a method has been developed to calculate the degree of 5-HT2 receptor occupancy in the frontal cortex [23]. In five clozapine treated patients is was confirmed that clozapine induce a very high 5-HT2 receptor occupancy (>80%). So far, few studies have been published regarding the degree of 5-HT2 receptor occupancy induced by classical neuroleptics. After haloperidol the 5-HT2 receptor occupancy is negligable whereas substantial 5-HT2 receptor occupancy has been reported in both patients treated with chlorpromazine (up to 69%) [24] and in thioridazine-treated patients (more than 50%) [25]. However, the combination of a low D2 receptor occupancy and high 5-HT2 is so far unique for clozapine.

According to a hypothesis put forward by Deutch et al [26], interaction with distinct isoforms of the D2 receptor may differentiate atypical from typical antipsychotics. In a study of the relationship between clozapine serum concentration and D2 receptor occupancy the relationship was best described by a saturation hyperbola giving a maximal D2 receptor occupancy of 61% [18]. One possible explanation for this finding is that clozapine interacts with distinct isoforms of the D2 receptor.

Olanzapine is a new antipsychotic drug. Binding studies of olanzapine *in vitro* have indicated a binding profile similar to that of clozapine. In a PET study of 3 healthy controls, 5-HT2 and D2 dopamine receptor occupancy was calculated after a single dose of 10 mg olanzapine [27]. Extrapolation from data obtained after a single dose suggested that at a lower dose range the D2 and 5-HT2 receptor occupancy may be similar to that induced by clozapine. Further PET examinations of patients treated with clinical doses of olanzapine under steaty state conditions are strongly suggested.

3. The dopamine hypothesis for the pathophysiology of schizophrenia

Different methods have been developed to quantify the D2 receptor density in schizophrenic patients. For all methods it is crucial that the patients who are examined are neuroleptic naive, i.e. that any increase in D2 receptor density cannot be explained by a drug induced up regulation. The methods can be devided into two categories. The first category includes single PET examinations of patients examined with a radioligand selective for D2 receptors. The striatum/cerebellum ratio is used as an index of the D2 receptor density (single S/C ratios). The second category includes two methods requiring two PET examinations in each patient and more complicated models for calculation of D2 receptor density (the equilibrium model and the kinetic compartment model).

3.1 SINGLE STRIATUM/CEREBELLUM RATIOS

The results from studies using single S/C ratios as an index of D2 receptor density in the striatum is summarized in table 1.

TABLE 1. Striatum/cerebellum ratios as an index of D2 receptor density

Author	Radioligand	Pat. No	Results
Wong et al 1985	[11C]NMSP	13	no elevation
Martinot et al 1990	76Br-bromospiperone	12	no elevation
Martinot et al 1991	76Br-bromolisuride	19	no elevation

No increase in D2 receptor density was found among schizophrenic patients in these studies. Martinot et al found a significant decline in the S/C ratio with age among controls, whereas no such correlation was found in groups of schizophrenic patients [29, 30]. However, in another study a significant decline in the S/C ratio with age was found in both healthy controls and schizophrenic patients [31].

3.2 EQUILIBRIUM MODEL

The equilibrium model includes an in vivo saturation procedure for quantitative determination of D2 dopamine receptor density (Bmax) and affinity (Kd) [32]. Two PET examinations are performed. In the first examination [11C]raclopride with a high specific activity is injected, and in the second examination [11C]raclopride with a low specific activity. Bmax and Kd values are obtained from individual Scatchard plots. Results from studies performed applying this method are summarized in table 2.

TABLE 2. Results from studies applying the equilibrium model

Author	Pat. No	Control No	Results
Farde et al 1987, 1991	18	20	no elevation
Hietala et al 1994	13	10	no elevation

The hypothesis of a generally elevated D2 receptor density among schizophrenic patients was thus not supported by these studies. However, in the study by Farde et al [33], a left to right assymetry with regard to the D2 receptor density in the putamen was found among patients but not among controls. Hietala et al [34] identified a subgroup of patients with relatively high striatal D2 receptor density.

3.3 KINETIC COMPARTMENT MODEL

A kinetic compartment model is applied to determine the uptake rate constant (k3) before and after an oral dose of 7.5 mg haloperidol. To calculate Bmax values, the reciprocals of the two k3 values are plotted versus brain haloperidol concentration in a linear graph [35]. Results from studies performed applying this method are summarized in table 3.

TABLE 3 Results from studies applying the kinetic compartment model

Author	Pat. No	Control No	Results
Wong et al 1986	10	11	2-3 fold elevation
Wong et al 1993 (extended series)	25	17	2 fold elevation (considerable overlap)
Nordström et al 1995	7	7	no elevation

The 2 fold elevation in D2 receptor density reported by Wong et al could thus not be replicated by our group applying a method similar to that described by Wong et al.

3.4 CONCLUSION

PET studies so far, have not revealed generally elevated D2 receptor densities in the striatum of schizophrenic patients. The results do not exclude the possibility that there are subpopulations of patients with elevated D2 receptor densities. The studies summarized above are limited to the D2 receptor density in the striatum. It cannot be exluded that schizophrenia is related to an overactivity in dopaminergic neurotransmission induced by other mechanisms or in extrastriatal brain regions.

4. References

1. van Rossum, J. M. (1967) The significance of dopamine-receptor blockade for the action of neuroleptic drugs, *5th Collegium Internationale Neuropsychopharmacologicum* 321-329.
2. van Rossum, J. M. (1966) The significance of dopamine receptor blockade for the mechanism of action of neuroleptic drugs, *Arch Int Pharmacodyn Ther* **160**, 492-494.
3. Connell, P. *Amphetamine Psychosis*, Vol. 5. Chapman & Hall, 1958.
4. Davis, K. L., Kahn, R. S., Ko, G., and Davidson, M. (1991) Dopamine in Schizophrenia: A Review and Reconceptualization, *Am J Psychiatry* **148**, 1474-1486.
5. Wagner, H. N., Jr., Burns, H. D., Dannals, R. F., Wong, D. F., Langstrom, B., Duelfer, T., Frost, J. J., Ravert, H. T., Links, J. M., Rosenbloom, S. B., Lukas, S. E., Kramer, A. V., and Kuhar, M. J. (1983) Imaging dopamine receptors in the human brain by positron tomography, *Science* **221**, 1264-1266.
6. Sedvall, G.,Farde, L., Persson, A., and Wiesel, F.-A. (1986) Imaging of neurotransmitter receptors in the living human brain, *Arch Gen Psychiatry* **43**, 995-1005.
7. Farde, L., Wiesel, F.-A., Halldin, C., and Sedvall, G. (1988) Central D2-dopamine receptor occupancy in schizophrenic patients treated with antipsychotic drugs, *Arch Gen Psychiatry* **45**, 71-76.
8. Farde, L., Hall, H., Ehrin, E., and Sedvall, G. (1986) Quantitative analysis of D2 dopamine receptor binding in the living human brain by PET, *Science* **231**, 258-261.
9. Cambon, H., Baron, J. C., Boulenger, J. P., Loc'h, C., Zarifian, E., and Mazière, B. (1987) In Vivo Assay for Neuroleptic Receptor Binding in the Striatum: Positron Tomography in Humans, *Br J Psychiatry* **151**, 824-830.
10. Smith, M., Wolf, A. P., Brodie, J. D., Arnett, C. D., Barouche, F., Shiue, C.-Y., Fowler, J. S., Russell, J. A. G., MacGregor, R. R., Wolkin, A., Angrist, B., Rotrosen, J., and Peselow, E. (1988) Serial [^{18}F]N-Methylspiroperidol PET Studies to Measure Changes in Antipsychotic Drug D-2 Receptor Occupancy in Schizophrenic Patients, *Biol Psychiatry* **23**, 653-663.
11. Baron, J. C., Martinot, J. L., Cambon, H., Boulenger, J. P., Poirier, M. F., Caillard, V., Blin, J., Huret, J. D., Loc'h, C., and Maziere, B. (1989) Striatal dopamine receptor occupancy during and following withdrawal from neuroleptic treatment: correlative evalutation by positron emission tomography and plasma prolactin levels, *Psychopharmacology* **99**, 463-472.
12. Farde, L., Nordström, A.-L., Wiesel, F.-A., Pauli, S., Halldin, C., and Sedvall, G. (1992) PET-analysis of central D1- and D2-dopamine receptor occupancy in patients treated with classical neuroleptics and clozapine - relation to extrapyramidal side effects, *Arch Gen Psychiatry* **49**, 538-544.

256

13. Nordström, A.-L., Farde, L., Wiesel, F.-A., Forslund, K., Pauli, S., Halldin, C., and Uppfeldt, G. (1993) Central D2-dopamine receptor occupancy in relation to antipsychotic drug effects - a double blind PET study of schizophrenic patients, *Biol Psychiatry* **33**, 227-235.

14. Farde, L. (1996) The advantage of using positron emission tomography in drug research, *TINS* **19**, 211-214.

15. Kapur, S., Remington, G., Jones, C., Wilson, A., DaSilva, J., Houle, S., and Zipursky, R. (1996) High Levels of Dopamine D2 Receptor Occupancy With Low-Dose Haloperidol Treatment: A PET study, *Am J Psychiatry* **153**, 948-950.

16. Casey, D. E. (1989) Clozapine: neuroleptic-induced EPS and tardive dyskinesia, *Psychopharmacology* **99**, S47-S53.

17. Kane, J., Honigfeld, G., Singer, J., and Meltzer, H. (1988) Clozapine for the treatment-resistant schizophrenic, *Arch Gen Psychiatry* **45**, 789-796.

18. Nordström, A.-L., Farde, L., Nyberg, S., Karlsson, P., Halldin, C., and Sedvall, G. (1995) D1-, D2 and 5-HT2 receptor occupancy in relation to clozapine serum concentration -a PET study in schizophrenic patients, *Am J Psychiatry*

19. Meltzer, H. Y. (1994) An overview of the Mechanism of Action of Clozapine, *J Clin Psychiatry (suppl B)* **55**, 47-52.

20. Farde, L., Nordström, A.-L., Wiesel, F.-A., Pauli, S., Halldin, C., and Sedvall, G. (1992) PET-analysis of central D1- and D2-dopamine receptor occupancy in patients treated with classical neuroleptics and clozapine - relation to extrapyramidal side effects, *Arch Gen Psychiatry* **49**, 538-544.

21. Karlsson, P., Smith, L., Farde, L., Härnryd, C., Wiesel, F.-A., and Sedvall, G. Lack of apparent antipsychotic effect of the dopamine D1-receptor antagonist SCH39166 in schizophrenia. Presented at conference, "XIX C.I.N.P. Congress." Washington, 1994.

22. Meltzer, H. Y. (1992) The Importance of Serotonin-Dopamine Interactions in the Action of Clozapine, *British Journal of Psychiatry* **160**, 22-29.

23. Nyberg, S., Farde, L., Eriksson, L., and Halldin, C. (1993) 5-HT2 and D2 dopamine receptor occupancy in the living human brain. A PET study with risperidone., *Psychopharmacology* **110**, 265-272.

24. Trichard, C., Paillère-Martinot, M.-L., Monfort, J. C., Blin, J., Dao-Castellana, M. H., Syrota, A., and Martinot, J. L. Cortical 5-HT2 receptors and antipsychotic drugs studied with PET in schizophreina: Preliminary Results. Presented at conference, "Association of European Psychiatrists. Sixth European Congress." Barcelona, 1992.

25. Nyberg, S., Nakashima, Y., Nordström, A.-L., Halldin, C., and Farde, L. (1996) PET studies of in vivo Binding Characteristics of Atypical Antipsychotic Drugs: review of D2- and 5-HT2-receptor Occupancy Studies and Clinical Response, *Br J Psychiatry (suppl)* **29**, 40-44.

26. Deutch, A. Y., Moghaddam, B., Innis, R. B., Krystal, J. H., Aghajanian, G. K., Bunney, B. S., and Charney, D. S. (1991) Mechanism of action of atypical antipsychotic drugs. Implications for novel therapeutic strategies for schizophrenia., *Schizophrenia Research* **4**, 121-156.

27. Nyberg, S., Farde, L., and Halldin, C. (1996) A PET study of 5-HT2 and D2 dopamine receptor occupancy induced by olanzapine in healthy subjects, *Neuropsychopharmacology* in press

28. Wong, D. F., Wagner, H. N., Pearlson, G., Dannals, D. F., Links, J. M., Ravert, H. T., Wilson, A. A., Suneja, S., Bjorvvinssen, E., Kuhar, M., J., and Tune, L. (1985) Dopamine receptor binding of C-11-3-N-methylspiperone in the caudate in schizophrenia and bipolar disorder: a preliminary report, *Psychopharm Bull* **21**, 595-598.

29. Martinot, J.-L., Peron-Magnan, P., Huret, J.-D., Mazoyer, B., Baron, J.-C., Boulenger, J.-P., Loc'h, C., Maziere, B., Caillard, V., Loo, H., and Syrota, A. (1990) Striatal D_2 Dopaminergic Receptors Assessed With Positron Emission Tomography and [^{76}Br]Bromospiperone in Untreated Schizophrenic Patients, *Am J Psychiatry* **147**, 44-50.

30. Martinot, J. L., Paillère-Martinot, M. L., Loc'h, C., Hardy, P., Poirier, M. F., Mazoyer, B., Beaufils, B., Mazière, B., Allilaire, J. F., and Syrota, A. (1991) The Estimated Density of D_2 Striatal Receptors in Schizophrenia -A Study with Positron Emission Tomography and ^{76}Br-Bromolisuride, *Br J Psychiatry* **158**, 346-350.

31. Nordström, A.-L., Farde, L., Pauli, S., Litton, J.-E., and Halldin, C. (1992) PET analysis of central [11C]raclopride binding in healthy subjects and schizophrenic patients- reliability and age effects, *Hum Psychopharmacol* **7**, 157-165.

32. Farde, L., Wiesel, F.-A., Stone-Elander, S., Halldin, C., Nordström, A.-L., Hall, H., and Sedvall, G. (1990) D_2 dopamine receptors in neuroleptic-naive schizophrenic patients, *Arch Gen Psychiatry* **47**, 213-219.

33. Farde, L., Wiesel, F., Hall, H., Halldin, C., Stone-Elander, S., and Sedvall, G. (1987) No D_2 receptor increase in PET study of schizophrenia, *Arch. Gen. Psychiat.* **44**, 671-672.

34. Hietala, J., Syvälahti, E., Vuorio, K., Någren, K., Lehikoinen, P., Ruotsalainen, U., Räkköläinen, V., Lehtinen, V., and Wegelius, U. (1994) Striatal D2 dopamine receptor characteristics in neuroleptic-naive schizophrenic patients studied awith positron emission tomography, *Arch Gen Psychiatry* **51**, 116-123.

35. Wong, D. F., Wagner, H. N., L.E., T., Dannals, R. F., Pearlsson, G. D., Links, J. M., Tamminga, C. A., Broussolle, E. P., Ravert, H. T., Wilson, A. A., Toung, J. K. T., Malat, J., Williams, F. A., O'Touma, L. A., Snyder, S. H., Kuhar, M. J., and Gjedde, A. (1986) Positron Emission Tomoghraphy reveals elevated D2-dopamine receptors in drug-naive schizophrenics, *Science* **234**, 1558-1563.

PET AND EPILEPSY IN ADULTS

Ivanka SAVIC
Department of Neuroscience, Karolinska Institute,
S-171 77 Stockholm, Sweden

1. Introduction

Positron Emission Tomography (PET) was introduced for studies on human epilepsy about two decades ago, and continues to be an important tool for localizing the epileptogenic region in patients with medically refractory epilepsy who are candidates for surgical treatment. Measurements of regional glucose metabolism (rCMRglu) using 18F-fluorodeoxyglucose (FDG) was the initial study of choice in epilepsy. Such studies require a steady state condition, and show a reduced metabolism within the epileptogenic region during the interictal state and elevated metabolism during the ictal state. In patients with temporal lobe epilepsy the interictal measurements identify the epileptogenic zone in 85% of cases. The area of hypometabolism, however, often exceeds the zone of seizure onset (defined from EEG), and although the underlying mechanisms are still unknown, it has been suggested that this area may denote both the epileptogenic region and the area of interictal dysfunction. Localized hypometabolism apart from areas of obvious lesions seen on MRI and CT is rare in patients with neocortical onset epilepsy, indicating a pathophysiologic difference between seizures generated in mesial temporal limbic structures and those generated in the neocortex.

Increasing interest has therefore been raised for PET studies on neuroreceptors, especially the opiate and benzodiazepine receptors, which have been vizualized with 11-C carfentanil and 11-C flumazenil. Interictal carfentanil PET studies of temporal lobe epilepsy have revealed increased mu-opioid receptor binding in neocortex of the epileptogenic temporal lobe, which could have relevance to opioid mechanisms of interictal behavioral disturbances. Interictal Flumazenil PET, on the other hand, reveals a reduction in benzodiazepine receptor binding in epileptogenic mesial temporal structures, which could reflect disturbances specific to GABA-mediated inhibitory mechanisms, or merely cell loss in this area. Of particular interest is that BZ receptor binding is also decreased in neocortical epileptogenic regions, which may become important for the presurgical localizations in these patients.

Results from different PET studies on generalized epilepsy are more difficult to evaluate because generalized epilepsy consists of a variety of syndromes. In general, ictal FDG PET measurements reveal increased, and interictal measurements a normal or diffusely reduced glucose metabolism. In contrast, results from Flumazenil studies are more variable, and related to the type of generalized epilepsy: whereas normal

B. Gulyás and H.W. Müller-Gärtner (eds.),
Positron Emission Tomography: A Critical Assessment of Recent Trends, 259–272.
© 1998 *Kluwer Academic Publishers.*

flumazenil binding has been demonstrated in patients with absence epilepsy, a reduced binding in the thalamus and elevated in the cerebellar nuclei has been reported in a pilot study on patients with generalized tonic clonic seizures.

PET will continue to be an important clinical and research tool in human epilepsy. Compared with other functional imaging techniques is has the advantage to measure different, and not only one specific neuronal function. PET should therefore become increasingly important in the future as a tool for understanding in vivo cerebral biochemical and pharmacological changes related not only to epileptogenesis, but to consequences of epilepsy, such as interictal behavioral disturbances, and to pharmacotherapy.

Epileptic seizures are paroxysmal behavioral manifestations caused by increased excitability and synchronous discharges of large neuronal populations. The underlying mechanisms are unknown, and changes in excitatory as well as inhibitory transmitter systems have been demonstrated in animal experiments (1-4).

Seizures represent a symptom complex rather than a disorder. In partial epilepsies, seizure manifestations reflect the specific function of the generating cerebral region. The area generating partial seizures is called the epileptogenic region (5). This abnormality is important to identify because its removal may lead to remission of seizures. The s c epileptogenic lesion, which may be gliosis, tumor, developmental abnormality, hemorrhage or focal infection, is often a part of epileptogenic region, and possibly but not necessarily, the cause of seizures. The epileptogenic region is characterized by an abnormality that persists between the ictal events. It is also characterized by reorganization of synapses, neuroreceptor and -transmitter changes, as well as cell loss (1). Since seizures are excitotoxic, synaptic reorganization may possibly exist also in the target areas for focal seizure activity, as a secondary phenomenon. This is especially valid for patients with partial seizures of limbic type, in whom a third zone is defined, the so called zone of interictal dysfunction (5). This zone may be related to the chronic behavioral disturbances which may be observed in patients with limbic epilepsy. The epileptogenic lesion, epileptogenic region, and the zone of interictal dysfunction are not necessarily overlapping, and are important to distinguish, especially during the evaluation for epilepsy surgery. The PET technique may here serve as an important tool.

2. 18F-FDG PET in partial epilepsy

2.1 LIMBIC SEIZURES

In patients with limbic seizures of mesial temporal lobe origin, which is the most frequent form of intractable epilepsy, the epileptogenic region shows interictal hypometabolism. This is observed in about 80% of cases (6-8). However, the hypometabolic changes usually exceed the size of the epileptogenic lesion, as well as the EEG-defined epileptogenic region (9). Furthermore, both the degree and the extent of hypometabolism varies, even in patients with identical pathology (10).

Over the past ten years attempts have been made to map the pattern of interictal regional hypometabolism in partial epilepsy and to associate it with various factors supposed to influence this pattern. Although some researchers suggest that patients with mesial temporal lobe tumors preferably have mesial temporal lobe hypometabolism, and those with hippocampal sclerosis an anterior and lateral temporal lobe hypometabolism (11), no consistent anatomical pattern has, up to now, been possible to demonstrate (12). Thus, among patients who have mesial temporal sclerosis as the sole pathology underlying their seizures, subcortical changes may, or may not be found. Furthermore, whereas some patients exhibit parietal and frontal lobe hypometabolism coexisting with temporal lobe hypometabolism others do not (13,14).

A hypothesis was therefore recently put forward that the hypometabolic pattern in patients with partial epilepsy may be related to the features of the particular seizure that preceded the PET scan (15).

A retrospective study was conducted on 53 pat with seizures of mesial temporal lobe origin, 48 with surgically verified hippocampal sclerosis. Patients were classified into four groups depending on semiology of the seizure preceding the PET scan as determined from EEG - video telemetry. In fourteen patients this seizure was focal limbic, (characterized by auras or staring spells), in eighteen the CPS was widespread limbic (including automatisms); ten patients had a CPS with posturing, and eleven a secondarily generalized CPS. Regions with hemisphere-normalized concentration of [18F]-FDG below the 95% confidence interval of values from eight controls were defined as hypometabolic. The location of these regions was then compared among the four groups, and the degree of hypometabolism related to the time from seizure to the [18F]-FDG PET scan.

The hypometabolic area was limited to the epileptogenic zone if the preceding seizure was focal limbic, whereas in patients with widespread limbic seizures it included additional limbic cortex (p=0.03). Patients with posturing differed from both previous groups by having hypometabolism in the extralimbic frontal lobe (p<0.0001), and subjects with secondarily generalized seizures differed from all others due to cerebellar (p<0.001) and parietal lobe (p<0.05) reductions (Fig 1). In ten patients the seizure preceding the [18F]-FDG-PET scan was not the habitual one, and the last habitual seizure occurred 4-10 days earlier. The regional pattern of hypometabolism in all these patients was compatible with the seizure preceding the PET scan rather than with their habitual seizure, which was more limited.

These observations emphasize the functional character of the hypometabolic patterns, indicating that pathological metabolism may be related not only to the generation, but also expression of seizures. At least some of the observed interictal depressions of rCMRglu may result from functional impairments in cortico-cortical and cortico-subcortical connections either directly related to epileptic activity or as a sequele of long-standing therapy refractory epilepsy. The areas with interictal hypometabolism may correspond to the zone of interictal dysfunction and be related to the behavioral and cognitive disturbances which are frequently observed in patients with limbic seizures. For example, Rausch et al. reported a lower verbal IQ and a lower score on recall of logical prose in patients with left hemispheric hypometabolism

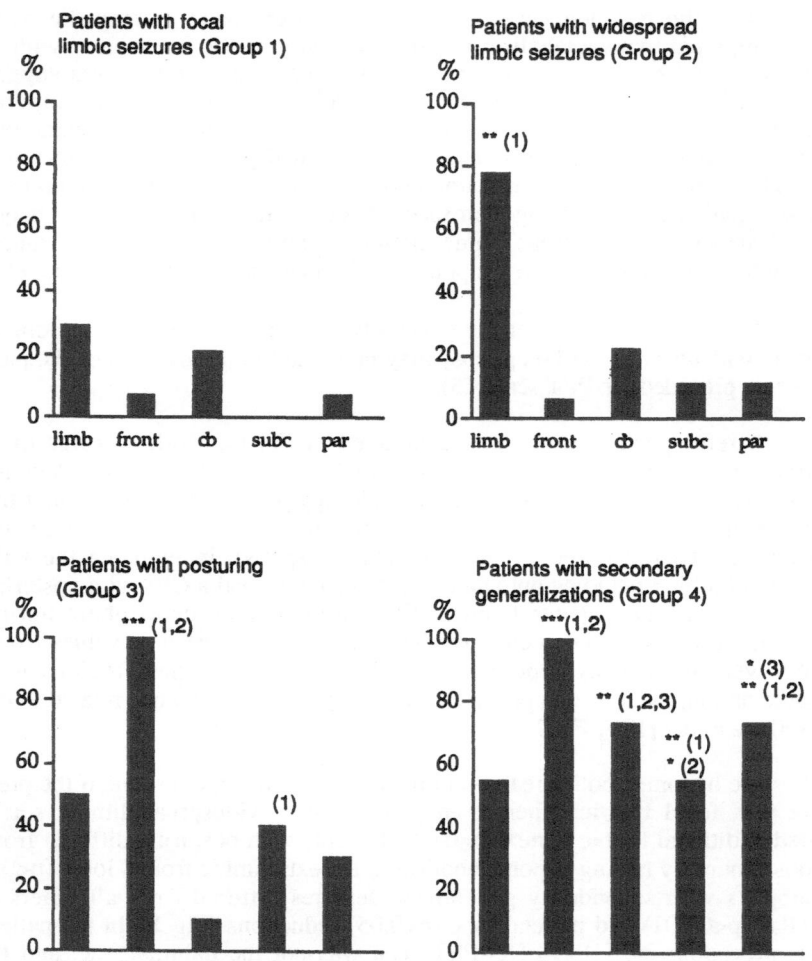

Figure 1. Percent of patients having one or several abnormal extrafocal regions within each of the five defined categories of regions (horisontal axis). Vertical axis: Percent patients within each group with one or several hypometabolic regions within the separate cathegory of regions.

Abbreviations: Limb = limbic regions; Front = frontal lobe cortex (extralimbic); Cb = cerebellum; Par = parietal cortex; Subcort = subcortical regions

Statistical significance (*= p<0.05; **= p<0.01, ***= p<0.001) is calculated in relation to the separate groups (specified under brackets).

From: Savic I, Altshuler L. Baxter L, Engel J Jr. Pattern of interictal hypometabolism in [18F]FDG PET reflects prior seizure semiology in patients with mesial temporal lobe seizures. Arch Neurol, in press. (With permission of the publisher.)

(16). Altshuler and col. found left inferior frontal lobe hypometabolism in depressed patients with mesial temporal lobe seizures, but not in the matched non depressed patients (17).

2.2 NEOCORTICAL SEIZURES

A localized interictal hypometabolism is less frequently observed in patients with neocortical partial seizures, and a correct detection has been reported in 50-91% of patients (18,19). Whether this difference in relation to limbic seizures is due to the underlying pathology, or to the fact that both ictal and postictal events are much shorter in neocortical epilepsies, is presently uncertain.

2. 3 MECHANISMS UNDERLYING METABOLIC CHANGES IN EPILEPSY

The mechanisms behind metabolic changes in epilepsy are fairly unknown. It has been suggested that the observed hypometabolic pattern during the interictal state may be a combined effect of lesion and deafferentation from the lesion. Such a concept is, however, incongruent with the observation that the extent of extrafocal hypometabolism decreases two years after epilepsy surgery (20). Alternatively, the synaptic connections may be re-organized within the epileptogenic region and its projection areas. A further possibility is that the area of hypometabolism reflects active mechanisms engaged to suppress subsequent ictal events, or, represents passive aftereffects of seizures. Such, seizure related mechanisms are suggested by recent observation of differentiated pattern of cerebellar hypometabolism in patients with partial seizures. In subjects with seizures of mesial temporal lobe origin there was a tendency towards ipsilateral cerebellar hypometabolism if the seizures were confined to the generating temporal lobe, whereas the cerebellar hypometabolism was contralateral if the seizures spread to the ipsilateral frontal lobe. Contralateral cerebellar hypometabolism was also found in patients whose seizures were generated by and confined to a frontal lobe (21).

3. 15O-H2O-PET in partial epilepsies

An other PET method that reflects dynamic events, and would be expected to illustrate regions involved in generation and spread of seizures is the measurement of cerebral blood flow (CBF), for example after a bolus injection of 15O-labeled water. Given the underlying assumptions of homogenous CBF and tracer blood:tissue partition coefficient, there is a nearly linear relation of tissue counts to CBF. Because of the short half-life of 15O, multiple injections can be given in a single experimental session, and the method is well suited for activation procedures. Interictal studies in patients with partial epilepsy seem, however, less localizing with respect to seizure onset than PET measurements of glucose metabolism (22). This discrepancy suggests that glucose metabolism and CBF may be uncoupled in partial epilepsy also during the interictal state, and that mechanisms underlying changes in interictal rCBF are not necessarily the same as those affecting the rCMRglu.

Ictal measurements with 15O-H2O-PET are difficult to conduct because O2 is a short lived tracer. In a recent study on 15 patients with complex partial seizures that were provoked with pentylenetetrazole Theodore et al. found bilateral temporal lobe increases in CBF, as well as bilateral increases in the thalamic and cerebellar blood flow (23), suggesting that these regions are actively involved in the spread of complex partial seizures.

Both rCBF and rCMRglu are state dependent parameters, and useful for imaging of seizure-effects, but perhaps less suitable for delineation of seizure generating regions. Neuronal reorganization characterizes severe partial seizures, and changes in both NMDA and GABA benzodiazepine (BZ) receptors have been found in surgically resected tissue (1). Therefore, it seems more appropriate to use tracers for the neuroreceptors involved in the epileptogenic process when trying to delineate the epileptogenic region with PET. Because synaptic reorganization theoretically should be more pronounced in patients with intractable seizures, receptor PET could be also be useful to predict this condition.

4. Flumazenil PET in partial epilepsies

Studies with flumazenil PET frequently show a reduced binding in the epileptogenic regions (24-32), both in neocortical and limbic epilepsies. The reduced 11-C flumazenil-binding is, according to quantitative models applied, caused by a reduction of receptor density rather than affinity (2).

Recent data suggest that patients with intractable seizures may have a reduced BZ receptor density (although less pronounced than the reduction in the epileptogenic region) also in the primary projection areas to the epileptogenic zone, which was not found in patients with pharmacologically controllable seizures. Furthermore, the BZ receptor reduction in the epileptogenic region was found to be more extensive and more pronounced in patients with intractable seizures, than in those with pharmacologically controllable seizures (30). Data from same study suggest a correlation between seizure frequency and degree of BZ receptor reduction in the epileptogenic zone, indicating that BZ receptor changes, at least in some patients with partial epilepsy, may be related to seizure generating mechanisms or, be an effect of seizures. This view is supported by the recent observation of specific pattern BZ receptor changes in the cerebellum, depends on the type of partial seizure. In a recent PET study a reduced density was found in the anterior and posterior cerebellum ipsilateral to the focus region in patients with mesial temporal lobe seizures, whereas subjects with frontal lobe seizures had reduced BZ receptor density in the anterior cerebellum contralateral to the epileptogenic zone (31), Fig 2. These differentiated changes are congruent with the available data on cortico-cerebellar projections, and suggest a seizure related phenomenon.

It has been suggested that the reduced flumazenil binding is a non specific finding, which merely may reflect loss of pyramidal cells. However, correlative studies on degree of BZ receptor reduction and cell loss indicate that the degree of BZ receptor reduction is relatively more pronounced than cell loss in area CA1 and CA3, and not

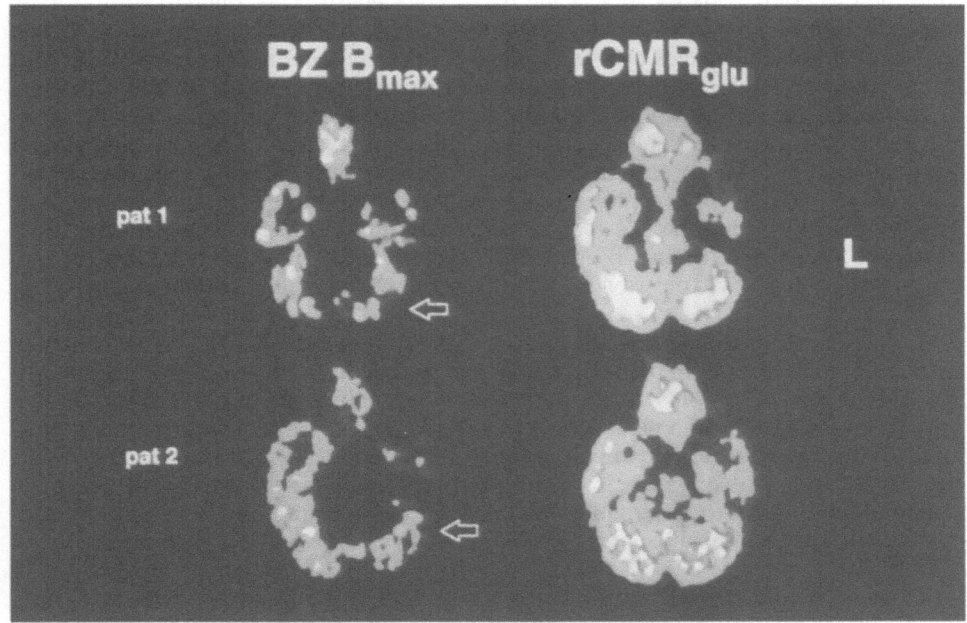

Figure 2. Two patients with seizure onsets in the left mesial temporal lobe. The BZ receptor density (left images) and the glucose metabolism (right images) is shown in each patient. Cerebellar BZ receptor density is reduced ipsilaterlly to the focus region in both patients. The scale (Sokoloff's scale) is arbitrary denoting as follows (pmol/cc): black 0, dark blue 15, light blue 30. green 40, yellow 65, and orange 80. (Colour figure: see colour section.)

From: Savic I, Thorell JO. Localized cerebellar changes of BZ receptor density in human partial epilepsy. Archives of Neurology, 1996;53:656-662. (With permission of the publisher.)

occurring in the dentate gyrus where the cell density is significantly decreased (33-34). Furthermore, in a recent pilot study on 11-C flumazenil binding before, and one year after the successful epilepsy surgery the BZ receptor density was found to normalize in the primary projection areas to the epileptogenic region, which presurgically showed reduced values. No changes between the two scans were seen in other cortical regions. All the four investigated patients had daily seizures before surgery, and became seizure free after. In one patient who had increased seizure frequency during the second scan, the reduction of BZ receptor density in the epileptogenic region became more pronounced (35). These data, although very preliminary, suggest a possibility of dynamic changes in BZ receptor density, perhaps in relation to seizures themselves. One possibility is that the BZ receptor exposing dendrites regenerate after seizure remission. An other is that BZ receptors were blocked by an endogenous ligand presurgery, or that the expression of BZ receptors may vary with seizure frequency.

Recent publication by Richardson and col. indicates that flumazenil binding may be increased in epileptogenic neocortex, especially if the underlying lesion is cortical dysplasia (36). This upregulation may reflect an increased cell density. It could also be compensatory to a depletion of GABA, suggesting that Vigabatrin could be a suitable treatment for these particular patients. A further evaluation of flumazenil PET in various forms of epilepsy is therefore highly encouraged. The technique may potentially be useful to differentiate between various seizure types, and perhaps also between those who probably are intractable vs. not.

5. PET studies on other receptors in partial epilepsies

5.1 OPIATE RECEPTORS

Opioid mechanisms in the brain play an important role in epilepsy, particularly in the mediation of postictal phenomena (37). Morphine and the opioid peptides elicit proconvulsant or anticonvulsant effects depending on the dose, rate of administration, and model system. Using a highly selective m opioide receptor agonist [11C]carfentanil, J Frost and col. demonstrated that patients with partial epilepsy of mesial temporal lobe origin and normal CT scans had a significantly elevated m opiate receptor binding in the temporal neocortex ipsilateral to the focus region (38). In the same area, the regional glucose metabolism, that was parallelly measured in these patients, was significantly reduced. Within the mesial temporal cortex, that according to EEG recordings was the seizure onset region, both the [11C]carfentanil binding and the glucose metabolism was slightly reduced.

In the next series of experiments Frost et al. compared the binding of [11C] diprenorphine and [11C]carfentanil in 11 patients with mesial temporal seizure onset in the surface EEG recordings. [11C]carfentanil, is a selective ligand for the m-rec, whereas [11C]diprenorphine binds non-specifically to m k and d opioide receptors. Therefore, it was of special interest to analyze if the binding pattern of the two ligands was different. The [11C]diprenorphine, displayed a symmetric binding in all temporal lobe regions. The selective increase in m-opioide receptor binding was interpreted to reflect a compensatory mechanism aimed to prevent the seizure spread from the mesial temporal cortex (39). Unfortunately, only anecdotal experiments with opiate receptor ligands have hitherto been conducted in patients with neocortical seizures, and it is presently uncertain whether the pattern of opiate receptor changes differs between patients with mesial temporal vs. neocortical seizures.

5.2 HISTAMINE RECEPTORS

Another compound that seems to be involved in the termination of seizures and may, therefore, function as an endogenous anticonvulsant is histamine. The drug doxepin is a well recognized antidepressant with high affinity for the histamine H1 receptors. Iinuma et al. used 11C-doxepin in nine patients with CPS of frontal or temporal lobe origin to study binding of H1 receptors. They found increased binding in the temporal neocortex ipsilateral to the focus in eight of nine patients. Like in the opiate studies of Frost and col., the authors speculated that the increased H1 receptor binding may represent a defensive mechanism in the brain serving to counteract the spread of epileptic discharges (40).

Taken together, the available PET data on partial epilepsy in adult patients indicate that whereas the interictal FDG PET often localizes both the epileptogenic zone and the zone of interictal dysfunction, Flumazenil PET seems to preferably indicate the epileptogenic region independently on its specific location. In contrast, the histamine and opiate PET can potentially demarcate the regions responsible to prevent seizure spread.

6. FDG PET in generalized epilepsies

Whereas numerous PET studies have hitherto been conducted on patients with partial epilepsy, relatively limited interest has been paid to patients with generalized epilepsies. Studies from generalized epilepsy patients are difficult to evaluate since generalized epilepsy consists of a variety of syndromes, and only data from similar and well defined patient populations may be compared.

In an initial study on four children with genuine petit mal epilepsy where FDG was injected just before 10 minutes of hyperventilation, Engel and col. reported a 2.0 to 3.5 times increased global cerebral metabolic rate over the control (41). In a follow-up experiment several months after successful drug therapy and when hyperventilation failed to induce absences no abnormalities in glucose metabolism were noted (41). A subsequent evaluation of ictal, postictal and control PET scan in a patient who underwent electroconvulsive chock treatments showed that the pattern of regional glucose metabolism varies depending on the state investigated: Whereas an increase in global cerebral metabolic rate was observed during the ictal scan, the global metabolic rate was postictally decreased (41).

7. PET studies on neuroreceptors in generalized epilepsies

Several other PET tracers have been used to identify possible anatomical substrates in primarily generalized seizures. Prevett and col. showed an increase in the thalamic cerebral blood flow during hyperventilation provoked absences in adult patients with petit mal epilepsy (42). They subsequently investigated same group of patients with 11-C diprenorphine (43), and 11-C flumazenil (44).

An increased wash out of 11-C dyprenorphine was found in the sensory cortices during absence seizures, but normal findings during the interictal state. Similar, but statistically insignificant changes were seen also in the thalamus. These data were interpreted to indicate a possible release of convulsive enkephalines during absences, which is in accordance with animal studies showing an induction of absences in cat after administration of enkefalines (37). In a follow up study on same category of patients and following the same design, the flumazenil binding was found to be normal (44) suggesting that the changes in 11-C diprenorphine binding were selective.

At variance to findings in subjects with absence seizures, a pilot study on nine patients with primarily generalized tonic and clonic seizures showed that the BZ receptor density was decreased in the thalamus and increased in the cerebellar nuclei (28). The observed changes are in accordance with the current theories about disturbed cerebello-thalamo-cortical connections, and a pathological recruitment of GABAergic neurons in patients with generalized seizures (45). It is, however, presently uncertain whether these changes represent the cause or result of seizures.

8. Summary

8.1 PET IN PRESURGICAL EVALUATION

FDG PET is useful in presurgical evaluation of pat with mesial temporal lobe seizures in whom the standardized anterior temporal lobe resection is planed. If PET is localizing, the intracranial recordings may be avoided.

FDG PET is also useful to exclude surgery, for example in an event of multiple hypometabolic areas. Some preliminary data indicate that it also may be predictive - the degree of focal hypometabolism is shown to be related to a better postsurgical outcome.

PET studies on neuroreceptors receptors are promising, and should be further explored. Especially the interaction between different receptor systems would be interesting to study since several receptor systems are altered in epilepsy.

8.2 FUTURE RESEARCH

Consecutive measurements of receptor binding in the same subject during different clinical conditions, for example before and after treatment are recommended in order to elucidate a possible state dependency of receptor changes.

Using multimethodological approaches, for example by combination of PET and magnetic resonance spectroscopy, it is now possible to study the relation between receptor and transmitter concentrations.

Partial epilepsy is a unique condition to study with PET, and may be regarded as a human model to study the correlation between behavioral changes and synaptic plasticity.

9. Acknowledgements

This study was supported by the Swedish Medical Research Council, Karolinska Institute, Wenner-Gren Center Foundations, Ture Nilsson's Foundation, The Swedish Institute, The Swedish Association for Neurologically Handicapped People, the Swedish Medical Society.

10. References

1. McDonald JW, Garofalo EA, Hood T, et al. Altered excitatory and inhibitory amino acid receptor binding in hippocampus of patients with temporal lobe epilepsy. Ann Neurol 1991;29:529-541.

2. Babb TL and Brown JW. Pathological findings in epilepsy. In: Engel J Jr. ed. Surgical treatment of the epilepsies. New York: Raven Press, 1987:511-540.

3. Krnjevic K. GABA mediated inhibitory mechanisms in relation to epileptic discharge. In: Jasper HH and van Gelder NM, eds. Basic Mechanisms of Neuronal Hyperexcitability. New York: Loss cop. 1983:49-80.

4. Schwartzkroin P. Hippocampal slices in experimental and human epilepsy. In: Delgado-Escueta AV, Ward AA Jr, Woddbury DM and Porter RJ, eds. Advances in Neurology. New York; Raven Press, 1986;44:813-847.

5. Engel Jr J. Basic mechanisms in epilepsy. In: Seizures and Epilepsy. Engel J Jr, editor. Philadelphia: F A Davis Company, 1989.

6. Engel J Jr, Henry TR, Risinger MW, et al. Presurgical evaluation for partial epilepsy: Relative contributions of chronic depth-electrode recordings versus FDG-PET and scalp-sphenoidal ictal EEG. Neurology 1990;40:1670-1677.

7. Theodore WH, Newmark ME, Sato S, et al. 18F fluorodeoxyglucose positron emission tomography in refractory complex partial seizures. Ann Neurol 1983;14:429-437.

8. Ryvlin P, Cinotti L, Froment J, et al. Metabolic patterns associated with non-specific magnetic resonance imaging abnormalities in temporal lobe epilepsy. Brain 1992;114:2363-2383.

9. Engel J Jr, Brown W, Kuhl D, et al. Pathological findings underlying focal

temporal hypometabolism in partial epilepsy. Ann Neurol 1982;12:518-528.

10. Henry TR, Mazziotta JC, Engel J Jr. Interictal metabolic anatomy of mesial temporal epilepsy. Arch Neurol 1993;50:582-589.

11. Hajek M, Antonini A, Leenders CL, Wieser HG. Mesiobasal versus lateral temporal lobe epilepsy: Metabolic differences in the temporal lobe shown by interictal 18F FDG positron emission tomography. Neurology 1993;43:79-86.

12. Henry TR, Babb TL, Engel J Jr, et al. Hippocampal neuronal loss and regional hypometabolism in temporal lobe epilepsy. Ann Neurol 1994;36:925-927.

13. Sperling MR, Gur R, Alavi A, Gur RE, Resnick S, O'Connor MJ, Reivich M. Subcortical metabolic alterations in partial epilepsy. Epilepsia 1990; 31(2):145-155.

14. Theodore WH, Fishbein D, Dubinsky R. Patterns of cerebral glucose metabolism in patients with partial seizures. Neurology. 1988;38:1201-1206.

15. Savic I, Altshuler L. Baxter L, Engel J Jr. Pattern of interictal hypometabolism in [18F]FDG PET reflects prior seizure semiology in patients with mesial temporal lobe seizures. Arch Neurol, in press.

16. Rausch R, Henry TR, Ary CM, et al. Asymmetric interictal glucose hypometabolism and cognitive performance in epileptic patients. Arch Neurol. 1994;51:139-144.

17. Altshuler L, Devinsky O, Post RM, Theodore W. Depression, anxiety, and temporal lobe epilepsy:laterality of focus and symptomatology. Arch Neurol 1990:47:284-290.

18. Henry TR, Sutherling WW, Engel J Jr, et al. Interictal cerebral metabolism in partial epilepsies of neocortical origin. Epilepsy Res 1992;10:174-182.

19. Swartz BE and Delgado-Escueta AV. Complex partial seizures of extratemporal origin. In: Wieser HG, Speckmann E-J, Engel J Jr, editors. The epileptic focus. London: John Libbey, 1987:137-175.

20. Hajek M, Wieser HG, Khan N, Antonini A, Schrott PR, Maquire P, Beer HF, Leenders KL. Preoperative and postoperative glucose consumption in mesio basal and lateral temporal lobe epilepsy. Neurology 1994; 44(11)2125-32.

21. Savic I, Altshuler L. Passaro E, Baxter L, Engel J Jr. Localized cerebellar hypometabolism in human partial seizures. Epilepsia 1996;37(8):781-787.

22. Gaillard WD, White S, Reeves P, Theodore WH. Comparison of 19-FDG-

PET with H2O (O-15)-PET and Tc-99m HMPAO-SPECT in patients with intractable partial seizures. Neurology 1992;42(suppl. 3):298.

23. Theodore WH, Balish M, Leiderman D, Bromfield E, Sato S, Herscovitch P. Effect of seizures on cerebral blood flow measured with 15O-H2O and positron emission tomography. Epilepsia 1996;37(8):796-802.

24. Savic I, Roland P, Sedvall G, Persson A, Pauli S, Widen L. In-Vivo demonstration of reduced benzodiazepine receptor binding in human epileptic foci. Lancet 1988; ii:864-66.

25. Savic I, Widen L, Thorell JO, Blomquist G, Ericson K, Roland P. Cortical benzodiazepine receptor binding in patients with generalized and partial epilepsy. Epilepsia 1990;31(6):724-30.

26. Savic I, Widen L, Stone-Elander S. Feasibility of reversing benzodiazepine tolerance with flumazenil. Lancet 1991;337:133-37.

27. Savic I, Ingvar M, Stone-Elander S. Comparison between 11-C flumazenil and [18F] FDG as PET markers of epileptic foci. J Neurol Neurosurg Psychiatry. 1993;5:615-621.

28. Savic I, Pauli S, Thorell JO. In vivo demonstration of altered Benzodiazepine receptor density in the cerebello-thalamo-cortical loop in human generalized epilepsy. J Neurol Neurosurg Psychiatry 1994;57:797-804.

29. Savic I, Thorell JO, Roland PE. [11C]Flumazenil PET specifically visualized frontal epileptogenic regions in humans. Epilepsia 1995;36:1225-1232.

30. Savic I, Svanborg E, J O Thorell. Cortical benzodiazepine receptor changes are correlated to seizure frequency in patients with intractable partial seizures. Epilepsia 1996;37:236-244.

31. Savic I, Thorell JO. Localized cerebellar changes of BZ receptor density in human partial epilepsy. Archives of Neurology, 1996;53:656-662.

32. Henry TR, Frey KA, Sackellares JC, et al. In vivo cerebral metabolism and central benzodiazepine-receptor binding in temporal lobe epilepsy. Neurology 1993;43:693-696.

33. Burdette DE, Sakurai SY, Henry TR, et al. Temporal lobe central benzodiazepine binding in unilateral mesial temporal lobe epilepsy. Neurology 1995;45:934-941.

34. Johnson EW, Lanerolle NC, Kim J, et al. "Central " and "peripheral" benzodiazepine receptors; opposite changes in human epileptogenic tissue. Neurology 1992;42:811-815.

35.   Savic I, Halldin C, Blomqvist G and Litton JE. Pre-and postsurgical 11-C
       flumazenil PET in temporal lobe epilepsy. Epilepsia 1995:36;6;S138.

36.   Richardson MP, Koepp MJ, Brooks DJ, Duncan JS. [11C]Flumazenil PET
       in extratemporal lobe epilepsy with normal MRI. Epilepsia 1996;37,Suppl
       4:123.

37.   Caldecott Hazard S, Shavit Y, Ackermann RF, Engel JJr., Fredrickson RCA
       and Liebeskind JC. Behavioural and electrographic effects of opioids on
       kindled seizures in rats. Brain Res. 1982;251:327-333.

38.   Frost JJ, Mayberg HS, Fisher RS, et al. M opiate receptors measured by
       Positron Emission Tomography are increased in temporal lobe epilepsy.
       Ann Neurol 1988;23:231-237.

39.   Mayberg HS, Sandzot B, Meltzer CC, Fisher RS, Lesser RP, Dannals RF,
       Lever JR, Wilson AA, Ravert HT, Wagner HN Jr, Bryan RN, Cromwell CC
       and Frost JJ. Quantification of mu and non-mu opiate receptors in temporal
       lobe epilepsy using positron emission tomography. Ann Neurol. 1991;30:3-
       11.

40.   Iinuma K, Yokoyama H, Otsuki T, et al. Histamine H1 receptors in complex
       partial seizures. Lancet 1993;238-241.

41.   Engel J Jr, Kuhl DE, Phelps ME. Patterns of local cerebral glucose
       metabolism during epileptic seizures. Science. 1982;218:64-66.

42.   Prevett MC, Duncan JS, Jones T, Fish D, Brooks D. Demonstration of
       thalamic activation during typical absence seizures using H2(15O) and PET.
       Neurology 1995:45(7) 1396-402.

43.   Bartenstein PA, Duncan JS, Prevett MC, Cunningham VJ, Fish DR, Jones
       AK. Investigation of the opiate system in absence seizures with positron
       emission tomography. JNNP 1993;56(12)1295-1302.

44.   Prevett MC. Lammertsma AA, Brooks DJ, Cunningham VJ, Fish D,
       Duncan JS. Benzodiazepine abd GABA receptor binding during absence
       seizures. Epilepsia 1995;36(6)592-9.

45.   Gloor P and Fariello RG. Generalized epilepsy: some of its cellular
       mechanisms differ from those of focal epilepsy. Trends in Neurosci.
       1988;1:63-68.

MAPPING CEREBRAL RESPONSES TO VOLATILE ANESTHETICS IN HUMANS

FERENC GYULAI
Department of Anesthesiology & Critical Care Medicine
University of Pittsburgh Medical Center
200 Lothrop Street, Room C-207
Pittsburgh, PA 15213-2582, U.S.A.

1. Introduction

The cerebral targets of volatile anesthetics have remained obscure despite abundant *in vitro* and *in vivo* animal studies addressing various aspects of anesthetic action. Recent advances in functional brain imaging, such as positron emission tomography (PET), offer an opportunity to investigate the net effects of a drug or stimulus on regional neuronal activity *in vivo*, in the entire human brain.

Drug effect on net cerebral metabolic activity can be mapped by measuring either regional cerebral blood flow (rCBF) or metabolic activity (rCMR), as reflections of neuronal activity.

Mapping cerebral neuronal activity by measuring absolute rCBF at the capillary level is based on the modified Kety-Schmidt technique [1], where flow-dependent accumulation or disappearance of a bolus-injected, freely diffusable positron-emitting radiotracer ^{15}O-water is used [2]. The rapid decay (half-life ~ 2 minutes) and favorable kinetic parameters of ^{15}O-water allow studies to be repeated in rapid succession with a relatively high temporal resolution. The alterations in rCBF, in turn, can be interpreted as reflecting changes in neuronal activity [3] provided rCBF and rCMR remain *coupled*. In order to take advantage of the ^{15}O-water technique for mapping the cerebral effects of volatile anesthetics, the assessment of rCBF/rCMR coupling is required, since it has been shown that flow and metabolism diverge in the presence of these agents [4].

The adaptation of the technique of Sokoloff et al. [5] allows the measurement of absolute rCMR with PET employing deoxyglucose labeled with the positron emitter ^{18}F (^{18}FDG) [6]; this substrate is trapped within neurons at a rate directly proportional to their metabolic activity. Since ^{18}F is taken up into neurons slowly, and has a half-life of 110 minutes, this technique has a poorer temporal resolution than ^{15}O-water PET scanning, and is not suitable for scanning during multiple experimental

273

B. Gulyás and H.W. Müller-Gärtner (eds.),
Positron Emission Tomography: A Critical Assessment of Recent Trends, 273–290.
© 1998 *Kluwer Academic Publishers.*

conditions in one session. However, this does not pose a limitation to rCBF - rCMR comparisons, where only a scan in the presence and another in the absence of the agent is obtained.

Since regional cerebral tracer concentration is linearly related to rCBF or rCMR, images obtained in the presence and absence of a drug can be directly compared (subtracted) to identify areas of activity changes following normalization of each scan to the same total number of counts [7]. Additional stereotactic normalization facilitates averaging of individual image differences which in turn increases signal-to-noise ratio and consequently sensitivity.

The feasibility of this technique will be demonstrated in elucidating the cerebral effects of nitrous oxide, the most frequently administered volatile anesthetic agent.

The antinociceptive effects of a low concentration of nitrous oxide (i.e., 20%) have been exploited for many years in the practice of dentistry, obstetrics, and emergency medicine. Recent work addressing the mechanisms of nitrous oxide analgesia suggests that some interaction may exist between nitrous oxide and endogenous opioid peptides. Animal studies have demonstrated nitrous oxide-induced met-enkephalin and β-endorphin release [8], and in humans at least some of nitrous oxide's antinociceptive effect has been shown to be reversible by naloxone [9]. Endogenous opioid peptide release, however, has not been confirmed in human studies, largely due to methodological difficulties posed by the invasiveness of the measurement technique [10]. Although such studies may suggest at least one neurochemical mediator for nitrous oxide analgesia, they do little to reveal the neural networks that participate in the antinociceptive response.

The complexity of the human pain experience emphasizes the importance of human studies focusing on the ultimate variable of interest - the perception of pain. According to contemporary pain theories, the various components of nociception, such as sensory-discriminative and affective-motivational, involve different brain areas [4]. PET has proven to be a sensitive tool for mapping the net functional effects of noxious peripheral thermal stimuli showing highly localized cerebral activation responses in the anterior cingulate cortex, thalamus, supplementary motor area, and primary and secondary somatosensory cortices [12,13]. Since nociceptive processing seems to involve distinct brain areas, it is reasonable to assume that analgesic drugs also exert their effect in a highly localized manner. PET has also been used to show that systemic morphine- and fentanyl-induced pain relief is associated with activation in the anterior cingulate, prefrontal cortex and caudate nucleus [14,15,16]. These observations suggest that a site of pain-opioid interaction is the anterior cingulate, the cortical projection of the medial pain system which is primarily responsible for processing the affective-motivational components of pain [11]. In contrast, more recent *in vivo* opioid receptor imaging by PET directly demonstrates high density of opiate binding sites in the thalamus, cingulate, and prefrontal cortices, with a relative lack of binding in the somatosensory cortices [17], where the lateral pain system projects [11].

To test the hypothesis that the medial pain system is the common site of action of the pharmacologically dissimilar systemic analgesics nitrous oxide and opioids, we used PET to compare changes in regional cerebral synaptic activity induced by a pain stimulus, to those accompanying the combined administration of a pain stimulus and 20% nitrous oxide inhalation.

2. Materials and Methods

2.1. SUBJECTS

Seventeen right-handed healthy volunteers (six female, three male) aged 20 to 46 years (mean, 27.4 years) gave written informed consent. All procedures were approved by the University of Pittsburgh Institutional Review Board and Radioactive Drug Review Committee. Subjects were screened for neurological, psychiatric and substance abuse disorders. All the women had negative serum pregnancy tests on the day of the PET scan.

2.2. EXPERIMENTAL DESIGN

Subjects underwent PET scanning in three groups, each under four experimental conditions (TABLE 1).

Five subjects (Pain-Nitrous Oxide Experimental Group) received a 48°C tonic, noxious, thermal pain stimulus ([PS] condition), produced by a thermostatically-controlled 2x2 cm aluminum hot plate (Omega Cn7600 Thermostat, Stanford, CT) applied to subject's left (nondominant) volar forearm. This thermal stimulator maintained the preset temperature throughout the stimulus duration of 2 min. During the control condition ([CONTROL]), subjects inhaled room air; during the nitrous oxide condition ([NITROUS OXIDE]), a mixture of 20% nitrous oxide and 30% oxygen was administered. The nitrous oxide-modulated pain stimulation condition ([PS + NITROUS OXIDE]) consisted of the simultaneous administration of the tonic pain stimulus and nitrous oxide. In each experimental condition all subjects fixed their gaze at a crosshair on an overhead computer screen (TABLE 1).

To assess whether nitrous oxide affected rCBF-rCMR coupling, both were separately assayed under control and nitrous oxide conditions in the initial four subjects (Nitrous Oxide Control Group [n = 8]). In each of the remaining four subjects, only two rCBF scans were obtained during the control and nitrous oxide conditions. Nitrous oxide is

TABLE 1. Experimental conditions and data analysis.

EXPERIMENTAL CONDITIONS	
PAIN-NITROUS OXIDE EXPERIMENTAL GROUP (n = 5)	
[CONTROL]	room air inhalation focusing on crosshair
[NITROUS OXIDE]	nitrous oxide inhalation focusing on crosshair
[PS]	room air inhalation focusing on crosshair pain stimulation
[PS + NITROUS OXIDE]	nitrous oxide inhalation focusing on crosshair pain stimulation
NITROUS OXIDE CONTROL GROUP (n = 8)	
[CONTROL]	room air inhalation focusing on crosshair
[NITROUS OXIDE]	nitrous oxide inhalation focusing on crosshair
VISUAL CONTROL GROUP (n = 4)	
[CONTROL]	room air inhalation focusing on crosshair
[NITROUS OXIDE]	nitrous oxide inhalation focusing on crosshair
[VS]	room air inhalation visual stimulation
[VS + NITROUS OXIDE]	nitrous oxide inhalation visual stimulation
DATA ANALYSIS	
COMPARISONS (SUBTRACTIONS)	ACTIVATED BRAIN AREAS RELATED TO:
PAIN-NITROUS OXIDE EXPERIMENTAL GROUP	
[PS] - [CONTROL]	pain stimulation
[[PS + NITROUS OXIDE] - [NITROUS OXIDE]	pain stimulation in the presence of nitrous oxide
[PS + NITROUS OXIDE] - [PS]	nitrous oxide inhalation during pain stimulation
NITROUS OXIDE CONTROL GROUP	
[NITROUS OXIDE] - [CONTROL]	nitrous oxide inhalation
VISUAL CONTROL GROUP	
[VS] - [CONTROL]	visual stimulation
[VS + NITROUS OXIDE] - [NITROUS OXIDE]	visual stimulation in the presence of nitrous oxide

PS, pain stimulation, VS, visual stimulation

known to increase global CBF in a dose-dependent manner [18]; such neuronal activity-independent global CBF increases could alter or obscure any activation-induced rCBF changes. Thus, to ensure that cerebral vascular responsiveness was not blunted during the nitrous oxide experimental condition, four subjects underwent PET scanning during a visual activation paradigm known to elicit a maximal cerebral activation response in the visual cortex (Visual Control Group) [19].The subjects' gaze remained fixed on the crosshair in each experimental condition except during visual stimulation. During the visual stimulation condition ([VS]), subjects viewed red and black annular segments reversed at a frequency of 6 Hz, which evokes an rCBF increase of maximal intensity in the occipital cortex [19]. A nitrous oxide-modulated visual stimulation condition ([VS + NITROUS OXIDE]) consisted of the combination of nitrous oxide administration and visual stimulation (TABLE 1).

rCBF and rCMR were measured using the positron labeled tracer ^{15}O-water and ^{18}FDG, respectively, as indicators of regional synaptic activity [3,6,20]. In each experimental group 2 rCBF scans, and in the Nitrous Oxide Control Group 1 rCMR scan were obtained in each condition. For both the Pain-Nitrous Oxide Experimental and Visual Control Groups the order of scans was randomized, and counterbalanced to control for any possible order effects. However, control conditions always preceded nitrous oxide conditions, to avoid residual drug effects. Quantitative psychophysical ratings of pain stimuli were made on a 100 mm visual analog scale (VAS) by each subject, both 5 seconds after application and again 5 seconds before removal of the pain stimulus. Zero denoted no sensation, 50 denoted the pain threshold, and 100 signified severe pain.

2.3. PET SCANNING PROCEDURES

PET scans were obtained on an ECAT 951R/31 system (Siemens Medical Systems, Hoffman Estates, IL), which provides 31 parallel slices over an axial field of view of 10.8 cm, with a transverse and axial image resolution of approximately 6 mm full-width half-maximum (FWHM). Subjects were placed, and the PET gantry rotated and tilted such that the lowest imaging plane was approximately parallel to, and 1 cm superior to the canthomeatal line. To correct for positron emission attenuation by the intra- and extracerebral tissues a transmission scan was obtained using an external positron emitting rod source (^{68}Ge/^{68}Ga). Reconstruction of PET images from the obtained data resulted in scans of 128x128 pixels, with dimensions of 2.05 x 2.05 mm.

For each rCBF scan, a 50 mCi intravenous bolus injection of ^{15}O-water was used. Twenty seconds after the injection a 60-second data acquisition was started [20] with at least 12 min between each scan to allow the tracer concentration to decay to background levels. In the Nitrous Oxide Control group, for each rCMR scan, 5 mCi of

[18]FDG was injected by intravenous bolus followed by a 40-minute period, during which uptake and metabolic trapping of [18]FDG in the brain are nearly complete [5,21]. A 20-minute scanning period was then begun. Pain stimulation ([PS] and [PS + NITROUS OXIDE]) conditions in the Pain-Nitrous Oxide Experimental Group and visual ([VS] and [VS + NITROUS OXIDE]) conditions in the Visual Control Group began 15 seconds after the tracer injection and continued for 65 seconds. Nitrous oxide administration began approximately 15 min before PET scanning in the [NITROUS OXIDE] and [PS + NITROUS OXIDE], and [NITROUS OXIDE] and [VS + NITROUS OXIDE] conditions, to establish a steady-state brain concentration. Nitrous oxide dilutions (10 L/min total gas flow) were delivered through an anesthesia machine (Modulus II Plus, Ohmeda, Madison, WI) via tight-fitting face mask and semiclosed breathing circuit. An end-tidal nitrous oxide concentration of 20% was confirmed with a multi-gas analyzer (Rascal II, Ohmeda, Salt Lake City, UT). Other physiologic monitors included noninvasive blood pressure cuff, electrocardiograph, pulse oximeter, and capnograph.

2.4. DATA ANALYSIS AND STATISTICS

All PET images were reconstructed and registered within subject to correct for any head movement that occurred during the collection of scans [27]. Resulting data were analyzed separately using the statistical parametric mapping software [23,24]. During this process images are averaged within and across subjects for each experimental condition, to facilitate the detection of experimental condition-related rCBF and rCMR changes with enhanced sensitivity. Intersubject image averaging, however, mandates the normalization of brain position, size, and shape differences. Therefore, image sets from each subject were transformed onto the stereotactic atlas of Talairach and Tournoux [25] by an automated routine that first determines the intercommissural (AC - PC) line, rotates the PET data to match the axis and then proportionately stretches the PET image sets to fit the atlas. Then, to reduce image noise and accomodate for intersubject differences in cerebral anatomy and function, a Gaussian filter was applied to all image data (20 mm FWHM). Differences in global ^{15}O-water and [18]FDG activity between subjects and conditions were normalized by analysis of covariance, with global activity as an independent, and regional activity as a dependent variable [23]. With this correction inter- and intrasubject variations in activity due to global CBF and rCMR cannot obscure regional changes. The result of this process is a mean regional activity which is linearly related to the actual rCBF and rCMR values and associated variance for every pixel during each experimental condition. Comparisons of rCBF and rCMR between conditions were performed on a pixel-by-pixel basis using t statistics. Regions where pixels reached P values of < 0.01 were considered significant. To reach significance pixels also had to be in a contiguous group extending over more than one transverse plane, in order to provide protection against type I error, as demonstrated by the lack of significant changes in same-state comparisons. z values were calculated by transforming the t statistic to a unit Gaussian distribution to facilitate comparisions of experimental groups of different size. The results were also displayed in coronal, transverse, and sagittal views of the brain.

The three groups of subjects were analyzed separately. In each group, to identify brain areas of significant rCBF increases (activation) related to pain stimulation; nitrous oxide inhalation; pain stimulation in the presence of nitrous oxide; and nitrous oxide inhalation during pain stimulation, the image comparisons (subtractions) listed in TABLE 1 were made, visual stimulation in the absence and presence of nitrous oxide. In the Nitrous Oxide Control Group, to detect whether there was rCBF-CMR uncoupling, in areas where significant nitrous oxide-associated blood flow increases were identified in the averaged rCBF image of all *eight* volunteers, the averaged rCBF and rCMR scans of the first *four* subjects were assessed for differences on a pixel-by-pixel basis using the point analysis tool of the statistical parametric mapping software [23,24]. Specifically, in the averaged rCBF and rCMR data sets of the initial *four* subjects, the mean percentage changes of rCBF scan pixels, that show the most robust changes in the averaged rCBF scan pixels of all *eight* volunteers, are compared with those of stereotactically corresponding rCMR scan pixels by t tests, using a P level of 0.05 to define significant. Then, to assess nitrous oxide's effect on cerebral vasculature responsiveness, in the Visual Control Group in the identified areas of visual activation, the magnitude of visual stimulation-related rCBF increases, measured in the absence and presence of nitrous oxide, was compared with point analysis. Specifically, in areas of visual activation-related rCBF changes, mean percentage rCBF increases in randomly selected pixels of the occipital cortex during nitrous oxide vs. room air inhalation, were analyzed for differences with t statistics at a significance level of 0.05.

Vital sign data were averaged for all subjects in each condition and processed using analysis of variance to detect any significant ($P < 0.05$) differences between experimental conditions. VAS scores were analyzed nonparametrically by the Wilcoxon rank sum test.

3. Results

The peripheral thermal stimulus was rated as painful by all subjects, as reflected in the mean VAS score of 67 ± 4 mm (SEM). Inhalation of 20% nitrous oxide during the thermal stimulus ([PS + NITROUS OXIDE] condition), resulted in a significant decrease in the mean VAS score to 54 ± 5 by Wilcoxon rank sum test ($P < 0.03$), although the stimulus was still rated as painful. Vital signs did not significantly alter during the experimental conditions.

Comparison of scans obtained during room air inhalation alone to those acquired during pain stimulation ([PS] - [CONTROL]) showed significant rCBF increases in the contralateral thalamus, anterior cingulate cortex (area 24), and supplementary motor area when compared to control ($P < 0.01$; *Figure 1*, TABLE 2). The [NITROUS OXIDE] - [CONTROL] comparison revealed significant rCBF increases in the anterior cingulate cortex bilaterally (areas 24, 32; vertical extent relative to AC-PC line: 0 to 20 mm) ($P < 0.01$; TABLE 3). Comparison of rCBF during simultaneous pain stimulation and nitrous oxide inhalation, to that during nitrous oxide inhalation alone (i. e.,[PS + NITROUS OXIDE] - [NITROUS OXIDE]), demonstrated significant rCBF increases in the infralimbic area (area 25) and orbitofrontal cortex (areas 10, 11), contralateral to the side of pain stimulus (*Figure 2*, TABLE 4). The previously demonstrated pain-related activations were not found in the presence of nitrous oxide. This is demonstrated by the nonsignificant percentage rCBF increases in pixels located in the thalamus, anterior cingulate, and supplementary motor area (TABLE 4) and in *Figure 2* where the nitrous oxide effect has been subtracted from the pain plus nitrous oxide condition. Comparison of rCBF scans obtained during pain stimulation from those during simultaneous pain stimulation and nitrous oxide inhalation ([PS + NITROUS OXIDE] - [PAIN]) revealed activation in the infralimbic area (area 25), orbitofrontal cortex (areas 10,

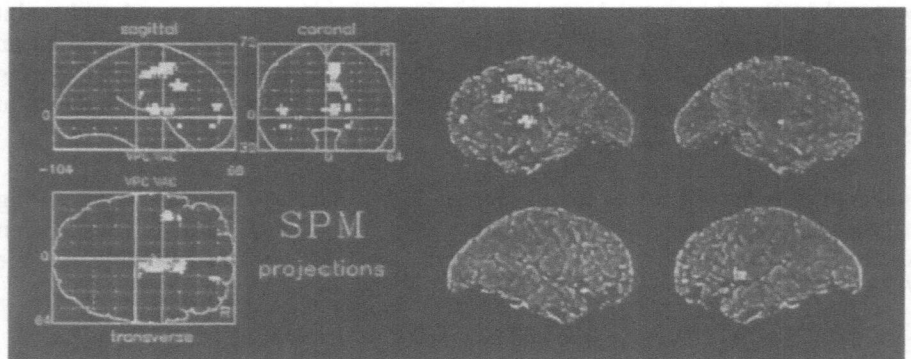

Figure 1. Statistical parametric maps of neuronal activation during pain stimulation alone ([PS] - [CONTROL]). Pixels that are significant at the given threshold of $P < 0.01$ are displayed on single sagittal, coronal, and transverse projections of the brain as lighter shades of gray, where the lightest shade indicates the greatest degree of activation (see TABLE 2 for anatomical locations). R, right hemisphere; L, left hemisphere; VAC, vertical line through the anterior commissure; VPC, vertical line through the posterior commissure.

TABLE 2. Areas of rCBF Increases During Pain Stimulation ([PS] - [CONTROL]).

Region	Brodmann's Area	Coordinates			Percentage rCBF Increase	z score
		x	y	z		
Thalamus (R)		8	-6	4	2.7	2.643
Thalamus (R)		10	-8	8	2.8	3.051
Thalamus (R)		4	-4	12	2.7	2.501
Anterior cingulate (R)	24	8	16	36	2.0	2.568
Anterior cingulate (R)	24	10	-14	40	2.7	2.415
Anterior cingulate (R)	24	4	4	40	2.3	2.451
Supplementary motor area (R)	6	6	14	48	2.7	2.381
Supplementary motor area (R)	6	10	6	52	2.0	2.372

Only areas with a z-score greater than 2.326 ($P < 0.01$) are listed. The coordinates in a standard stereotactic space (Talairach and Tournoux, 1988) are given (in millimeters) for the maximally significant pixel in each SPM-identified area. X is the lateral displacement from the midline (- for the left hemisphere), y is the anteroposterior displacement relative to the anterior commissure (- for positions posterior to the latter), and z is the vertical position relative to the AC-PC line (- if below this line). rCBF, regional cerebral blood flow; PS, pain stimulation; R, right.

TABLE 3. Areas of rCBF Increases (n = 8) and Comparisons of Percentage rCBF and rCMR Increases (n = 4) in the Same Areas During Nitrous Oxide Inhalation Alone ([NITROUS OXIDE] - [CONTROL]).

Region	Brodmann's Area	Coordinates			% Change rCBF	% Change rCMR
		x	y	z		
Anterior cingulate (L)	24	-6	28	20	2.21 ± 0.91	2.32 ± 1.12
Anterior cingulate (R)	24	6	34	16	2.35 ± 0.84	2.86 ± 1.46
Anterior cingulate (L)	24	-2	8	32	2.91 ± 0.96	3.20 ± 1.51
Anterior cingulate (R)	24	4	0	36	2.56 ± 1.25	2.18 ± 1.36

The magnitude of nitrous oxide-related rCBF and rCMR increases did not differ significantly ($P > 0.05$). Details regarding the stereotactic coordinates and abbreviations are as in TABLE 2. L, left; rCMR, regional cerebral metabolic rate.

282

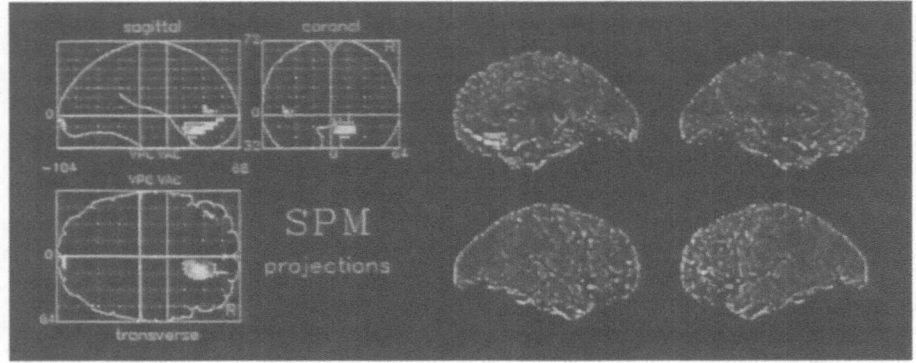

Figure 2. Statistical parametric maps of neuronal activation during pain stimulation in the presence of nitrous oxide ([PS + NITROUS OXIDE] - [NITROUS OXIDE]). The activation pattern reflects nitrous oxide-modulated pain responses. Regional brain activation is represented as described in *Figure 1*, and the relevant anatomical assignments are specified in detail in TABLE 4. Views and abbreviations are the same as for *Figure 1*.

11), and anterior cingulate cortex (area 24; vertical extent relative to AC-PC line: 0 to 20 mm). Areas 25 and 11 were activated contralateral to the pain stimulus (*Figure 3*, TABLE 5). Activation of area 24 in the right hemisphere did not overlap with the pain-activated subregion of area 24.

Point analysis of rCBF and rCMR scans revealed no significant ($P < 0.05$) differences between the 20% nitrous oxide-induced percentage increases in randomly selected pixels in the anterior cingulate cortex images (TABLE 3). Visual stimulation-related percentage rCBF increases in randomly selected pixels of the occipital cortex during nitrous oxide inhalation ([VS + NITROUS OXIDE]) did not differ significantly from those measured during room air ([VS]) ($P > 0.05$; TABLE 6).

4. Discussion

The pain activation ([PS] - [CONTROL]) results of this study (*Figure 1*, TABLE 2) are in agreement with the findings of previous investigations. [12,13] Activation in the thalamus and anterior cingulate cortex (area 24) in response to peripheral pain stimulation, as we report, has been a consistent finding in these studies. [12,13] Electrophysiological recordings indicate that nociceptive information reaches the anterior cingulate via direct projections from the medial thalamic nuclei [26]. Clinical observations of chronic pain patients following anterior cingulotomy underscore its importance in the affective-motivational aspects of pain [27].

TABLE 4.   Areas of rCBF Increases During Pain Stimulation in the Presence of Nitrous Oxide ([PS + NITROUS OXIDE] - [NITROUS OXIDE]).

Region	Brodmann's Area	Coordinates			Percentage rCBF Increase	z score
		x	y	z		
Infralimbic area (R)	25	12	36	-16	4.2	3.743*
Orbitofrontal cortex (R)	11	10	36	-12	4.1	3.576*
Orbitofrontal cortex (R)	10	18	48	-8	3.2	2.539*
Orbitofrontal cortex (R)	10	16	52	-4	3.0	2.480*
Thalamus (R)		8	-6	4	0.2	NS
Thalamus (R)		10	-8	8	0.0	NS
Thalamus (R)		4	-4	12	-0.2	NS
Anterior cingulate (R)	24	8	16	36	0.4	NS
Anterior cingulate (R)	24	10	-14	40	-0.3	NS
Anterior cingulate (R)	24	4	4	40	0.8	NS
Supplementary motor area (R)	6	6	14	48	0.6	NS
Supplementary motor area (R)	6	10	6	52	0.9	NS

* $P < 0.01$. Coordinates and abbreviations are as in TABLES 2 and 3. NS, nonsignificant.

Another brain region where significant pain-related activation was found is the supplementary motor area contralateral to the noxious stimulus. Human PET studies have shown significant supplementary motor area activation during movement planning [28]. Thus, the activation observed in our study is consistent with pain-related initiation of withdrawal from the noxious stimulus, although none of the subjects did actually move during the study.

The [PS + NITROUS OXIDE] - [NITROUS OXIDE] (*Figure 2*, TABLE 4) comparison revealed that pain stimulation, in the *presence* of nitrous oxide, was not associated with the activation observed in the thalamus, anterior cingulate cortex, and supplementary motor area during pain stimulation in the *absence* of nitrous oxide

284

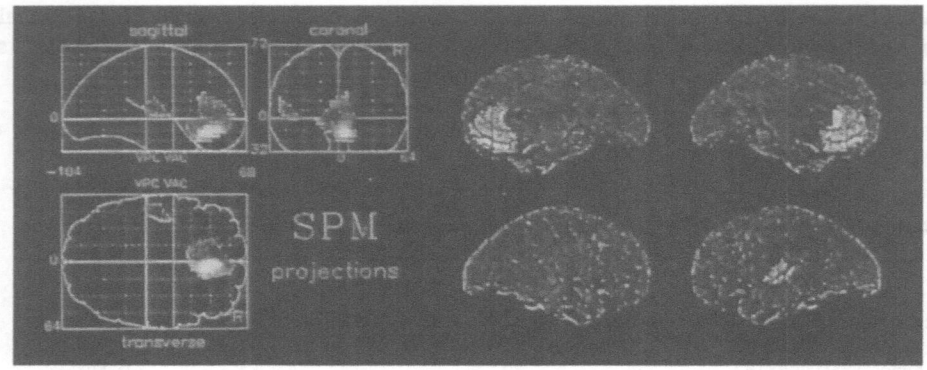

Figure 3. Statistical parametric maps of neuronal activation during nitrous oxide inhalation in the presence of pain stimulation ([PS + NITROUS OXIDE] - [PS]). The activation pattern reflects pain-modulated nitrous oxide responses. Regional brain activation is represented as described in *Figure 1*, and the relevant anatomical assignments are specified in detail in TABLE 5. Views and abbreviations are the same as for *Figure 1*.

([PS] - [CONTROL]; *Figure 1*, TABLE 2). The [PS + NITROUS OXIDE] - [NITROUS OXIDE] comparison, however, showed new areas of activation in the infralimbic (area 25) and orbitofrontal cortices (areas 10, 11) contralateral to the pain stimulus. Additional brain areas activated during simultaneous pain stimulation and nitrous oxide inhalation are revealed in the [PS + NITROUS OXIDE] - [PS] comparison, and include bilateral anterior cingulate activation in the same small regions as in the [NITROUS OXIDE] - [CONTROL] comparison (*Figure 3*, TABLES 3, 5). This indicates that pain stimulation during nitrous oxide inhalation does not alter the activation pattern evoked by nitrous oxide alone. The different pain activation pattern in the presence versus absence of nitrous oxide, however, suggests that nitrous oxide affects cerebral nociceptive processes.

Alternatively, the lack of pain-related response in the thalamus, anterior cingulate, and supplementary motor area during nitrous oxide inhalation, could be due to the fact that nitrous oxide, administered in the absence of pain, might have evoked activation in the same brain areas that pain evoked activation, thus the [PS + NITROUS OXIDE] - [NITROUS OXIDE] subtraction would have subtracted out apparent pain-related activations. However, this is not the case, because our [NITROUS OXIDE] - [CONTROL] comparison shows that the nitrous oxide-related activations are in different areas from those associated with pain (TABLE 3).

TABLE 5. Areas of rCBF Increases During Nitrous Oxide Inhalation in the Presence of Pain ([PS + NITROUS OXIDE] - [PS]).

Region	Brodmann's Area	Coordinates			Percentage rCBF Increase	z score
		x	y	z		
Infralimbic area (R)	25	4	32	-16	4.2	4.402
Orbitofrontal cortex (R)	11	4	34	-12	4.9	4.498
Orbitofrontal cortex (R)	10	8	44	-8	5.0	3.783
Anterior cingulate (L)	24	-8	28	0	4.2	2.896
Anterior cingulate (R)	24	8	32	0	5.0	3.368
Anterior cingulate (L)	24	-10	42	4	4.5	2.420
Anterior cingulate (R)	24	6	40	4	3.1	3.196
Anterior cingulate (L)	24	-2	34	12	3.6	3.170
Anterior cingulate (R)	24	2	28	20	3.8	2.775

Only areas with a z score greater than 2.326 ($P < 0.01$) are listed. Coordinates and abbreviations are as in TABLES 2 and 3.

The likelihood that the pain activation pattern observed during nitrous oxide inhalation is related to this agent's antinociceptive effect is supported by the association with a decreased subjective pain experience, as reflected by the significantly lower VAS scores. Limitations to this interpretation are posed by effects of nitrous oxide other than analgesia, and the nonrandom order of PET scans. Although they were not measured, subjective effects of nitrous oxide, such as mild sedation, and euphoria reported by each volunteeer, could result in cerebral responses that interfered with the pain-related activation during nitrous oxide inhalation contributing to the observed activation pattern. Subjects were blinded as to whether room air or nitrous oxide was being administered, but since room air conditions (e.g., [CONTROL], [PS]) always preceded nitrous oxide conditions ([NITROUS OXIDE], [PS + NITROUS OXIDE]), the possibility of some order effect cannot be ruled out.

Although nitrous oxide inhalation significantly decreased subjective pain experience, the stimulus was still rated as painful. The contradiction that *total* elimination of activation in the thalamus, anterior cingulate, and supplementary motor area is associated only with *partial* pain relief may be explained by the possibility that during nitrous oxide inhalation, pain-related net metabolic responses only decreased to a level undetectable by PET.

TABLE 6. Comparisons of Percentage rCBF Changes Obtained by Point Analysis, In Areas of Occipital Cortex Activated by Visual Stimulation During Nitrous Oxide vs. Room Air Inhalation ([VS], [VS + NITROUS OXIDE]).

Region	Brodmann's Area	Coordinates			% Change rCBF [VS]	% Change rCBF [VS + NITROUS OXIDE]
		x	y	z		
Gyrus occipitalis inferior (L)	17	-12	-90	-8	6.3 ± 1.56	7.4 ± 2.12
Gyrus occipitalis inferior (R)	17	20	-92	-8	6.5 ± 2.71	8.9 ± 1.70
Gyrus occipitalis inferior (L)	18	-24	-90	-4	5.8 ± 0.82	5.9 ± 1.85
Gyrus occipitalis inferior (R)	18	18	-88	-4	5.9 ± 1.60	7.2 ± 0.96
Cuneus (L)	17	-8	-96	4	2.6 ± 0.90	2.3 ± 1.56
Cuneus (R)	17	4	-94	4	2.7 ± 1.10	2.4 ± 1.69
Cuneus (L)	18	-18	-88	4	3.7 ± 1.37	4.7 ± 1.14
Cuneus (R)	18	22	-90	4	2.1 ± 1.13	2.9 ± 1.60
Cuneus (L)	18	-18	-92	8	3.8 ± 1.52	3.3 ± 1.66
Cuneus (R)	18	12	-92	8	2.1 ± 1.85	2.5 ± 1.69

The magnitude of visual activation did not show significant difference between [VS] and [VS + NITROUS OXIDE] conditions ($P > 0.05$). Coordinates and abbreviations are as in TABLES 2 and 3. VS, visual stimulation.

In principle, the use of rCBF as an indicator of regional neuronal activity could be confounded in several different ways. Any observed rCBF changes could be due either to nitrous oxide-induced alterations in arterial carbon dioxide content [29], or to rCBF/rCMR uncoupling [30]. Another potential confounding variable could be nitrous oxide's effects on the responsiveness of the cerebral vasculature [18]. The first possibility is ruled out by the lack of change detected by capnography. To assess the second possibility, we measured and compared the magnitude of nitrous oxide-related rCBF and rCMR increases and found that the two parameters remained tightly coupled in the presence of 20% nitrous oxide (TABLE 3).

To control for the possibility that nitrous oxide directly affects cerebral vasculature responsiveness independent of neuronal activation, we compared occipital rCBF increases evoked by visual stimulation, in the presence and absence of nitrous oxide. The response characteristics of visual stimulation are well described under physiological circumstances showing an ~ 40% rCBF increase in the occipital cortex

[19], therefore, providing a large and sensitive signal. Since 20% nitrous oxide has been shown not to perturb the metabolic activity of the primary visual cortex (TABLE 3), significant differences between the magnitude of visual activation in the presence and absence of nitrous oxide would have to be attributed to nitrous oxide-related perturbation of cerebral vascular reactivity. There were no significant differences in the visual activation whether nitrous oxide was present or not, even with the relaxed criteria for significance of $P < 0.05$ (TABLE 6). This indicates that rCBF reactivity remains intact during exposure to 20% nitrous oxide permitting us to interpret our pain-nitrous oxide interaction data directly.

Careful analysis of our findings in light of known neuroanatomical and electrophysiological data can be used to map the sites of nitrous oxide's analgesic action. Nitrous oxide could suppress medial thalamic neuronal activity directly, by influencing intrathalamic circuits, or indirectly, via affecting thalamic afferent mechanisms. Electrophysiological studies indicate that nitrous oxide increases the firing of thalamic reticular inhibitory neurons thereby increasing the inhibitory influence on thalamic somatosensory relay nuclei and consequently decreasing net corticopetal transmission [31]. Alternatively, the depression of thalamic activation could be the result of nitrous oxide's modulatory effect on ascending pain pathways, or other thalamic afferents.

Since infralimbic and orbitofrontal activation occurs exclusively during pain stimulation in the presence of nitrous oxide ([PS + NITROUS OXIDE]), and this activation pattern is associated with significantly decreased pain perception, it is likely that these areas contribute to the antinociceptive effect of nitrous oxide. Infralimbic and orbitofrontal stimulation have been shown to produce analgesia [32] possibly via their connections with ascending and descending pain modulatory pathways [33]. Another possibility is that bilateral anterior cingulate activation, produced by nitrous oxide both in the absence ([NITROUS OXIDE]) and presence of pain stimulation ([PS + NITROUS OXIDE]), alone or together with the infralimbic and orbitofrontal responses, represents the neuroanatomical substrate of the observed antinociceptive effect.

Comparing our results with human PET findings addressing the cerebral sites of opioid analgesia, may reveal antinociceptive mechanisms common to systemic analgesics. The available data suggest that the primary loci of action are in the thalamus and in the cortical projections of the medial pain system. Jones et al. [14] have demonstrated activation of discrete areas of the anterior cingulate and prefrontal cortices when morphine was administered systemically in one chronic pain patient. However, only activation associated with morphine analgesia was reported; no data were presented on the effects of pain or morphine alone, which would allow a more complete assessment of pain-opioid interactions. We have reported anterior cingulate activation, as well as bilateral thalamic deactivation, following fentanyl systemically administered during a peripheral pain stimulus [15]. The thalamus seems to be a common target for both fentanyl and nitrous oxide. But since fentanyl is associated with neither depression in the anterior cingulate cortex, nor activation in the infralimbic and orbitofrontal areas when administered in the presence of pain, these two analgesics appear to modulate pain perception via different neural circuits.

In summary, evidence is presented that nitrous oxide's antinociceptive effects in humans are associated with the abolition of pain-evoked activation in the thalamus, area 24 of the anterior cingulate cortex, and supplementary motor area, as well as activation of the infralimbic and orbitofrontal cortices. Responses in the latter areas raise the possibility that nitrous oxide activates ascending and descending antinociceptive modulatory pathways in addition to depressing pain-induced cerebral responses. These findings strongly suggest that analgesic concentrations of nitrous oxide modulate cerebral pain processing, at least in part, via the medial pain system. Furthermore, the data demonstrate the potential of PET in elucidating hitherto inaccessible aspects of anesthetic action under the most relevant circumstances – in the living human brain.

5. References

1. Kety, S.S. and Schmidt, C.F. (1948) The nitrous oxide method for the quantitative determination of cerebral blood flow in man; theory, procedures and normal values. J. Clin. Invest. 27, 476-483.
2. Herscovitch, P., Markham, J., Raichle, M.E. (1983) Brain blood flow measured with intravenous $H_2^{15}O$. I. Theory and error analysis. J. Nucl. Med. 14, 782-789.
3. Meyer, E. (1991) ^{15}O studies with PET. In: Diksic, M. and Reba, R.C. eds. Radiopharmaceuticals and brain pathology studies with PET and SPECT. CRC Press, Boca Raton, 166-197.
4. Bendo, A.A., Kass, I.S., Hartung, J., Cottrell, J.E. (1992) Neurophysiology and neuroanesthesia. In: Barash P.G., Cullen B.F., Stoelting R.K., eds. Clinical Anesthesia, J.B. Lippincott Company, Philadelphia, 871-919.
5. Sokoloff, L., Reivich, M., Kennedy, C., Des Rosiers, M.H., Patlak, C.S., Pettigrew, K.D., Sakurada, O., Shinohara, M. (1977) The [C-14] deoxyglucose method for the measurement of local cerebral glucose utilization: Theory, procedure, and normal values in the conscious and anesthetized albino rat. J. Neurochem. 28, 897-916.
6. Phelps, M.E., Huang, S.C., Hoffman E.J., Selin, C., Sokoloff, L., Kuhl, D.E. (1979) Tomographic measurement of local cerebral glucose metabolic rate in humans with 2-(F-18)fluoro-2-deoxy-D-glucose: validation of method. Ann. Neurol. 6, 371-88.
7. Raichle, M.E. (1987) Circulatory and metabolic correlates of brain function in normal humans. In: Mountcastle, V.B., Plum, F., eds. Handbook of Physiology, Section I, The Nervous System, Volume 5, Higher Functions of the Nervous System, Oxford University Press, New York, 643-674
8. Zuniga, J.R., Joseph, S.A., Knigge, K.M. (1987) The effect in nitrous oxide on the central endogenous pro-opiomelanocortin system in the rat. Brain Res. 420, 56-65.
9. Chapman, C.R., Benedetti, C. (1979) Nitrous oxide effects on cerebral evoked potential to pain. Anesthesiology 51, 135-138.
10. Way, W.L., Hosobuchi, Y., Johnson, B.H., Eger, E.I. II, Bloom, F.E. (1984) Anesthesia does not increase opioid peptides in cerebrospinal fluid of humans. Anesthesiology 60, 43-45.

11. Albe-Fessard, D., Berkley, K.J., Kruger, L., Ralston, H.J., Willis, W.D. (1985) Diencephalic mechanism of pain sensation. Brain Res. Rev. 9, 217-296.

12. Casey, K.L., Minoshima, S., Berger, K.L., Koeppe, R.A., Morrow, T.J., Frey, K.A. (1994) Positron emission tomographic analysis of cerebral structures activated specifically by repetitive noxious heat stimuli. J. Neurophysiol. 71, 802-807.

13. Coghill, R.C., Talbot, J.D., Evans, A.C., Meyer, E., Gjedde, A., Bushnell, M.C., Duncan, G.H. (1994) Distributed processing of pain and vibration by the human brain. J. Neurosci. 14, 4095-4108.

14. Jones, A.K., Friston, K.J., Qi, L.Y., Harris, M., Cunningham, V.J., Jones, T., Feinman, C., Frackowiak, R.J. (1991) Sites of action of morphine in the brain. Lancet 338, 825.

15. Adler, L., Firestone, L.L., Mintun, M., Winter, P. (1994) Central mechanisms of pain and opioid analgesia elucidated by Positron Emission Tomography (PET). Anesthesiology 81, A917.

16. Firestone, L.L., Gyulai, F., Mintun, M., Adler, L.J., Urso, K., Winter, P. (1996) Human brain activity response to fentanyl imaged by positron emission tomography. Anesth. Analg. 82, 1247-1251.

17. Jones, A.K., Qi, L.Y., Fujirawa, T., Luthra, S.K., Ashburner, J., Bloomfield, P., Cunningham, V.J., Itoh, M., Fukuda, H., Jones, T. (1991) In vivo distribution of opioid receptors in man in relation to the cortical projections of the medial and lateral pain systems measured with positron emission tomography. Neurosci. Lett. 126, 25-28.

18. Field, L.M., Dorrance, D.E., Kreminska, E.K., Barsouam, L.Z. (1993) Effect of nitrous oxide on cerebral blood flow in humans. Br. J. Anaesth. 70, 154-159.

19. Fox, P.T., Raichle, M.E. (1984) Stimulus rate dependence of regional cerebral blood flow in human striate cortex, demonstrated by positron emission tomography. J. Neurophysiol. 51, 1109-1120.

20. Raichle, M.E., Martin, W.R.W., Herscovitch, P., Mintum, M.A., Markaham, J. (1983) Brain blood flow measured with intravenous $H_2^{15}O$. II. Implementation and validation. J. Nucl. Med. 24, 790-798.

21. Huang, S.C., Phelps, M.E., Hoffman, E.J., Sideris K., Selin, C.J., Kuhl, D.E. (1980) Noninvasive determination of local cerebral metabolic rate of glucose in man. Am. J. Physiol. 238, E69-E92.

22. Woods, R.P., Cherry, S.R., Mazziotta, J.C. (1992) Rapid automated algorithm for aligning and reslicing PET images. J. Comp. Assist. Tomogr. 16, 620-633.

23. Friston, K.J., Frith, C.D., Little, P.F., Dolan, R.J., Lammerstma, A.A., Frackowiak, R.S. (1990) The relationship between global and local changes in PET scans. J. Cereb. Blood Flow Metab. 10, 458-466.

24. Friston, K.J., Passingham, R.E., Nutt, J.G., Heather, J.D., Sawle, G.V., Frackowiak, R.S. (1991) Localization of PET images: direct fitting of the intercommissural (AC-PC) line. J. Cereb. Blood Flow Metab. 9:690-5

25. Talairach, H. and Tournoux, P. (1988) Co-planar stereotactic atlas of the human brain. 3-dimensional proportional system: an approach of cerebral imaging, Thieme Medical Publishers, New York.

26. Sikes, R.W. and Vogt, B.A. (1992) Nociceptive neurons in area 24b of rabbit anterior cingulate cortex. J. Neurophysiol. 68, 1720-1732.

27. Corkin, S. (1980) A prospective study of cingulotomy. In: Valenstein ES. ed. The Psychosurgery Debate, Freeman, San Francisco, 164-204

28. Roland, P.E., Larson, B., Lassen, N.A., Skinhoj, E. (1980) Supplementary motor area and other cortical areas in the organization of voluntary movements in man. J. Neurophysiol. 43, 118-136.

29. Dubbink, D.A. (1991) Physiologic effects of hyper- and hypocarbia In: Faust RJ. ed. Anesthesiology Review, Churchill Livingstone Publishing Co., New York, 24-26

30. Pelligrino, D.A., Miletich, D.J., Hoffman, W.E. (1984) Nitrous oxide markedly increases cerebral cortical metabolic rate and blood flow in the goat. Anesthesiology 60, 405-412.

31. Angel, A. (1991) The G. L. Brown lecture. Adventures in anaesthesia. Exp. Physiol. 76, 1-38.

32. Hardy, S.G.P. (1985) Analgesia elicited by prefrontal stimulation. Brain Res. 339, 281-284.

33. Hardy, S.G.P., Leichnetz, G.R. (1981) Frontal cortical projections to the periaqueductal gray in the rat: a retrograde and orthograde horseradish peroxidase study. Neurosci. Lett. 23, 13-17.

ACKNOWLEDGMENTS

The author wishes to thank Drs. Leonard L. Firestone, Mark A. Mintun, Peter M. Winter for their collaboration in the project. This work was supported by a Clinical

Scholar's Research Award from the International Anesthesia Research Society, the University Anesthesiology and Critical Care Medicine Foundation, and the UPMC PET Facility.

THE EFFECT OF A SINGLE-DOSE INTRAVENOUS VINPOCETINE ON CHRONIC STROKE PATIENTS. A PET STUDY

Balázs GULYÁS[1], László CSIBA[2], Levente KERÉNYI[2], László GALUSKA[3] and Lajos TRÓN[4]

[1]Department of Neuroscience, Karolinska Institute, S-171 77 Stockholm, Sweden, [2]Department of Neurology, [3]Central Laboratory for Nuclear Medicine, [4]PET Centre, Debrecen University Medical School, H-4012 Debrecen, Hungary

1. Abstract

With the purpose of evaluating the effect of vinpocetine (Cavinton®), given in a single-dose infusion, on the cerebral metabolism and perfusion of chronic stroke patients, we measured in 12 patients with focal cerebrovascular lesions the regional and global cerebral metabolic rates of glucose with positron emission tomography (PET) before and after a single dose intravenous vinpocetine treatment. The global and regional glucose metabolic rates showed no significant changes before and after Vinpocetine treatment, however, the kinetic constants (k_1, k_2), related to the facilitated transport of glucose through the blood-brain-barrier in both directions, significantly increased in the whole brain. This increase could be observed in the non-affected hemisphere and in the affected hemisphere outside the infarcted area.

The findings indicate that a single-dose vinpocetine treatment significantly improves the transport of glucose through the blood-brain-barrier in both directions (uptake and release of glucose) in the non-affected hemisphere and in the peri-infarct region of the affected hemisphere.

2. Introduction

Vinpocetine, a vinca alkaloid, has been known to be a clinically useful dilator of the cerebral vasculature, selectively increasing the cerebral blood flow (Mchedlishvili and Ormotsadze, 1981; Tokiwa et al., 1982, Imamoto et al., 1983, 1984; Sugawa et al., 1986; Kiss and Kárpáti, 1996). This effect is facilitated by the positive haemorheological effects of vinpocetinee, including the decrease of blood and plasma viscosity, platelet aggregability and intravascular coagulation, and positive influence on erythrocyte deformability (Kuzuya, 1985, Szobor and Klein, 1992; Schmid-Schönbein et al., 1988; Hayakawa, 1992). In addition, vinpocetine has proved to be neuroprotective by increasing cerebral oxygen extraction and utilisation (Bíró et al., 1976; Tohgi et al., 1990) as well as by its anti-oxidative effect (Miyamoto et al., 1989). It has also been

291

B. Gulyás and H.W. Müller-Gärtner (eds.),
Positron Emission Tomography: A Critical Assessment of Recent Trends, 291–306.
© 1998 Kluwer Academic Publishers.

shown that vinpocetinee increases the cerebral glucose uptake, however, it does not increase the cerebral glucose metabolism (Jucker et al., 1988).

Vinpocetine has been recommended for the treatment of cerebrovascular diseases, including stroke and post-stroke states (cfr. Nagy, 1994). Indeed, during the past two decades world-wide thousands of patients with focal cerebral ischemia have been treated with either *per os* or intravenous vinpocetine products. Against the fact that the *in vitro* effects of vinpocetine have been studied in great detail and several clinical studies aimed at the assessment of its *in vivo* effects on neurological patients, no detailed study has been performed to quantify the *in vivo* effects of vinpocetine on the regional and global cerebral blood flow and metabolic rates in humans both under physiological and pathological conditions, including stroke and post-stroke states.

With the advance of functional imaging techniques, including PET, it has become possible to measure quantitatively (PET) the functional parameters related to cerebral blood flow and metabolism, and, in combination with anatomical imaging modalities (CT, MRI) and computerised brain atlas systems, to correlate these findings with anatomical structures in the human brain. Several studies in the past have dealt with the global and regional metabolic and functional changes in the brain in acute, sub-acute and chronic stroke patients (for reviews, see e.g. Heiss and Herholz, 1994, Heiss and Graf, 1994, Heiss and Podreka, 1993,). However, no functional imaging study has until now dealt with the effects of a commonly used therapeutic agent, vinpocetine, on cerebral metabolism and flow in chronic stroke patients.

The main objectives of the present study were to assess in chronic stroke patients the effects of vinpocetine on
♦ (i) the global and regional cerebral blood flow and
♦ (ii) the global and regional cerebral glucose uptake and metabolism.

In the study we used a combinations of imaging techniques, including CT, MR, SPECT, PET and Transcranial Doppler in 12 neurological patients in a post-stroke status to assess the effects of a single dose vinpocetine infusion on the parameters of global and regional cerebral blood flow and glucose metabolism. The present analysis focuses on the findings obtained with positron emission tomography. A description of the quantification (Balkay et al., 1997) as well as the analysis of the effects of vinpocetine on the kinetic constants (Trón et al., 1997) is summarised in this volume.

3. Materials and Methods

3.1 SUBJECTS

The study was performed in the PET Centre of the Debrecen University Medical School on twelve patients in a post-stroke status (8 men, 4 woman; age: 55-70 years, mean ± 1 s.d.: 62.4 ± 4.9 years). The patients and their relatives were fully informed about the objectives, details, and risks of the study and they gave a written consent, in agreement with the Helsinki Declaration (1964) and the OPRR Reports (1989). The study was approved by the Ethical Committee of the Debrecen University Medical School. The chronic ischemic stroke-patient population was selected so that they represent an average patient population including individuals with various duration and origin of their chronic cerebral vascular disease. The clinical status was assessed by different scales (Orgogozo, Unified NS, SNS - see later). Five patients had a unilateral infarction in the region of the middle cerebral artery (3 on the right and 2 on the left side), whereas in 7 cases the infarction was bilateral in the territories of the middle cerebral arteries. The duration between the ischemic stroke and the present study was 13.4 ± 11.9 months (range: 5.5 - 41 months).

3.2 PATIENT PROTOCOL

All the patients were previously hospitalised with the diagnosis of acute stroke in the Department of Neurology of the Debrecen University Medical School. The diagnosis was established on the basis of clinical symptoms as well as by CT or MRI scans; the patients' clinical status was assessed by the Orgogozo Scale (Orgogozo, 1989), the Scandinavian Neurological Scale (1985), and the Unified Stroke Scale (Edwards et al., 1991). During the acute phase of the disease, the patients were treated by conventional medication, and their post-acute recovery phase was regularly monitored by neurologists of the Department. In patients with bilateral infarctions the more seriously infarcted hemisphere was indicated as "affected" side.

During the present study, repeated physical, neurological, psychological and neuroimaging investigations as well as laboratory tests were performed. The study included a three-day hospitalisation period. On the first day, clinical status was assessed, an MRI scan and an individually moulded head-fixation helmet was made for the patient. During the second day, the subject received in the form of intravenous infusion 500 ml physiological solution (Salsol), followed by (i) a TCD (trans-cranial Doppler) investigation, (ii) a SPECT, and (iii) a PET investigation. During the third day of the study, the subject received intravenously 20 mg vinpocetine dissolved in 500 ml Salsol. This infusion was followed again by (i) a TCD investigation, (ii) a SPECT, and (iii) a PET investigation. As the present paper focuses only on the results of the PET investigation, we henceforth concentrate on this imaging modality.

3.3 BRAIN SCANNING PROCEDURES

CT and MRI: The clinical diagnosis on each patient was confirmed by CT (8 cases) or MRI (4 cases) investigation during the acute phase of the disease. The present investigation, however, included an MRI scan on each patient with a Shimamatzu SMT-100X 1.0 T scanner (T1 and T2 weighted images). The infarct area, affected by the stroke, was determined on each MRI scans by two independent physician-observers. The regions, identified by the observers, were later used to define volumes-of-interest on the PET images.

First PET scanning: During the PET investigations, the patients were equipped with an individually moulded plastic head fixation helmet (Greitz et al., 1980, Bergström et al., 1981), which held the patient's head in identical position in both scans (before and after vinpocetine treatment). Two intravenous cannulae were placed in the

294

subjects' cubital veins, one in the right and one in the left side, for the purposes of administering the physiological solution (with and without vinpocetine), for injecting the tracer, as well as for taking venous blood samples for measuring the time activity curves of the tracer in the blood and plasma.

The patients were at rest during the PET measurements, as defined by Roland and Friberg (1985), the investigation was conducted under sensory suppression. The eyes were closed and covered by dark eye occluders. Beside the ambient noise of the camera fans, there was no noise in the camera room. The PET tracer administration started upon finishing the administration of the i.v. solution.

The second PET scanning took place on the subsequent day. Before the scan the subjects were received in a 45-minute intravenous infusion 20 mg Vinpocetine dissolved in 500 ml Salsol. Similarly to the first scan, the second PET scan commenced upon finishing the administration of the i.v. solution. In addition to taking blood samples at regular intervals for determining the time activity curve of the tracer in the blood, further blood samples were taken to measure the blood glucose and vinpocetinee concentrations.

The quantitative PET measurements were made on a GE 4096 Plus whole body positron camera with 5 mm in-plane resolution and 6.5 mm inter-slice distance (Rota-Kops et al., 1990). The axial field-of-view of the camera is 103 mm. The camera produced 15 transaxial slices of the brain. ^{18}F-deoxy-D-glucose (FDG) was used as tracer, which was given in a bolus injection (10 sec) in the right or left cubital vein parallel with the start of data acquisition (administered activity: 0.15±0.02 mCi/kg in 5 ml physiological solution). Data acquisition with the PET camera and blood sampling started parallel with the bolus injection. In the emission scans, data were sampled for a total of 30 min. The PET images were reconstructed with a Hanning filter to an effective image resolution (FWHM) of 4.5 mm using attenuation correction obtained from a separate transmission scan. The time activity curves of the tracer in the blood were calculated on the basis of regular measurements of tracer activity in blood samples taken manually from the patients.

3.4 PET DATA PROCESSING AND ANALYSIS

On the basis of the PET measurements and the time activity curve of the tracer in the blood, regional cerebral radioactivity concentrations were transformed into quantitative measurements of regional cerebral glucose metabolic rates and related kinetic constants in volumes-of-interest (VOI) (Balkay et al., 1997; Trón et al., 1997). The cerebral glucose metabolic rate (CMR$_{glu}$) and the kinetic constants were determined inside the following VOI's: (A) whole brain, (B) non-affected hemisphere, (C) affected hemisphere, (D) stroke region, (E) affected hemisphere, cortical region outside the stroke region. The stroke-regions were determined on the basis of the patients MR images by two independent physician-observers. For further analysis, this latter region was divided into two components: (F) a "para-stroke" region, the cortex adjacent to the infarct region within the supply territory of the medial cerebral artery, and (G) the "peri-stroke" region: everythin else in the affected hemisphere, outside the stroke and para-stroke regions. (Fig. 1.)

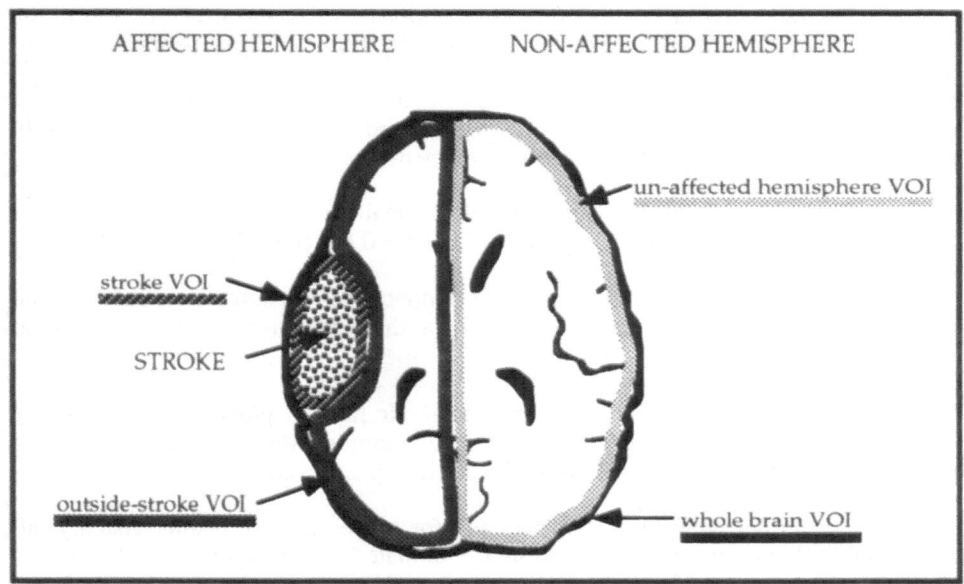

Figure 1.
A. (Above) The determination of the VOI's, shown in one horizontal brain slice.
B. (Below) The "structure" of the VOI's in the affected hemisphere.

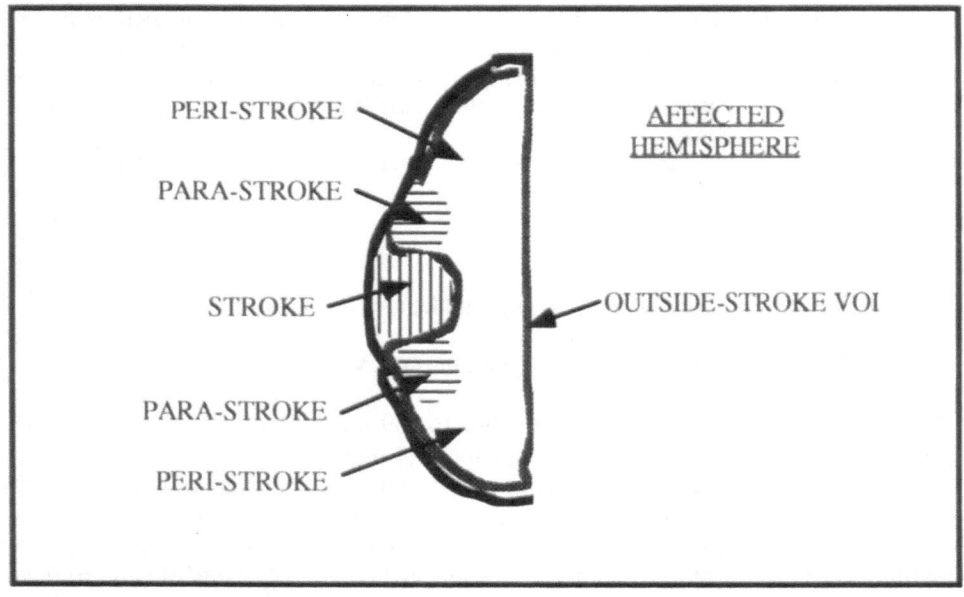

The rationale for creating these two regions outside the stroke region in the affected hemisphere comes from earlier PET investigations on stroke patients (Heiss et al., 1994) indicating the functionally different tissue compartments of the affected hemisphere, including a rim of viable tissue with dilated vasculature around the infarct region (which we term here "para-stroke" region) and the more or less intact tissue in the rest of the hemisphere (which we term here "peri-stroke" region).

Consequently, the VOI's covered all the main compartments of the diseased brain in the following manner: (A) = (B) + (C); (C) = (D) + (E); (E) = (F) + (G).

PET measurements before and after Vinpocetine infusion were obtained from all 12 patients. For the final analysis of the data, three patients were excluded due to technical reasons.

The global and regional cerebral metabolic rates of glucose and the kinetic constants within the VOI's before and after the administration of Vinpocetine were compared with each other using a Student t statistics (paired, one tailed).

For further analysis of the data and for expressing the results visually, all individual PET images were processed in the computerised Human Brain Atlas (HBA) system of the Karolinska Institute (Roland et al., 1994). The voxel size of the HBA is 1 x 1 x 1 mm. In the first step, the individual PET images were adjusted in both size and shape to the standard HBA brain contours (stereotactic standardisation). The stereotactically standardised individual brain images of global glucose metabolic rates and kinetic constants were averaged across the patient population, giving rise to averaged and variance images representing global metabolic rate and kinetic constant images before and after Vinpocetine treatment. Subtraction images as well as three-dimensional statistical images (Roland et al., 1993) were created to indicate and visualise those loci in the brain for which the regional cerebral glucose metabolic rates and kinetic constants differed significantly before and after Vinpocetine administration.

Figure 2. (See opposing page) Image processing in the Human Brain Atlas (HBA).

4. Results

The VOI based values for CMR_{glu}, k_1, k_2, and k_3 are shown in Table 1.

4.1 METABOLIC AND KINETIC PARAMETERS BEFORE VINPOCETINE TREATMENT

The average $gCMR_{glu}$ value in the patients (6.239 mg/100 g/min) was close to the physiological value of the average population (5.4 mg/100 g/min; Roland 1993). As expected, the average $rCMR_{glu}$ value in the non-affected hemisphere (6.861 mg/100 g/min) exceeded that in the affected hemisphere (5.617 mg/100 g/min), the difference was significant (p=0.006). The decreased metabolic rate in the affected hemisphere was not homogeneous: the stroke region had low metabolic rate (5.472 mg/100 g/min) then the region outside the infarct area (5.790 mg/100 g/min), the difference between them being not significant. However, as the stroke region mainly comprised of cortical grey

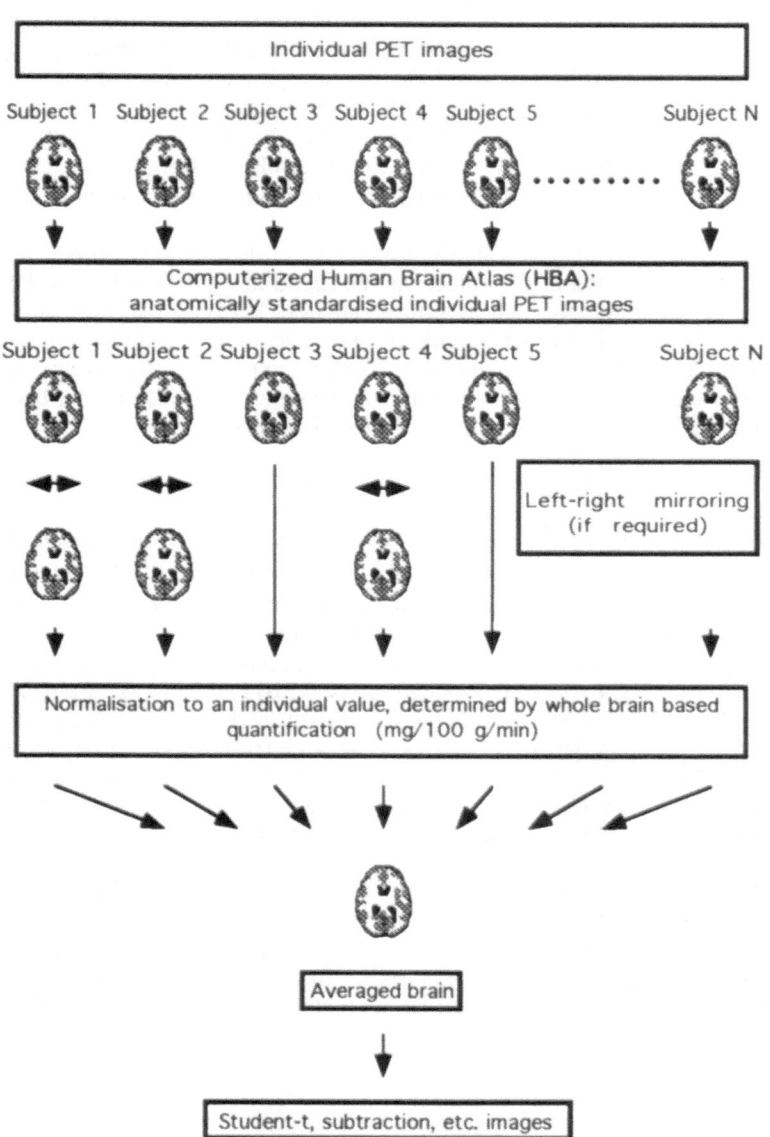

Figure 2.

matter, whereas the overall region outside the infarct area comprised of cortical and subcortical grey matter, white matter and ventricles. It is therefore more appropriate to compare the rCMR$_{glu}$ values between the stroge region and the "para-stroke region" that we defined as the cortical grey matter region within the supply territory of the medial cerebral artery outside the stroke region. In this respect, the para-stroke region had significantly higher rCMR$_{glu}$ value than the stroke region (5.790 mg/100 g/min and 8.729 mg/100 g/min, respectively, the difference being significant [p=0.0007]).

The changes of the kinetic constants followed the above trends. A comprehensive description of the values can be found in Table 1, whereas the relative proportions of the metabolic and kinetic parameters between the various VOI's are displayed in Table 2. The most remarkable facts, as expected and as shown in the Tables, are the reduction in the value of the kinetic constants in the stroke region and their increase in the region neighbouring the stroke region. The differences between the values in these VOI's were significant for all three kinetic constants.

4.2 METABOLIC AND KINETIC PARAMETERS AFTER VINPOCETINE TREATMENT

A single dose Vinpocetine infusion did change neither the gCMR$_{glu}$ nor the rCMR$_{glu}$ within the various VOI's significantly. It did not result in significant change in the k3 values, either. It, however, resulted in significant changes in the k1 and k2 parameters. The k1 and k2 kinetic constants increased significantly in the whole brain (from 0.101 to 0.121 [+20%] and from 0.201 to 0.255 [+27%], respectively), in the non-affected hemisphere from 0.108 to 0.128 [+19%] and from 0.210 to 0.258 [+23%], respectively) and in the affected hemisphere, with special regard to the region falling outside the stroke region (from 0.101 to 0.121 [+20%] and from 0.262 to 0.319 [+22%], respectively). The greatest changes took place in the "peri-stroke" regions, i.e. in the relatively healthy tissue compartments of the affected hemisphere (from 0.080 to 0.121 [+51%] and from 0.195 to 0.295 [+51%], respectively). (Table 1)

A single dose vinpocetine treatment did not alter significantly the relative proportions of the gCMR$_{glu}$ or rCMR$_{glu}$ between the VOI's. It, however, changed significantly the relative proportions of the k1 and k3 constants between the peri-stroke region VOI and most of the other VOI's, indicating that the above described effect of vinpocetine is more selective in the peri-stroke region than in other regions. (Table 2)

	Before	After	Δ%	p	sign.
	Vinpocetin infusion			*(Student-t)*	

CMRglu *(mg/min/100g ± 1 SEM)*

	Before	After	Δ%	p	sign.
Whole brain:	6.239±0.295	6.213±0.305	0	0.477	-
Un-affected hemisphere:	6.861±0.371	6.799±0.386	-1	0.442	-
Affected hemisphere:	5.617±0.338	5.636±0.293	0	0.482	-
Stroke region:	5.472±0.427	4.996±0.257	-9	0.062	-
Region outside stroke:	5.790±0.317	5.894±0.370	+2	0.350	-
Para-stroke region:	8.729±0.650	7.600±0.815	-13	0.083	-
Peri-stroke region:	4.602±0.042	5.148±0.047	+12	0.179	-

k1

	Before	After	Δ%	p	sign.
Whole brain:	0.101±0.001	0.121±0.011	+20	**0.036**	+
Un-affected hemisphere:	0.108±0.007	0.128±0.010	+19	**0.024**	+
Affected hemisphere:	0.093±0.007	0.113±0.012	+22	0.099	-
Stroke region:	0.084±0.007	0.093±0.010	+11	0.107	-
Region outside stroke:	0.101±0.009	0.121±0.016	+20	0.088	-
Para-stroke region:	0.146±0.009	0.153±0.021	+5	0.333	-
Peri-stroke region:	0.080±0.009	0.121±0.016	+51	**0.015**	+

k2

	Before	After	Δ%	p	sign.
Whole brain:	0.201±0.015	0.255±0.027	+27	**0.044**	+
Un affected hemisphere:	0.210±0.016	0.258±0.020	+23	**0.030**	+
Affected hemisphere:	0.198±0.016	0.251±0.021	+27	**0.009**	+
Stroke region:	0.181±0.018	0.218±0.025	+20	**0.036**	+
Region outside stroke:	0.214±0.023	0.303±0.028	+42	**0.017**	+
Para-stroke region:	0.262±0.021	0.319±0.040	+22	0.087	-
Peri-stroke region:	0.195±0.029	0.295±0.036	+51	**0.042**	+

k3

	Before	After	Δ%	p	sign.
Whole brain:	0.106±0.008	0.105±0.012	-1	0.488	-
Unaffected hemisphere:	0.115±0.010	0.112±0.010	-3	0.393	-
Affected hemisphere:	0.099±0.072	0.108±0.015	+9	0.220	-
Stroke region:	0.102±0.010	0.096±0.011	-6	0.216	-
Region outside stroke:	0.099±0.012	0.115±0.018	+16	0.069	-
Para-stroke region:	0.131±0.013	0.126±0.016	-4	0.345	-
Peri-stroke region:	0.086±0.008	0.108±0.017	+26	0.141	-

Table 1. Cerebral glucose metabolic rates and kinetic constants in selected VOI's before and after vinpocetine tratment.

CMR

	Whole brain	Un-affected hemisphere	Affected hemisphere	Stroke	Outside stroke	Para-stroke	Peri-stroke
Whole brain	-	0	+1	-8	+2	-18	+9
Un-affected hemisphere	n.s.	-	+1	-7	+3	-15	+9
Affected hemisphere	n.s.	n.s.	-	-8	+2	-20	+9
Stroke	n.s.	n.s.	n.s.	-	+12	-8	+19
Outside stroke	n.s.	n.s.	n.s.	n.s.	-	-22	+8
Para-stroke	n.s.	n.s.	n.s.	n.s.	n.s.	-	+15
Peri-stroke	n.s.	n.s.	n.s.	n.s.	n.s.	n.s.	-

k1

	Whole brain	Un-affected hemisphere	Affected hemisphere	Stroke	Outside stroke	Para-stroke	Peri-stroke
Whole brain	-	-1	+1	-6	0	-19	+21
Un-affected hemisphere	n.s.	-	+2	-5	+1	-15	+21
Affected hemisphere	n.s.	n.s.	-	-8	-2	-22	+31
Stroke	n.s.	n.s.	n.s.	-	+10	-9	+35
Outside stroke	n.s.	n.s.	n.s.	n.s.	-	-19	+21
Para-stroke	n.s.	n.s.	n.s.	n.s.	n.s.	-	+24
Peri-stroke	*	*	*	*	*	n.s.	-

k2

	Whole brain	Un-affected hemisphere	Affected hemisphere	Stroke	Outside stroke	Para-stroke	Peri-stroke
Whole brain	-	-3	-1	-5	+13	-5	+19
Un-affected hemisphere	n.s.	-	+3	-2	+15	-1	+21
Affected hemisphere	n.s.	n.s.	-	-4	+13	-5	+20
Stroke	n.s.	n.s.	n.s.	-	+21	+1	+27
Outside stroke	n.s.	n.s.	n.s.	n.s.	-	-17	+6
Para-stroke	n.s.	n.s.	n.s.	n.s.	n.s.	-	+18
Peri-stroke	n.s.	n.s.	n.s.	n.s.	n.s.	n.s.	-

k3

	Whole brain	Un-affected hemisphere	Affected hemisphere	Stroke	Outside stroke	Para-stroke	Peri-stroke
Whole brain	-	-1	+10	-5	+17	-14	+22
Un-affected hemisphere	n.s.	-	+10	-3	+17	-1	+21
Affected hemisphere	n.s.	n.s.	-	-14	+6	-15	-13
Stroke	n.s.	n.s.	n.s.	-	+23	+3	+19
Outside stroke	n.s.	n.s.	n.s.	n.s.	-	-22	+7
Para-stroke	n.s.	n.s.	n.s.	n.s.	n.s.	-	+20
Peri-stroke	*	*	*	*	n.s.	*	-

Table 2. The proportions of metabolic rates and kinetic rates between VOI's. The changes between the two experimental conditions (before and after vinpocetine treatment) are expressed in % and are shown right of the diagonal. The corresponding statistical significance of the changes are shown left of the diagonal (n.s. = not significant; * = $p \leq 0.05$). (See page after next one.)

4.3 POPULATION AVERAGES

In the later phase of the analysis, in order to visualise these changes, the global metabolic and kinetic parameters were used to normalise the anatomically standardised individual images and, in the next step, to create averaged images across the patient population (see Methods). The above observations were expressed in a pictorial way as shown in Figures 3 and 4. In Figure 3 a horizontal slice of the brain, averaged across the patient population and corresponding to global $gCMR_{glu}$ values, is displayed. As shown in the images, the highest changes in local $gCMR_{glu}$ values are present in the non-affected hemishere as well as in the regions falling outside the center stroke and its the cortical regions flanking it. In Figure 4 thresholded global $gCMR_{glu}$ images with three dimensional surface rendering are displayed: the images are thresholded at $gCMR_{glu}$ = value. As shown by the images before and after vinpocetine treatment, the region around the spatially averaged center of the stroke (no colour) diminishes following by the vinpocetine treatment.

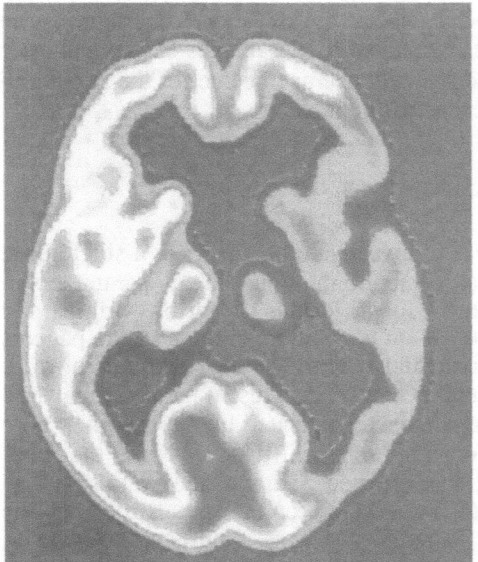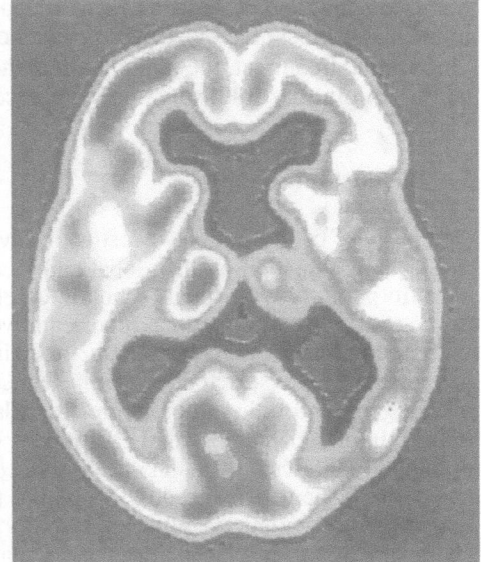

Figure 3. Averaged CMR_{glu} images of the population. A horizontal slice of the anatomically standardised population brain, averaged across the patient population, is displayed. The individual brains had their respective global CMR_{glu} values, the present average image corresponds to their mean. HBA convention, z = +20 mm (i.e. 20 mm above the AC-PC plane). The affected hemispheres are on the right side in the brain. Left panel: global mean CMR_{glu} image before vinpocetine treatment, right panel: global mean CMR_{glu} image after vinpocetine treatment. Colour coding: red: > 34

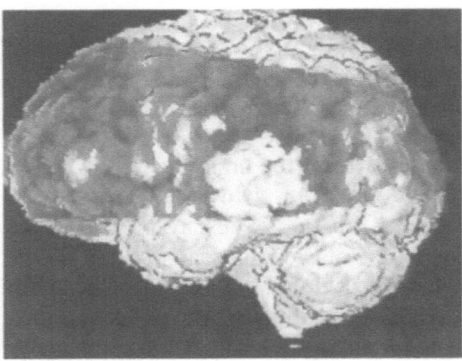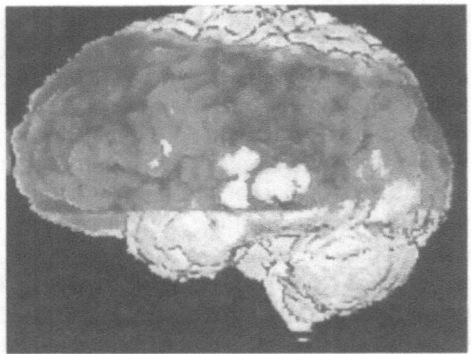

Figure 4: Thresholded global CMR$_{glu}$ images with three dimensional surface rendering are displayed. Lateral view, affected hemisphere. The coloured section of the brain represents the field of view of for which each subject's PET image was overlaping. The images are thresholded at CMR$_{glu}$= value. As shown by the images before and after vinpocetine treatment, the region around the spatially averaged center of the stroke (no colour) diminishes following by the vinpocetine treatment.

5. Discussion

The purpose of the present investigation was to investigate the effect of a single dose vinpocetine injection on the global and local metabolic rates and kinetic parameters in chronic patients with focal ischemic cerebrovascular lesions of various origins and durations. In order to gather a representative patient population, chronic stroke patients with focal cerebrovascular lesion in their medial cerebral artery in either or both hemispheres were selected for the present study. Clearly, the present patient population had in general a rather poor cerebrovascular status, as indicated by the fact that more than half of the patients had infarctions on both sides (in these patients the more dominantly infarcted hemisphere was termed the "affected" side) and the overall average age of the patients (62.4 ± 4.9). Similarly, the period between the acute stroke and the present study varied largely (range: 5.5 - 41 months). These features are, however, characteristic for the chronic cerebrovascular patient population treated with vinpocetine products orally or, sometime, intravenously. Consequently, against the fact that the present investigation included a relatively limited patient material, the sample represented the "usual" patient population undergoing Vinpocetine treatment.

The quantification of the present study is based upon a tracer kinetic model of cerebral glucose up-take and utilisation, and for the analysis the cerebral metabolic rates (CMR) and the k_1, k_2, k_3 kinetic constants were used (Balkay et al., 1997). Generally, the first two constants characterise the transfer velocity of glucose between the intravascular and intracellular compartments (k_1: brain up-take, k_2: brain release), whereas k_3 characterises the intracellular phosphorylation rate of glucose (the fourth component, k_4, characterising dephosphorylation, is considered to be zero in brain tissues). The parameter k1 depends on regional cerebral blood flow, up-take of glucose through the blood-brain barrier, and and the regional activity of a glucose transporter protein, glucose permease. The parameter k_3 depends on the regional expression and activity of the hexokinase enzyme (Szabó, 1997).

A single dose vinpocetine infusion did not change significantly the global and regional glucose metabolic rates in the brain. It, however, increased significantly the k_1 and k_2 kinetic constants in the whole brain, in the non-affected hemisphere and in the affected hemisphere, with special regard to the region falling outside the stroke region. The greatest changes took place in the "peri-stroke" regions, i.e. in the relatively healthy tissue compartments of the affected hemisphere. On the other hand, there were no significant changes with respect to the k_3 parameter.

The composition of the VOI's was non-homogeneous and the contribution of the partial volume effect was not assessed during the calculations. Whereas the whole brain VOI and the VOI of the non-affected hemisphere are composed of a combination of cortical and sub-cortical grey matter, white matter and ventricular space, the stroke VOI mainly consists of cortical grey matter and, to some extent, subcortical white matter. The dominant component of the VOI's of the para-stroke regions was cortical grey matter. Finally, the VOI comprising the rest of the affected hemisphere (i.e. the peri-stroke region) comprises cortical grey matter (within the supply territories of the anterior and posterior cerebral arteries), subcortical grey matter, white matter and ventricles. Not only the anatomical, and, consequently, metabolic and kinetic components, of the VOI's are inhomogeneous, but there are also differences in their metabolic and kinetic components due to the pathophysiology of the disease. The stroke region mainly comprises metabolically non- or dys-functioning grey matter, whereas the cortical regions flanking it (para-stroke region) consist of grey matter with both hypo- and hyper-functioning tissue and dilated vessels (Heiss et al.,   ). As in each patient the stroke preceded the present PET investigations on the average almost by one year, it is reasonable to believe that in this cortical region the anatomical consequences of tissue regeneration (e.g. increased capillary generation) and the physiological compensating mechanisms (e.g. dilated vessels, relatively increased regional blood flow) have already been stabilised. These regions, on the other hand, proved to be relatively "inert" with respect to the effects of a single dose vinpocetine infusion.It is therefore not surprising that (i) both the metabolic rates and kinetic constants within the stroke region were significantly lower that those in the surrounding cortical regions and (ii) a single infusion of vinpocetine has not exerted any significant effect within the stroke region and its surrounding cortical regions within the territory of the medial cerebral artery (para-stroke region).

The robust and significant effects of vinpocetine were found in the relatively "healthy" brain tissue: in the non-affected hemisphere and the regions in the affected hemisphere falling outside the stroke and para-stroke regions. A single dose vinpocetine infusion significantly increased both the k_1 and k_2 constants in these regions, and this effect was the strongest in the affected hemisphere's "healthy" regions. In these regions, in addition, the k_3 constant and, consequently, the regional cerebral glucose metabolic rates increased, though these changes were not significant per se. Nevertheless, when the changes of both the k1 and k3 constants within this region were compared to those in other regions, it became apparent that the single dose vinpocetine treatment more selectively affected the peri-stroke region than other regions.

These observations indicate a clear tendency: glucose up-take and, to a certain extent, glucose metabolism, in the relatively healthy brain tissue, spared by ischemic attacks in chronic stroke patients, can be positively influenced by a single dose administration of vinpocetine.

6. References

Balkay, L., Molnár, T., Boros, I. and Lehel, S. Quantification of FDG uptake using kinetic models. In: Gulyás, B. and Müller-Gärtner, H. W. (eds.) Positron emission tomography: A critical assessment of recent trends. Dordrecht: Kluwer, 1997.

Bergström M, Boëthius J, Eriksson J, Greitz T, Ribbe T, Widén L (1981): Head fixation device for reproducible positron alignment in transmission CT and positron emission tomography. J Comput Assist Tomogr 5: 136-141.

Biro K. Karpati E. Szporny L. Protective activity of ethyl apovincaminate on ischaemic anoxia of the brain. Arzneimittel-Forschung. 26(1976):1918-1920.

British Medical Journal, 18 July 1964. p. 177.

De Haan, R., Horn, J., Limburg, M., Van Der Meulen, J. and Bossuyt, P. A comparison of five stroke scales with measures of disability, handicap and quality of life. Stroke 24(1993):1178-1181.

Edwards, D. F., Chen, Y.-W. and Diringer, M. N. Unified neurological stroke scale is valid in ischemic and hemorrhagic stroke. Stroke 26(1995):1852-1858.

Greitz T, Bergström M, Boëthius J, Kingsley D, Ribbet T (1980) Head fixation device for integration of radiodiagnostic and radiotherapeutic procedures. Neuroradiology 19: 1-6.

Hayakawa M. Comparative efficacy of vinpocetinee, pentoxifylline and nicergoline on red blood cell deformability. Arzneimittel-Forschung. 42(1992):108-110.

Hayakawa M. Effect of vinpocetinee on red blood cell deformability in stroke patients. Arzneimittel-Forschung. 42(1992):425-427.

Hayakawa M. Effect of vinpocetinee on red blood cell deformability in vivo measured by a new centrifugation method. Arzneimittel-Forschung. 42(1992):281-283.

Heiss, W. D. and Herholz, K. Assessment of pathophysiology of stroke by positron emission tomography. Eur. J. Nucl. Med. 21(1994):455-465.

Heiss, W. D. and Graf, R. The ischemic penumbra. Curr. Op. Neurol. 7(1994):11-19.

Heiss, W. D. and Podreka, I. Role of PET and SPECT in the assessment of ischemic cerebrovascular disease. Cerebrovasc. Brain Metab. Rev. 5(1993):235-263.

Heiss, W. D., Graf, R., Wienhard, K., Lottgen, J., Saito, R., Fujita, T., Rosner, G., and Wagner, R. Dynamic penumbra demonstrated by sequential multitracer PET after middle cerebral artery occlusion in cats. J. Cerebr. Blood Flow Metab. 14(1994):892-902.

Imamoto T. Tanabe M. Shimamoto N. Kawazoe K. Hirata M. Cerebral circulatory and cardiac effects of vinpocetinee and its metabolite, apovincaminic acid, in anesthetized dogs. Arzneimittel-Forschung. 34(1984):161-169.

Jucker, M., Meier-Ruge, W., Baettig, K. Psychopharmacology 96(1988):29.

Kiss, B. and Kárpáti, E. Vinpocetine hatásai, hatásmechanizmusa: legújabb eredmények. 1996.

Kuzuya F. Effects of vinpocetinee on platelet aggregability and erythrocyte deformability. Therapia Hungarica. 33(1985):22-34.

Mchedlishvili, G. I. and Ormotsadze, L. G. The effect of ethyl apovincaminate on vasospasm of the circulatory isolated internal carotid artery in dogs. Arzneimittel-Forschung. 31(1981):414-418.

Nagy, Z. (ed.) Stroke Ellátás. (The Treatment of Stroke. *In Hungarian*) Budapest, Springer Hungarica, 1994.

Nagy, Z., Bönöczk, P., Pánczél, G., and Pataky, I. Vinpocetine review. Praxis 5(1996):1-6).

OPRR Reports (1989): Protection of Human Subjects. Code of Federal Regulations. Part 46. Washington, Department of Health and Human Services.

Orgogozo, J. M. Evaluation of treatments in ischemic-stroke patients. In: Amery, W. K. (ed.) Clinical Trial Methodology in Stroke. London, Mallière Tindall, 1989. pp. 35-53.

Roland, P. E. Brain Activation. New York: Wiley, 1993.

Roland, P. E. and Friberg, L. Localization of cortical areas activated by thinking. J.Neurophys. 53(1985):1219-1243.

Rota Kops E. Herzog H. Schmid A. Holte S. Feinendegen LE., Institute of Medicine, Nuclear Research Center, Juelich, Performance characteristics of an eight-ring whole body PET scanner. J.Comp. Ass. Tomogr. 14(3):437-45, 1990.

Scandinavian Stroke Study Group. Multicenter trial of hemodilution in ischemic stroke: Background and study protocol. Stroke 16(1985):885-890.

Szabó, Z. Membrane transporters. In: Gulyás, B. and Müller-Gärtner, H. W. (eds.) Positron emission tomography: A critical assessment of recent trends. Dordrecht: Kluwer, 1997.

Szobor A. Klein M. Examinations of the relative fluidity in cerebrovascular disease patients. Therapia Hungarica. 40(1991):8-11.

Tohgi H. Sasaki K. Chiba K. Nozaki Y. Effect of vinpocetinee on oxygen release of hemoglobin and erythrocyte organic polyphosphate concentrations in patients with vascular dementia of the Binswanger type. Arzneimittel-Forschung. 40(1990):640-643.

Tokiwa, T., Senoo, T., Narita, S., Minobe, Y., Sakihama, M., Tokiwa, S., Kaneko, S. Clin. Rep. 16(1982)2179.

Trón, L., Szakáll, S., Veress, G., Németh, F. and Galuska, L. Vinpocetine affects the kinetic constants of FDG accumulation: an application of registration and kinetic modelling. In: Gulyás, B. and Müller-Gärtner, H. W. (eds.) Positron emission tomography: A critical assessment of recent trends. Dordrecht: Kluwer, 1997.

COUPLING BETWEEN CEREBRAL BLOOD FLOW AND METABOLISM IN THE PRIMATE: METHODOLOGICAL AND PHARMACOLOGICAL ISSUES

Pascale SCHUMANN, Eric T. MacKENZIE
University of Caen, CNRS UMR 6551
Cyceron, Cyclotron Biomedical Unit
Bd H. Becquerel - BP 5229
F-14074 CAEN Cedex, France

The concept of coupling between cerebral blood flow (CBF), metabolism and function was enunciated as early as 1890 by Roy and Sherrington, as follows: "We conclude then that the chemical products of cerebral metabolism contained in the lymph which bathes the walls of the arterioles of the brain can cause variations of the calibre of the cerebral vessels: that in this re-action the brain possesses an intrinsic mechanism by which its vascular supply can be varied locally in correspondence with local variations of functional activity." However, this hypothesis remained to be tested until techniques for measurements of regional CBF and metabolism were available. A major step in this issue was achieved by the development of autoradiographic methods which allows the measurement of local CBF with tracers such as $[^{131}I]$trifluoroiodomethane [1] or $[^{14}C]$iodoantipyrine [2]. The tight relationship between regional CBF and function was demonstrated by measuring specific increases in CBF after experimentally induced local increases in functional activity that are restricted to a few defined neuroanatomical areas [3]. As an example, olfactory stimulation specifically enhanced the cerebral glucose metabolism (CMRGlu), as measured by the $[^{14}C]$deoxyglucose technique, in the brain regions from the olfactory system [3]. With the advent of non invasive techniques such as positron emission tomography (PET) and, more recently, functional magnetic resonance imaging (MRI), such studies have become increasingly possible in man [4, 5]. From this succint historical description, it is evident that advances in our understanding of the relationships between CBF, metabolism and function are completely dependent on the technological progress in the measurement of each of these parameters. Such developments include improvement not only in their spatial but also in their temporal resolution.

This article reviews the advances that have been made in the understanding of the relationships between flow, metabolism and function, their different pharmacological bases and the relevance of these findings to the functional imaging.

1. Coupling and uncoupling: what are the definitions?

If the term "coupling" designates the capacity of the cerebral circulation to provide the appropriate amount of energetic substrates that are required by the brain, this condition may be uniquely answered at rest when local CBF is distributed in proportion to the

B. Gulyás and H.W. Müller-Gärtner (eds.),
Positron Emission Tomography: A Critical Assessment of Recent Trends, 307–316.
© 1998 *Kluwer Academic Publishers.*

resting activity of brain structures as reflected by their metabolic rates. One pictorial representation of this coupling may be provided by the tight correlation that exists between CBF and CMRGlu over brain regions [5, 6] (Figure 1), and also seen in the striking homogeneity of the parametric images for cerebral oxygen extraction fraction (OEF). Indeed, at rest, the local CBF displays a heterogenous and regiospecific pattern that is distributed among brain regions in a manner similar to the local cerebral metabolism [6]. This correlation between regional flow and metabolism may correspond to the strict definition of the coupling and, accordingly, any alteration of this linear relation, i.e. modification of the slope or disparition of the correlation, may be called "uncoupling". Capillary density may play an major role in this coupling since the number of perfused capillaries is tightly correlated with the CBF and the CMRGlu [7] (Figure 2).

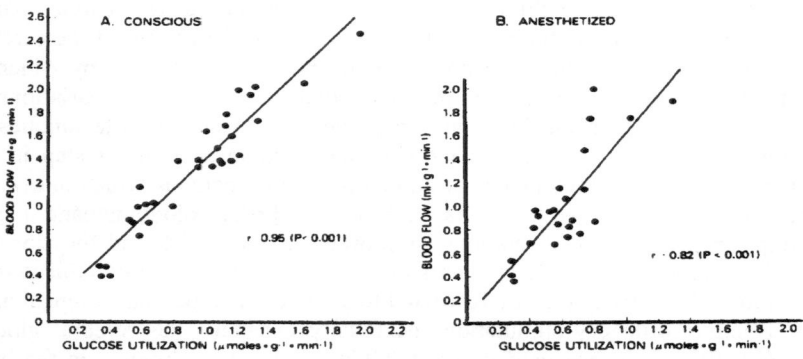

Figure 1: Correlation between cerebral blood flow and glucose utilization measured with [^{14}C]iodoantipyrine and [^{14}C]deoxyglucose techniques in either conscious (A) or thiopental anaesthetized (B) rats (after [6]).

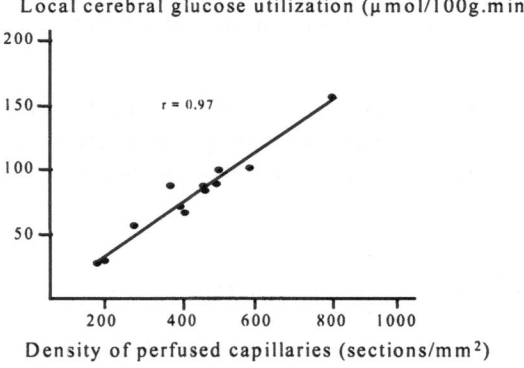

Figure 2: Correlation between the local cerebral glucose utilization and the density of perfused capillaries in the rat brain (adapted from [7]).

Based on this definition, functional activation may lead to a diversity of situations, given the variety of the flow/metabolism relationships observed in the literature. Indeed, coupling between CMRGlu and CBF seems to be preserved in some activation paradigms such as sensory stimulation in the rat [3] or mental activation in man [8] whereas it has been reported that experimental stimulation of some intracerebral structures may induce an increase in CBF without any change in CMRGlu, thereby leading to an "uncoupling" [9]. However, the physiological occurrence and, therefore, the physiological role, of the latter type of stimulation remains to be tested. This observation shows undoubtly that CBF and CMRGlu may be, in some cases regulated by distinct, parallel pathways in response to neuronal activation. This constitutes an interesting alternative of the classical view of coupling "in series" where the CBF is directly related to the CMRGlu of the activated neurone. The aim of the increase in CBF in face of unmodified metabolic requirements of the brain needs further explanation, as will be discussed later.

If the term "coupling" usually refers to the relationships between CBF and CMRGlu, it should also refer to the cerebral oxidative metabolism. As such, functional activation, by inducing a disproportionately greater increase in CBF than in $CMRO_2$, provokes a focal physiological "uncoupling" [10, 11]. For example, it has been shown that visual stimulation in man causes increases in CBF and CMRGlu by approximately 50% whereas $CMRO_2$ is enhanced by only 5% [12]. Accordingly, the molar ratio for the increase of oxygen to glucose consumption is only 0.4:1 during neural activity. In other words, the major part of the glucose uptake increase is not oxidized. Since the amount of ATP produced by glycolysis is far less than that by oxidative phosphorylation, one might consider that the energetic advantage of this activity-induced change in metabolism is quite low (around 8%). Does this mean that the energy requirements of the brain during neural activity are far less than it may be deduced from the substantial increase in CMRGlu? If so, then what function is served by the CBF and CMRGlu increases during functional activation? The hypothesis of a predominant glycolytic pathway during cerebral activation has been experimentally affirmed by the observation that lactate production is increased during somatosensory stimulation [13]. If the energy yield of glycolysis is relatively inefficient, although apparently sufficient, its preferential use might be explained by a faster availability of ATP than that which would be provided by oxidative phosphorylation. The brain energetic yield should then be calculated not only in terms of quantity but also in terms of rapidity in ATP production. Recently, Magistretti and co-workers (1996) [14] have proposed a cellular model of the relationships between neuronal activity and CMRGlu during functional activation that is particularly relevant to the nonoxidative glucose utilization reported by functional imaging studies. In physiological conditions, glucose is the near exclusive energetic susbtrate for the brain and is, at rest, metabolized at 95% *via* oxidative phosphorylation. This constitutes the main biochemical pathway for oxygen consumption in the adult. However, certain metabolic intermediates such as lactate or ketone bodies can constitute alternative substrates for cerebral energy metabolism. Their poor permeability across the blood brain barrier limits their contribution in physiological situations. At the cellular level, Magistretti and co-workers (1996) [14] have postulated that lactate could serve as an energetic substrate for neurons if produced from glucose in the astrocytes and then transferred into neurons. Furthermore, the concept of increased energy demand during functional activation may not be in accordance with the findings of the cerebral ischaemia studies. Indeed, it has been shown in rats subjected to transient global ischaemia that somatosensory stimulation, although still efficient as shown by persistent evoked potentials, is not accompanied by any change in CBF or CMRGlu [13]. However, since the recovery of electrophysiological function does not necessarily imply normal neurological performance [15], the physiological relevance of this uncoupling between electrical activity *versus* cerebral metabolism and blood flow is not completely resolved nor, indeed, understood.

To summarize, the term "coupling" may refer to a situation where the energetic supply is intimately linked to the brain's metabolic requirements, a situation that is achieved at rest. However, since only 50 % of the oxygen and 10% of the glucose available in the circulation are extracted by the brain, one might advance the hypothesis that the supply of energetic substrates at rest is already in excess of the requirements of the resting consumption of the brain.

2. The pharmacological bases of the flow/metabolism coupling

Two major hypotheses have been advanced to elucidate the mechanisms of the vascular response to functional activation, one of them based on the release of vasoactive substances by locally activated neurons and/or astrocytes, the other involving a neurogenic control by fibers arising from intracerebral structures and directly innervating the cerebral vessels [16]. There is no doubt that these two classical views of the CBF regulation could coexit in the brain although their respective roles are far from being well characterized.

2.1. NEUROVASCULAR COUPLING MEDIATORS: THE METABOLIC HYPOTHESIS

The metabolic hypothesis states that the vasoactive catabolites of neurons may regulate CBF in proportion to their metabolism, itself a function of their activity. Vasoactive factors can also be released by the endothelium and contribute to the maintenance of a basal cerebrovascular tone. It is not as clear whether at rest, flow/metabolism coupling and the maintenance of the basal tone are two phenomena that may be considered distinct or if coupling is one, amongst others, of the mechanisms by which cerebrovascular resistance is regulated. Nitric oxide (NO), endothelin, prostaglandins are some of the vasoactive agents that are synthesised by the endothelium in response to various stimuli (shear stress, neurotransmitters..) [17]. The simultaneous implication of neuronal and endothelial factors in several features of cerebrovascular regulation makes the analysis of the biological implications difficult. Moreover, the role of glial cells has long been neglected but has been recently brought to light by the study of Magistretti and co-workers (1996) [14], who suggested a tight cooperation between neurones and astrocytes during functional activation. This concept gives a prominent role to the astocytes in the regulation of cerebral metabolism during neuronal activation. Indeed, these cells occupy a strategic position between neurons, at the synaptic level, and capillaries; the glia, through the astrocytic endfeets, could then constitute a major element in the neurovascular coupling. However, the role of astrocytes in the vascular response to neural activation remains to be determined.

The classical metabolic candidates for coupling factors between CBF and cerebral activity are pCO_2, extracellular pH, potassium, adenosine and prostaglandins. More recently, neuronal and glial NO has been added to the list of potential mediators of the flow/metabolism coupling.

2.1.1. *CO₂, protons and potassium*

CO_2 is the final metabolite of glucose and, therefore, occupys a place of choice in the metabolic hypothesis [18]. Cerebrovascular resistance is clearly decreased by a rise in arterial pCO_2 and, conversely, hypocapnia leads to vasoconstriction. It is widely assumed that the cerebrovascular effects of CO_2 are mediated by the protons that are generated from CO_2 by carbonic anhydrase in the extracellular space of the brain. Indeed, extracellular acidosis induces a strong dilatatory action on cerebral vessels [19]. For this reason and based on the production of lactate during activation, it has been postulated that protons may be involved in the coupling. However, measurements of cerebral pH during neuronal stimulation are still lacking [20]. Likewise, it has been though attractive to consider potassium (K^+) as a mediator released by active neurones, taken up by astrocytes and, finally released by glial endfeet in the vicinity of cerebral arterioles, leading to a vasodilatation [21].

2.1.2. *Adenosine*

The attenuation of the CBF response to sensory stimulation by adenosine receptors antagonists has led to the "adenosine hypothesis" of metabolic coupling [22]. According to this concept, an early mismatch between oxygen consumption and availability in the brain would lead to a dephosphorylation of energy-rich adenosine nucleotides, thereby increasing the concentration of adenosine. The hypothesis of hypoxia, even transient is however unlikely [20]. Likewise, the microdialysis studies of adenosine release during sensory stimulation have never been conclusive [23].

2.1.3. *Prostaglandins*

The vasoactive properties of the products of the cyclo-oxygenase pathway, prostaglandins and thromboxanes, are well documented in the literature and this has led some authors to postulate an active role for these products in the coupling between flow and metabolism [24]. Moreover, prostaglandin synthesis takes place in various brain compartments including neurons although the major prostaglandin, PGI_2 is predominantly synthesised in endothelium and choroid plexus; PGI_2, is a potent vasodilatatory agent. It is commonly assumed that thromboxane A_2, PGE_2 and $PGF_{2\alpha}$, synthesised in vascular, glial and, to a lesser extent, neuronal cells, are essentially vasoconstrictive. Synthesis of PGD_2, taking place at the same sites, leads to a vasodilatatory response (for review, see [25]). Finally, inhibitors of cyclo-oxygenase are reputed to decrease resting CBF and inhibit the CBF response to hypercapnia [26, 27]. Although it is well established that indomethacin, a potent inhibitor of cyclo-oxygenase, dramatically decreases CBF in various species, without affecting the cerebral glucose metabolism [28], the effects of indomethacin on the cerebral oxygen consumption remained controversial [29]. An unchanged CMRGlu associated with a decreased oxidative metabolism by indomethacin would suggest a shift towards the anaerobic metabolism of glucose. The energetic yield being less than that of oxidative phosphorylation, a reduction in ATP production may cause a decrease in cerebral perfusion.

We recently addressed the effects of indomethacin on the coupling between CBF and $CMRO_2$ as measured by PET in the anaesthetized baboon [30]. Experiments were performed in 5 animals with a 7-slice Leti TTV 03 PET camera according to the ^{15}O

312

steady-state technique [31]. Intravenous administration of indomethacin (bolus 20 mg/Kg followed by perfusion 10 mg/Kg.hr) resulted in a marked and homogenous decrease in CBF of approximately 30% in all cortical and subcortical regions studied (Figure 3). In contrast, the $CMRO_2$ remained unchanged under indomethacin perfusion, suggesting that the potent cerebrovascular effects of this drug are not related to a decreased cerebral metabolism.

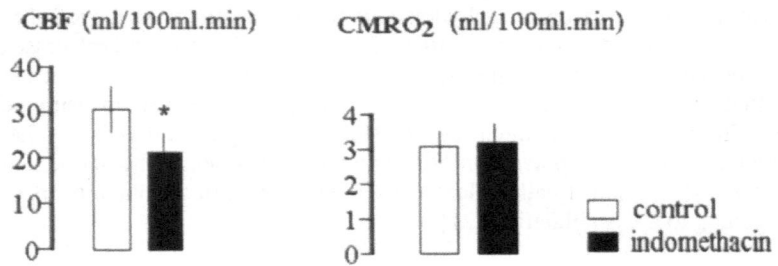

Figure 3: Effect of indomethacin (20 mg/Kg.hr, i.v.) on the resting cortical cerebral blood flow (CBF) and cerebral oxygen metabolism ($CMRO_2$) measured by PET in five anaesthetized baboons.

This study would then speak for an action of indomethacin on the coupling between flow and metabolism although more experimental data are still required to attribute its mechanism of action to an inhibition of prostaglandin synthesis [24]. Moreover, these results demonstrate the feasibility of such PET investigations in which the cerebrovascular and metabolic effects of a compound are concomitantly analysed. Such paradigms may be performed under strict physiological, biochemical and pharmacokinetic control in the non-human primate. The evident limitation for the measurement of $CMRO_2$ by PET is the somewhat low temporal resolution for the [15]O steady-state technique.

2.1.4. Nitric oxide
The gaseous bioradical NO constitutes the "ideal" molecule for coupling in that it is a diffusible, short-lived and a potent vasodilatator agent synthesised by astrocytes, neurones and endothelial cells [32] (Figure 4).

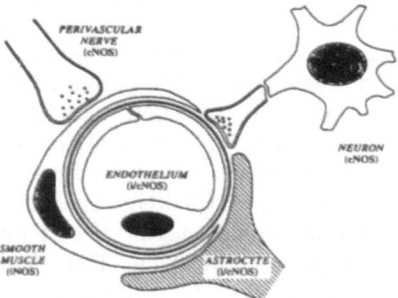

Figure 4: Cellular localization of the NO-synthase (NOS) at the microcirculatory level in the cerebrovascular bed (After [33]).

NO has been implicated in numerous responses of the cerebral circulation, including the basal cerebrovascular tone and the vasodilatation that follows hypercapnia and somatosensory stimulation (for review, see [33]). Further, it has been shown that inhibitors of the NO synthase partly inhibit these responses. The discovery of novel specific inhibitors of NO synthase has confirmed the fact that neuronal NO is effectively involved in cerebrovascular regulation [34, 35]. However, since inhibitors of NO synthase do not completely abolish these responses, it is unlikely that NO is the sole agent responsible for the coupling. Since NO is implicated in mediating or modulating several responses to putative agents of the coupling such as CO_2 or H^+ [19], it is more likely that a complex mixture of molecules are simultaneously and interactively involved in the adjustement of flow to metabolism in the active brain.

2.2. NEUROVASCULAR COUPLING MEDIATORS: THE NEUROGENIC HYPOTHESIS

The neurogenic hypothesis is essentially based on the two following arguments (for review, see [36]. First, the rapidity of coupling-related events is hard to reconcile with the timing of an accumulation of metabolites. The vascular response may then be more related to the removal of metabolites than to the supply of oxygen and glucose. However, as already mentioned, the temporal resolution given by the techniques available may themselves be little compatible with the biological processes. Second, the observed mismatch between flow and metabolism during certain activation studies may not milite in favour of the prominent role of metabolites in the vascular response to neuronal function. Briefly, there is an abundant innervation of cerebral vessels and stimulation of the central pathways of these systems (for example, cholinergic or serotoninergic pathways) leads to changes in CBF with inconsistent modifications in CMRGlu [9]. While the finality of these cerebrovascular responses are unknown, several hypotheses have been advanced. For example, an increase in CBF in response to a stimulation may have a favorable consequence on the cerebral energetic metabolism [37]. Indeed, the CO_2 washout enhanced by the increase in CBF leads to a rise in intracellular pH, which in turn disinhibits phsophofructokinase, the glycolytic regulatory enzyme. The lactate produced by glycolysis leads to an acidosis that can be rapidly buffered by the alkalosis due to the elimination of CO_2. This constitutes an advantage since acid buffering is a biological process that would otherwise consume energy [37]. The cerebrovascular response would then have an initiatory role in neuronal function. It has also been suggested that the CBF response to neurogenic stimulation could be directed to the anticipation of an additional stimulus. In other terms, this vascular response could constitute one aspect of the attentional processes of the brain to differing stimuli.

3. The flow/metabolism coupling: relation to functional activation studies

From the pharmacological studies of coupling discussed above, it becomes evident that any pathological condition or pharmacological intervention that interferes with the mechanisms responsible for coupling may affect the results of activation studies based on CBF measurements. For example, a widely consumed adenosine receptor antagonist such as caffeine may influence the CBF response to functional activation. Likewise, it

314

would be important to study the effects of indomethacin, a drug that is widely prescribed to a population of patients with not only articular disorders but also neurological diseases, on the cerebrovascular responses to somatosensorial or even cognitive stimuli. It is also clear that one must analyze the implication of NO in this vascular response. For these purposes, PET is ideally suited in that it corresponds to the methodological conditions in which activation studies are performed in man. Additionally, as already mentioned, global cerebral ischaemia illustrates a situation where the CBF response does not accurately reflect the underlying neuronal activity [13]. This situation could be particularly important for the understanding of those patients with cerebral diseases in which the flow/metabolism relationships may also be disturbed.

In conclusion, although the mechanisms that couple perfusion to metabolism and activity in the brain are particularly complex and multifactorial, it is of crucial importance to better characterize their pharmacological bases in order to take into account drugs or diseases that may interfere with the coupling and, accordingly, the interpretation of activation studies.

References

1. Landau, W.M., Freygang, W.H., Rowland, L.P., Sokoloff, L. and Kety, S.S. (1955) The local circulation of the living brain: values in the unanaesthetized and anaesthetized cat, *Trans Am Neurol Assoc* **80**, 125-129.
2. Sakurada, O., Kennedy, C., Jehle, J., Brown, J.D., Carbin, G. and Sokoloff, L. (1978) Measurement of local cerebral blood flow with iodo-^{14}C-antipyrine, *Am J Physiol* **234**, H59-H66.
3. Sokoloff, L. (1977) Relation between physiological function and energy metabolism in the central nervous system, *J Neurochem* **29**, 13-26.
4. Raichle, M.E., Grubb, R.L., Gado, M.H., Eichling, J.O. and Ter-Pogossian, M.M. (1979) Correlation between regional cerebral blood flow and oxidative metabolism. In vivo studies in man, *Arch Neurol* **33**, 523-526.
5. Prichard, J.W., and Rosen, B.R. (1994) Functional study of the brain by NMR, *J Cereb Blood Flow Metab* **14**, 365-372.
6. Sokoloff, L. (1981) Relationships among local functional activity, energy metabolism, and blood flow in the central nervous system, *Federation Proc* **40**, 2311-2316.
7. Klein, B., Kuschinsky, W., Schröck, H. and Vetterlein, F. (1986) Inter-dependency of local capillary density, blood flow and metabolism in rats brain? *Am J Physiol* **251**, H1333-H1340.
8. Lund Madsen, P., Hasselbalch, S.G., Hagemann, L.P., Skovgaard Olsen, K., Bülow, J., Holm, S., Wildschiodtz, G., Paulson, O.B. and Lassen, N.A. (1995) Persistent resetting of the cerebral oxygen /glucose uptake ratio by brain activation: evidence obtained with the Kety-Schmidt technique. *J Cereb Blood Flow Metab* **15**, 485-491.
9. Kimura, A., Sato, A. and Takano, Y. (1990) Stimulation of the nucleus basalis of Meynert does not influence glucose utilization of the cerebral cortex in anaesthetized rats. *Neurosci Lett* **119**, 101-104.
10. Fox, P.T. and Raichle, M.E. (1986) Focal physiological uncoupling of cerebral blood flow and oxidative metabolism during somatosensory stimulation in human subjects, *Proc Natl Acad Sci USA* **83**, 1140-1144.

11. Seitz, R.J., and Roland, P.E. (1992) Vibratory stimulation increases and decreases the regional cerebral blood flow and oxidative metabolism: a positron emission tomography (PET) study, *Acta Neurol Scand* **86**, 60-67.
12. Fox, P.T., Raichle, M.E., Mintum, M.A. and Dence, C. (1988) Nonoxidative glucose consumption during focal physiologic neural activity, *Science* **241**, 462-464.
13. Ueki, M., Linn, F. and Hossmann, K. (1988) Functional activation of cerebral blood flow and metabolism before and after global ischemia of rat brain, *J Cereb Blood Flow Metab* **8**, 486-494.
14. Magistretti, P.J. and Pellerin, L. (1996) Cellular bases of brain energy metabolism and their relevance to functional brain imaging: evidence for a prominent role of astrocytes, *Cereb Cortex* **6**, 50-61.
15. Hossmann, K., Schmidt-Kastner, R. and Grosse-Ophoff, B. (1987) Recovery of integrative central nervous function after one hour global cerebro-circulatory arrest in normothermic cat, *J Neurol Sci* **77**, 305-320.
16. Lou, H.C., Edvinsson, L. and MacKenzie, E.T. (1987) The concept of coupling blood flow to brain function: revision required?, *Ann Neurol* **22**, 289-297.
17. Wahl, M., and Schilling, L. (1993) Regulation of cerebral blood flow - A brief review, *Acta Neurochir* **59**, 3-10.
18. Lassen, N.A. (1959) Cerebral blood flow and oxygen consumption in man, *Physiol Rev* **39**, 183-238.
19. Niwa, K., Lindauer, U., Villringer, A. and Dirnagl, U. (1993) Blockade of nitric oxide synthsesis in rats strongly attenuates the CBF response to extracellular acidosis. *J Cereb Blood Flow Metab* **13**, 535-539.
20. Villringer, A., and Dirnagl, U. (1995) Coupling of brain activity and cerebral blood flow: basis of functional neuroimaging, *Cerebrovasc Brain Metab Rev* **7**, 240-276.
21. Paulson, O.B., and Newman, E.A. (1987) Does the release of potassium from astrocyte endfeet regulate cerebral blood flow?, *Science* **237**, 896-898.
22. Dirnagl, U., Niwa, K., Lindauer, U. and Villringer, A. (1994) Coupling of cerebral blood flow to neuronal activation: role of adenosine and nitric oxide, *Am J Physiol* **267**, H296-H301.
23. Northington, F.J., Matherne, G.P., Coleman, S.D. and Berne, R.M. (1992) Sciatic nerve stimulation does not increase endogenous adenosine production in sensory-motor cortex, *J Cereb Blood Flow Metab* **12**, 835-843.
24. Pickard, J.D. (1981) Role of prostaglandins and arachidonic acid derivatives in the coupling of cerebral blood flow to cerebral metabolism, *J Cereb Blood Flow Metab* **1**, 361-384.
25. Schaad, N.C., Magistretti, P.J. and Schorderet, M. (1991) Prostanoids and their role in cell-cell interactions in the central nervous system, *Neurochem Int* **18**, 303-322.
26. Pickard, J.D., and MacKenzie, E.T. (1979) Inhibition of prostaglandin synthesis and the response of baboon to cerebral circulation to carbon dioxide, *Nature (New Biol)* **245**, 187-188.
27. Wang, Q., Paulson, O.B.and Lassen, N.A. (1993) Indomethacin abolishes cerebral blood flow increase in response to acetazolamide-induced extracellular acidosis: a mechanism for its effect on hypercapnia?, *J Cereb Blood Flow Metab* **13**, 724-727.
28. McCulloch, J., Kelly, P.A.T., Grome, J.J. and Pickard, J.D. (1982) Local cerebral circulatory and metabolic effects of indomethacin, *Am J Physiol* **243**, H416-H423.
29. Nowicki, J.-P., Jourdain, D. and MacKenzie, E.T. (1987) NADH fluorescence in vivo: changes in cerebral oxidative metabolism and perfusion induced by pentobarbital, indomethacin and salicylate. *J Cereb Blood Flow Metab* **7**, 280-288.

316

30. Schumann, P., Touzani, O., Young, A.R., Verard, L., Morello, R. and MacKenzie, E.T. (1996) Effects of indomethacin on cerebral blood flow and oxygen metabolism: a positron emission tomographic investigation in the anaesthetized baboon, *Neurosci Lett* **220**, 1-5

31. Frackowiak, R.S.J., Lenzi, G.L., Jones, T. and Heather, J.D. (1980) Quantitative measurement of regional cerebral blood flow and oxygen metabolism in man using ^{15}O and positron emission tomography: theory, procedure and normal values, *J Comput Assist Tomogr* **4**, 727-736.

32. Goadsby, P.J., Kaube, H. and Hoskin, K.L. (1992) Nitric oxide couples cerebral blood flow and cerebral metabolism, *Brain Res* **595**, 167-170.

33. Iadecola, C., Pelligrino, D.A., Moskowitz, M.A. and Lassen, N.A. (1994) Nitric oxide synthase inhibition and cerebrovascular regulation. *J Cereb Blood Flow Metab* **14**, 175-192.

34. Kelly, P.A.T., Ritchie, I. and Arbuthnott, G.W. (1995) Inhibition of neuronal nitric oxide synthase by 7-nitroindazole: effects upon local cerebral blood flow and glucose use in the rat, *J Cereb Blood Flow Metab* **15**, 766-773.

35. Cholet, N., Bonvento, G. and Seylaz, J. (1996) Effects of neuronal NO synthase inhibition on the cerebral vasodilatory response to somatosensory stimulation, *Brain Res* **708**, 797-200.

36. Reis D.J. and Iadecola C. (1989) Central neurogenic regulation of cerebral blood flow, in J. Seylaz and P. Sercombe (eds.), *Neurotransmission and Cerebrovascular Function II,* Elsevier Science,.Amsterdam, pp.369-390.

37. Woods, B.T. (1988) Metabolic advantage of neuronal coupling of regional brain activity and blood flow. *Ann Neurol* **23**, 629-630.

Part five:
PET in neuroscience research

BRAINMAP™: ELECTRONIC INTEGRATION OF MIND AND BRAIN

PETER T. FOX, M.D. and JACK L. LANCASTER, Ph.D.

University of Texas Health Science Center, Research Imaging Center

7703 Floyd Curl Drive, San Antonio TX 78284-6240 USA

Mapping the Mind

The human brain has been described as the most complex system in the known universe. In our normal, daily activity of perceiving, understanding and moving within the physical world, our brains are constantly mapping that world. Each property of every object that we perceive -- the shape, color, location, motion, sound, and the like -- is extracted by a set of highly specialized brain areas. While some areas perform perception, other areas plan movements and envision consequences. While we do not yet know the total number of functional areas in the brain, Michael has hypothesized that each elementary, information-processing operation of the mind is enabled by a distinct functional area of the brain. Complex mental operations (like reading this page) require the integrated activity of tens, possibly hundreds, of localized populations of neurons. In the face of such overwhelming complexity, the availability of accurate, detailed, up-to-date maps of the brain and its functions is an absolute necessity for the brain scientist or clinician.

319

B. Gulyás and H.W. Müller-Gärtner (eds.),
Positron Emission Tomography: A Critical Assessment of Recent Trends, 319–329.
© 1998 *Kluwer Academic Publishers.*

Functional neuroimaging has made dramatic advances over the past decade. Only a few years ago, imaging even the gross anatomy of the human brain in living persons was impossible. Today, anatomical imaging shows structural details previously seen only with a microscope. Of even greater consequence, the functioning of the brain can be mapped in both space and time. As many of the articles in this volume illustrate, cartographers of the human brain now have a variety of powerful brain-imaging methods from which to choose. Positron-emission tomography (PET), functional magnetic resonance imaging (fMRI), magneto-encephalography (MEG), event-related potentials (ERPs), and other less evolved but no-less-promising techniques can all map the workings of the human mind. As a consequence, the influx of the knowledge about the functional organization of the human brain has become enormous. We have learned more about the human brain in the past decade than in the previous century. To cope with this enormous influx of information, a continually updated -- and, therefore, electronic -- map of the human the brain has been developed. This electronic atlas of the brain is called BrainMap™.

The BrainMap™ Database

BrainMap™ is an electronic environment for collecting and comparing functional maps of the human brain that are produced by the current generation of non-invasive, neuroimaging research methods. Developed by Peter Fox and Jack Lancaster at the University of Texas Health Science Center at San Antonio, Texas, BrainMap™ is a service chiefly targetting scientists actively researching the workings of the human brain. Nevertheless, BrainMap™ can be accessed by clinicians, educators and those merely curious, seeking to learn a bit more about their own brains. BrainMap's™ electronic environment consists of a relational database and a suite of tools for contributing, retrieving, viewing and interpreting data about brain-imaging studies. Recent developments even allow detailed models of activation patterns in as-yet-untested paradigms to be created.

BrainMap's™ database resides at the Research Imaging Center in San Antonio, on a UNIX workstation (Sun SparcStation 20) connected to the internet. BrainMap's™ database is accessed via the internet using Brainmap™ interface tools, also developed by Fox and Lancaster. The easiest way for interested readers to access BrainMap™ is with a World Wide Web browser (http://ric.uthscsa.edu/services).

Brain images, as produced by modern neural imaging methods, are actually matrices of numbers recording measurements of in the three spatial dimensions, i.e., 3-D number matrices. The amount of detail present in a brain image is determined by the size of individual volume elements (voxels) of which the 3-D image (number matrix) is composed. As the brain occupies a volume of approximately 20 cm (long) x 15 cm (wide) x 15 cm (tall), a brain image with a resolution of 1 mm^3 contains 4.5 million individual voxels. Functional brain mapping is performed by comparing images during complex tasks with those during simpler control states; the need for task and control images doubles the total amount of data needed. If multiple task and control states are reported (as is often done), the amount of data is proportionately larger. For reliability, brain-mapping studies are often repeated many times in each subject and then replicated across tens of persons, greatly increasing the total amount of data. Thus, a typical brain-imaging research article is based on many gigabytes (1×10^9 bytes) of data. Given the enormous volume of data underlying any brain-imaging study, sharing of raw data via the internet is impractically slow. BrainMap™ solves this problem by restricting its content to meta-data: data about data.

BrainMap's™ meta-data includes in-depth descriptions of the behavioral conditions (cognitive tasks) studied as well as the control states to which the test condition were contrasted. The imaging methods used are described in detail. The subject population is fully described, including age, sex, handedness, race and diagnosis. The authors and laboratory from which the data originated, as well as citations of published reports of the data, are documented. Finally, the results of the study are reported, often in greater detail than can be permitted by conventional print media. The results can be reported in great detail both because of the efficiency of electronic storage and because of a very concise anatomical-coding scheme developed by a Parisian neurosurgeon, Jean Talairach.

The Talairach Space

Talairach was among the first to recognize that brain anatomy, although admittedly variable from person to person, is sufficiently regular that it can be practically addressed using Cartesian coordinates. Anatomical coordinates are created by first placing the brain images of the individual subjects into a standard orientation, thereby referencing them all to a common origin and to common major axes. The brain images are then scaled to the same shape and size, using an arbitrary standard published by Talairach. Once the brains are spatially normalized, each anatomical location in any brain can be identified by 3 numbers: the x, y and z addresses of that point. Prior to Talairach's coordinate method, anatomical location was determined solely by visualizing structural boundaries and physical relations to neighboring brain structures, such as the sulci (grooves) and gyri (bumps) of the surface of the brain. When constrained to use surface "bump and groove" anatomy, the most effective way of communicating location is to provide a picture. As discussed above, modern brain images are simply too large to be efficiently shared on-line. On the other hand, Talairach's strategy for addressing the entire brain with coordinates -- removing the need to visualize the gross anatomy of the individual person -- allows function-location relationships to be communicated with incredible concision: three numbers are sufficient to describe an exact location within the brain.

The accuracy with which Talairach coordinates localize brain structures has been a matter of some debate. For many years, some brain scientists have argued that individual variations in brain shape (e.g., the number and size of specific gyri and sulci) was too great to be accommodated by Talairach's geometrical approach. While this objection still persists, the power of modern "morphing" algorithms is steadily convincing all but the most traditional anatomists. These are the same type of algorithm used for special effects in cinema and television, where one face or object is "morphed" into another. The reader can judge from his own observations at the movies whether modern computer science has the power to warp all brains into a standard shape and size. Even more exciting is the prospect of applying this transformation in reverse, thereby apply to individual patients the knowledge derived from morphing and merging the observations from thousands of prior studies.

Although quite useful for database development and efficient communications, Talairach's "anatomical geometry" can also be viewed as a generalized modeling construct. The modeling construct is a Cartesian space, bounded to create an addressing scheme for the brain. Within Talairach's idealized "brain space", we place observations about structure and function. These observations can be group-mean effects (e.g., activated locations within a grand-mean image), individual effects (e.g., activated locations within the images of individual subjects), or even probabilities that a specific structure or function will reside at a specific coordinate. As all observations are referenced to the same spatial framework, they collectively constitute a three-dimensional model of the human brain that is continuously updated by the research community itself. BrainMap™ (above) was the first community-based, electronic, human-brain model to be developed. A complementary electronic tool -- the Structure Probability Map (SPMap), describing structure-location probabilities within this space, is well underway (Mazziotta et al., 1995). Jointly, BrainMap™ and SPMap revise the entire concept of a brain atlas. Rather than a printed atlas illustrating a single subject or an author's conception of an "idealized brain", these tools provide locations and bounds -- both functional and anatomical -- derived from neuroimaging studies in thousands of participants. Thus, the concept of a cumulative brain model, implicit in the use of the Talairach space and clearly enunciated by Talairach nearly three decades ago, has become an electronic reality.

For the many among us who still find gross morphology the most intuitive conceptualization of brain anatomy, a labeling tool -- the "Talairach Daemon" -- that instantly defines the subset of Talairach's space bounding any nucleus, gyrus, tract or cytoarchitectonic area has been developed. Although not yet openly available, the "Talairach Daemon" will soon be made available for interactive use with a Java applet and with a c-code function for incorporation in user-developed applications. With these electronic guides, anyone will be able navigate the Talairach space. Equally as important, these tools will more effectively integrate traditional terminologies into Talairach's space, hopefully minimizing conflicts with neuroanatomical traditionalists, while still reaping the many benefits of Talairach's profound concept.

Integrating the Many Dimensions of Brain Research

However numerous and detailed as the brain's maps of function prove to be, structure-function mapping is only a first step to understanding the complexities of the brain. Structure-function mapping is one view of the brain. The brain can be considered from many other perspectives. Only when all possible views are considered and reconciled will we truly understand the human brain.

Time, for example, is a fundamental aspect of cognition and of brain function that imaging studies have only begun to explore. Although our thoughts may seem instantaneous, they actually are played out over time, like pieces of music. From the instant that you direct your eyes to this WORD until you understand its meaning, about 300 milliseconds will elapse. Even this simple thought unfolds through the sequential performance of dozens of specialized brain areas. When imaging these mental processes with PET or functional MRI, the sequence is compressed: every area is seen as if active at the same instant. Imagine hearing an entire score played in an instant. All the beauty and order of the music would be lost! The chronometry of thought -- the order of our mental music -- can be detected by event-related potentials (ERP) and magnetoencephalography (MEG). These temporally precise methods, unfortunately, lack the spatial precision of PET, MRI and other tomographic imaging methods. A attractive solution is to integrate the findings of these several modalities.

Leading brain-imaging laboratories throughout the world are beginning to make four-dimensional (time plus space) functional maps, by combining the use of techniques optimized for temporal mapping (ERP and MEG) with those optimized for spatial mapping. As each of these methods places different constraints on the physical environment needed for data acquisition (at least, for optimal acquisition), scanning with each of the two (or more) methods generally is performed sequentially, rather than simultaneously. While this optimizes the data obtained with each method, it raises several hurdles that must be overcome in the process of creating 4D maps. Two of the chief problems are: 1) registration of the multiple images types with one another; and 2) development of cognitive experimental paradigms that can be performed under the varied conditions of the different modalities while activating the same brain areas in a reliable way. Despite these difficulties, the challenge is being met, as can be seen in the recent study of Heinse and colleagues from the University of California, San Diego. The BrainMap™ development team has responded to these developments by expanding the database to include studies from ERP and MEG, as well as the use of multiple modalities within a single study.

The integration process is only begun by the consideration of four dimensions (time and space). Each brain area is connected to hundreds of other areas. In any task, some of these connections are active, while others are not. Connectivity both anatomical and effective (functional) is another view of the brain that has only just begun to be explored. Brain areas differ in the type of neurons present as well as their local (within-area) wiring. Post-mortem studies of the human brain are providing detailed maps of micro-anatomy, that must also be included in a comprehensive view of the brain. Finally, the brain codes and transfers information using a great variety of chemical mediators. PET and other radiotracer methods make regional maps of neurotransmitters, which eventually must be transformed into the Talairach space and integrated into the BrainMap™ environment.

The prospects for understanding the workings of the human brain and its product, the human mind, have never been more promising. Computer-based methods for imaging, mapping and modeling the brain have created a revolution in the neurosciences. The amount known about the workings of the brain is increasing exponentially. The next great challenge is to synthesize this wealth of information into maps and models that convert data into knowledge.

Acknowledgments

The work described above is supported by an EJLB Foundation grant and and National Institute of Mental Health grants P20 DA52176-01 (Human Brain Project).

References

1. Fox, PT (1995) Spatial Normalization: Origins, Objectives, Applications and Alternatives. *Human Brain Mapping* 3: 161-164.

2. Fox, PT, Lancaster, J.L. (1995) Neuroscience on the Net. *Science* 266: 994-995.

3. Fox PT, Mikiten S, Davis G, Lancaster JL (1994): BrainMap: A database of human functional brain mapping. In, Thatcher RW, Hallett M, Zeffiro T, Roy John E, Huerta M (eds): *Functional Neuroimaging: Technical Foundations*. San Diego: Academic Press, pp. 95-106.

4. Fox, PT, Woldorff MG (1994) Integrating Human Brain Maps. *Current Opinions in Neurobiology* 4:151-156.

5. Mazziotta JC, Toga AW, Evans A, Lancaster JL, Fox PT (1995): A probabilistic atlas of the human brain: theory and rational for its development. *NeuroImage* 2: 89-101.

6. Posner MI, Petersen SE, Fox PT, Raichle ME. (1988) Localization of Cognitive Operations in the Brain. *Science* 240: 1627-1631.

7. Talairach J, Szikla G (1967): Atlas D'anatomie stereotaxique du telencephale. *Etudes Anatomo-Radiologiques.* Paris: Masson & Cie.

8. Talairach J, Toumoux, P (1988): *Coplanar stereotaxic atlas of the human brain.* NY: Thieme Medical.

DESIGNING ACTIVATION EXPERIMENTS

K.J. FRISTON, C.J. PRICE & C. BUECHEL

The Wellcome Department of Cognitive Neurology, Institute of Neurology, Queen Square, London WC1N 3BG, UK

I. Introduction

In this chapter we review different sorts of experimental design and examine some of the assumptions about the relationship between cognitive function and neurobiology. We start with a simple taxonomy of experimental design that includes (i) categorical, (ii) parametric and (iii) factorial designs. Each of these designs is discussed from a conceptual viewpoint and illustrated using exemplar activation studies. The remaining sections of the chapter focus on parametric and factorial designs by considering specific examples of their use. For parametric designs we have chosen nonlinear regression and show how curvilinear relationships between task parameters and evoked hemodynamic responses can be characterized. The example of factorial designs was chosen to address cognitive subtraction and the validity of pure insertion.

2. A taxonomy of experimental design

2.1. CATEGORICAL DESIGNS

The tenet of this approach is that the difference between two tasks can be formulated as a separable cognitive or sensorimotor component and that the regionally specific differences in brain activity identify the corresponding functionally specialized area. Early applications of subtraction range from the functional anatomy of word processing [1] to functional specialization in extrastriate cortex [2]. The latter studies involved presenting visual stimuli with and without some specific sensorial attribute (e.g. colour, motion *etc*). The areas highlighted by subtraction are identified with homologous areas in monkeys that show selective electrophysiological responses to equivalent visual stimuli. Cognitive subtraction is conceptually simple and is a very effective device in mapping functional anatomy. It does however depend on the assumption that cognitive states differ in components that can be *purely inserted* or removed with no interaction between them, both at the level of the function and its neural implementation. The possible fallibility of this assumption (see below) has prompted the exploration of other experimental designs.

2.2. PARAMETRIC DESIGNS

The premise here is that regional physiology will vary monotonically and systematically with the amount of cognitive or sensorimotor processing. Examples of this approach include the experiments of Grafton *et al* [3] who demonstrated significant correlations between rCBF and the performance of a visually guided motor tracking task (using a pursuit rotor device) in the

B. Gulyás and H.W. Müller-Gärtner (eds.),
Positron Emission Tomography: A Critical Assessment of Recent Trends, 331–342.
© 1998 *Kluwer Academic Publishers.*

primary motor area, supplementary motor area and pulvinar thalamus. The authors associated this distributed network with early procedural learning. On the sensory side Price *et al* [4] have demonstrated a remarkable linear relationship between rCBF in peri-auditory regions and frequency of aural word presentation. Significantly this correlation was not observed in Wernicke's area, where rCBF appeared to correlate, not with the discriminative attributes of the stimulus, but with the presence or absence of words. This nonlinear relationship between stimulus presentation frequency and evoked response speaks to the importance of modelling nonlinear associations explicitly in the analysis. This theme will be taken up again below.

Time dependant changes in physiology are clearly central in studies of learning and memory. Many animal models of procedural learning depend on habituation and adaptation, either at a behaviourial or electrophysiological level. In the context of functional imaging, physiological adaptation to a challenge is simply the change in rCBF activation with time. This is an interaction.

2.3. FACTORIAL DESIGNS

At its simplest an interaction is basically a change in a change. Interactions are associated with *factorial* designs where two *factors* are combined in the same experiment. The effect of one factor, on the effect of the other, is assessed by the interaction term (two factors interact if the level of one factor affects the effect of the other). Factorial designs have a wide range of applications. The first PET experiment of this sort was perhaps the simplest imaginable and examined the interaction between motor activation (sequential finger opposition paced by a metronome) and time (rest - performance pairs repeated 3 times) [5]. Significant adaptation was seen in the cerebellar cortex (ipsilateral to hand moved) and cerebellar nuclei. These results are consistent with the electrophysiological studies of Gilbert and Thach [6], who demonstrated a reduction in simple and complex spike activity of Purkinje cells in the cerebellum during motor learning in monkeys. Psychopharmacological activation studies are examples of a factorial design [7]. In these studies subjects perform a series of baseline-activation pairs before and after the administration of a centrally acting drug. The interaction term reflects the modulatory drug effect on the task-dependent physiological response. A further example of factorial designs would include experiments designed to examine the interaction between cognitive processes, for example dual task interference paradigms. An example of this approach [8] involved analysis of encoding of episodic verbal material (using paired associates). Memory and control tasks were performed under two conditions, an easy and a difficult manual distracter task. Encoding is generally confounded with priming. However, due to the differential impact of the distracter task on encoding and priming, the authors were able to make inferences about encoding *per se* using the interaction effects.

The examples cited above all involve the use of factorial designs where each factor has a number of discrete or categorical levels. These designs have facilitated the study of adaptation, neuromodulation and interference at a cognitive level. It is of course possible to combine parametric and factorial designs, wherein a task or stimulus parameter is varied under two or more conditions. For example the frequency of stimuli can be manipulated under different forms of attentional set (Frith and Friston, in press). The interaction effects in this context can be thought of as a change in the slope of the regression of hemodynamic response on the task parameter under different conditions. This sort of design has clear applications in examining the changes in sensitivity of a particular area to stimulation under different conditions. We will return to factorial designs and their relationship to 'cognitive interactions' below.

2.4. SUMMARY

Experimental design has been briefly reviewed using a taxonomy of activation studies that distinguishes between categorical (subtractive), parametric (dimensional) and factorial (interaction) designs. Subtraction designs are well established in functional mapping but are predicated on possibly untenable assumptions about the relationship between brain dynamics and the functional processes that ensue. For example even if, from a functionalist perspective, a cognitive component can be added without interaction among pre-existing components the brain's implementation of these processes is almost certainly going to show profound interactions. This follows from the observation that neural dynamics are nonlinear. Parametric approaches avoid many of the shortcomings of 'cognitive subtraction'; in testing for systematic relationships between neurophysiology and sensorimotor, psychophysical, pharmacologic or cognitive parameters. These systematic relationships are not constrained to be linear or additive and may show very nonlinear behaviour. The fundamental difference between subtractive and parametric approaches lies in treating a cognitive process, not as a categorical invariant, but as a dimension or attribute that can be expressed to a greater or lesser extent. It is anticipated that parametric designs of this type will find an increasing role in psychological and psychophysical activation experiments. Finally factorial experiments provide a rich way of assessing the affect of one manipulation on the effects of another. The designs can be used to look at a variety of effects; we have mentioned adaptation, neuromodulation and dual task interference as three important examples. The assessment of differences in activations between two or more groups represents a further question about regionally specific interactions. The limiting case of this example is where one group contains only one subject, This may be one way to proceed with single subject analyzes; in that the interesting things about an individual's activation profile are how it relates to some normal profile or a profile obtained from the same subject in different situations or at a different time. These differences in activations are interactions.

3. Nonlinear parametric approaches

Parametric study designs can reveal information about the relationship between a study parameter (e.g. word presentation rate) and regional cerebral blood flow (rCBF). The brain's responses in relation to study parameters might be non-linear, therefore linear regressions, as often used in the analysis of parametric studies, might not represent a sufficient characterization. We now introduce a method that fits nonlinear functions of stimulus or task parameters to rCBF responses, using second order polynomials. This technique is implemented in the context of the general linear model and statistical parametric mapping. We also consider the usefulness of statistical inferences, based on statistical parametric maps of the F statistic - SPM{F}. These points will be illustrated with a PET activation study using an auditory paradigm of increasing word presentation rate.

3.1. BACKGROUND

Digital signal processing utilises many different techniques to characterize discrete signals by a combination of a number of basis functions. Well known examples are the Fourier expansion and the polynomial expansion. We adapt this technique to characterize rCBF responses in terms of a set of basis functions of a given parameter (e.g. study parameter or response) and show how rCBF responses can be approximated by a small number of these functions. In the current example we use a polynomial expansion of the task parameters. This is effected by modelling the responses in terms of a 'design matrix' whose columns contain the parameter of interest, the parameter of interest squared, and so on. The resulting parameter estimates correspond to the coefficients of a polynomial expansion of the response variable in terms of the original parameter. To test the overall significance of the regression we test the null hypothesis that these coefficients are zero in the usual way, using the F statistic. This analysis is performed for every voxel. The ensuing SPM{F} can be interpreted as an image of the significance of the non-linear relationship between the parameter and rCBF response.

This general approach to parametric studies can also be extended to compare the non-linear responses of different groups. To test for differences, the polynomials appear twice in the design matrix. First the columns are replicated in both groups; and in the second partition the polynomials are inverted for the second group. This second partition models differential responses and effectively represents a non-linear model of the interactions. These interactions can be designated as the effects of interest, treating the common responses as confounds, and the resulting SPM{F} depicts the significance of the differential response.

3.2 AN EXAMPLE

A patient, who had recovered from severe aphasia after an infarction, that was largely confined to the left temporal lobe and involving the whole of the superior temporal gyrus, was scanned 12 times while listening to words presented at different rates.. To illustrate a comparison between subjects, a normal subject was studied using the same paradigm. The SPM{F} reflecting the significance of a nonlinear relationship between word presentation rate and evoked responses, in the patient, is shown in Figure 1 (upper panel). As an example, of this non-linear relationship, the adjusted and fitted responses from a voxel (at -34, 40, 24 mm) is shown in the lower left panel. This region evidences a highly non-linear (inverted U) rCBF response in relation to word rate.

Figure 2 demonstrates statistical inference about nonlinear differences in rCBF responses between the two subjects. The first two columns of the design matrix are the effects of interest (i.e. differences). Confounding effects are the commonalities (columns 3 and 4), global blood, subject effects (column 5 and 6) and global activity (column 7). There are two maxima apparent in the SPM{F}. To demonstrate a non-linear interaction we have chosen a voxel in the left hippocampus (-18, -26, -12 mm). Note the decrease of rCBF in relation to increasing word rates in the normal subject whereas the patient shows increasing rCBF.

3.3. SUMMARY

Non-linear fitting techniques allow detection of rCBF responses in brain regions that might not be so evident using simple (i.e. linear) regression. This approach can model a family of nonlinear rCBF responses, without specifying the exact form of the relationship. Different brain areas can show different responses to a study parameter, which can then be used to characterize each area involved in this task. Although we have restricted our model to a second order polynomial regression, other basis functions could be used; However the interpretation of polynomial coefficients (i.e. a decomposition into linear and non-linear effects) is more intuitive than interpreting the coefficients of some basis functions. This may be important in some experiments where the introduction of non-linearity (the second order terms) considerably improves the fit. In general questions pertaining to the number, and type, of basis functions are questions of model selection.

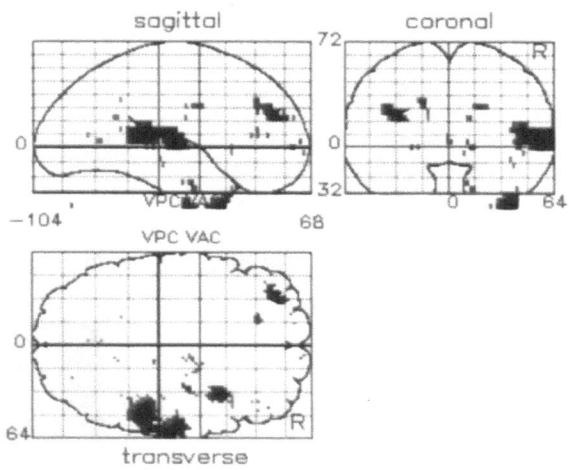

SPM{F}: p < 0.002 {uncorrected}

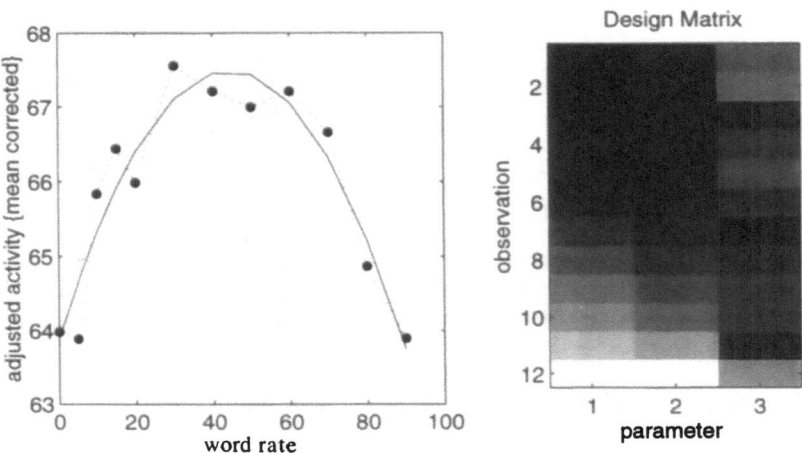

Figure 1: SPM{F} for a polynomial regression in a patient who had recovered from an ischaemic infarction in the territory of the left middle cerebral artery. Voxels over a threshold of F = 15 are shown. The regression for a voxel in a left frontal region (x=-34, y=40 and z=24 mm). F=20, df 2,8, p<0.001 (uncorrected) is also shown.

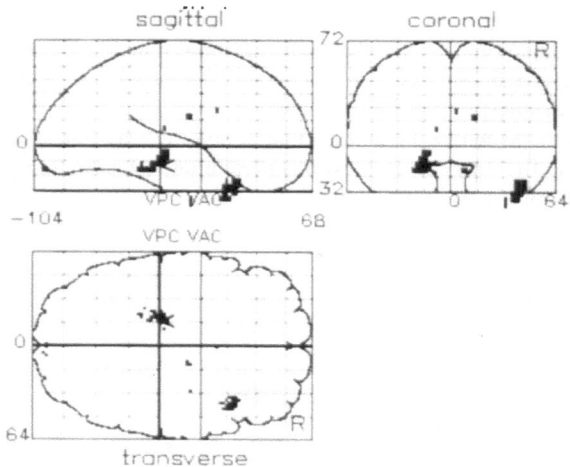

SPM{F}: p < 0.001 {uncorrected}

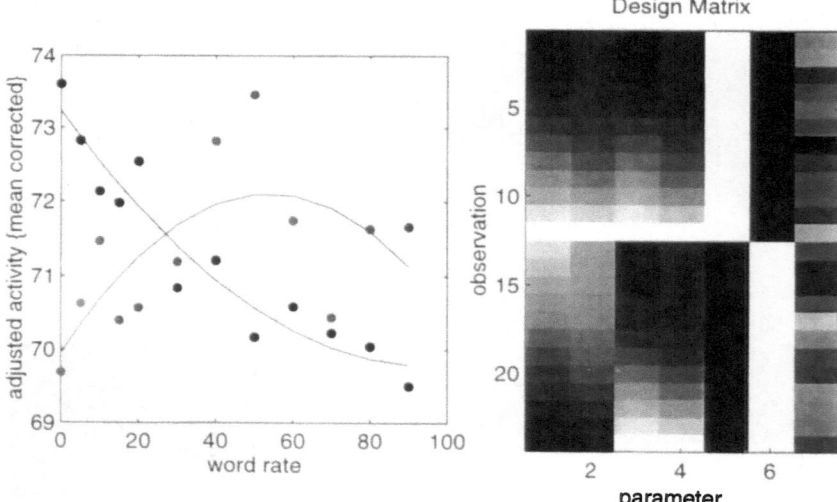

Figure 2: SPM{F}, regression and design matrix for differential rCBF responses in the patient and the control subject. Voxels over a threshold of F = 10 are shown. Regression for differential rCBF responses at a voxel x:-18, y:-26 and z:-12 mm are shown. F=17, df 2,17, p<0.0005 (uncorrected)

4. Subtraction, factorial designs and pure insertion

This section represents a critique of *cognitive subtraction* as a conceptual framework used in the design of brain activation experiments and allows us to demonstrate the potential usefulness of factorial designs. Subtraction designs are well established and powerful devices in mapping cognitive neuroanatomy but are predicated on possibly untenable assumptions about the relationship between brain dynamics and the functional processes that ensue (and where these assumptions may be tenable they are seldom demonstrated to be so). Concerns with cognitive subtraction can be formulated in terms of the relationship between cognitive processes and their neuronal implementation. In this section we suggest that nonlinear systems like the brain do not behave in a fashion that is consistent with pure insertion and illustrate our point with a simple example - the functional anatomy of phonological retrieval during object naming.

4.1. SUBTRACTION, PURE INSERTION AND ADDITIVE FACTORS LOGIC

Cognitive subtraction involves the successive elaboration of a task by adding separable cognitive components and measuring the resulting increases in neuronal activity elicited by these tasks. The physiological activations, that obtain on serial subtraction of these measurements, are then identified with the added cognitive components. The approach, predicated on *pure insertion* assumes that each cognitive component evokes an 'extra' physiological activation that is the same irrespective of the cognitive or physiological context. Pure insertion is an idea that underlies the original Donders subtractive method and has proven itself in many situations; for example in the psychophysics of reaction time measurements during the detection of visual targets embedded in a background of distracters. However pure insertion in the context of brain activation experiments is an *a priori* assumption that has not been validated in any physiological sense. We present here an evaluation, in physiological terms, of cognitive subtraction by focusing on pure insertion: This evaluation follows Sternberg's proposal [9] to use additivity and *interaction* within factorial designs (the additive factor method) to address this issue.

Pure insertion is implicit in cognitive subtraction. The idea is that as a new cognitive (A) component is added to a task, the implementation of the pre-existing components (e.g. B) remains unaffected. Otherwise the difference between the two asks would comprise component A and the interaction between A and B i.e. AxB. I.e. pure insertion requires that one cognitive component does not affect the effect of another. In factorial designs pure insertion is another way of saying that the interaction terms are negligible. The fact that interactions can be measured, using functional imaging [5], means that the validity of pure insertion can be addressed empirically. Here we use a simple factorial design to demonstrate that the physiological brain does not conform to pure insertion.

4.2. THE NON-LINEAR BRAIN AND INTERACTIONS

Even if a cognitive component can be added without interacting with pre-existing components, the brain's implementation of these processes is almost certainly going to show profound interactions. This follows from the observation that neural dynamics are nonlinear. Nearly all theoretical and computational neurobiology is based on this observation. The point being made here is that although a cognitive science model, describing the functions, may be framed in terms of separable, non-interacting elements, the implementation of those functions is not. Consequently the structure of the cognitive components (functional model) and the brain's physiological implementation are not isomorphic and the mapping of one onto the other is problematic. Cognitive subtraction makes some strong assumptions about this relationship which are difficult to reconcile with basic neurobiology.

One of the innumerable examples of nonlinear brain dynamics, that confound cognitive subtraction, is modulation; from classical neuromodulation to large-scale modulatory interactions between different cortical areas. A particularly relevant example here is the modulatory role of attention on perceptual processing; for example the responsiveness of V5 to motion in the visual field. It is likely that this responsiveness is enhanced by selectively attending to motion. V5 activation therefore represents an interaction between visual motion and selective attention. Consider now an experiment in which visual motion is presented with and without selective attention for motion. The resulting difference in physiological activation of V5 would, in terms of cognitive subtraction, be attributed to selective attention for motion. This would be a fallacious conclusion because the differential responses of V5 represents an interaction between visual analysis of a particular attribute and selective attention for that attribute. In neuronal terms this might be described as a modulation of V5 responsiveness to motion in the visual field mediated by afferents from some higher order area. The fallacy would be revealed by repeating the experiment in the absence of visual motion. In this instance 'selective attention for motion' should not activate V5 because there are no motion-dependent responses to modulate. This second experiment would demonstrate an interaction between 'selective attention for motion' and 'visual motion' using a factorial experimental design. The V5 example highlights the close relationship between functional interactions (between different cognitive or sensorimotor components) and statistical interactions that can be inferred using factorial designs. It should be said that we do not consider 'attention' a cognitive component (although the 'control' of attention can be) but our point is clearly illustrated by this example.

We suggest that a potentially more powerful approach to cognitive neuroanatomy is to consider interactions explicitly, both in terms of the cognitive model and in terms of the experimental design and analysis. This acknowledges that the conjunction or integration of two or more cognitive processes may require an active physical instantiation. For example naming an object involves object recognition, phonological retrieval and the integration of the two, where that integration calls upon distinct brain processes. Put more simply phonological retrieval influences object recognition, and *vice versa,* and these effects are physiologically measurable. This perspective requires a factorial design where the interaction term represents the non-additive (i.e. nonlinear) physiological concomitants of naming a recognised object, that is independent of the activations produced by recognising or naming alone. This is the example used below.

4.3. AN EXAMPLE

The example chosen uses the same data to address the question, "is the inferotemporal region implicated in phonological retrieval during object naming?" from a subtraction perspective and from a factorial perspective. Considerable evidence from neuroanatomy and unit-electrode recordings suggests that neurons, in the inferotemporal cortex of animals, respond selectively to specific objects in the visual field, or have the appropriate responses to do so [10, 11]. On the basis of this and other evidence it might be hypothesised that inferotemporal regions are functionally specialised for object recognition in man. Furthermore lesion studies in man have shown that the ability to name objects is impaired, when the inferotemporal regions are damaged. For example De Renzi *et al.*[12] studied the neuropsychological correlates of inferior temporal ischaemic damage. As well as showing alexia subjects were impaired on naming objects and photographs. This sort of evidence suggests that the integrity of the inferotemporal cortex may be necessary for phonological retrieval (among other things) in object naming.

Consider the problem of designing an experiment to identify brain areas selectively activated by phonological retrieval during object naming. The cognitive processes involved in this task include visual analysis, object recognition, phonological retrieval and speech. Suppose that we are not concerned with the sensorimotor aspects of the task but wish to test the hypothesis that the inferotemporal regions are involved in both object recognition and phonological retrieval. In this case we want a series of tasks that, on successive subtraction, isolate these two cognitive components (i.e. three tasks). The tasks used were:

A Saying 'yes' when presented with a coloured shape (**visual analysis, and speech**)

B Saying 'yes' when presented with a coloured object (**visual analysis, speech and object recognition**).

C Naming a visually presented coloured object (**visual analysis, speech, object recognition and phonological retrieval**).

Subtraction of task A from task B should identify brain regions associated with object recognition and subtraction of task B from task C should identify regions implicated in phonological retrieval. The hypothesis here is that both subtractions should activate the left inferotemporal regions (among other regions). The data were obtained from six subjects scanned 12 times (every eight minutes) whilst performing one of four different tasks (the three tasks A, B, C and a further task D to be described below). The subjects were instructed to respond with either 'yes' (tasks A and B) or a name (tasks C and D). In this analysis we are only concerned with differential responses in the left inferotemporal region.

Significant (p < 0.05 uncorrected) voxels in the subtraction of task A from task B included, as predicted, a small focus in the left inferotemporal region. No significant activation were detected in the second subtraction, comparing tasks B and C. From the current perspective the key thing to note is that the inferotemporal regions showed no further activation due to phonological retrieval. Figure 3 shows the activity of a voxel in the inferotemporal region [-48, -24, -32mm BA 20 - according to the atlas of Talairach and Tournoux (1988)] during the three tasks (A, B and C). On the basis of these results one might conclude that the inferotemporal region are specialised for (implicit) object recognition and that this cognitive component is sufficient to explain the activations even in the context of naming that object. In other words there is no evidence for differential responses in the inferotemporal regions due to phonological retrieval. The conclusion is that the inferotemporal regions cannot be implicated in phonological retrieval (as far as can be measured with PET).

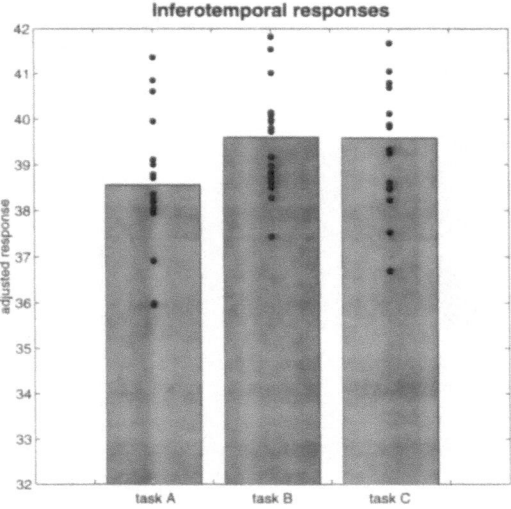

Figure 3: Task-dependent activities: Adjusted (for the confounding effects of global activity and the subject or block effect) rCBF equivalents for the three tasks (A, B and C). The bars represent mean activity and the dots are the individual data points from each scan. Note that there is activation in passing from task A to B but not in going from task B to C. These data were taken from a voxel in the left inferotemporal regions (-48,-18, -24mm).

340

4.3.1   A re-evaluation using the factorial perspective

The forgoing conclusion is wrong because it assumes pure insertion: Namely that the activation due to object recognition is the same, irrespective of whether phonological retrieval is present or not. In order to say that phonological retrieval does not activate the inferotemporal region (according to the second subtraction) one has to assume that the activation due to object recognition is the same as in the first subtraction (i.e. object recognition in the absence of phonological retrieval). To validate this assumption we need to measure activations due to object recognition in the presence of phonological retrieval. This can be effected by comparing tasks that involve phonological retrieval with and without object recognition. This comparison required the fourth condition:

D Name the colour of a presented shape (**visual analysis, speech, and phonological retrieval**).

Subtraction of task A from task B gives the activations due to object recognition in the absence of phonological retrieval and subtraction of task D from C gives the recognition-dependent activation in the context of phonological retrieval. Pure insertion requires these activations to be the same and this is not the case. Figure 4 (upper panel) shows that the activation in the context of phonological retrieval is far greater than under conditions without phonological retrieval. In other words phonological retrieval can be thought of as modulating the recognition-dependant responses of the inferotemporal region and in this sense the inferotemporal regions contribute to phonological retrieval.

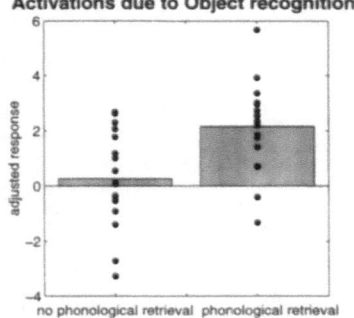
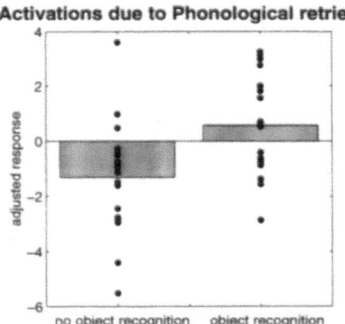

Figure 4: Modulation of task-dependent activations: Upper panel: The activations due to object recognition with (right) and without (left) phonological retrieval. Lower panel: The same but comparing the activation due to phonological retrieval with (right) and without (left) object recognition. As in the previous Figure the bars represent mean activity and the dots are the individual data points from each scan.

There is an alternative and equivalent perspective on this interaction, that considers the inferotemporal activations induced by phonological retrieval with and without object recognition. Figure 4 (lower panel) shows that, in the absence of object recognition, phonological retrieval *deactivates* the inferotemporal regions, whereas in the context of object recognition, this effect is nullified if not reversed. These data come from a voxel in BA 20 (-48, -18, -24mm). In summary inferotemporal responses do discriminate between situations where phonological retrieval is present or not and can be directly implicated in this cognitive process. These differential responses are expressed at the level of interactions and were revealed only with a factorial experimental design.

Using the same statistical model as in the previous analysis we tested for the main effect of object recognition, the main effect of phonological retrieval and the interaction between them using the appropriate contrasts. The SPM reflecting the interaction effects is shown in Figure 5 where significant voxels are rendered onto an MRI scan (white areas are significant at $p < 0.05$ uncorrected). This effect depicts the interaction (augmented activation), due to the conjunction of phonological retrieval and object recognition and allows us to confirm that the inferotemporal region is implicated in phonological retrieval during object naming.

saggital

coronal

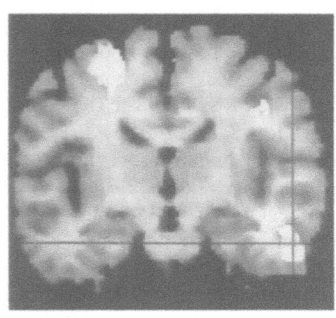

transverse {left = right}

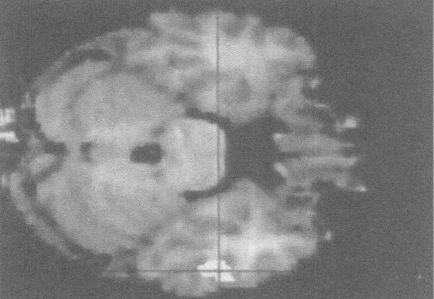

Figure 5: Factorial SPM. White areas correspond to voxels that showed a significant interaction [i.e. (C - D) - (B - A)] at $p < 0.05$ (uncorrected) rendered onto a MRI scan.

4.4. SUMMARY

We have presented a critique of cognitive subtraction and in particular the notion of pure insertion upon which subtraction relies. The main contention is that pure insertion may, or may not be, a valid cognitive science-level description but it is almost certainly not valid in relation to the brain's implementation of cognitive functions. Pure insertion disallows any interactions and yet these interactions are evident, even in the simplest experiments. To illustrate this point we have used a factorial experiment designed to elucidate the functional anatomy of object recognition and phonological retrieval.

In conclusion we suggest that the effect of a cognitive component (this is independent of all other components) is best captured by the main effect of that component and that the integration of various components (i.e. the expression of one cognitive process in the context of another) is embedded in the interaction terms. Brain regions can be functionally specialised for this integration in the sense that they can demonstrate significant interactions in terms of their physiological responses. If we are right then brain *activations* are only part of the story in mapping cognitive anatomy. Regionally specific *interactions* may hold the key for a more complete and richer characterization.

References

1. Petersen SE, Fox PT, Posner MI, Mintun M, Raichle ME (1989) Positron emission tomographic studies of the processing of single words *J Cog. Neurosci* 1:153-170
2. Lueck CJ Zeki S Friston KJ Deiber NO Cope P Cunningham VJ Lammertsma AA Kennard C and Frackowiak RSJ. (1989) The colour centre in the cerebral cortex of man. *Nature* 340:386-389
3. Grafton S Mazziotta J Presty S Friston KJ Frackowiak RSJ and Phelps M (1992) Functional anatomy of human procedural learning determined with regional cerebral blood flow and PET. *J. Neuroscience* 12:2542-2548
4. Price CJ Wise RJS Ramsay S Friston KJ Howard D Patterson K and Frackowiak RSJ (1992) Regional response differences within the human auditory cortex when listening to words. *Neurosci. Letters* 146:179-182
5. Friston, K.J Frith, C Passingham, R.E Liddle, P and Frackowiak, RSJ (1992) Motor Practice and neurophysiological adaptation in the cerebellum: a positron tomography study. *Proc. Royal Soc. Lond. Series B.* 248:223-228.
6. Gilbert PFC and Thach WT. (1977) Purkinje cell activity during motor learning *Brain Res.* 128:309-328
7. Friston, K.J Grasby, P Bench,C Frith, C Cowen, P Little, P., Frackowiak, R.S.J and Dolan, R (1992) Measuring the neuromodulatory effects of drugs in man with positron tomography. *Neuroscience Letters* 141:106-110.
8. Fletcher PC Frith CD Grasby PM Shallice T Frackowiak RSJ and Dolan RJ (1995) Brain systems for encoding and retrieval of auditory-verbal memory. *Brain* 118:401-416
9. Sternberg S (1969) The discovery of processing stages: Extension of Donders method. *Acta Psychologica* 30:276-315
10. Ungerleider LG and Mishkin M (1982) Two cortical visual systems: In Ingle DJ, Goodale MA and Mansfield RJW (eds.) *Analysis of Visual Behaviour* MIT Press, Cambridge pp. 549-586
11. Desimone R, Albright TD, Gross CG, Bruce C (1984) Stimulus selective properties of inferior temporal neurons in the macaque. *J Neurosci.* 4:2051-2062
12. De Renzi E, Zambolin A and Crisi G (1987) The pattern of neuropsychological impairment associated with left posterior cerebral artery territory infarcts. *Brain* 110:1088-1116

THE HUMAN MOTOR SYSTEM: PRINCIPLES VERSUS PLASTICITY

R. J. SEITZ, G. SCHLAUG, M. F. SCHÜLLER
Department of Neurology,
Heinrich-Heine-University Düsseldorf,
Moorenstraße 5
D-40225 Düsseldorf

ABSTRACT: The cerebral representations of normal human brain functions and their changes in diseases can be mapped by measurements of the regional cerebral blood flow with positron emission tomography. It was shown that motor cortical representations can be modified by learning and during recovery from neurological diseases. Also, skill learning activates premotor and parietal cortical areas probably related to processing of feed-forward and feed-back information. In neonatal lesions of the cerebral cortex the cortical representations of motor functions may occur with abnormal topography outside their original gyral locations or even in the contralateral hemisphere. In adulthood motor representations appear to be less plastic but have been described to evade even across gross anatomical borders in slowly progressive disorders. Evidence suggests that cerebral reorganization apparently develops gradually over time. It differs from short-term recovery after acute brain lesions with only partial system damage. Data will be provided showing that normal skill learning and compensatory relearning in brain diseases engage identical areas in frontomesial and inferior premotor cortex.

1. Introduction

Imaging of the regional cerebral blood flow (rCBF) with positron emission tomography (PET) is a powerful tool to map the brain structures participating in motion, sensation, and cognition. The rCBF increases occur in those cerebral structures in which populations of neurons have a high energy metabolism due to intense and sustained activation [2, 58]. Dynamic recording methods provided evidence that the activation-related hemodynamic response reaches a first maximum after approximately 3 s being significantly raised as long as activation continues [32, 33, 57]. It was shown that the rCBF changes in specifically activated brain regions have a lower inter-subject variance than in non-activated areas and that the local magnitude of the mean rCBF changes across a number of subjects is determined by the residual anatomical variability of the spatially standardized PET images [36, 48, 60].

PET studies during different types of motor activation paradigms demonstrated the somatotopic organization of the human motor cortex (MI), the involvement of the

343

B. Gulyás and H.W. Müller-Gärtner (eds.),
Positron Emission Tomography: A Critical Assessment of Recent Trends, 343–355.
© 1998 *Kluwer Academic Publishers.*

contralateral basal ganglia, of the anterior part of the cerebellum on the side of limb movement, and the regular activation of the supplementary motor area (SMA) even in simple finger movements [8, 14, 15, 18, 42, 55, 67]. Already in children, these activations can be shown owing to the stability of the underlying neuronal system (Fig. 1). Basically, however, these studies demonstrated the feasibility of rCBF measurements to detect the salient anatomical structures regularly participating in motor activity in human subjects. The scientific novelty of such observations was modest, since it had been known from clinical neurology since the time before Wolrd War II that these structures of the human brain are critical for normal movement generation [12].

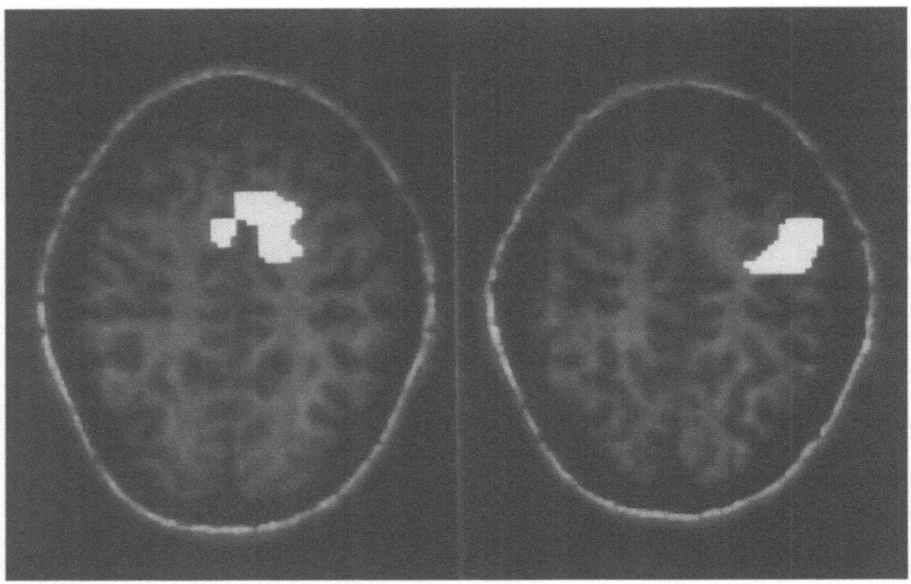

Figure 1: Significant rCBF increases (p < 0.005) in left motor cortex and bilateral in frontomesial cortex in a right-handed eight-years old boy during self-initiated performance of a simple right-hand thumb-finger-opposition sequence at a mean movement rate of 1.7 Hz. The details of the rCBF measurements, of the statistical data processing, and the anatomical coregistration have been described elsewhere [23, 31, 60].

2. Sensorimotor integration

Anatomical and physiological evidence from laboratory studies in non-human primates have revealed the participation of a large number of brain areas during movement initiation and execution, intense connectivity patterns between the participating areas, and a multifold parcellation of gross anatomical structures such as the premotor and parietal cortex [26, 43, 65]. One of the intriguing questions one would like to learn about is, whether similar functional units can also be identified in the human brain and how they are orchestrated for producing complex movements, such as trajectorial movements used in pointing and writing, or fractionated finger movements used in movement sequences, grasping or somatosensory exploration, or in bimanual manipulations.

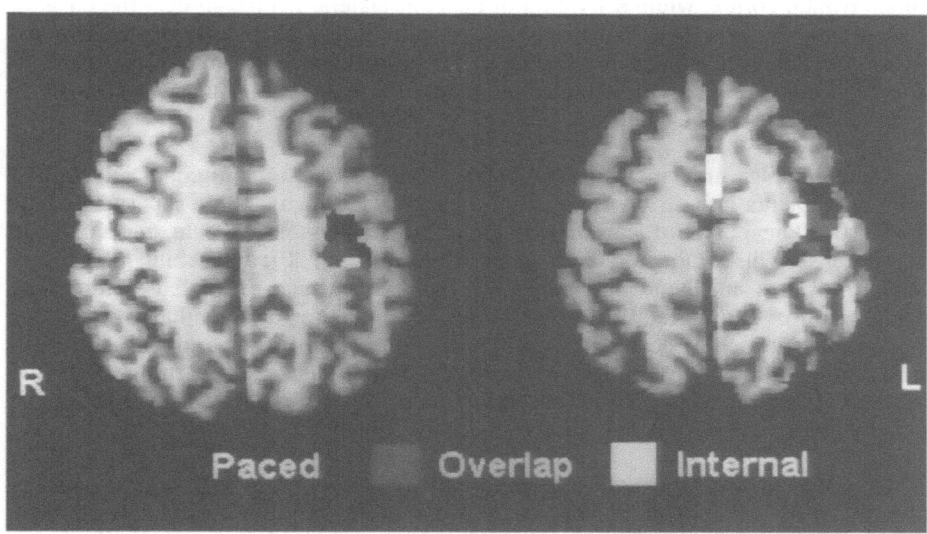

Figure 2: Significant mean rCBF increases (p < 0.05) during index flexion movements in MI (grey, 38 %). Internal pacing activated the SMA (white, 26 %), external pacing the dorsolateral PMC (dark, 33%). Significant mean rCBF increases consisted of clusters of least 11 pixels of the pixels exceeding 2.365 in t-map calculation of each stimulation condition compared to rest.

PET studies have revealed that subtle differences in the instructions and performance characteristics of movements critically affect the rCBF activation patterns in motor, premotor, and frontomesial cortical areas. Recently, evidence accumulated that movement execution is associated with rCBF increases in the posterior portion of MI in the depth of the central sulcus, while preparation to move activates the anterior portion of MI at the cortical surface [28]. Similarly, the SMA was demonstrated to be more active during more difficult than during simple motor tasks [47, 56]. In our laboratory, we obtained evidence in a group of eight healthy, right-handed volunteers (age range 21 to 29 years) that internally paced index-finger extensions activated in addition to MI the SMA, while the same movements being acoustically driven activated MI and the dorsolateral premotor cortex (PMC). This is illustrated for the study group after spatial normalization of the rCBF-images and superimposition of the significant mean rCBF increases on a spatially aligned MR image (Fig. 2). Recently, evidence was obtained showing that repetitive self-paced index extension or thumb-finger opposition movements performed in a continuous repetitive manner are associated with strong contralateral MI and SMA activations, while intermittent performance with an average inter-movement interval of 7.4 s resulted in a marked activation of the dorsal frontal cortex, the anterior part of the SMA, the cingulate motor area, and a markedly attenuated mean rCBF change in MI due to the low movement rate [34]. These data were substantiated by electrophysiological recordings showing that the premovement potentials were dominated by dipoles in the mesial dorsal frontal cortex, while activation of the contralateral MI began near the onset of muscle activity [25, 34]. The "command to move" that is underlying the internal pacing can thus be localized to activations in prefrontal and mesiofrontal cortical areas. The function of the cingulate motor area is still unknown, but may serve a role in force control [11] and also in bimanual actions [61].

More complex movements activated a large number of brain regions. Selection of hand movements into different directions was shown to activate the superior parietal lobe [10]. The posterior parietal cortex was also demonstrated to participate in somatosensory discrimination of objects and surfaces [39, 50], in reaching [29], and in trajectorial arm and hand movements [7, 20]. These data reavealed a functional parcellation of the parietal areas in the human brain for differentiated aspects of movement guidance. In accordance with this interpretation, the superior parietal cortex was demonstrated to participate during the initial phase of learning of finger movement sequences [27, 51]. These activations tapered as learning proceded and sensory guidance was no longer employed by the subjects.

In contrast to these consistent observations, significant mean rCBF changes during motor activity in the prefrontal cortex were more variable across laboratories. Prefrontal cortex activations were reported for self-initiation of simple index movements [25] and in the initial phase of sequence learning [27] after data analysis based on large image reconstruction filters and the SPM-software. We did not observe such changes in similar conditions using quantitative measurements of rCBF with freely diffusible rCBF tracers such as $[^{11}C]$-fluoromethane or $[^{15}O]$-butanol, small image reconstruction filters, and image evaluation software taking into account the locally varying variability of rCBF changes across different subjects in pixel-by-pixel t-map analysis. Apart from methodological differences, however, subtle differences in tasks design have also to be appreciated as was outlined above.

3. Skill learning

Of particular intest is to what extent learning or task specific aspects of skill acquisition affect the patterns of rCBF changes. As was pointed out above, a number of different cortical areas, the basal ganglia and the cerebellum show differential patterns of significant mean rCBF changes during motor learning. The results are summarized in Table 1. Remarkable were the significant mean rCBF changes in the inferior frontal cortex and in the temporo-parietal region in the right cerebral hemisphere, even in right-sided movements and in movement imagery. In location, they represented the right-hemispheric homologues to Broca's and Wernicke's areas that are major nodes for language processing [35].

Table 1: Significant mean rCBF changes during motor learning

	Finger movement sequences		Hand trajectories	
	SEITZ+ROLAND 1992 [51]	JENKINS et al. 1994 [27]	GRAFTON et al. 1992 [19]	SEITZ et al. 1994 [52]
Pacing	int. 1,8 bis 3,4 Hz	ext. 0,33 Hz	ext.1/min	40-160mm/s
MI	+	=	+	+
dPMC	=	=	n.e.	+
vPMC	-	-	n.e.	=
SMA	=	+	+	+
Sup. parietal	-	-	+	+
Inf. parietal	-	-	n.e.	-
Prefrontal	=	-	=	=
Striatum	+	=	=	=
Ant. cerebellum	=	-	+	+
Dentate nucleus	=	-	n.e.	-

MI: Motor Cortex, dPMC: dorsal premotor cortex, vPMC: ventral premotor cortex, SMA: supplementary motor area. =: no changes, +: rise of the rCBF-increases, -: reduction of the rCBF, int: internal pacing, ext.: external pacing, n.e.: not examined.

In accordance with earlier observations [51] the right-sided inferior frontal cortex was activated during implicit learning of a sequence of key pressing in a stimulus reaction paradigm [44]. Since in these paradigms the movements themselves were well known to the subjects, the new task was to select the next finger movement in the correct temporal order. This process may be termed the "command how to move" and be interpreted as realization of the feed-forward model of action [70]. Interestingly, this area was also activated during imagery of hand position [4, 40]. Most recent observations from our laboratory show that voluntary, self-paced performance of finger movements against a low force of 0.25 Nm specifically also activated this area (Fig. 3). Taking the evidence from Gallese et al. [16], this area probably corresponds to the inferior premotor cortex and codes action recognition. Probably, this function is mediated by so-called mirror-neurons that were shown in monkeys to fire during hand movements and also during observations of the same movements made by another monkey or the experimenter. Thus, there is good evidence to suggest that the right inferior frontal cortex, being located in a homologue location in the right hemisphere as Broca's area in the left hemisphere represent the human homologue of subarea F5 in the lower premotor cortex of monkeys.

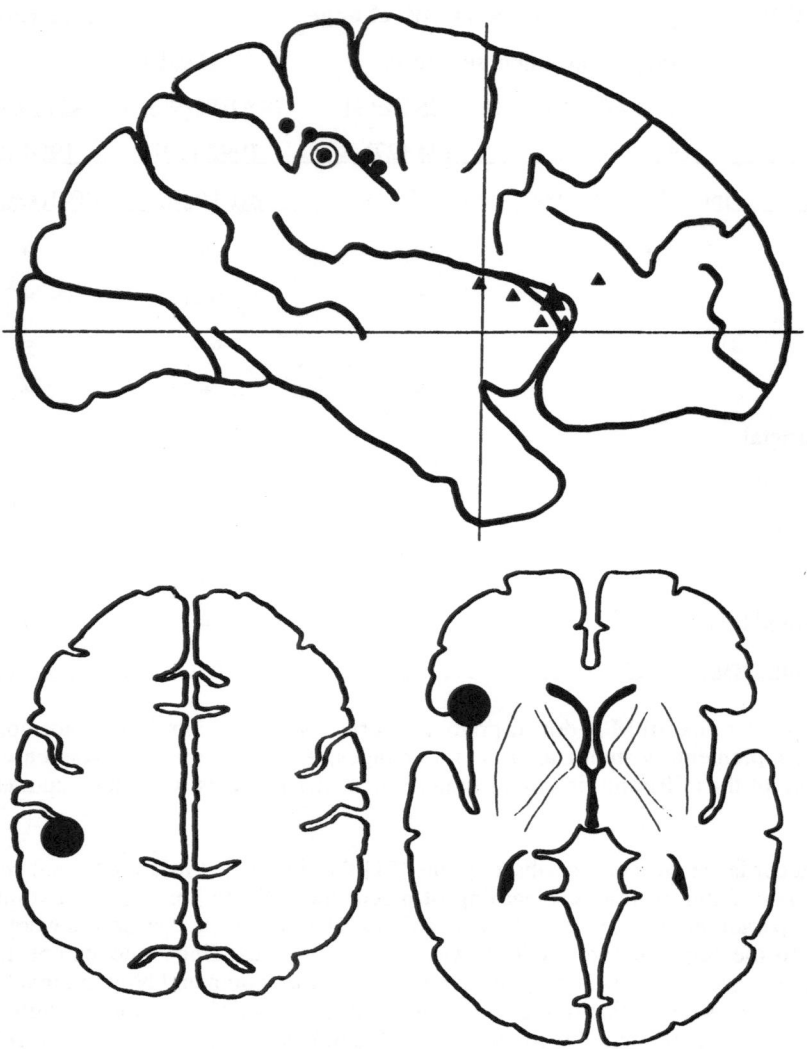

Figure 3: Activations of right inferior frontal and inferior parietal in stereotactic coordinates (x, y, z) of Talairach and Tournoux [64] in lateral and transaxial view:

Supramarginal activations		Lower premotor cortex activations	
Deiber et al. 1991 [10]:	40, -46, 48	Seitz, Roland 1992 [51]: 31, 20, 3	
Pardo et al. 1991 [41]:	44, -26, 36	Bonda et al. 1995 [4]: 39, 15, 3	
Seitz, Roland 1992 [51]:	35, -27, 38	Parsons et al. 1995 [40]: 56, 24, 12	
Bonda et al. 1996 [5]:	58, -38, 38	Rauch et al. 1995 [44]: 45, -1, 10	
Canavan et al. 1995 [7]:	34, -41, 46	Scholle et al. 1996 [48]: 39, 12, 7	
Scholle et al. 1996 [48]:	49, -28, 27		
Mean (◉):	43, -34, 39	Mean (▲): 42, 14, 7	

During learning of a new skill, there is ongoing processing of incoming reafferent information probably used for error correction. Continuously, the aimed movement is compared with the movement actually made in a feed-back loop. Focussed attention and selective sensory movement control have been suggested to be engaged during learning of finger movement sequences and trajectorial movements used for hand writing [51, 52]. In these studies an important activation was present initially during learning and during accurate task performance in the right inferior parietal cortex. In both tasks it faded after learning. Interestingly, the same area was also activated during the perception of biological hand motion [5]. Pardo et al. [41] demonstrated that this area is involved during sustained somatosensory attention. Deiber et al. [10] showed that this area is particularly active during performance of directed hand movements with random direction selection. Recently, we were able to show that the ventral part of this area revealed an activation related to compensation of passive right-hand finger perturbation movements (Fig. 3). Finally, imagery of movement trajectories activated a dorsal part of this area [7]. Altogether, these data demonstrate that the temporo-parietal cortical area around the supramarginal gyrus extending from the superior temporal to the intraparietal sulcus is engaged during attentive sensory movement control. This area and the lower premotor area are intensely interconnected [9].

4. Relearning in disease

A scientifically challenging and clinically important question is, whether the cerebral areas participating in normal motor learning become similarly engaged in motor recovery after brain lesions. Recent results from our laboratory suggest that functional restitution is determined by a number of different factors as summarized below.

The most critical factor for spontaneous restitution of funtion in the adult human brain is the degree of damage of the dedicated neuronal substrate as for example the motor cortex or the pyramidal tract [2]. In this respect a damage of the central nervous system is equivalent to that of the peripheral nerves. Neonatal lesions can induce large-scale remapping of function beyond the normal boundaries of gross anatomical structures and even with abnormal engagement of the contralateral cerebral hemisphere even in large hemispheric lesions [46]. In the adult brain this has never been shown. However, in partial damage of a system, within-system recovery can occur. Two mechanisms could be demonstrated to be relevant in this situation. First, there is remapping of cortical representations particularly in long standing diseases of the nervous system [53, 54]. This remapping may exceed the anatomical border of the motor cortex and appears to be related to the location of the lesion within the affected anatomical structure. Evidence from animal experiments suggests that muscles have overlapping cortical representations and are highly synaptically interconnected [37, 63]. These cortical units probably provide the underlying substrate for such changes, since they were shown to expand and to increase their overlap after specific training in the ischemic brain [38]. The other factor is the preservation of an intact thalamic circuitry [2]. By multiple regression analysis we could demonstrate that patients who recovered from stroke have an enhanced functional interrelation of the ipsilesional thalamus with the contralesional cerebellum and the frontomesial cortex [1]. Taking the evidence that passive right-hand finger perturbations induce an rCBF increase in the anterior lobe of the ipsilateral cerebellum indicating the processing of afferent information [48], the data by Azari et al. [1] would mean that the plastic change in the cerebello-thalamo-frontomesial connection is related to preparation of the next movement in response to the preceeding one. In other words the data suggested that the recovered patients successfully established the "command to move". This process might be subserved by direct corticospinal neurons [15] and by the corticoreticulospinal tract. This tract was shown to orginate in a large proportion in the SMA and to provide the machinery for coordinated axial and limb movements on both sides of the body [30]. Indeed, the SMA was shown to be particularly active in patients with lesions in the motor system who had recovered from hemiparesis [53, 68, 69].

In addition, lower premotor activation in the right cerebral hemisphere was observed in MI lesions [53]. These data support the conclusion that the patients were relearning the finger movements in terms of feed-forward movement generation. As suggested above for normal motor learning, the patients appeared to establish the "command how to move". Interestingly, the lower premotor cortex and the supramarginal gyrus both showed significant mean metabolic depressions in patients suffering from motor neglect, although these areas were in no case comprised the infarctions [66]. Thus, voluntary motor activity appears to employ these areas. In this context, it should be mentioned that also the anterior cingulate gyrus was suppressed in the patients with motor neglect, while it was activated during imagery of movements [7, 61]. The predeliction of the right hemispheric activations is in accordance with the known clinical phenomenon that intentional neglect, be it termed anasognosia for hemiplegia, directional hypokinesia or motor impersistance, can usually be found in patients with right hemispheric lesions [17, 21, 22]. The lack of activation in the posterior region in our patients suggests the predominant need of voluntary generation of action relative to attentive movement guidance. This was substantiated by the behavioral finding that recovery of motor functions was not correlated with somatosensory recovery [1]. In addition, motor recovery was shown to improve significantly as a result of voluntary repetitions of active movements but not after passive cutaneous stimulation [6, 24].

5. References

1. Azari NP, Binkofski F, Pettigrew KD, Freund H-J, Seitz RJ (1996) Altered cerebral functional (rCMRGlc) interactions in thalamic circuitry predicts motor recovery following hemiparetic stroke. Neuroimage 2, S373

2. Binkofski F, Seitz RJ, Arnold S, Claßen J, Benecke R, Freund H-J (1996) Thalamic metabolism and integrity of the pyramidal tract determine motor recovery in stroke. Ann Neurol, 39, 460-470

3. Blomqvist G, Seitz RJ, Sjögren I, Halldin C, Stone-Elander S, Widen L, Solin O, Haaparanta M (1994) Regional cerebral oxidative and total glucose consumption during rest and activation studied with positron emission tomography. Acta Physiol Scand 151, 29-43

4. Bonda E, Petrides M, Frey S, Evans A (1995) Neural correlates of mental transformations of the body-in-space. Proc Natl Acad Sci USA 92, 11180-11184

5. Bonda E, Petrides M, Ostry D, Evans A (1996) Specific involvement of human parietal systems and the amygdala in the perception of biological motion. J Neurosci 16, 3737-3744

6. Bütefisch C, Hummelsheim H, Denzler P, Mauritz K-H. Repetitive training of isolated movements improves the outcome of motor rehabilitation of the centrally paretic hand. J Neurol Sci 130, 59-68 (1995)

7. Canavan AGM, Seitz RJ, Yágüez L, Knorr U, Tellmann L, Herzog H, Hömberg V (1995) Imagery of trajectorial movements activates parietal but not motor cortex. Neuroscience Abstracts, J Neurosci 21, 269

8. Colebatch JG, Deiber M-P, Passingham RE, Friston KJ, Frackowiak RSJ (1991) Regional cerebral blood flow during voluntary arm and hand movements in human subjects. J Neurophysiol 65, 1392-1401

9. Deacon TW (1992) Cortical connections of the inferior arcuate sulcus cortex in the macaque brain. Brain Res 573, 8-26

10. Deiber M-P, Passingham RE, Colebatch JG, Friston KJ, Nixon PD, Frackowiak RSJ (1991) Cortical areas and the selection of movement: a study with positron emission tomography. Exp Brain Res 84, 393-402

11. Dettmers C, Fink GR, Lemon RN, Stephan KM, Passingham RE, Silberzweig D, Holmes A, Ridding MC, Brooks DJ, Frackowiak RSJ (1995) Relation between cerebral activity and force in the motor areas of the human brain. J Neurophysiol 74, 802-815

12. Foerster O (1936) Motorische Felder und Bahnen. In: Bumke O, Foerster O (eds) Handb Neurol, Vol 6, Allgemeine Neurologie, Julius Springer Verlag, Berlin. pp 1-357

13. Fox PT, Raichle ME, Thach WT (1985) Functional mapping of human cerebellum with positron emission tomography. Proc Natl Acad Sci USA 82, 7462-7466

14. Fox PT, Fox JM, Raichle ME, Burde RM (1985b) The role of cerebral cortex in the generation of voluntary saccades: a positron emission tomographic study. J Neurophysiol 54, 348-369

15. Galea MP, Darian-Smith I (1994) Multiple corticospinal neuron populations in the macaque monkey are specified by their unique cortical origins, spinal terminations, and connections. Cereb Cortex 4, 166-194

16. Gallese V, Fadiga L, Fogassi L, Rizzolatti G (1996) Action recognition in the premotor cortex. Brain 119, 593-609

17. Gold M, Adair JC, Jacobs DH, Heilman KM (1994) Anosognosia for hemiplegia: an electrophysiologic investigation of the feed-forward hypothesis. Neurology 44, 1804-1808

18. Grafton ST, Woods RP, Mazziotta JC, Phelps ME (1991) Somatotopic mapping of the primary motor cortex in humans: activation studies with cerebral blood flow and positron emission tomography. J Neurophysiol 66, 735-743

19. Grafton ST, Mazziotta JC, Presty S, Friston KJ, Frackowiak RSJ, Phelps ME (1992) Functional anatomy of human procedural learning determined with regional cerebral blood flow and PET. J Neurosci 12, 2542-2548

20. Grafton ST, Fagg AH, Woods RP, Arbib MA (1996) Functional anatomy of pointing and grasping in humans. Cereb Cortex 6, 226-237

21. Heilman KM, Bowers D, Coslett B, Whelan H, Watson RT (1985) Directional hypokinesia: prolonged reaction times for leftward movements in patients with right hemispheric lesions and neglect. Neurology 35, 855-859

22. Heilman KM, Watson R (1996) Motor neglect: what do we mean? Neurology 46, 1492-1493

23. Herzog H, Seitz RJ, Tellmann L, Müller-Gärtner H-W (1996) Quantification of regional cerebral blood flow with ^{15}O-butanol and positron emission tomography in humans. Cereb Blood Flow Metab 16, 645-649

24. Hummelsheim H, Hauptmann B, Neumann S (1995) Influence of phsyiotherapy facilitation techniques on motor evoked potentials in centrally paretic hand extensor muscles. Electroencephal Clin Neurophysiol 97, 18-28

25. Jahanshahi M, Jenkins H, Brown RG, Marsden CD, Passingham RE, Brooks DJ (1995) Self-initiated versus externally triggered movements. I. An investigation using measurement of regional cerebral blood flow with PET and movement-related potentials in normal and Parkinson's disease subjects. Brain 118, 913-933

26. Jeannerod M, Arbib MA, Rizzolatti G, Sakata H (1995) Grasping objects: the cortical mechanisms of visuomotor transformation. TINS 18, 314-320

27. Jenkins IH, Brooks DJ, Nixon PD, Frackowiak RSJ, Passingham RE (1994) Motor sequence learning: a study with positron emission tomography. J Neurosci 14, 3775-3790

28. Kawashima R, Roland PE, O'Sullivan BTO (1994) Fields in human motor areas involved in preparation for reaching, actual reaching, and visuomotor learning: a positron emission tomography study. J Neurosci 14, 3462-3474

29. Kawashima R, Roland PE, O'Sullivan BTO (1995) Functional anatomy of reaching and visuomotor learning: a positron emission tomography study. Cereb Cortex 2, 111-122

30. Keizer K, Kuypers HGJM (1989) Distribution of corticospinal neurons with collaterals to the lower brain stem reticular formation in monkey (macaca fascicularis) Exp Brain Res 74, 311-318

31. Knorr U, Weder, B, Kleinschmidt A, Wirrwar A, Huang Y, Herzog H, Seitz RJ (1993) Identification of task-specific rCBF changes in individual subjects: validation and application for positron emission tomography. J Comput Assist Tomogr 17, 517-528

32. Krüger G, Kleinschmidt A, Frahm J (1996) Dynamic MRI sensitized to cerebral blood oxygenation and flow during sustained activation of human visual cortex. MRM 35, 797-800

33. Malonek D, Grinvald A (1996) Interactions between electrical activity and cortical microcirculation revealed by imaging spectroscopy: implications for functional brain mapping. Science 272, 551-554

34. MacKinnon CD, Kapur S, Hussey D, Verrier MC, Houle S, Tatton WG (1996) Contributions of the mesial frontal cortex to the premovement potentials associated with intermittent hand movements in humans. Human Brain Mapp 4, 1-22

35. Mesulam MM. Large-scale neurocognitive networks and distributed processing for attention, language, and memory. Ann Neurol 28, 597-613 (1990)

36. Missimer G, Knorr U, Seitz RJ, Maguire RP, Schlaug G, Leenders KL (1996) Comparative analysis of rCBF increases in voluntary index flexion movements. In: Quantification of Brain Function using PET. Jones T, Cunningham V, Meyers R, Bailey D (eds), Academic Press, pp 393-397

37. Nudo RJ, Jenkins WM, Merzenich MM, Prejan T, Grenda R. Neurophysiological correlates of hand preference in primary motor cortex of adult Squirrel monkeys. J Neurosci 12, 2918-2947 (1992)

38. Nudo RJ, Milliken GW, Jenkins WM, Merzenich MM (1996) Use-dependent alterations of movement representations in primary motor cortex of adult squirrel monkeys. J Neurosci 16, 786-807

39. O'Sullivan BT, Roland PE, Kawashima R (1994) A PET study of somatosensory discrimination in man. microgeometry versus macrogeometry. Eur J Neurosci 6, 137-148

40. Parsons LM, Fox PT, Downs JH, Glaas T, Hirsch TB, Martin CC, Jerabek PA, Lancaster JL (1995) Use of implicit motor imagery for visual shape discrimination as revealed by PET. Nature 375, 54-58

41. Pardo JV, Fox PT, Raichle ME (1991) Localization of a human system for sustained attention by positron emission tomography. Nature 349, 61-64

42. Petit L, Orssaud C, Tzourio N, Salamon G, Mazoyer B, Berthoz A (1993) PET study of voluntary saccadic eye movements in humans: basal ganglia-thalamocortical system and cingulate cortex involvement. J Neurophysiol 69, 1009-1017

43. Passingham RE (1996) Anatomical connectivity of the thalamus and SMA, cortex, and motor cortex. In: Neurophysiology of the human supplementary motor area: positron emission tomography. In: Adv Neurol. 70 Lueders HO (ed), p 478

44. Rauch SL, Savage CR, Brown HD, Curran T, Alpert NM, Kendrick A, Fischman AJ, Kosslyn SM (1995) A PET investigation of implicit and explicit sequence learning. Human Brain Mapp 3, 271-286

45. Roland PE Meyer E, Yamamoto YL, Thompson CJ (1982) Regional cerebral blood flow changes in cortex and basal ganglia during voluntary movements in normal human volunteers. J Neurophysiol 48, 467-480

46. Sabatini U, Toni D, Pantano P, Brughitta G, Padovani A, Bossao L, Lenzi GL (1994) Motor recovery after early brain damage. A case of brain plasticity. Stroke 25, 514-517

47. Schlaug G, Knorr U, Seitz RJ (1994) Inter-subject variability in acquiring a motor skill. A study with positron emission tomography (PET). Exp Brain Res 98, 523-534

48. Scholle HC, Bradl U, Hefter H, Martin P, Tellmann L, Herzog H, Seitz RJ (1996) Cerebral network related to torque-load compensation and torque generation. EJN (Suppl 9): 133

49. Seitz RJ, Bohm C, Greitz T, Roland PE, Eriksson L, Blomkvist G, Rosenkvist G, Nordell B (1990) Accuracy and precision of the computerized brain atlas programme (CBA) for localization and quantification in positron emission tomography. J Cereb Blood Flow Metab 10, 443-457

50. Seitz RJ, Roland PE, Bohm C, Greitz T, Stone-Elander S (1991) Somatosensory discrimination of shape: tactile exploration and cerebral localization. Eur J Neurosci 3, 481-492

51. Seitz RJ, Roland PE (1992) Learning of finger movement sequences: a combined kinematic and positron emission tomography study. Eur J Neurosci 4, 154-165

52. Seitz RJ, Canavan AGM, Yagüez L, Herzog H, Tellmann L, Knorr U, Huang Y, Hömberg V (1994) Successive roles of the cerebellum and premotor cortices in trajectorial learning. Neuroreport 5, 2541-2544

53. Seitz RJ, Huang Y, Knorr U, Tellmann L, Herzog H, Freund H-J (1995) Large-scale plasticity of the human motor cortex. Neuroreport 6, 742-744

54. Seitz RJ, Freund H-J (1996) Plasticity of the human motor cortex. In: Freund H-J, Sabel BA, Witte OW (eds) Brain Plasticity. Adv. Neurol. 73: 323-335

55. Seitz RJ, Schlaug G, Knorr U, Steinmetz H, Tellmann L, Herzog H (1996) Neurophysiology of the human supplementary motor area: positron emission tomography. In: Adv Neurol. 70 Lueders HO (ed), pp 167-175

56. Shibasaki H, Sadato N, Lyshkow H, Yonekura Y, Honda M, Nagamine T, Suwazono S, Magata Y, Ikeda A, Miyazaki M, Fukuyama H, Sato R, Konishi J (1993) Both primary motor cortex and supplementary motor area play an important role in complex finger movement. Brain 116, 1387-1398

57. Sitzer M, Knorr U, Seitz RJ (1994) Cerebral hemodynamics during sensorimotor activation in humans. J Appl Physiol 77, 2804-2811

58. Sokoloff L (1986) Cerebral circulation, energy metabolism, and protein synthesis: general characteristics and principles of measurement. In: Phelps ME, Mazziotta JC, Schelbert HR (eds) Positron emission tomography and autoradiography. Principles and applications for the brain and heart. Raven Press, New York, pp 1-72

59. Steinmetz H, Seitz RJ (1991) Functional anatomy of language processing: neuroimaging and the problem of individual variability. Neuropsychologia 29, 1149-1161

60. Steinmetz H, Huang Y, Seitz RJ, Knorr U, Herzog H, Hackländer T, Kahn T, Freund H-J (1992) Individual integration of positron emission tomography and high-resolution magnetic resonance imaging. J Cereb Blood Flow Metab 12, 919-926

61. Stephan KM, Fink GR, Passingham RE, Silberzweig D, Ceballos-Baumann AO, Frith CD, Frackowiak RSJ (1995) Functional anatomy of the mental representation of upper extremity movements in healthy subjects. J Neurophysiol 73, 373-386

62. Stephan KM, Binkofski F, Tellmann L, Tass P, Wunderlich G, Knorr U, Herzog H, Müller-Gärtner HW, Sturm V, Seitz RJ, Freund H-J (1996) Anterior cingulate activation during bimanual coordination: kinematic and functional imaging data in acquired lesions. Neuroimage 2, S418

63. Stepniewska I, Preuss TM, Kaas J. Architectonics, somatotopic organization, and ipsilateral cortical connections of the primary motor area (M1) of Owl monkey. J Comp Neurol 330, 238-271 (1993)

64. Tailarach J., Tournoux P. (1988) Co-planar stereotaxic atlas of the human brain. 3-dimensional proportional system: an approach to cerebral imaging. Thieme Verlag, Stuttgart, New York

65. Tanji J (1994) The supplementary motor area in the cerebral cortex. Neurosci Res 19, 251-268

66. von Giesen HJ, Schlaug G, Steinmetz H, Benecke R, Freund H-J, Seitz RJ (1994) Cerebral network underlying unilateral motor neglect: evidence from positron emission tomography. J Neurol Sci 125, 29-38

67. Walter H, Kristeva R, Knorr U, Schlaug G, Huang Y, Steinmetz H, Nebeling B, Herzog H, Seitz RJ (1992) Individual somatotopy of primary sensorimotor cortex revealed by intermodal matching of MEG, PET, and MRI. Brain Topography 5, 183-187

68. Weder B, Knorr U, Herzog H, Nebeling B, Kleinschmidt A, Huang Y, Steinmetz H, Freund H-J, Seitz RJ (1994) Tactile exploration of shape after subcortical ischemic infarction studied with positron emission tomography. Brain 117, 593-605

69. Weiller C, Chollet F, Friston KJ, Wise RJS, Frackowiak RSJ (1992) Functional reorganization of the brain in recovery from striatocapsular infarction in man. Ann Neurolo 31, 463-472

70. Wolpert DM, Ghahramani Z, Jordan MI (1995) An internal modal for sensorimotor integration. Science 269, 1880-1882

Acknowledgements: This work was supported by the Sonderforschungsbereich 194 of the Deutsche Forschungsgemeinschaft and grant 01KO95093 of the Bundesministerium für Bildung, Wissenschaft, Forschung und Technologie.

ACTIVATION OF THE VISUAL VENTRAL STREAM IN HUMANS: AN FMRI STUDY

N. J. SHAH, C. AINE*, H. SCHLITT, B. J. KRAUSE, D. RANKEN*, J. GEORGE*, AND H.- W. MÜLLER-GÄRTNER
Institute for Medicine
Forschungszentrum Jülich GmbH
52425 Jülich
Germany

and

Biophysics Group
Los Alamos National Laboratory
Los Alamos, New Mexico 87545
U. S. A

Abstract

Invasive studies in non-human primates and non-invasive functional imaging studies in humans suggest the existence of two streams for the processing of visual information, the dorsal stream and the ventral stream. Evidence from anatomical and physiological studies indicates differences in sensitivity of the two streams to stimulus parameters which include luminance, spatial and temporal frequency, and chromatic cues. In this fMRI study, we have attempted to preferentially excite the ventral stream using a paradigm which combines stimulus features designed to favour the ventral stream, which include stimulus position (e.g., foveal versus peripheral). These results were compared with responses evoked by a stimulus designed to preferentially excite the dorsal stream.

In all subjects, activation of the primary visual cortex was more anterior when the stimulus was presented in a more peripheral location as compared with the activation elicited by a stimulus in a more foveal location. The differences in location of the activated region as a function of stimulus position reflect retinotopic organisation. As predicted, our results revealed more regions of activation for the foveal stimulus. These were located in the occipital temporal region and were lateral and ventral of the response evoked by the more peripheral stimulus containing luminance cues. The results demonstrate the feasibility of activating the visual ventral stream with this paradigm and subsequent detection of the activation patterns with fMRI.

357

B. Gulyás and H.W. Müller-Gärtner (eds.),
Positron Emission Tomography: A Critical Assessment of Recent Trends, 357–369.
© 1998 *Kluwer Academic Publishers.*

1. Introduction

Recent non-invasive imaging studies in humans (e.g., PET, fMRI and MEG/EEG) have shown that selective manipulation of visual stimulus properties (e.g., colour, shape, faces, spatial location and motion) can preferentially activate different cortical foci (1). These results provide support for the existence of two streams of processing visual information ("ventral" versus "dorsal" processing streams), first noted in invasive non-human primate studies. Ungerleider and Mishkin (2) originally proposed the existence of these two streams of processing as mediating "object" versus "spatial" vision (i.e., the "what" versus "where" systems), based on lesion data in monkeys. For example, lesions of the central but not peripheral visual field representation of area V1 (the primary visual area) impaired pattern discrimination, similar to monkeys that had lesions in inferior temporal cortex (ITC), which is part of the "ventral" stream. Later studies (3-5) suggested that peripheral field representations remain rather distinct and separate throughout cortex and send "preferential input" into the dorsal pathway, ending in parietal cortex. Therefore it was hypothesised that the "dorsal" stream serves as an alerting function for orienting attention or eyes to peripheral stimuli. They also suggested that the response latencies to peripheral stimulation may be faster compared with response latencies in ITC (ventral stream) which reflects primarily central field input. In contrast, neurons in ITC respond more strongly to foveal stimuli than peripheral ones, and are presumably related to pattern recognition.

Livingstone and Hubel (6, 7) proposed that "parvocellular" and "magnocellular" systems may mediate the differences noted between the two processing streams. These cell types remain relatively separate and independent from levels of the retina throughout cortex and they differ in their sensitivities for colour selectivity, contrast sensitivity, temporal properties and spatial resolution. The end-points of the "magno" and "parvo" streams are parietal cortex and ITC which were noted previously to be associated with a predominance of peripheral field and central field representations, respectively. Consequently, it appears that there is some relationship between these two characterisations of the two streams of processing, but the exact relationship is unknown. However, recent evidence from invasive monkey studies suggests that the "magnocellular" and "parvocellular" streams are not strictly segregated and controversy regarding these two streams of processing has ensued (8-11).

Early primate and cat studies have shown: (a) cell diameters in the retina are smaller in the fovea and show a progressive increase in size as they extend into the periphery—there is a tendency for receptive field (RF) size (for both magno and parvo cell types) to increase from central to more peripheral locations (14); (b) there is also a progressive tendency for magno-type cell concentration to increase from central to peripheral locations (15); (c) the magno-type and parvo-type afferent projections from the retina to lateral geniculate nucleus (LGN) remain largely segregated as they course through cortex and project along fast and slow pathways, respectively, to distinct and non-overlapping layers of visual cortex (15-20); and (d) cell diameter is inversely proportional to the spatial frequency (SF) that the cell is most responsive to; smaller cell diameters are tuned to higher SFs and slower temporal frequencies and vice versa (21).

Tolhurst (22) also noted that smaller RFs have longer integration times (i.e., longer response times) and have a preference for moving stimuli of low speeds (1-5 degrees per second). The RF properties noted above provide a neural basis for differentiating between "magno" and "parvo" streams in general: 1) the "magno" system, usually associated with motion analysis, is highly sensitive to luminance differences, relatively insensitive to colour, and is more sensitive to lower spatial frequencies and higher temporal frequencies than the "parvo" system. In contrast, the "parvo" system, usually associated with colour/form analysis, is sensitive to chromatic contrast, higher spatial frequencies and lower temporal frequencies. If we examine the RF properties of ITC and parietal cortex, we find that ITC neurons are larger than striate or extrastriate areas and invariably include central vision and extend well into both hemifields (23). The visual RFs of parietal neurons, in contrast, are unusually large and the fovea is not included, although some neurons extended to within 2 degrees of the fovea and most represent the contralateral half-field of vision (24).

Preliminary magnetoencephalography (MEG) data using stimulus conditions similar to those outlined in this paper, suggest that the foveal, isoluminant stimulus tends to evoke activity in areas more ventral and lateral than the peripheral isochromatic stimulus. In the latter case, activity was more medial and dorsal (25). MEG studies also indicate that responses to peripheral field stimulation are earlier than responses to foveal stimulation (12, 13). These MEG findings have been used here to design stimuli, to be used in a functional magnetic resonance imaging (fMRI) study with a combination of features which preferentially excite one stream or the other.

Magnetic resonance imaging methods which are sensitised to changes in cerebral blood oxygenation level (BOLD) (26, 27) may be used to measure regional changes of the MR signal associated with visual stimulation of localised regions in the brain. Functional imaging using MRI exploits the fact that changes in blood flow and metabolism in brain tissue resulting from sudden functional challenge are uncoupled. Local increases in cerebral blood flow are not paralleled by an equivalent increase in oxygen consumption (28). Hence, a relative increase of diamagnetic oxyhaemoglobin results in decreased intravoxel dephasing and therefore an increased MR signal. These signal changes form the basis of the BOLD method and can be measured with a number of MR methods including echo planar imaging (EPI) (27), our current method of choice.

In this fMRI study, we attempt to separate the two streams of processing by comparing responses to a stimulus containing a combination of features which should preferentially excite the "ventral" stream with responses to a stimulus containing features which should preferentially excite the "dorsal" stream. The stimulus parameters manipulated for the fMRI study were: (a) spatial position; (b) spatial frequency; (c) temporal frequency; and (d) chromatic contrast (i.e., isoluminant versus isochromatic). Due to the length of subject time required in these fMRI studies, the number of stimulus conditions which can be acquired in one experimental setting is limited and only two stimulus conditions were examined.

2. Methods

2.1. IMAGING METHODS AND SUBJECTS

A Siemens 1.5T Vision scanner, equipped with an EPI booster, was used to acquire the data. Time series images were acquired using the gradient-echo, EPI technique with the following sequence parameters optimised to show activity elicited by the stimulation paradigm described below: repeat time (tr) = 3s; echo time (te) = 66ms; flip angle (θ) = 90°; field of view (FOV) = 200mm; image matrix = 32 x 32; slice thickness = 6mm; interslice spacing = 0.6mm; 16 slices. Given these parameters, the voxel size was approximately isotropic (6 x 6 x 6mm). Following optimisation of the stimulus and sequence parameters, the signal-to-noise ratio (SNR) of the acquired images was improved by using the whole-body coil for transmission of radiofrequency and a quadrature surface coil for reception of the signal. In order to help visualise differences between the two stimulus conditions, the acquired 16 slices were aligned with the bottom of the temporal lobe, which favoured regions active in the "ventral" stream.

Ten informed subjects, with no history of neurological illness, fixated on a small cross-hair for the duration of the experimental conditions. Six additional subjects were presented with our optimised stimulation paradigm and also participated in the PET and MEG counterparts of this study, the results of which will be presented elsewhere. For each stimulus condition the paradigm comprised a 51s baseline followed by a 30s stimulation period and a 51s rest period with the stimulation-rest conditions being repeated 4 times. Additional high-resolution anatomical images were acquired for some subjects using the MP-RAGE sequence (magnetisation-prepared, rapid acquisition gradient echo) with the following parameters: tr = 40ms; te = 5ms θ = 40°; matrix = 256 x 256; FOV = 200mm; slice thickness = 6mm. A full 3D acquisition using the MP-RAGE sequence, with a standard head coil in transmit/receive mode, was also acquired with the following sequence parameters: tr = 40ms; te = 5ms θ = 40°; matrix = 256 x 256; FOV = 250mm; slice thickness = 1.41mm; 128 sagittal slices.

2.2. OPTIMISATION OF THE STIMULATION PARADIGM

The experimental set-up described above was determined after a thorough preliminary study in which stimulus and image-acquisition parameters were varied. The data thus acquired were analysed and, given the various limitations, an optimum set of conditions derived. Variation of sequence parameters included the following: matrix size (32 x32 and 64 x 64); flip angle of the radiofrequency excitation (θ = 90° and 30°); a standard radiofrequency head coil was used in transmit/receive mode. The standard head coil and the 64 x 64 matrix were both found to be sub-optimal since subsequent data analysis revealed small activation ratios and (mean activated signal divided by the mean rest signal) noisy signal-time curves. The number of time points in the paradigm were also varied to find an optimum; the number of baseline and rest images ranged from 5 to 20, the number of activated images ranged from 5 to15 and total number of images acquired for each experimental condition ranged from 50 to 125. The optimised paradigm, described above in section 2.1., was chosen on the basis of clear and prominent activation in the calculated maps of the correlation coefficient. Given the limitations of the stimulus projection hardware and the image acquisition software, (noted below) it was clear at the outset that our experimental set-up was better suited for imaging regions associated with the "ventral" system.

The primary stimulus parameters manipulated during the preliminary studies were: (a) stimulus sizes. The goal was to keep stimulus size as small as possible whilst maintaining acceptable activation ratios; (b) rate of alternation (2, 4, 6, 8 Hz) between bands of circular sinusoidal bulls-eye stimuli; and (c) chromatic content of the stimulus. fMRI and PET investigations of vision often use large stimuli in order to evoke a strong haemodynamic response. As a consequence, it is difficult to determine whether multiple regions of activation are present in the image or whether the responses are "extended." Although most PET and fMRI studies report that 8 Hz alternation rates for checkerboard reversals are optimal, these stimuli typically do not contain chromatic content which could affect the optimal rate. In addition, most PET and fMRI studies have used checkerboard reversals as a stimulus of choice, which contain higher spatial frequencies due to the sharp edges. Because spatial frequency is a parameter which can help differentiate between the two streams of processing, we limited the spatial frequency content of the stimulus by using circular sinusoids. Isoluminance was determined for each subject by having them psychophysically match the analog video output levels for red and green while in the experimental setting (6). Subjects adjusted the intensity of the red/green circular sinusoids which were alternating at 30 Hz (i.e., red-to-green and green-to-red) until the "flicker" appeared to disappear or was at a minimum. The experimenter stepped through the intensity levels while the subject responded that the flicker was better or worse. The matched values of intensities of red and green were stored in the output look-up tables created for each subject. The isochromatic stimulus had bands ranging from black to a blend of red/green (i.e., it was the average of the red/green bands and was luminance modulated). Achromatic, luminance modulated sinusoids were also examined, initially.

The following descriptions represent the parameters which we found to be "optimal" for differentiating between the two streams of processing, given the projection hardware limitations. Both stimuli (Figure 1) were: (a) circular with radially symmetric sinusoidal variation in either colour contrast or luminance (i.e., bulls-eye patterns); (b) 100% contrast against a black background; and (c) were presented in the lower right visual field. The stimulus designed to activate the "ventral" stream was 3.5° in diameter and placed foveally (centred at 1.8° and 1.5° from the horizontal and vertical meridia, respectively). The foveal stimulus was isoluminant (i.e., the red/green bands were perceptually equated for luminance by the subject), had a spatial frequency of 1.3 cycles per degree (cpd) with the red/green bands alternating at 2Hz. The stimulus placed peripheral to the foveal stimulus (centred at 5.6° and 4.5° from the horizontal and vertical meridia, respectively) was larger in diameter (7.3° in diameter, in order to compensate for cortical magnification effects) and was a luminance modulated isochromatic stimulus (it appeared as alternating yellow/black bands). The spatial frequency of the stimulus was 0.5 cpd and the bands alternated at 4 Hz. Due to hardware limitations, the latter stimulus could not be placed as peripheral as desired, in order to preferentially activate the "dorsal" stream, but it was included as a parafoveal stimulus for comparison.

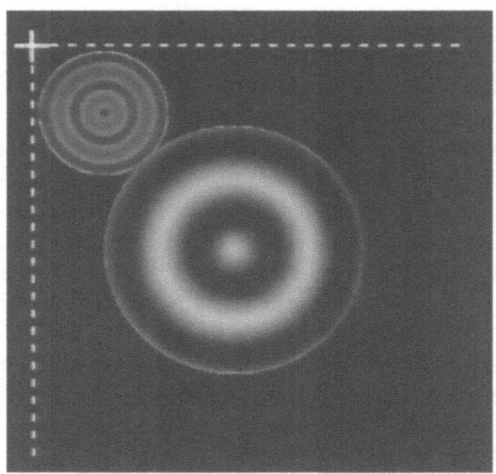

Figure 1:   A schematic representation of the stimuli used to excite the ventral stream (foveal location) and the dorsal stream (more peripheral location). The radially symmetric sinusoidal variation in either colour or contrast should be noted. Subjects were instructed to fixate on the small white cross during the baseline and activation conditions. During one block of trials (e.g., foveal, isoluminant), the red/green bands alternated at 2Hz, against a black background. During a second block of trials (e.g., parafoveal, isochromatic), the yellow/black bands alternated at 4Hz, against a black background.

Stimuli were rear-projected (LCD projector) onto a screen placed at the subjects' feet, while lying in the scanner. A small mirror was placed above the forehead, attached to the head-holder, permitting subjects to view the stimuli along the bore of the scanner. The distance from the subjects' eyes to the display was 192 cm. Foam was placed around the head to help prevent involuntary head motion.

2.3. DATA ANALYSIS

The first five images of each time-series, during which the MR signal reaches equilibrium, were discarded from all analysis. In particular the first image, if included in the data analysis, has the potential to be a serious confound, since subjects often jerk their heads in response to the sudden noise of the scanner. In order to remove the influence of gross head movement and also physiologically related motion, all data sets were motion corrected, using SPM96b software (30-32). Given the relatively low spatial resolution of the EPI images, bilinear interpolation was employed during the motion correction procedure; using a more time-intensive sinc interpolation method yielded similar motion corrections. Significantly activated voxels were found by using SPM96b software based on the "General Linear Model" approach for the analysis of time-series data (29-31). The time-course of each voxel was compared with a user-specified reference vector (a box-car with exponential rise and decay, time constant = 3s, to model the haemodynamic response). Data were not spatially or temporally smoothed or high-pass filtered. Given the number of scans and the repeat time, no correction for spin-history effects was performed. It is well-known that the images acquired with the EPI sequence suffer from geometrical distortions (33). For this reason, the resultant Z-maps (which have the same distortions as the EPI images) are depicted on the native EPI images.

3. Results

Figure 2 reflects the type of variability seen across subjects for one stimulus condition (foveal isoluminant). For some of the 6 subjects displayed here, regions of activation appear more extended while others appear more focal. All 6 of these subjects indicate multiple foci in the ventral occipital-temporal regions.

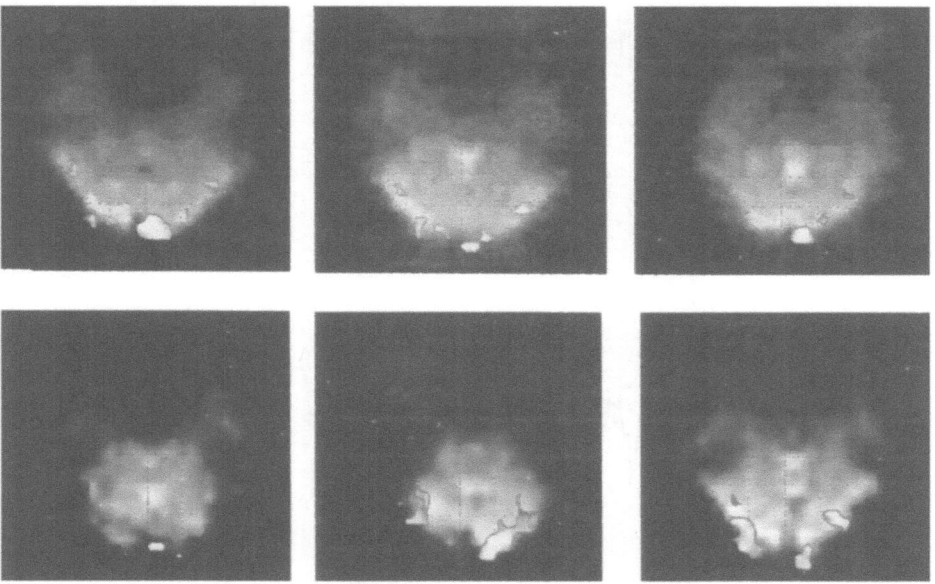

Figure 2: A ventral slice showing responses to the foveal stimulus for 6 subjects; the same approximate anatomical slice is depicted from each subject. The inferior regions of the occipital and temporal cortex are displayed at the bottom of these axial images. Because a surface coil was used, anterior regions seem to disappear into the background. The left hemisphere is displayed on the right of this figure and all remaining figures.

Figure 3 shows multiple focal regions of activation evoked by an isoluminant, chromatic stimulus located foveally and alternating at 2Hz for a single, representative

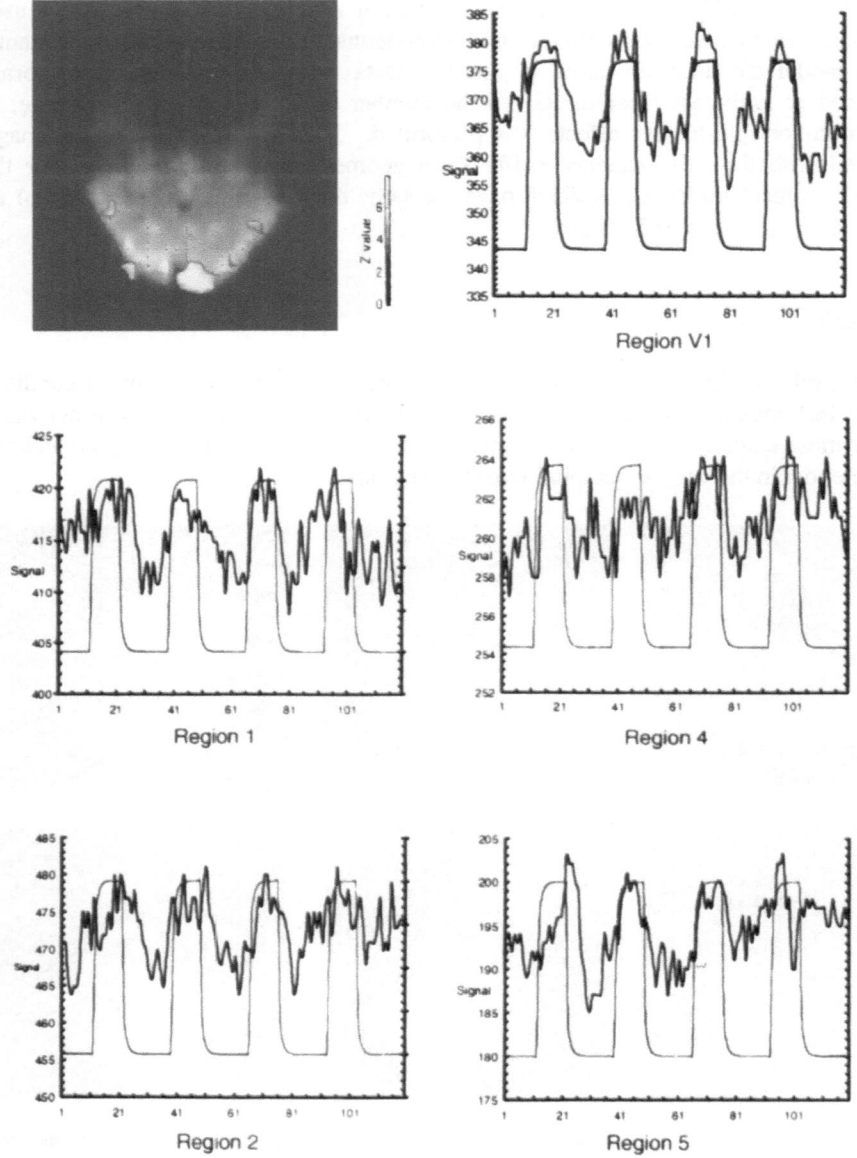

Figure 3: A ventral slice showing responses to the foveal stimulus. The activated regions and their corresponding signal-time curves (in arbitrary units) are also shown. The activation ratios (mean activated signal divided by the mean rest signal) ranged from 1.2% to 5.8%. The highest activation ratio is from the primary visual cortex. The regions are numbered counter-clockwise, starting from the left (as depicted above) of the slice. The subject shown in this figure is the same subject shown in the top left of Figure 2. Note the change in scale across the different time curves.

subject. Signal-time curves are shown for each of the activated regions of primary visual and lateral occipital-temporal cortex with Z-scores ranging from 5.44 to 7.98 and activation ratios ranging from 1.2% (region of smallest extent in the left hemisphere) to 5.8% for V1. These unsmoothed and unfiltered time-courses show a clear pattern of activation for each of the foci.

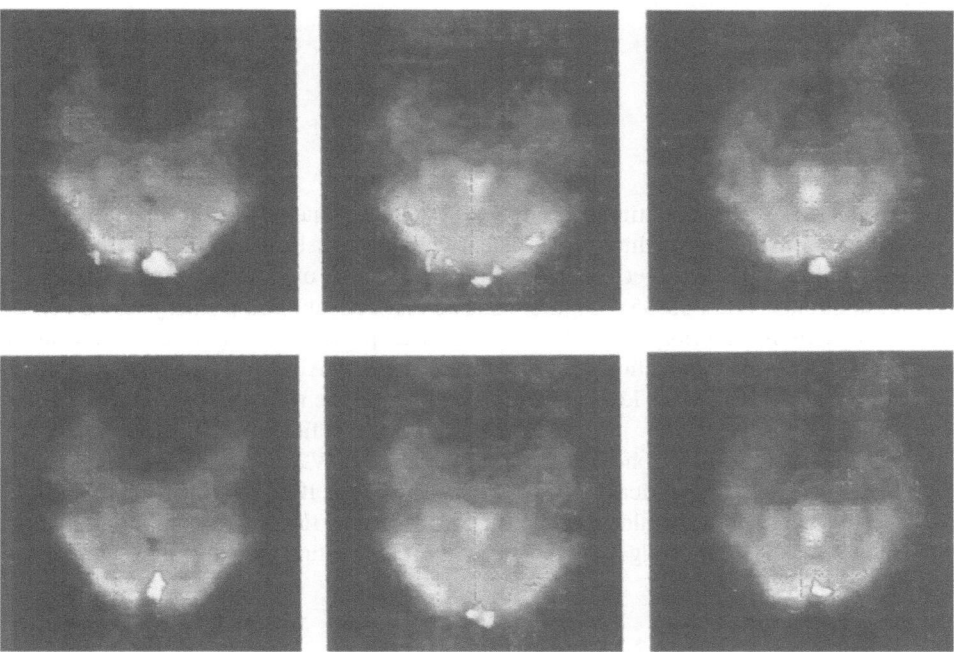

Figure 4:   A comparison of the activation for both stimulus conditions for a ventral slice oriented along the inferior occipital and temporal cortex for three representative subjects.   The top row reveals regions of activation in response to the foveal, isoluminant stimulus and the bottom row shows responses to the parafoveal, isochromatic stimulus.   More regions of activation are evident for the foveal stimulus than for the more peripheral stimulus.   Retinotopy is also evident from this result.   The three subjects depicted in this figure correspond to those in the top row of Figure 2.

Figure 4 depicts the results obtained from three representative subjects for both stimulus conditions. Activations in response to both stimulus conditions are shown for a ventral slice oriented along the inferior occipital and temporal cortex for one representative subject. All of these ventral slices were oriented approximately parallel to the bottom of the temporal lobe. The same slice is shown for both conditions. In the upper row of images, the foveal stimulus (isoluminant, chromatic stimulus, alternating at 2Hz) shows activated regions lateral and ventral to regions active in response to the more peripheral stimulus (bottom row of Figure 4), the luminance-modulated isochromatic stimulus which alternated at 4Hz. Almost no activity can be seen in lateral ventral regions in these slices. Only primary visual cortex activation (area V1) contralateral to the presented stimulus is evident (note that the left hemisphere is displayed on the right side of the figure). Slices adjacent to ones shown here did not reveal ventral/lateral activity either. In all of the six subjects investigated under optimised conditions, the activation evident in the primary visual cortex was located more anterior when the stimulus was presented in a more peripheral location, as compared with activations evoked by the stimulus located more foveally. This result reveals retinotopic organisation.

4. Discussion

4.1. RETINOTOPY

The foveal and parafoveal stimuli produced systematic changes in the location of V1 activation, for all subjects, which is consistent with the classical model of V1 retinotopy (i.e., the point-to-point projection from the visual field onto striate and extrastriate cortex) in humans (e.g., 33-35). Figure 4 shows that activation in primary visual cortex was located more anterior when the stimulus was presented to parafoveal locations compared with foveal locations in the visual field. Also, primary visual cortex activation was evident in the left hemisphere (shown on the right side of Figure 4) since the stimuli were presented in the lower right visual field. Callosal transfer of information between hemispheres occurs along the V1/V2 border, which represents field locations along the vertical meridian (36). Therefore, it is not surprising that lower right field stimulation would activate a focal region limited to the contralateral hemisphere. It is interesting that the regions of activation outside area V1 appear bilateral.

4.2. VENTRAL / DORSAL STREAMS

Figures 2, 3, and 4 reveal multiple focal regions of activation primarily in response to the foveal isoluminant stimulus. The stimulus size used to evoke these responses (3.5° in diameter) was smaller than normally used in fMRI and PET studies and was presented at a much slower rate (2Hz) than that used in other studies. Kleinschmidt and colleagues (37) also found 2Hz an optimal rate of stimulation for isoluminant chromatic stimulation. Judging from our preliminary studies, optimal rates are likely to vary depending upon the stimulus eccentricity and other stimulus parameters. For example, the optimal rate appeared to increase as the stimulus became more peripheral but this rate depended upon the chromatic content of the stimulus.

In the ventral slices, up to six focal regions were identified. The time-courses of each of these regions (see Figure 2) indicate a temporal pattern which is consistent with the activation paradigm with haemodynamic changes ranging from ~1.2% to 5.8%. As predicted, more regions of activation were evident in the ventral slices for the foveal, isoluminant chromatic stimulus. In comparison, the isochromatic parafoveal stimulus evoked primarily V1 activation in the same ventral slice. These results corroborate findings from a recent MEG study which has also revealed regions of activation, in response to an isoluminant, foveal stimulus, which were ventral and lateral compared to responses to a peripheral isochromatic or achromatic stimulus (25). This MEG study revealed regions of activation, in response to the peripheral achromatic stimulus, which were more medial and dorsal (e.g. stronger parietal activity was present and areas V1 and lateral lingual cortices were pronounced). Taken together, these results suggest that careful manipulation of stimulus parameters, including visual field position or eccentricity, can aid in identifying cortical regions associated with the ventral and dorsal streams of processing.

The current fMRI study was unsuccessful in showing dorsal activation for the following reasons: (a) Due to limitations in the projection system and the larger stimulus size necessary for evoking haemodynamic changes, we were unable to achieve peripheral stimulation. (b) The volume covered by 16 slices was inadequate for imaging changes ranging from the bottom of the temporal lobe to superior parietal lobe. We hope to investigate this issue further in the near future. Using a somewhat different paradigm, Kleinschmidt and colleagues (37) were also unable to demonstrate dorsal activity.

5. Acknowledgements

We are indebted to Mr. E. Müller (Siemens AG, Germany) for generously providing us with pre-release versions of his echo planar sequences. Sequence modifications by, and discussions with, Dr. S. Posse are also gratefully acknowledged. We thank Mrs. M.- L. Grosse-Ruyken for her assistance and Dr. C. C. Woods for helpful discussions.

6. References

(1) Haxby, J. V., Horwitz, B., Ungerleider, L.G., Maisog, J. Ma., Pietrini, P. and Grady, C.L. (1994). The functional organization of human extrastriate cortex: A PET-rCBF study of selective attention to faces and locations, *J. of Neuroscience*, **14**, 6336-6353.

(2) Ungerleider, L. G. and Mishkin, M. (1982) Two cortical visual systems. In D. J. Ingle, R. J. W. Mansfield and M. A. Goodale (Eds.). *The Analysis of Visual Behaviour* (pp. 549-586). Cambridge: MIT Press.

(3) Desimone, R. and Ungerleider, L.G. Neural mechanisms of visual processing in monkeys. In F. Boller and J. Grafman (Eds.). *Handbook of Neuropsychology. Vol. 2*. Elsevier, pp. 267-299, 1989.

(4) Boussaoud, D., Ungerleider, L.G. and Desimone, R. (1990) Pathways for motion analysis: Cortical connections for the medial superior temporal and fundus of the superior temporal visual areas in the Macaque, *J. of Comp. Neurol.* **296**, 462-495.

(5) Baizer, J.S., Ungerleider, L.G. and Desimone, R. (1991) Organization of visual inputs to the inferior temporal and posterior parietal cortex in macaques, *J. of Neuroscience*. **11**, 168-190.

(6) Livingstone, M. S., and Hubel, D. H. (1987) Psychophysical evidence for separate channels for the perception of form, color, movement, and depth. *J. of Neuroscience* **7**, 3416-3468.

(7) Livingstone, M. S., and Hubel, D. H. (1988) Segregation of form, color, movement, and depth: anatomy, physiology, and perception. *Science*, **240**, 740-749.

(8) Malpeli, J.G., Schiller, P.H. and Colby, C.L. (1981) Response properties of single cells in monkey striate cortex during reversible inactivation of individual lateral geniculate laminae. *J. Neurophysiol.* **46**, 1102- 1119.

368

(9) Maunsell, J.H.R., Nealey, T.A. and DePriest, D.D. (1990) Magnocellular and parvocellular contributions to responses in the middle temporal visual area (MT) of the Macaque monkey. *J. of Neuroscience*, **10**, 3323-3334.

(10) Lachica, E.A., Beck, P.D. and Casagrande, V.A. (1992) Parallel pathways in macaque monkey striate cortex: Anatomically defined columns in layer III. *Proc. Natl. Acad. Sci.* **89**, 3566-3570.

(11) Ferrera, V.P., Nealey, T.A. and Maunsell, J.H.R. (1994) Responses in macacque visual area V4 following inactivation of the parvocellular and magnocellular LGN pathways. *J. of Neuroscience*. **14**, 2080-2088.

(12) Aine, C., George, J., Supek, S., Ranken, D., Best, E., Tiee, W., Flynn, E., Lewine, J. and Wood, C. Differences in the temporal dynamics of visual-evoked neruomagnetic activity for central versus peripheral stimulation. Human Brain Mapping. Supplement 1, 25, 1995. First International Conference on Functional Mapping of the Human Brain, Paris, France, June 27-30, 1995

(13) Aine, C.J., Supek, S. and George, J.S. (1995) Temporal dynamics of visual-evoked neuromagnetic sources: Effect of stimulus parameters and selective attention. *International J. of Neuroscience*, **80**, 79-104.

(14) Stone, J. and Johnston, E. (1981) The topography of primate retina: A study of the human, bushbaby, and new and old-world monkeys. *J. Comp. Neurol.* **196**, 205-23.

(15) Breitmeyer, B.G. and Ganz, L. (1976) Implications of sustained and transient channels for theories of visual pattern masking saccadic suppression, and information processing. *Psychol. Review*, **83**, 1-36.

(16) Lund, J.S. and Boothe, R.G. (1975) Interlaminar connections and pyramidal neuron organization in the visual cortex, area 17, of the macaque monkey. *J. Comp. Neurol.* **159**, 305-334.

(17) Ferster, D. and LeVay, S. (1978) The axonal arborization of lateral geniculate neurons in the striate cortex of the cat. *J. of Comp. Neurol.* **182**, 923-944.

(18) Singer, W. (1977) Control of thalamic transmission by corticofugal and ascending reticular pathways in the visual system. *Physiol. Review*, 57:386-420

(19) Bullier, J. and Henry, G.H. (1979) Laminar distribution of first order neruons and afferent terminals in cat striate cortex. *J. Neurophysiol.* **42**, 1271-1281.

(20) Rodieck, R.W. (1979) Visual pathways. *Ann. Rev. Neurosci.* **2**, 193-225.

(21) Enroth-Cugell, C. and Robson, J.G. (1966) The contrast sensitivity of retinal ganglion cells of the cat. *J. Physiol.* **187**, 517-551.

(22) Tolhurst, D.J. (1975) Reaction times in the detection of gratings by human observers: A probabilistic mechanism. *Vision Res.* **15**, 1143-1149.

(23) Gross, C.G., Rocha-Miranda, C.E., and Bender, D.B. (1972) Visual properties of neurons in inferotemporal cortex of the macaque. *J. Neurophysiology*, **35**, 96-111.

(24) Yin, T.C.T. and Mountcastle, V.B. (1977) Visual input to the visuomotor mechanisms of the monkeys parietal lobe. *Science*. **197**, 1381-1383.

(25) Aine, C., Chen, H.-W., Ranken, D., Mosher, J., Best, E., George, J., Lewine, and Paulson, K. (1997) An examination of chromatic/achromatic. stimuli presented to central/peripheral visual fields: An MEG study. Presented at the Third International Conference on Functional Mapping of the Human Brain, Copenhagen, Denmark, May 19-23, 1997.

(26) Ogawa, S., Tank, D.W., Menon, R., Ellermann, J.M., Kim, S.-G., Merkle H. and Ugurbil, K. (1992) Intrinsic signal changes accompanying sensory stimulation: Functional brain mapping with magnetic resonance imaging. *Proc. Natl. Acad. Sci.* **89**, 5951-5955.

(27) Kwong, K.K., Belliveau, J.W., Chesler, D.A., Goldberg, I.E., Weisskoff, R.M., Poncelet, B.P., Kennedy, D.N., Hoppel, B.E., Cohen, M.S., Turner, R., Cheng, H.-M., Brady, T.J. and Rosen, B.R. (1992) Dynamic magnetic resonance imaging of human brain activity during primary sensory stimulation. *Proc. Natl. Acad. Sci.* **89**, 5675-5679.

(28) Fox, P.T. and Raichle M.E. (1986) Focal physiological uncoupling of cerebral blood flow and oxidative metabolism during somatosensory stimulation in human subjects. *Proc. Natl. Acad. Sci.* **83**, 1140-1144.

(29) Friston K.J., Holmes A.P., Worsley K.J., Poline J.-B., Frith C.D. and Frackowiak R.S.J. (1995) Spatial parametric mapping in functional imaging: A general linear approach. *Human Brain Mapping* **2**, 189-210.

(30) Friston K.J., Jezzard P. and Turner R. (1994) The analysis of functional MRI time-series. *Human Brain Mapping* **2**, 69-78.

(31) Friston K.J., Ashburner J., Frith C.D., Poline J.-B., Heather J.D. and Frackowiak R.S.J. (1995) Spatial registration and normalisation of images. *Human Brain Mapping* **2**, 165-189.

(32) Jezzard P. and Balaban R.S. (1995)   Correction of geometric distortion in echo planar images from B_0 Field Variations *Mag. Res. Med.* **34**, 65-73.

(33) Holmes, G. (1918) Disturbances of vision by cerebral lesions. *Br. J. Ophthalmol.* **2**, 353-384.

(34) Horton, J.C. and Hoyt, W.F. (1991) The representation of the visual field in human striate cortex. *Arch. Ophthalmol.* **109**, 816-824.

(35) Holmes, G. (1945) The organisation of the visual cortex in man. *Proc. R. Soc. Lond. Series B (Biol.),* **132**, 348- 361.

(36) Zeki, S.M. (1978) Functional specialisation in the visual cortex of the rhesus monkey. *Nature,* 274, 423-428,

(37) Kleinschmidt, A., Lee, B.B., Requardt, M., and Frahm, (1996) *J. Exp. Brain Res.* **110**, 279-288.

PET AND FMRI STUDIES OF CEREBELLAR FUNCTION IN SENSATION, PERCEPTION, AND COGNITION

LAWRENCE M. PARSONS AND PETER T. FOX

Research Imaging Center

University of Texas Health Science Center at San Antonio

1. Introduction

The cerebellum contains half of our neurons [99], has reciprocal anatomical connections with many non-motor areas in cerebral cortex [61, 84], and in hominids evolved in size more in proportion to the extent of cognitive adaptations than of motor ones [23, 56, 73]. Yet the cerebellum has been missing throughout much of the history of the functional neuroanatomy of mental processes. It was absent from the influential functional neuroanatomy of, e.g., Wernicke [98], Brodmann [16], Penfield [75], and Talairach [94]. The cerebellum has instead been assumed for more than a century to be dedicated solely to fine coordinated motor control [4, 38, 46, 47, 57, 95]. This conclusion was made first on the basis of lesion-deficit observations in non-human animal experiments [34] and in clinical neurology [44, 95], and later confirmed by electrophysiological studies in rat and monkey [24, 47, 95]. Findings consistent with this hypothesis have also recently been produced by neuroimaging studies that use PET and fMRI to record brain activation of healthy humans [26,32].

These two classical dogmas, that (a) the cerebellum does not participate in mental processes such as cognition and perception and that (b) the cerebellum only serves to support fine motor control, have recently become increasingly challenged by results in

B. *Gulyás and H.W. Müller-Gärtner (eds.),*
Positron Emission Tomography: A Critical Assessment of Recent Trends, 371–391.
© 1998 *Kluwer Academic Publishers.*

neuropsychology [13, 18, 27, 41, 45, 50]. These dogmas have also been challenged by neuroimaging studies and this chapter is concerned with such new provocative data produced specifically by investigations of sensory, perceptual, and cognitive tasks. We will not review the neuroimaging studies of cerebellar function during language processing and skill learning (see [85]).

We will briefly present some fundamental aspects of PET and fMRI, review the principal empirical data in this area, and conclude with a discussion of current proposals of cerebellar function. To preview, new PET and fMRI studies provide striking evidence that the cerebellum is intensely and selectively active during sensory and cognitive tasks, even in the absence of explicit or implicit motor behavior. Focal activity is observed in the lateral cerebellar hemispheres during the processing of auditory, visual, cutaneous, spatial, and tactile information, and in anterior medial cerebellar regions during somatomotor behavior. Moreover, there is a double dissociation between (a) cerebellar activity and sensory processing and (b) motor behavior and activity in known motor areas in cerebral cortex. The findings contradict the classical motor coordination theory of cerebellar function but are predicted by, or are at least consistent with, new alternative theories.

2. Functional Neuroimaging

The studies discussed here employed either PET, which detects regional cerebral blood flow (rCBF) [29, 30], or fMRI, which is sensitive to changes in blood oxygenation [54, 62]. Both measures can be used as indirect correlates of neural activity [7, 31, 63, 82, 93]. In addition, each neuroimaging modality yields data typically consistent with, or complementary to, data from lesion-deficit and electrophysiological studies of neural function [79, 80, 96].

Functional neuroimaging investigations have a brief, decade long, history [77, 78], yet they have yielded data that appear to be having considerable impact on hypotheses of cerebellar function. The impact is by virtue of increasing the variety of data implicating cerebellar function in non-motor behavior and reflects two important virtues of PET and fMRI, as compared to neurophysiology and neurology. First, neuroimaging allows for paradigms in which subjects (healthy humans) can be trained to perform more intricately controlled and more cognitive tasks than is possible with other animals. Second, whereas researchers in neurology and neuropsychology wait for accidents of nature to yield "pure" cases of selective brain damage, a neuroimaging study can be designed to efficiently localize components of the neural substrate of a mental process and be conducted in a relatively short amount of time.

For various reasons however, there are at present in fact just a few functional neuroimaging studies examining cerebellar involvement in cognition and perception. Many studies are constrained by their imaging instruments (either PET or fMRI) to have a limited field of view of subjects' brains, often excluding all or much of the cerebellum. Many studies use regions of interest (ROI) analyses in brain areas other than cerebellum and do not record or analyze data outside of those regions. Moreover, amongst those studies that have observed and analyzed activity in the cerebellum, often the researchers were not prepared to interpret the unexpected cerebellar activation and consequently de-emphasized it in reporting their findings. Lastly, in the Talairach and Tournoux [94] atlas, which is the stereotaxic coordinate space used in nearly all published reports for standardized reference of neuroimaging activation, the cerebellum is depicted in quite imprecise anatomical detail.

There will likely be many more PET and fMRI studies exploring the role of the cerebellum in sensory, perceptual, and cognitive tasks. This is because new instruments with wider fields of view will soon be in more common use [6]; stereotaxic coordinate atlases that include more precise representations of cerebellum will soon be available [86]; and there is likely to be heightened interest in cerebellar participation in non-motor processing as a result of new publications [85]. Furthermore, the number of neuroimaging studies of all kinds published each year is now doubling annually, and there is an increasing sophistication of experimental designs, higher resolution measurement instruments, and more powerful analysis methods [28].

3. Early Study of Cerebellar Function in Cognition

Perhaps the earliest attempt to use neuroimaging methods to specifically evaluate the possible role of the cerebellum in non-motor sensory or cognitive activity was the 1994 study by Kim, Ugurbil, and Strick [52]. This experiment used fMRI to assess whether or not deep lateral cerebellar nuclei, which are the only output for the large cerebellar hemispheres, are active during cognitive activity. Healthy subjects performed two tasks that were assumed to have comparable motor activity but different extents of cognitive processing. In the control task, subjects performed a visually-guided task in which they reached for and grasped a peg in a hole, then placed the peg in the next in a series of holes, eventually moving four pegs from one end of the series to the other. In the experimental task, the same subjects performed the same visually-guided peg movement, but were required to move the pegs under the constraints of three rules. This task was presumed to elicit cognitive activity (i.e., problem solving) as subjects decided in which sequence the pegs should be moved, in order to comply with the rules. The extent of activity imaged in a single transverse plane containing the dentate nuclei was three to four times greater during the experimental task than during the control. Because similar overt motor behavior was observed in the two tasks, the authors concluded that the increase in dentate activity was associated with cognitive processing.

A sensible alternative interpretation of the Kim *et al.* experiment is that the increased dentate activity may occur if subjects used a strategy of imagining moving their hands to rearrange the pegs as an aid in planning a sequence of peg movements. In the experimental condition, because subjects were not allowed to move pegs backwards and future moves depend on prior moves, sequences of moves must be planned to complete the task. It is well documented in cognitive science that in solving physical problems similar to this task, subjects often imagine possible manipulations and configurations [2, 36, 51, 89]. If the Kim *et al.* subjects used this strategy, then there would be a considerable amount of implicit motor behavior occurring during the experimental task and none during the control task. A PET study by Parsons, Fox *et al.* [71] showed that when subjects imagine reaching into a visually-presented target hand orientation [65, 68], even in the absence of any motor behavior, there is strong activity in the cerebellum, including the lateral hemispheres, and, as well, in many motor areas in cerebral cortex. (Other neuroimaging studies have also observed cerebellar activation during various imagined or observed motor performances [19, 20, 83]). Without data that decouple implicit motor behavior and any other, more abstract, problem solving cognitions that may be present, the most accurate conclusion from the Kim *et al.* data is probably that deep cerebellar lateral nuclei are activated by imagined motor behavior. So, these early data do not appear to demonstrate cerebellar activation for cognitive processing unrelated to motor control.

4. Dissociating Perceptual-Cognitive and Somatomotor Functions Within Cerebellar Regions

Another strategy is to design an experiment capable of yielding a regional dissociation within cerebellum between somatomotor processing and perceptual/cognitive processing. This approach is illustrated by three recent studies, each employing different experimental designs. Each study implicates posterior lateral cerebellar regions in perceptual processing (of either visual, spatial, or auditory information) and implicates medial and anterior cerebellar regions in somatomotor processing.

One such investigation [33] used PET to explore the neural systems underlying the representation of music, a complex rule-based communication informed by harmonic structure, melody, rhythm, timbre, emotional expression, and poetic semantics. This experiment recorded brain activation of expert pianists performing either scales or the third movement of J. S. Bach's Italian Concerto. Both performances were executed from memory without reading a score or seeing the keyboard or hands and both required two hands, each hand playing similar sequences of notes. Brain states during Scales and during Bach were contrasted in order to outline the neural substrate of music representation. Both the Scales and Bach conditions produced equally extensive bilateral activation in premotor and motor cortex and in anterior medial cerebellar areas that correspond to sensory-motor representations of each hand. More significantly however, the laterality of activations produced by Scales and Bach were dramatically different in other brain areas. Whereas strong multifocal activations were observed only during Scales in left auditory temporal cortex (Brodmann Area (BA) 22), comparable activations were present only during Bach in right auditory temporal cortex (BA 21). These regions in auditory cortex were the sole cortical activations showing a double dissociation between hemisphere and performed task, but they mirror an observed pattern of dissociation between cerebellum and task in which intense focal activations were observed in an intermediate lateral posterior area in right cerebellum for Scales only and in an homologous area in left cerebellum for Bach only. Because right cerebellum has its primary connectivity with left cortex [17], we proposed that during Bach, activation in right auditory cortex is related to that in left cerebellum. (Indeed, across eight subjects and three trials, there was a +0.77 correlation between activation in the right auditory areas and that in the left lateral cerebellar ones.) By the same token, we suggested that during Scales, activation in left auditory cortex is related to that in right cerebellum.

These PET data suggest that lateral cerebellar areas are selectively involved in supporting the auditory and cognitive (but non-motor) representations of music and scales. At least one other neuroimaging study has observed cerebellar activation (also via PET) during an auditory processing task [43]. Subjects in this task attended either to a moving sound or a stationary sound, without producing motor behavior. Significant cerebellar activation occurred only during the moving sound task, implying that the cerebellum was involved in supporting auditory-spatial information processing.

A second example of this regional dissociation approach is an fMRI study of visual selective attention tasks [1]. Subjects performed either (a) a visual attention condition which possessed no motor behavior component, (b) a motor behavior condition which possessed no selective visual attention, or (c) both the visual selective attention and motor response tasks together. Prior to the experiment, the researchers selected two regions of interest (ROIs) in a coronal image plane passing through the cerebellum. An ROI in the superior region of the posterior (left) cerebellum was assumed to be involved in visual attention and an anterior ROI in the right cerebellar hemisphere was deemed likely to be associated with somatomotor processing involving the right hand. The authors' predictions were confirmed. In the visual attention (no motor response) condition, the superior posterior ROI was activated, but the anterior ROI was not. In the motor (no selective visual attention) condition, the anterior ROI was activated, but the superior posterior ROI was not. In the conjoint condition, both of these regions were active to about the same extent as in their separate conditions. The authors concluded that the cerebellum is involved in selective attention operations which are anatomically differentiated from cerebellar somatomotor activations.

A related conclusion was drawn by a third recent study using a different approach. These researchers [92] metanalyzed PET investigations of various visual information processing tasks, some with motor components, some without motor components. The analysis located activated areas which generalized across tasks, while isolating other areas that were differentially sensitive to selected task components. Left medial cerebellar areas, which were active for somatomotor processes, were dissociated from right lateral cerebellar areas, which were active for non-motor, apparently visual processing functions. The latter regions were active and unmodulated by the presence or absence of motor responses during either active or passive conditions. Furthermore, increases in these areas were sensitive to experimental variables that held the motor response constant. These findings are further support for a regional dissociation within the cerebellum between (a) perceptual or cognitive processing, present in posterior and lateral areas and (b) somatomotor processing, occurring in medial and anterior regions.

5. Double Dissociation of Cerebellar Function and Motor Processing Per Se

Strong evidence of cerebellar participation in perceptual and cognitive tasks without motor components is revealed by a PET study of the mental rotation of abstract objects [72]. Subjects discriminated between identical and mirror image pairs of simple Shepard-Metzler objects visually presented [69, 90]. Each subject performed two tasks (see Figure 1): (a) when the objects in a pair were separated by a rotation about their long central segment and (b) when corresponding pairs were not separated by a rotation. Subjects made only covert judgment responses. Activations specific to the mental rotation process were dissociated from activations for the other components of the task (the encoding, comparison, and judgment processes) by subtracting (b) from (a).

Figure 1. From Parsons and Fox [72]:
(top) Examples of the mental rotation task stimuli: a pair with identical shape and a pair with different (mirror image) shape.
(bottom) Examples of the shape judgment control task stimuli: a pair with identical shape and a pair with different (mirror image) shape.

Observed activations specifically associated with mental rotation were strongest and most extensive in the cerebellum (see Figure 2). Net cerebellar activation was 4.5 times that for parietal cortex, two times that for occipital cortex, and 3.5 times that for temporal cortex. Cerebellar activations in superior vermis, deep nuclei, and inferior and superior lateral areas were observed bilaterally but were twice as strong on the right. The subjects made no gross movements, apart from eye movements, activations for which were subtracted out by the statistical contrast with the shape judgment control task. Indeed, there was no significant rotation-specific activation detected in areas known to be involved in the execution of eye movements (e.g., the frontal eye fields or the superior colliculus). Moreover, there is strong evidence from psychophysical studies that observers do not perform this task by implicit motor activity such as imagining manipulating the stimulus with one's hand [65, 66, 67, 68]. There was in fact no significant activation detected in the other areas involved either in imagined or implicit body movement or in the planning or execution of motor behavior. In sum, these data show that the cerebellum can participate intensely in a cognitive and perceptual task, without the presence of motor behavior and without the participation of the known motor areas in cerebral cortex. These results indicate that the cerebellum is involved in some way in supporting the visual spatial processing performed during mental rotation, but they do not indicate specifically what function it performs (see discussion below).

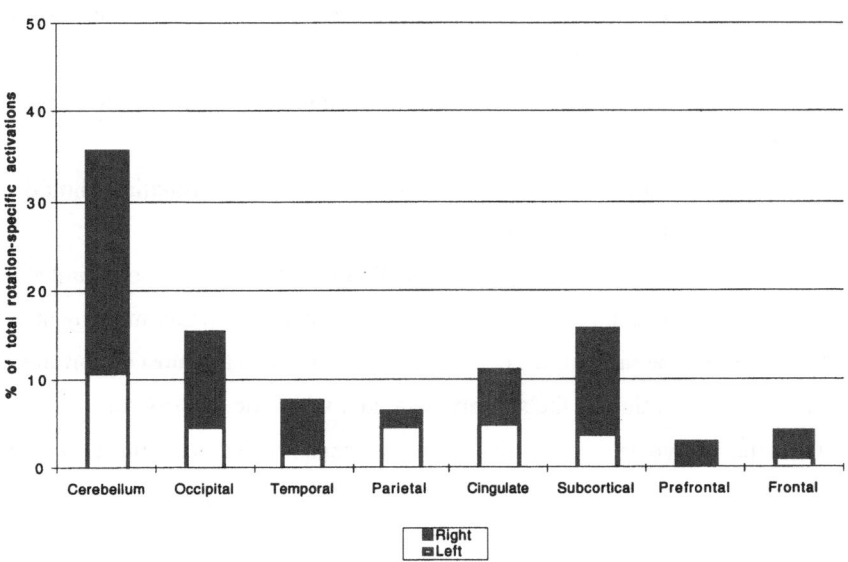

Figure 2. From Parsons and Fox [72]: The percentage of total rotation-specific activation detected by PET in left and right cerebellum, cortical, and subcortical structures.

Thus, four studies [1, 33, 72, 91], particularly [72], indicate a single dissociation between (a) cerebellar activity (specifically, in the lateral hemispheres) and (b) motor behavior and activity in motor areas in cerebral cortex. These studies also suggest a positive association between (c) cerebellar activity and (d) perceptual or cognitive processing.

The most systematic examination yet of such possibilities was conducted in an fMRI study by Gao, Parsons, Bower, Xiong, Li, and Fox [35]. This study monitored activity in the deep lateral cerebellar (dentate) nucleus in humans performing tasks involving both passive and active sensory tasks involving their fingers. The dentate nucleus provides the sole output for the large lateral cerebral hemispheres of the cerebellum and its activity is most often associated with finger movements. This experiment was designed to test the *a priori* implications of the hypothesis that the cerebellum monitors and adjusts sensory acquisition to optimize processing in the rest of the brain and is not engaged by motor control per se [10, 11]. This hypothesis was suggested by results in electrophysiology in rat cerebellum during tactile tasks [12]. We tested four implications of this hypothesis in humans:

(a) dentate nuclei should respond to sensory stimuli even when there are no accompanying overt finger movements;

(b) finger movements not associated with tactile sensory discrimination should not induce substantial dentate activation;

(c) the requirement to make a sensory discrimination with the fingers should induce an increase in dentate activation, with or without accompanying finger movements;

(d) the dentate should be most strongly activated when there is the most opportunity to modulate the acquisition of the sensory data, i.e., when the sensory discrimination involves the active repositioning of tactile sensory surfaces through finger movements.

Subjects performed each of four tasks (see Figure 3). In the Cutaneous Stimulation task, they passively experienced sandpaper rubbed against the immobilized pads of the fingers of each hand. In the Cutaneous Discrimination task, subjects were asked to actively compare (without making an overt response) whether the coarseness of the sandpaper on one hand matched that on the other. In the Grasp Objects task, they used each hand to repeatedly reach for, grasp, raise and then drop an object (an imperfect sphere). In the Grasp Objects Discrimination task, they used each hand independently to repeatedly grasp one object with one hand, while using the other hand to grasp another object, and noticed (without making an overt response) whether the two objects had the same shape or not. In neither task could a subject see the objects being manipulated or discriminated. During these tasks, a transverse plane containing the dentate nucleus was imaged in each subject.

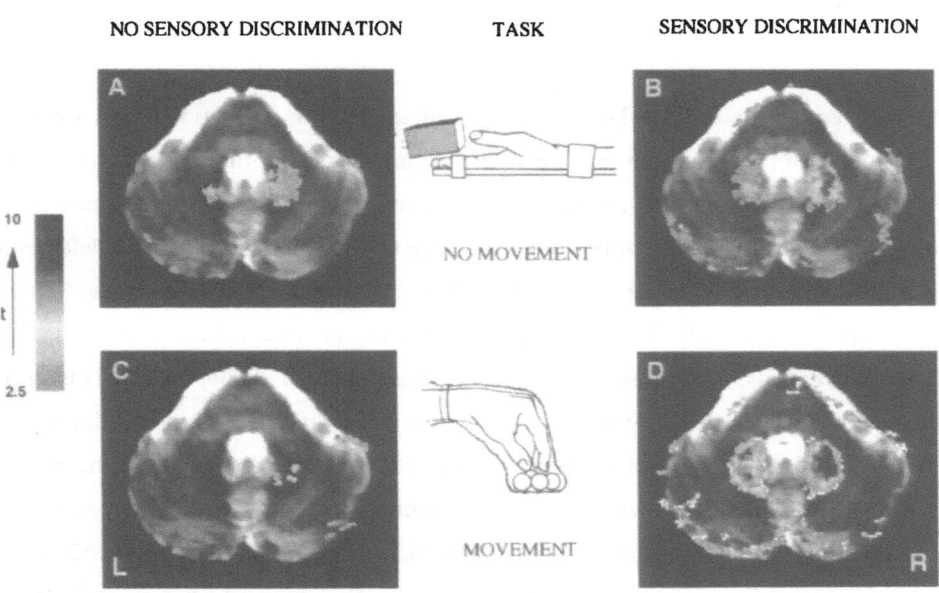

Figure 3. From Gao, Parsons, Bower, Xiong, Li, and Fox [35]: Functional MRI (color) overlaid on anatomical MRI (gray) showing dentate activations (P < .05) for (A) Cutaneous Stimulation, (B) Cutaneous Discrimination, (C) Grasp Object, and (D) Grasp Object Discrimination tasks. **(Reprinted with permission, © 1996 AAAS)**

Each of the four implications described above were confirmed. The cerebellar output nuclei (dentate) showed significant task-induced blood flow increases when participants experienced cutaneous stimulation alone (see upper left panel in Figure 3). Thus, there are responses in dentate to purely sensory stimuli, i.e., those unaccompanied by overt motor behavior of the type classically associated with dentate activation. This finding confirms PET results that show cerebellar activation during hand vibration [32]. When the same stimuli were presented under identical conditions, but a covert discrimination of the cutaneous stimuli was required, the dentate nuclei exhibited activation more than twice as intense and three times as extensive (see upper right panel in Figure 3). The enhanced activity could be a consequence of the cerebellar connectivity with prefrontal cortex which supports the working memory processes that may be necessary for comparing and discriminating sensory experiences [5, 39, 61, 87]. The activations observed in these two cutaneous tasks are thus consistent with the hypothesis that the cerebellum is engaged during the acquisition and processing of sensory information, and is even more strongly engaged during discrimination.

In the second pair of tasks, we compared cerebellar activation in a sensory discrimination task requiring rapid, coordinated movements of the fingers to a control task with similar finger movements but no discrimination. The control task produced very slight, statistically insignificant dentate nucleus activation (see the lower left panel in Figure 3). The slight activation likely reflected cutaneous stimulation of the fingers that touched the stimuli. The lack of activation confirms that rapid, coordinated, fine finger movements, in the absence of a sensory discrimination, do not detectably engage the dentate nucleus. This slight activation is consistent with the slight activation in the Kim, Ugurbil, and Strick control condition (see above) which also required fine motor control. This finding is evidence that the primary role of the cerebellum is not coordination of motor behavior for its own sake. By far the strongest and most extensive activation for any of the tasks studied here was observed when subjects made a covert discrimination of object shape using their fingers (see lower right panel of Figure 3). The latter finding is consistent with earlier PET experiments reporting posterior cerebellar activity during tactile exploration tasks [88]. The striking difference between the degree of dentate activation in the two grasping tasks (lower panels of Figure 3) provides evidence of strong cerebellar support for sensory discrimination.

In a PET study imaging the whole brain during these four tasks [70], we found that the anterior cerebellar regions were active during the grasping tasks. We predicted this activity on the basis of the hypothesized role of the cerebellum in the acquisition of proprioceptive information. In addition, these regions were equally active during the grasping tasks whether or not there was a task requirement to discriminate. This finding was predicted because the motor behavior during the Grasp Object and Grasp Object Discrimination tasks were comparable. However, in somatosensory areas of cerebral cortex during the two grasping tasks, there was greater activity when there was a requirement to discriminate object shape than when there was no such requirement. This result was predicted by the *a priori* hypothesis that activity in lateral cerebellar areas associated with support of sensory discrimination would be regulated by somatosensory regions in the cerebral cortex. There was no detectable activity in anterior cerebellar regions during the two cutaneous tasks. This result is consist with the hypothesis that those regions are specialized for sensory acquisition of proprioceptive information which is not being acquired during those tasks because there was no motor behavior.

In sum, results from these last two studies indicate that the lateral cerebellar hemispheres are active to the degree that a sensory discrimination is required. Also, independently, the anterior cerebellar regions are active only to the extent to which there is motor behavior. The latter neural activity is hypothesized to be due to sensory acquisition of proprioceptive information accompanying motor behavior.

These results are not inconsistent with data from studies that implicate the cerebellum in motor behavior. In both neurophysiological studies of awake animals and other neuroimaging studies, the sensory and motor components of task performance have not been well dissociated. Determining whether a brain area has a motor or sensory function is a subtle problem. Motor behavior is guided by ongoing sensory acquisition of information about the object toward which action is directed. In addition, continuously updated sensory data is necessary for accurate, coordinated, and smooth motor behavior. The two studies just described appear to have successfully unconfounded sensory and motor components in behavior. Because the earlier imaging and animal studies did not decouple the sensory and motor processing, it is quite plausible that cerebellar activity in the other studies [26] was also due to acquisition of proprioceptive or tactile information rather than to motor behavior per se. Additional studies separating the influences of these factors are necessary to resolve these issues.

The data from these experiments provide the second part of a double dissociation. Results described above [1, 31, 72, 91] show that there is a single dissociation such that cerebellum may be intensely active, without the presence of motor behavior or of activity in motor areas in cerebral cortex. The current data [35, 70] show the oppositely-directed single dissociation: that fine motor behavior can occur without the participation of the lateral hemispheres of the cerebellum. In combination, there is a double dissociation between (a) cerebellar processing and sensory processing and (b) motor behavior and cerebral cortical motor system activity. Furthermore, these data strongly implicate cerebellum in sensory acquisition.

6. Implications for Hypotheses About Cerebellar Function

In summary, the neuroimaging results reviewed here indicate that regions in the posterior lateral cerebellum are activated selectively by tasks involving auditory, auditory-spatial, visual, visual-spatial, cutaneous, and tactile-spatial information processing. The results also indicate that anterior and medial cerebellar areas are active during explicit and implicit somatomotor behaviors. Other recent neuroimaging studies which we have not reviewed report cerebellar activation in support of other cognitive processes, such as semantic association [58, 68, 76], attention [55], working memory [21, 53], and verbal learning and memory [3, 42]. There are also early indications from PET studies that vermal cerebellar activation is associated with emotional states such as panic, sadness, depression, and fear [8, 22, 37, 59, 81]. It is likely that these mental conditions have few or no motor components but possess various sensory and cognitive components [25].

Thus, the neuroimaging data showing cerebellar activation during non-motor processing are quite varied and numerous. However, the findings do not indicate precisely what function(s) the cerebellum performs. With this in mind, we consider the data reviewed here in light of various current hypotheses about cerebellar function (excluding those pertaining to linguistic processing and skill learning).

These hypotheses can be very briefly described as follows. (a) The cerebellum is dedicated only to fine motor control [38, 46, 47, 57, 95]. (b) The cerebellum models dynamic change in objects formed by perceptual and cognitive processes [48]. (c) The cerebellum tracks dynamic changes in perceived states of objects [74]. (d) The cerebellum performs timer functions, operating multiple, independent timer processes for each potential response being prepared for in a situation [49]. (e) The cerebellum binds interval-timing and sequence-coding processes to other forms of encoding and representation [40]. (f) The cerebellum monitors and adjusts sensory acquisition [11]. (g) The cerebellum modulates attention by learning to predict and prepare for imminent information acquisition, analysis, or action [18]. (h) The cerebellum modulates cerebral cortical processes, dampening oscillation, maintaining function steadily around a homeostatic baseline, and smoothing performance [84].

The classical motor control theory of cerebellar function (a) is refuted by the neuroimaging data reviewed here. The apparent falsification of that theory marks the start of a new era in the understanding of cerebellar physiology. As researchers now reconceptualize cerebellar function, it is natural that new alternative hypotheses ((b) - (h) above) are still in early stages of development, and expressed in general and similar terms. It is also unsurprising that the hypotheses have not yet been tested by neuroimaging or other studies in ways designed to discriminate among them. Nor do any neuroimaging data fortuitously invalidate any of the new hypotheses. Proposals assuming cerebellar functions that occur in many different tasks, such as attention, sensory acquisition control, event timing, event sequencing, and process modulation, can appear to be consistent with cerebellar activation during the sort of sensory and cognitive tasks discussed here. However, certain neuroimaging findings may be better described by hypotheses proposing more specialized functions. For example, the predominance of cerebellar activation during the operation of mentally rotating abstract objects, which may involve kinematic modeling, seems more closely described by hypotheses (b) and (c).

The sensory acquisition controller hypothesis (f) appears overall to be the most promising current alternative because of its power to predict unanticipated neuroimaging results, its clear neurobiological support, and its parsimony (for details, see [11]). This proposal is an example of an hypothesis which assumes a single general function for all cerebellar regions, rather than assuming multiple functions (e.g., [9, 74]. An important argument for a single general function is that the structure of the cerebellum is uniform [64] and structure is known to have a strong influence on computational function [91]. This hypothesis assumes that cerebellar circuitry performs the same function everywhere but the kind of information on which the computation is performed varies with regional variation in efferent and afferent projections. Thus, the information operated upon could vary in the sense modality involved, since it is known that the cerebellum has reciprocal projections with nearly every sensory system [14, 15, 97]. It is important to note however, that scientific understanding of the principles of distributed computation is still rudimentary. As we deepen our grasp of these principles, hypotheses about cerebellar function are likely to have substantially different structure, and to have quite different supporting arguments, than at present.

In conclusion, we note an incipient trend for known facts, explicit hypotheses, and effective research strategies from neuroimaging, cerebellar neurobiology, neurology, and experimental psychology to become better integrated. This allows us to look forward to a more comprehensive and accurate basis of knowledge about the cerebellum. This knowledge will surely have important implications for our understanding of brain function in general and for our practice of clinical neuroscience.

7. References

1. Allen, G., Courchesne, E., Buxton, R. B., and Wong, E. C. (1996) Dissociation of attention and motor opera in the cerebellum, *Proc. of the Third Annual meeting of the Cognitive Neuroscience Society*, pp. 28.
2. Anderson, J. R. (1995) *Cognitive psychology and its implications*, 4th edition. W. H. Freeman, New York.
3. Andreasen, N. C., O'Leary, D. S., Arndt, S., Cizadlo, T., Hurtig, R., Rezai, K., Watkins, G. L., Ponto, L. L. Hichwa, R. D. (1995) Short-term and long-term verbal memory: a positron emission tomography s *Proceedings of the National Academy of Sciences USA* **92**, 5111-5115.
4. Babinski, J. (1899) De l'asynergie cerebelleuse, *Review Neurologie* **7**,806-816.
5. Baddeley, A. (1986) *Working memory*, Oxford University Press, New York.

6. Bailey, D. (in press) Recent trends in PET camera designs, in H. W. Muller-Gartner and B. Gulyas (eds.), *PET: Critical assessment of recent trends*, Kluwer Academic Press, Dordrecht.

7. Bandettini, P. A., Jesmanowicz, A., Wong, E. C., and Hyde, J. S. (1993) Processing strategies for time-course data sets in functional MRI of the human brain, *Magnetic Resonance in Medicine* **30**, 161-173.

8. Bench, J. C., Friston, K. J., Brown, R. G., Scott, L C., Frackowiak, R. S. J., and Dolan, R. J. (1992) The anatomy of melancolia: Focal abnormalities of cerebral blood flow in major depression, *Psychological Medicine* **22**, 607-615.

9. Bloedel, J. (1992) Functional heterogeneity with structural homogeneity--how does the cerebellum operate?, *Behavioral and Brain Sciences* **15**, 666-678.

10. Bower, J. M. (1995) The cerebellum as a sensory acquisition controller, *Human Brain Mapping* **2**, 255-256.

11. Bower, J. M. (in press) Is the cerebellum sensory for motor's sake, or motor for sensory's sake: The view from the whiskers of a rat?, *Progress in Brain Research*.

12. Bower, J. M., and Kassell, J. (1990) Variability in tactile projection patterns to cerebellar folia crus-IIa of the Norway rat, *J. Comparative Neurology* **302**, 768-778.

13. Bracke-Tolkmitt, R., Linden, A., Canavan, G. M., Rockstroh, B., Scholz, E., Wessell, K., and Diener, H. C. (1989) The cerebellum contributes to mental skills, *Behavioral Neuroscience* **103**, 442-446.

14. Brodal, A. (1981) *Neurological anatomy in relation to clinical medicine*, Oxford University Press, New York.

15. Brodal, P. (1978) The corticopontine projection in the rhesus monkey: Origin and principles of organization, *Brain* **101**, 251-283.

16. Brodmann, K. (1909) *Vergleichende lokalisationlehre der grosshirnrinde in ihren prinzipien dargestellt auf grund des zellenbaues*, J. A. Barth, Leipzig.

17. Carpenter, M. B. (1991) *Core text of neuroanatomy*, 4th edition, Williams and Wilkins, Baltimore, MD.

18. Courchesne, E., Townsend, J, Akshoomoff, N. A., Saitoh, O., Yeung-Courchesne, R., Lincoln, A. J., James, H.E., Haas, R. H., Schreibman, L., and Lau, L. (1994) Impairment in shifting attention in autistic and cerebellar patients, *Behavioral Neuroscience* **108**, 848-865.

19. Decety, J., Sjohom, H., Ryding, E., Stenberg, G., and Ingvar, D. H. (1990) The cerebellum participates in mental activity -- tomographic measurements of regional cerebral blood flow, *Brain Research* **535**, 313-317.

20. Decety, J., Perani, D., Jeannerod, M., Bettinardi, V., Tadary, B., Woods, R., Mazziotta, J. C., and Fazio, F. (1994) Mapping motor representations with PET, *Nature* **371**, 600-602.

21. Desmond, J. E., Gabrieli, J. D. E., Ginier, B. I., Demb, J. B., Wagner, A. D., Enzmann, D. R., and Glover, G. H. (1995) A functional MRI (fMRI) study of cerebellum during motor and working memory tasks, *Society for Neurosciences Abstracts*, 1210.

22. Dolan, R. J., Bench, C. J., Scott, R. G., Friston, K. J., and Frackowiak, R. S. J. (1992) Regional cerebral blood flow abnormalities in depressed patients with cognitive impairment, *J. Neurology, Neurosurgery, and Psychiatry* **55**, 768-773.

23. Dow, R. S. (1942) The evolution and anatomy of the cerebellum, *Biological Review of Cambridge Philosophical Society* **17**, 179-220.

24. Dow, R. S. (1988) Contributions of electrophysiological studies to cerebellar physiology, *J. Clinical Neurophysiology* **5**, 307-323.

25. Ekman, P., and Davidson, R. J. (1994) *The nature of emotion*, Oxford University Press, New York.

26. Ellerman, J. M., Flament, D., Kim, S.-G., Fu, Q-G, Merkle, H. Ebner, T. J., Ugurbil (1994) Spatial patterns of functional activation of the cerebellum investigated using high-field (4-t) MRI, *Magnetic Resonance Imaging in Biomedicine* **7**, 63-68.

27. Fiez, J. A., Petersen, S. E., Cheney, M. K., and Raichle, M. E. (1992) Impaired nonmotor learning and error-detection associated with cerebellar damage--a single case study, *Brain* **115**, 155-178.

388

28. Fox, P.T., Lancaster, J. L., and Friston, K. J. (in press) *Mapping and modeling the human brain*, John Wiley and Sons, New York.

29. Fox, P. T., Mintun, M. A., Raichle, M. E., and Herscovitch, P. (1984) A non-invasive approach to quantitative functional brain mapping with $H_2^{15}O$ and positron emission tomography, *J. Cerebral Blood Flow and Metabolism* **4**, 329-333.

30. Fox, P. T., Mintun, M. A., Reiman, E. A., and Raichle, M. E. (1989) Enhanced detection of focal brain responses using intersubject averaging and distribution of subtracted PET images, *J. Cerebral Blood Flow and Metabolism* **8**, 642-653.

31. Fox, P. T., Raichle, M. E., Mintun, M. A., and Dence, C. (1988) Nonoxidative glucose consumption during focal physiological neural activity, *Science* **241**, 462-464.

32. Fox, P. T., Raichle, M. E., and Thach, W. T. (1985) Functional mapping of the human cerebellum with positron emission tomography.,*Proceedings of the National Academy of Science USA* **82**, 7462-7466.

33. Fox, P. T., Sergent, J. S., Hodges, D. A., and Parsons, L. M. (1996) Neural systems underlying musical performance, *Proceedings of the VI International MusicMedicine Symposium on Music, Physiology, and Medicine*.

34. Flourens, P. (1824) *Recherches experimentales sur les proprietes et les fonctions du systeme nerveux, dans les animaux vertegres*, Cervot, Paris.

35. Gao, J.-H., Parsons, L. M., Bower, J. M., Xiong, J., Li, J., and Fox, P. T. (1996) Cerebellum implicated in sensory acquisition and discrimination rather than motor control, *Science* **272**, 545-547.

36. Gentner, D., and Stevens, A. (1983) *Mental models*, Erlbaum, Mahwah, NJ.

37. George, M. S., Ketter, T. A., Parekh, P. I., Horwitz, B., Hersocovitch, P. and Post, R. M. (1995) Brain activity during transient sadness and happiness in healthy women, *American J. Psychiatry* **152**, 341-351.

38. Ghez, C. (1991) The cerebellum, in E. R. Kandel, J. H. Schwartz, and T. M. Jessell (eds.),*Principles of neural sciences*, 3rd edition, Elsevier, New York, pp. 626-646.

39. Goldman-Rakic, P. S. (1995) Toward a circuit model of working memory and the guidance of voluntary motor action, in J. C. Houk, J. L. Davis, and D. G. Beiser (eds.), *Models of information processing in the basal ganglia*, Massachusetts Institute of Technology Press, Cambridge, Massachusetts, pp. 131-148.

40. Grafman, J. (1996) Does the cerebell um contribute to complex cognitive processing, Symposium at the Annual Meeting of the International Society for Behavioral Neuroscience, Albequerque, New Mexico.

41. Grafman, J., Litvan, I., Massaquoi, S., Stewart, M., Sirigu, A., and Hallett, M. (1992) Cognitive planning deficits in patients with cerebellar atrophy, *Neurology* **42**, 1493-1496.

42. Grasby, P. M., Firth, C. D., Friston, K. J., Bench, C., Frackowiak, R. S. J., and Dolan, R. J. (1993) Functional mapping of brain areas implicated in auditory-verbal function, *Brain* **116**, 1-20.

43. Griffiths, T. D., Bench, J. C., and Frackowiak, R. S. (1994) Human cortical areas selectively activated by apparent sound movement, *Current Biology* **4**, 892-895.

44. Holmes, G. (1939) The cerebellum of man, *Brain* **62**, 1-30.

45. Horak, F. B., and Diener, H. C. (1994) Cerebellar control of postural scaling and central set in stance, *J. Neurophysiology* **72**, 479-493.

46. Houk, J. C., and Wise, S. P. (1995) Distributed modular architectures linking basal ganglia, cerebellum, and cerebral-cortex--their role in planning and controlling action, Cerebral Cortex **5**, 95-110.

47. Ito, M. (1984) *The cerebellum and neural control*, Appleton-Century-Crofts, New York.

48. Ito, M. (1993) Movement and thought: Identical control mechanisms by the cerebellum, *Trends in Neuroscience* **16**, 448-450.

49. Ivry, R. B. (1996) The timing hypothesis, Symposium at the Annual meeting of the International Society for Behavioral Neuroscience, Albequerque, New Mexico.

50. Ivry, R. B. and Keele, S. W. (1989) Timing functions of the cerebellum, *J. Cognitive Neuroscience* 1, 136-152.

51. Johnson-Laird, P. N., and Byrne, R. M. J. (1991) *Deduction*, Erlbaum, Mahwah, NJ.

52. Kim, S.-G., Ugurbil, and Strick, P. L. (1994) Activation of a cerebellar output nucleus during cognitive processing, *Science* 265, 949-951.

53. Klingberg, T., Roland, P. E., and Kawashima, R. (1995) The neural correlates of the central executive function during working memory--A PET study, *Human Brain Mapping* Supplement 1, 414.

54. Kwong, K. K., Belliveau, J. W., Chesler, D. A., Goldberg, I. E., Weiskoff, R. M., Poncelet, B. P., Kennedy, D. N., Hoppel, B. E., Cohen, M. S., and Turner, R. E., (1992) Dynamic magnetic resonance imaging of the human brain activity during primary sensory stimulation, *Proceeding of the National Academy Science USA* 89, 5951-5955.

55. Le, T. H, and Hu, X. (1996) Involvement of the cerebellum in intramodality attention shifting, *Neuroimage* 3, 246.

56. Leiner, H. C., Leiner, A. L., and Dow, R. S. (1993) Cognitive and language functions of the human cerebellum, *Trends in Neuroscience* 16, 444-447.

57. Llinas, R., and Sotelo, C. (1992) The cerebellum revisited, Springer-Verlag, New York.

58. Martin, A., Haxby, J. V., LaLonde, F. M., Wiggs, C. L., and Ungerleider, L. G., (1995) Discrete cortical regions associated with knowledge of color and knowledge of action, *Science* 270, 102-105.

59. Mayberg, H. S., Liotti, M., Jerabek, P. A., Martin, C.C., and Fox, P. T. (1995) Induced sadness: a PET model of depression, *Human Brain Mapping* Supplement 1, 396.

60. Mellet, E., Crivello, F., Tzourio, N., Joliot, M., Petit, L., Laurier, L. Denis, M., and Mazoyer, B. (1995) Construction of mental images based on verbal description: Functional neuroanatomy with PET, *Human Brain Mapping* Supplement 1, 273.

61. Middleton, F. A., and Strick, P. L. (1994) Anatomical evidence for cerebellar and basal ganglia involvement in higher cognitive function, *Science* 266, 458-461.

62. Ogawa, S., Menon, R. S., Tank, D. W., Kim, S.-G., Merkle, H., Ellermann, J. M., and Ugurbil, K. (1992) Functional brain mapping by blood oxygenation level-dependent contrast magnetic resonance imaging--a comparison of signal characteristics with a biophysical model, *Biophysics J.* 64, 309-397.

63. Orrison, W. W., Lewine, J. D., Sanders, J. A., and Hartshorne, M. F. (1995) *Functional brain imaging*, Mosby, St. Louis, Missouri.

64. Palay, S. L., and Chan-Palay, V. (1974) *Cerebellar cortex: Cytology and organization*, Springer, Berlin.

65. Parsons, L M. (1987a) Imagined spatial transformation of one's hands and feet, *Cognitive Psychology* 19, 176-241.

66. Parsons, L. M. (1987b) Imagined spatial transformation of one's body, *J. Experimental Psychology: General* 116, 172-191.

67. Parsons, L. M. (1987c) Visual discrimination of abstract mirror-reflected three-dimensional objects at many orientations, *Perception and Psychophysics* 42, 49-59

68. Parsons, L . M. (1994) Temporal and kinematic properties of motor behavior reflected in mentally simulated action, *J. Experimental Psychology: Human Perception and Performance* 20, 709-730.

69. Parsons L. M. (1997) New psychophysical constraints on theories of how an object's orientation and rotation are mentally represented. (Manuscript submitted for publication)

70. Parsons, L. M., Bower, J. M., Xiong, J., and Fox, P. T. (in preparation) Neuroimaging studies of whole brain activity during active and passive sensory tasks: Evidence for sensory acquisition control functions localized within cerebellum.

71.   Parsons, L. M., Fox, P. T., Downs, J. H., Glass, T., Hirsch, T. B., Martin, C. C., Jerabek, P. A., and Lancaster, J. L. (1995) Use of implicit motor imagery for visual shape discrimination as revealed by PET, *Nature* **375**, 54-59.

72.   Parsons, L. M., and Fox, P. T. (1995) Neural basis of mental rotation, *Society for Neuroscience Abstracts* **21**, 272.

73.   Passingham, R. E. (1975) Changes in the size and organization of the brain in man and his ancestors, *Behavioral Brain Evolution* **11**, 73-90.

74.   Paulin, M. G. (1993) The role of the cerebellum in motor control and perception, *Brain Behavioral Evolution* **41**, 39-50.

75.   Penfield, W., and Jasper, H. (1950) *Epilepsy and the functional anatomy of the human brain*, Little, Brown, and Co., Boston.

76.   Petersen, S. E., Fox, P. T., Posner, M. I., Mintun, M., and Raichle, M. E. (1989) Positron emission tomographic studies of the processing of single words, *J. Cognitive Neuroscience* **1**, 153-170.

77.   Posner, M. I., and Raichle, M. E. (1994) *Images of mind,* Scientific American Library (W. H. Freeman Co.), New York.

78.   richard, J. W., and Rosen, B. R. (1994) Functional study of the brain by NMR, *J. Cerebral Blood Flow and Metabolism* **14**, 365-372.

79.   Puce, A., Constable, R. T., Luby, M. L., McCarthy, G., Nobre, A. C., Spencer, D. D., Gore, J. C., and Allison, T. (1995) Functional magnetic resonance imaging of sensory and motor cortex: comparison with electrophysiological localization, *J. Neurosurgery* **83**, 262-270.

80.   Ramsey, N. F., Kirkby, B. S., Van Gelderen, P., Berman, K. F., Duyn, J. H., Frank, J. A., Mattay, V. S., Van Horn, J. D., Esposito, G., Mooner, C. T. W., and Weinberger, D. W. (1996) Functional mapping of human sensorimotor cortex with 3D BOLD fMRI correlates highly with H_2-^{15}O PET rCBF, *J. Cerebral Blood Flow and Metabolism* **16**, 755-764.

81.   Reiman, E. M., Raichle, M. E., Robins, E., Mintun, M. A., Fusselman, M. J., Fox, P. T., Price, J. L., and Hackman, K. A. (1989) Neuroanatomical correlates of a lactate-induced anxiety attack. *Archives of General Psychiatry* **24**, 493-500.

82.   Roland, P. E. (1993) *Brain activation*, Wiley-Liss, New York.

83.   Ryding, E., Decety, J., Sjohom, H., Stenberg, G., and Ingvar, D. H. (1993) Motor imagery activates the cerebellum regionally-- a SPECT rCBF study with ^{99m}Tc-HMPAO, *Cognitive Brain Research* **1**, 94-99.

84.   Schmahmann, J. D. (1996) From movement to thought: Structural and functional correlates of the cerebellar contribution to cognitive processing, *Human Brain Mapping* **4**, 174-198.

85.   Schmahmann, J. D. (in press) (ed.) *The cerebellum and cognition*, Academic Press, New York.

86.   Schmahmann, J. D., Doyon, J., Makris, N., Petrides, M., Holmes, C., Evans, A., and Kennedy, D. (1996) An MRI atlas of the human cerebellum in Talairach space, *Neuroimage* **3**, 122.

87.   Schmahmann, J. D., and Pandya, D. N. (1995) Prefrontal cortex projections to the basilar pons: implications for the cerebellar contribution to higher function, *Neuroscience Letters* **199**, 175-178.

88.   Seitz, R. J, Roland, P. E., Bohm, C., Greitz, T., and Stoneelander, S. (1991) Somatosensory discrimination of shape--tactile exploration and cerebral activation, *European J. Neuroscience* **3**, 481-492.

89.   Shepard, R. N., and Cooper, L. A. (1982) *Mental images and their transformations*, MIT Press, Cambridge, Massachusetts.

90.   Shepard, R. N., and Metzler, J. (1971) Mental rotation of three-dimensional objects, *Science* **171**, 701-703.

91.   Shepherd, G. (1989) The significance of real neuron architectures for neural network simulations, in E. Schwartz (ed.), *Computational Neuroscience*, MIT Press, Cambridge, Massachusetts, pp. 82-96.

92. Shulman, G. L., Corbetta, M., Buckner, R. L., Fiez, J. A., Miezin, F. M., Raichle, M. E., and Petersen, S. E. (in press) Common blood flow changes across visual tasks: I. Increases in subcortical structures and cerebellum, but not in non-visual cortext, *J. Cognitive Neuroscience*.

93. Shulman, R. G., Blamire, A. M., Rothman, D. L., and McCarthy, G. (1993) Nuclear magnetic resonance imaging and spectroscopy of human brain function, *Proceedings of the National Academy of Science USA* **90**, 3127-3133.

94. Talairach, J., and Tournoux, P. (1988) *Co-planar stereotaxic atlas of the human brain*, Thieme Medical, New York.

95. Thach, W. T., Goodkin, H. P., and Keating, J. G. (1992) The cerebellum and the adaptive coordination of movement, *Annual Review of Neuroscience* **15**, 403-442.

96. Wang, G., Tanaka, K., and Tanifuji, M. (1996) Optical imaging of functional organization in the monkey inferotemporal cortex, *Science* **272**, 1665-1668.

97. Welker, W. (1987) Spatial organization of somatosensory projections to granule cell cerebellar cortex: functional and connectional implications of fractured somatotopy, in J. S. King (ed.), *New concepts in cerebellar neurobiology*, Liss, New York, pp.239-280.

98. Wernicke, C. (1874) *Der aphasische symptomenkomplex*, Cohn and Weigert, Breslau.

99. Williams, R. W., and Herrup, K. (1988) The control of neuron number, *Annual Review of Neuroscience* **11**, 423-453.

HOW TO USE NEUROIMAGING TO STUDY VISUAL ATTENTION

Maurizio CORBETTA
Department of Neurology and Neurological Surgery,
Washington University School of Medicine
St. Louis, MO 63110, USA

1. Introduction

The last decade has witnessed an explosion in the field of human brain imaging. Several methodologies, most notably Positron Emission Tomography (PET) and functional magnetic resonance imaging (fMRI), allow us to image brain activity in vivo during sensory, motor, and cognitive behavior. The combination of brain imaging with task paradigms derived from cognitive psychology and psychophysics represent today one of the most effective ways to correlate in vivo cognition and brain activity (Raichle, 1994).

This paper describes two examples of how to use neuroimaging to study 'selective attention', namely the brain's ability to select certain stimuli over others, based on their ecological relevance and its own current behavioral goals. Consider a person sitting in front of a computer screen, who knows in advance that something interesting will appear in the left upper corner of the screen. His ability to detect a stimulus, even as simple as a flash of light, is greatly improved by the foreknowledge of the correct spatial location (Posner, 1980). It is now know that the allocation of attention to a spatial location or feature of an object (e.g. its color) produces very robust changes in the neuronal firing of several areas of the monkey's brain (Desimone and Duncan, 1995; Maunsell, 1995). Goals of current neuroimaging research on attention are to establish whether similar changes are observed in the living human brain, define the rules governing these modulations, i.e. whether attention is allocated to feature or objects, or whether visual attention is mediated by the same brain regions as auditory attention, define the neural sources of such attentional modulations, i.e. regions controlling the allocation of attention. Imaging studies can be particularly helpful in this regard since they allow to image activity in the whole brain, i.e. image both the site and source of attentional effects.

The first experiment localizes regions of the parietal and frontal human cortex that are activated when attention is explicitly directed toward visual locations to detect behaviorally relevant visual stimuli (Corbetta et al., 1993). The second experiment shows these regions become active, even in the context of an object identification task in which attention is cued to other features, if the task demands involve spatial selection (Corbetta et al., 1995). We propose that the parietal cortex is necessary to shift the focus of processing from one location/object to another in the visual field.

B. Gulyás and H.W. Müller-Gärtner (eds.),
Positron Emission Tomography: A Critical Assessment of Recent Trends, 393-399.
© 1998 *Kluwer Academic Publishers.*

2. Parietal and frontal regions control visuospatial attention

In one experiment we scanned normal volunteers while they were looking at a computer generated display consisting of eleven small boxes, displayed just below the horizontal meridian across both hemifields (0°, $+1^O$, $+3^O$, $+6^O$, $+10^O$, $+15^O$). In one condition, subjects maintained fixation on a central cross-hair, attended to different peripheral box locations in a predetermined sequence, and detected on each trial the onset of a brief flash within each box by a speeded key-press (shifting attention task). Subjects were explicitly instructed to shift attention to the next location to anticipate the occurrence of the flash. The shifting attention task was performed separately in the left or right visual hemifield in different PET scans. In a second condition, subjects maintained central fixation, detected flashes presented in the central fixation box, and ignored flashes that were presented in temporal asynchrony at random peripheral box locations. Subjects were explicitly instructed to concentrate on the central flashes, and ignore the peripheral flashes which were present in the left or right hemifield in separate scans (central detection task). In a third condition, subjects maintained central fixation while flashes were presented at random peripheral box locations either in the left or right hemifield in separate scans (passive task). Under this condition, attention was automatically attracted to various peripheral locations by the onset of the flash (Yantis and Jonides, 1990). The shifting attention and passive tasks, therefore, contrast two modes of cueing attention to a location: one cognitive, driven by task instructions about the predictive sequence of locations, and one automatic, driven by the abrupt onset of the sensory stimulus. It has been postulated that these different attentional modes are implemented in different neural pathways. A final control condition was run in which subjects simply fixated on the array of boxes, and no flashes (central or peripheral) were presented (fixation).

Figure 1, center, shows a group-averaged subtracted sagittal PET brain slice, in which activity obtained during the fixation condition is subtracted from activity obtained during the shifting task. The activity is averaged across a group of normal volunteers (n=24) participating to the study. The image subtraction isolates task-relevant regional blood flow change differences between the two experimental conditions, in this case higher activity in the shifting attention than fixation task. The foci of activation map in the superior parietal (near the intraparietal sulcus) and frontal cortex (near the superior middle frontal gyrus and precentral gyrus), and can be related to any number of differences in sensory, attentional, or motor processes.

Figure 1, left, shows that the parietal and frontal activations remain when the central detection task is subtracted as control condition. Since the two tasks have similar sensory and motor requirements, these activations can be more confidently explained by the attentional differences between tasks, namely the greater degree of attention to the periphery of the visual field in the shifting task. From these data we conclude that both parietal and frontal cortex mediate the allocation of attention to peripheral locations of the visual field. The two regions, however, may differently contribute to this function. Figure 1, right, shows significant activity in parietal cortex, but not in frontal cortex, when flashes suddenly occur in the periphery of the visual field during the passive task, and attention is automatically drawn to the periphery. Therefore, parietal cortex is active for both cognitive (i.e. instruction driven) and automatic (i.e. sensory driven) attention, while frontal cortex seems more related to the cognitive voluntary component of attention . Parietal activity may reflect various neuronal signals including preparatory oculomotor activity (although no eye movements actually occurred), attentional enhancement of visual responses and/or a pure cognitive signal encoding shifts of attention to various field locations. The involvement of parietal cortex is spatial selection is consistent with a large body of clinical and experimental literature (Mesulam, 1990), and more recent imaging studies (Nobre et al, in press). The presence of frontal activity during the shifting task is more surprising. Activations in nearby locations have been obtained during the execution of saccadic eye movements (Petit et al., 1993) and spatial short-term memory tasks (Smith et al., 1995; Courtney et al., 1995).

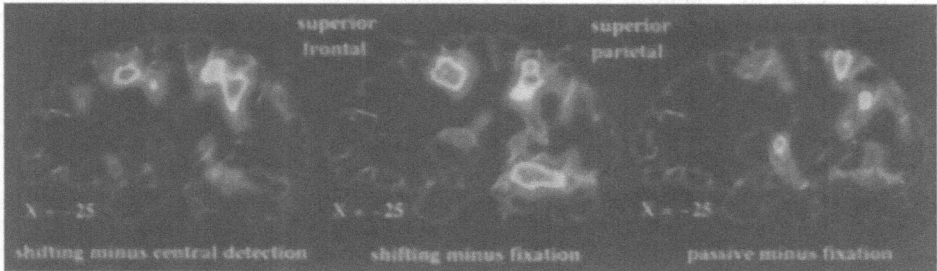

Figure 1. Parietal and frontal cortex activations (arrows), 25 mm left of midline, during peripheral spatial attentional tasks.

3. Parietal activity during visual search conjunction tasks

The establishment of a functional relationship between spatial selection and parietal activity is helpful because it allows to establish whether spatial selection is necessary for other visual behaviors in which its role is only hypothetical or controversial, based on current on psychophysical evidence. The parietal activity in other terms become a 'marker' of spatial selection, that can be used to ask more physiologically-oriented questions about attention. For example, a long-standing controversy in psychology is the one between 'parallel vs. serial' models of visual search. The basic phenomenon is well known: the search time for a low saliency target, e.g. a red triangle among red squares and green triangles, increases with the number of distractors, whereas the search time for a high saliency target, e.g. a red square among green distractors, is independent of the number of distractors. The reasons for this difference are unclear. Some models propose that the visual analysis of the display proceeds in parallel (parallel search) across the field in both cases, but that its efficiency declines in conditions of low target discriminability when more noise (distractors) is added to the system (Duncan and Humphreys, 1989). Others models propose that in conditions of low discriminability a spatial attention mechanism is additionally recruited to serially inspect each item in the field and discriminate between target and distractors (Treisman and Gelade, 1980).

In a recent report (Corbetta et al., 1995), we presented subjects with a display containing four windows arranged as a square around a fixation point. Each window contained colored moving dots which were presented for 0.5 seconds on each trial. The subjects searched in different scans for targets defined by either 'color' (e.g. a red target among non-red distractors), 'motion' (e.g. a fast moving target among slow moving distractors), or by a 'conjunction' of color and motion (e.g. a red and fast moving target among red and slow or non-red and fast moving distractors). Subjects pressed one key when the target was present and another key when the target was absent. The control condition was a 'passive viewing' condition in which the same stimuli were shown, but subjects alternated key-presses without discrimination.

Expectedly, search times increased linearly with the number of distractors in the field during the conjunction task, but remained flat during the color or motion task. The pattern of results is consistent with parallel analysis in the color or motion (feature) task, and either with a noisy parallel analysis or serial analysis by a spatial mechanism in the conjunction task.

Figure 2 shows sagittal and coronal group-averaged subtracted PET slices through the right superior parietal cortex, respectively in the shifting attention, conjunction, color, and motion tasks. The control condition is fixation for the shifting task (see above), and passive viewing for the other conditions. Across the whole brain the three visual search tasks are best differentiated by activity in the right superior parietal cortex (near intraparietal sulcus), which is significantly higher in the conjunction than in color or motion task. This right superior parietal region overlaps with regions related to spatial selection in the shifting task.

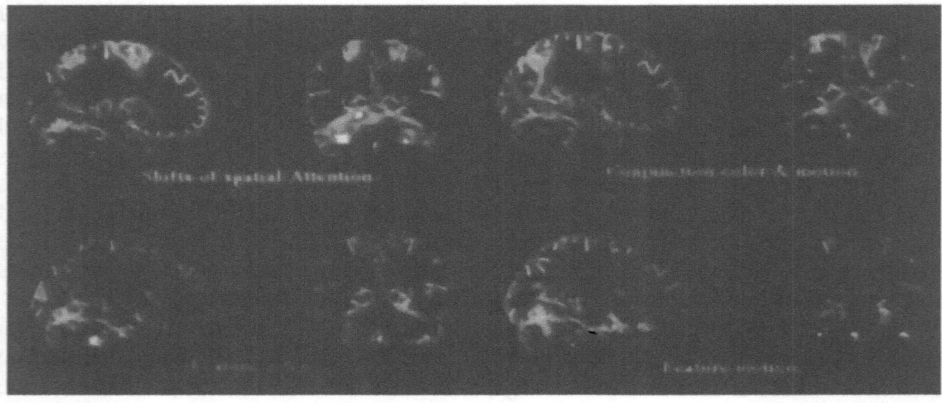

Figure 2. Right superior parietal activity (arrow) for shifts of attention and conjunction visual search.

The co-localization of activity suggests that some form of spatial selection uniquely occurred in the conjuction task, and therefore supports serial over parallel psychological models of visual search. Interestingly, the conjunction task did not explicitly cue location, as the targets were defined by color and motion. These findings, however, suggest that the brain had to 'turn on' selection by location once the discriminability in the display was too low to distinguish target from distractors.

The nature of the spatial selection signal implemented in parietal cortex is unknown. Single unit data indicate that neurons in posterior parietal cortex respond more strongly when attention is directed away from the neuron's receptive field (Robinson et al., 1995; Steinmetz and Constatidinis, 1995). These data are consistent with a model in which parietal cortex encode a shifting signal to realign the focus of attention toward novel behaviorally relevant stimuli. Another possibility is that superior parietal cortex code for the binding of color and motion information at a single location in the conjunction task. Binding activity would be again higher in the conjunction than feature task. This interpretation is consistent with a recent report of a patient with bilateral occipito-parietal lesions and severe difficulties in binding features of visually presented objects (Friedman-Hill et al, 1995). Explanations in terms of shifts of attention and feature binding are not necessarily exclusive if the superior parietal cortex contains a spatial map, topographically connected (directly or indirectly through other structures, e.g. the pulvinar) to various feature maps (color, motion, orientation). Parietal activity will therefore occur in tasks that emphasize processing at various locations as in the shifting task, or tasks involving more than one feature at one location as in the conjunction task.

4. Conclusions

Functional neuroimaging represents a powerful tool to ask questions about the functional neuroanatomy of cognitive functions such as attention. These functional anatomical maps can be used to neurobiologically constrain psychological theories of cognitive functions, and solve issues which cannot be addressed on pure behavioral grounds. Experiments exploring the interactions between large scale neural systems are likely to yield the most innovative information as they best exploit the whole brain coverage uniquely offered by neuroimaging over more traditional approaches (e.g. lesion or single unit recordings).

5. Acknowledgements

Supported by the Charles A. Dana Foundation and the McDonnell Center for Studies of Higher Brain Function.

6. References

Corbetta, M., Miezin, F. M., Shulman, G. L., and Petersen, S. E. (1993). A PET study of visuospatial attention. J. Neurosci., 13(3), 1202-1226.

Corbetta, M., Shulman, G. L., Miezin, F. M., and Petersen, S. E. (1995). Superior parietal cortex activation during spatial attention shifts and visual feature conjunction. Sci., 270, 802-805.

Courtney, S. M., Ungerleider, L. G., Keil, K., and Haxby, J. V. (1995). Object and spatial visual working memory activate separate neural systems in human cortex. Cer. Cortex, in press.

Desimone, R., and Duncan, J. (1995). Neural mechanisms of selective visual attention. Annu. Rev. Neurosci., 18, 193-222.

Duncan, J., and Humphreys, G. W. (1989). Visual search and stimulus similarity. Psychol. Rev., 96, 433-458.

Friedman-Hill, S. R., Robertson, L. C., and Treisman, A. (1995). Parietal contibutions to visual feature binding: evidence from a patient with bilateral lesions. Sci., 269, 853-855.

Maunsell, J.H.R. (1995) The brain's visual world: representation of visual targets in cerebral cortex. Sci., 270, 764-768.

Mesulam MM (1990) Large-scale neurocognitive networks and distributed processing for attention, language, and memory. Ann. Neurol. 28:597-613.

Nobre AC, Sebestyen GN, Gitelman DR, Mesulam MM, Frackoviack R.S.J., Frith C.D. (in press) Functional localization of the systems for visuospatial attention using positron emission tomography. Brain.

Petit, L., Orssaud, C., Tzourio. N., Salamon. G., Mazoyer, B., Berthoz, A. (1993). PET study of voluntary saccadic eye movements in humans: Basal ganglia-thalamocortical system and cingulate cortex involvement, 69, 1009-1017.

Posner, M. I. (1980). Orienting of attention. Quart. J. Exp. Psychol., 32, 3-25.

Raichle, M. E. (1994). Visualizing the mind. Sci.Am., 270 (4), 36-42.

Robinson, D.L., Bowman, E.M., Kertzman (1995) Covert orienting of attention in macaques. II. Contributions of parietal cortex. J.Neurophysiol., 74: 608-712.

Smith, E. E., Jonides, J., Koeppe, R. A., Awh, E., Schumacher, E. H., and Minoshima, S. (1995). Spatial versus object working memory: PET investigations. J. Cogn. Neurosci., 7(3), 337-356.

Steinmetz, M.A., Constatidinis, C. Neurophysiological evidence for a role of posterior parietal cortex in redirecting visual attention. Cerebr.Cortex, 5: 448-456.

Treisman, A. M., and Gelade, G. (1980). A feature-integration of theory of attention. Cog. Psychol., 12, 97-136.

Yantis S , Jonides J (1990) Abrupt visual onsets and selective attention: voluntary versus automatic allocation. J. Exp. Psych.: Hum. Percept. Perf. 16:121-134.

CHARACTERISING SELECTIVE ATTENTION WITH POSITRON EMISSION TOMOGRAPHY

Neurophysiological investigations

Geraint Rees, Karl Friston, Richard Frackowiak
and Chris Frith
The Wellcome Department of Cognitive Neurology,
Institute of Neurology,
Queen Square,
London WC1N 3BG

Abstract

Selection is the essence of attention, whether of a train of thought, an object or a spatial location. The most general way of characterising selectivity is as task-dependent modulation of neural activity. Such task-dependent modulation by attention of sensory input or motor output can take two forms. Activity in an area involved in a sensorimotor or cognitive task may be modulated in either a stimulus-dependent (phasic) or a stimulus independent (tonic) fashion. Conventional PET methodology cannot distinguish between these two types of modulation, as all neural activity within the scanning window is lumped together. However, by varying the rate of stimulus presentation orthogonally with task, these two types of modulation by attention can be distinguished. This is illustrated with a PET study of selective attention. In a non-spatial selective attention task, subjects identified visual targets defined by colour, orientation, or the conjunction of colour and orientation. A region in right dorsolateral prefrontal cortex was the only area specifically activated by the conjunction task, in a location that has previously been implicated in the performance of conjunction tasks in monkeys. A distributed network of cortical areas including prestriate cortex, motor and premotor cortex and inferior temporal cortex shows activity correlated with increasing stimulus presentation rate. Within this network of areas we demonstrate that inferior temporal cortex, premotor cortex and cerebellum are modulated by selective attention to

B. Gulyás and H.W. Müller-Gärtner (eds.),
Positron Emission Tomography: A Critical Assessment of Recent Trends, 401-413.
© 1998 *Kluwer Academic Publishers.*

visual conjunctions in the conjunction task. Moreover the manner of modulation differs in different areas, being stimulus related in premotor cortex and left cerebellum ("phasic") and stimulus-independent ("tonic") in inferior temporal cortex and right cerebellum. This is the first demonstration in humans of two physiologically distinct types of modulation of cortical activity by selective attention. We suggest that this technique can distinguish between the site and source of attentional modulation.

Introduction

This paper investigates the fundamental nature of selective attention. We characterise attention as a modulatory influence on neural processing, and describe a novel PET methodology that makes use of this insight to allow direct inference about the physiological form of these modulatory influences. This methodology represents a way of using PET to characterise not only the anatomical distribution of functionally distinct areas, but also to characterise the physiological nature of their interaction.

Selection is the essence of attention, whether of a train of thought (1) an object or a particular spatial location (2). Selectivity at such a phenomenal level of experience implies mechanisms at the neural level for selecting among different types of incoming information. Perhaps the broadest way to characterise attention is therefore as a modulatory influence on sensorimotor processing. Modulatory influences could be of exogenous or endogenous origin, and this is reflected in many current theoretical models of selective attention (3). These suggest that attention can be captured by bottom up exogenous factors related to stimulus saliency, but is also under the influence of top-down, voluntary control. Top-down control is a way of biasing stimulus-related signals in favour of current behaviour or according to task demands. It is this latter form of attention that we address in this paper.

Different types of attentional modulation

Presentation of a visual stimulus evokes transient neural activity. It is a reasonable assumption that the activity evoked is stereotyped and therefore invariant across successive presentations of the same stimulus. Repeated stimulus presentations therefore evoke a train of neural activity, illustrated in figure 1. The neural implementation of selective attention (top-down influences on sensory processing) that we have discussed implies a modulation of that transient activity. This modulatory influence could in principle take two distinct forms (figure 1).

First, the transient activity associated with processing of the visual stimulus may be enhanced (or suppressed) directly. This means that the neural activity associated with the processing of individual stimuli is altered. It has been suggested that this sort of modulation is active in prestriate cortex during selective attention to visual motion (4).

Note that this characterisation of attention suggests that without stimuli, there is no observable effect of attentional modulation. An intuitive way of characterising this type of attentional modulation might be as a 'gain control', and a perusal of the literature suggests that this is the conceptual approach to attention implicitly or explicitly adopted by many authors (2,3,4). However the modulation could act a second way, by increasing the tonic neural activity in a cortical area even without stimulus related neural signals. This type of modulation takes the form of a 'bias' signal that is task dependent, but stimulus independent. However it has its effect on processing by virtue of the coexpression of both attention related and stimulus related activity in a given cortical location. Although this presupposes that the effects on neural activity of stimulus and attention are independent, this does not imply that the expression of attention can have no modulatory influence on stimulus processing in that area. If the first form of attentional modulation is analogous to a gain control on an amplifier, one might think of this kind of modulation as analogous to altering the tone or balance controls. The absolute quantity of signal is not altered, but the balance of its intrinsic characteristics is changed.

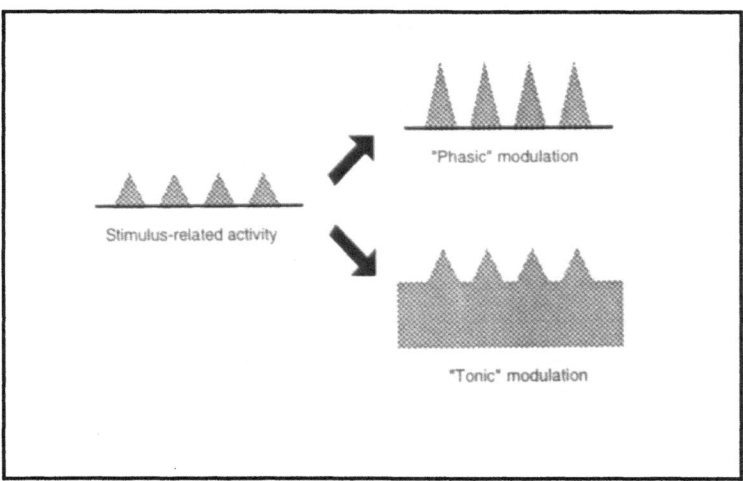

Figure 1 . Stereotyped transient activity evoked by a stimulus may
be modulated in a stimulus-dependent manner ("phasic"; upper right)
or in a stimulus-independent manner ("tonic"; lower right)

Theoretical precedents for this type of "tonic" modulation exist; the biased competition model (5) of selective attention suggests that visually responsive inferior temporal cortex neurons are thought to receive a tonic 'bias' signal altering their processing of visual information according to task demands. Note that attentional modulation might have a negative, rather than positive, influence; that is, the "phasic" modulation might result in a lesser amount of neural activity per stimulus, and the "tonic" modulation might results in a baseline activation lower than the original level of activity. The important distinction between these two types of modulation is not whether they increase or decrease overall neural activity, but whether the modulation is stimulus-dependent or stimulus-independent.

Functional imaging techniques such as positron emission tomography (PET) integrate activity within a scanning frame, and are unable to discriminate between neural activity of the form shown on the right hand side of figure 1. However, such techniques can be used to detect the transient changes in activity evoked by a stimulus if the rate of presentation of the stimulus is varied across scans. Several studies with PET have shown that the rate of stimulus presentation or response production is strongly correlated with blood flow in the relevant brain areas (6,7,8). This finding leads us to suggest that the slope relating activity to presentation rate is an index of the amount of transient activity associated with the presentation and subsequent processing of a single stimulus. We can use this relationship to characterise those functionally segregated areas associated with performance of the task. Phasic and tonic (or bias) modulatory effects, as we have defined them, can now be distinguished (Figure 2).

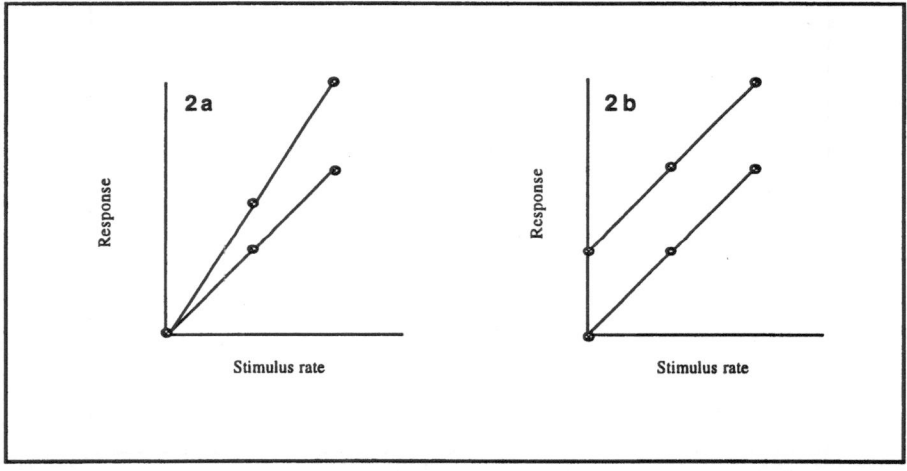

Figure 2. Diagram to illustrate how integrated stimulus or response related phasic activity produces a correlation between rCBF and stimulus presentation rate. A modulatory influence on the processing of each stimulus or response can have an effect on the phasic activity, increasing the slope of the line (2a); that is to say, the rate dependency is determined by the task at hand. Alternatively, a modulatory influence can have a tonic effect that results in no change of the slope describing rate dependency, but a shift due to the task at hand that can be quantified by a change in intercept (2b). This implies that there may be activity in the absence of stimulus related signals.

An increase in the transient activity evoked by a stimulus will cause an increase (or decrease) in the slope relating activity to presentation rate (Figure 2a), whereas a tonic elevation or bias will result in a change of intercept with no change in slope (ie a main effect that is independent of presentation rate, Figure 2b).

This is an important relationship to understand, and it should be emphasised that we are using the terms "phasic" and "tonic" to characterise *changes* in the relationship between activity and presentation rate. Quite separately, we could distinguish areas with only a task related effect, or with only a rate related effect. We **do not** characterise such areas as tonic or phasic, as our terms refer only to modulatory influences on areas which also show rate dependency. Again, note that figure 2 shows a positive relationship between response and stimulus presentation rate. Our theoretical analysis still holds if the relationship between activity and presentation rate is negative, or that the modulatory influence is negative in nature.

To what extent is this technique valid? It presupposes that there will be a linear, or near-linear relationship between stimulus and response related activity and stimulus presentation rate. If the relationship is highly nonlinear, then the effects of modulatory influence may be obscured or exaggerated. In visual cortex, the relationship between flicker rate of a reversing checker board and rCBF is linear up to 7.8Hz (6). In primary auditory cortex, a linear correlation between presentation rate and words (7) or tones (8) and rCBF has been demonstrated. It therefore seems reasonable to assume that the relationship between stimulus presentation rate and rCBF is linear for primary sensory cortex. This assumption is empirically validated by the data analysis we present below.

Example Data

We now show how this methodology can be used to characterise functionally segregated brain areas modulated by attention in two physiologically distinct ways in a non-spatial attention demanding task. We chose non-spatial attention as there is already an extensive literature on spatial attention implicating the parietal lobe in shifts of spatial attention and modulation of prestriate cortical activity according to task demands (9). We sought evidence for frontotemporal interactions in the context of attention to different aspects of simple objects on the background of compelling theoretical and experimental evidence in monkeys for this interaction (10,11). The experiment is described in detail elsewhere (12) and here we focus not on the biological aspects of the experiment, but on its methodology which illustrates the theoretical approach discussed here.

METHODS

Participants were presented with a stream of visual information in the form of coloured ellipses. The ellipses were presented at four different rates (between 30 and 90 ellipses per minute) and could be coloured red, green or blue and oriented horizontally or vertically. Participants responded to every ellipse by pressing a button to indicate whether it was a target or not. There were three conditions; the 'conjunction' condition with targets defined by the conjunction of red colour and horizontal orientation, and two 'feature' conditions where targets were either red, or horizontal respectively. The proportion of targets were adjusted in each condition such that 50% of the stimuli were targets; in other respects, the stimuli and presentation were identical across all conditions. Six subjects performed these tasks during a series of twelve PET regional cerebral blood flow scans with $H_2^{15}O$ as the tracer. The experiment is therefore characterised as a 4x3 factorial design with subjects undergoing one scan for each combination of four presentation rates and three task conditions.

RESULTS

Statistical analysis of the functional imaging data allows us to distinguish cortical areas on the basis of their response characteristics (13,14). We were interested in the comparison between conjunction task and the weighted average rCBF of the feature tasks to characterise areas that would be selectively activated in the conjunction task. We identified only one area, in dorsolateral prefrontal cortex (Brodmann area 8), that was significantly activated in the conjunction tasks (figure 3). The activity in this area showed no rate dependency. In contrast to this single area of activation, considering all tasks together we identified a network of areas that showed increases and decreases in activity covarying with stimulus presentation rate. These areas represent a distributed network responsible for performance on this task (table 1; areas negatively correlated with presentation rate shown in figure 4).

To identify the loci of attentional modulation within this network, we performed two further analyses. In the first analysis, we identified areas that both showed a rate effect and an interaction with task. This is equivalent to the interaction term in our factorial design, and identifies areas corresponding to our figure 2a; this is the form of modulation that we have termed "phasic." The second analysis identified areas that showed both and main effect of task *and* a rate dependency (this is termed a conjunction analysis). This is equivalent to the effect shown in our figure 2b, and that we have termed "tonic." In effect, we were identifying areas with four types of response characteristics in our data: -

I	Effect of task, no effect of rate
II	Effect of rate, no effect of task
III	Effect of rate that interacts with the task effect (interaction term in the factorial design)
IV	Effect of rate *and* of task (but no interaction)

Areas of types III and IV we interpret as showing one of the two types of attentional modulation that we have described above. Again we emphasise that we are using the terms phasic and tonic in this context to refer to modulation by attention. Of course, it is also possible to conceive of activity in areas II-IV as modulated in a "phasic" manner by presentation rate; this is not the manner in which we use the term phasic in this paper. Finally, it is possible that there are areas that show both tonic attentional modulation *and* phasic attentional modulation, although we found no such areas in our data.

Figure 5 illustrates the differing types of modulatory activity by plotting the adjusted rCBF values from voxels in the cerebellar, premotor and inferotemporal areas identified in table 2 as a function of stimulus presentation rate. This graphically illustrates not only the distinction between positive and negative correlation of rCBF with presentation rate, but also the clear distinction between tonic and phasic modulation of this rate-related activity by selective attention to visual conjunctions.

Discussion

The biological interpretation of our results is discussed elsewhere (12). Here we will focus on the distinction between the types of modulation seen in our data. Our methodological approach is unique within the functional imaging literature, and has now been used to successfully study auditory attention as well as visual attention (8). There are two features of the data we will discuss; the presence of multiple cortical areas that show a negative correlation of activity with presentation rate, and the theoretical implications of identifying areas with distinct types of modulation by attention.

DECREASES

Positive correlations of rCBF with increasing presentation rate were seen in areas of visual, premotor and motor cortex. This might be thought biologically plausible as increasing presentation rate resulted in presentation of more targets, and produced more responses from the subjects. However the most striking and statistically significant changes in rCBF were in fact decreases with increasing stimulus presentation rate. Decreases in activity, especially in medial prefrontal structures, have been seen frequently in our laboratory. However, in a categorical design it is difficult to unambiguously attribute decreases to an active condition. Decreases in rCBF in an active condition may represent activations (relative to global normalised rCBF) in the rest condition. But the factorial parametric design of our experiment has no rest condition, and we can unambiguously attribute the decreases in rCBF seen with increasing presentation rate to stimulus related activity. The spatial distribution of the areas we identify has interesting parallels with decreases seen in other functional neuroimaging studies (15) and in single cell recording studies in monkeys (16). This suggests that it may be important to specifically investigate decreases in activity in addition to increases. This has implications for experimental design, in that a categorical design is limited in the theoretical interpretation of such effects, and implies that factorial parametric experimental designs may become increasingly important (17). The rCBF decreases observed probably correspond to decreases of synaptic activity in relatively large neuronal populations (18). It should be remembered that PET measures predominantly presynaptic activity, and the decreases may represent decreases in activity of either excitatory or inhibitory (or mixed) populations of neurons. So it is difficult to make direct inferences about the nature of the underlying neural activity. Nevertheless it is striking that decrements in activity with a similar anatomical distribution in inferior temporal cortex have been recorded previously in visually responsive neurons in monkeys performing a match-to-sample task.

TYPES OF MODULATION

Modulation of sensory processing according to task demands has been shown in previous functional imaging studies in prestriate cortex (9). The categorical nature of the design in these studies did not allow inference about the type of modulation, but evidence from convergent ERP studies suggests that the effects may be of the "phasic" kind (although we note in passing that it is not clear that ERP can distinguish between the two types of modulation that we have characterised). Our data are strikingly different in their spatial distribution. We find no modulation of activity in prestriate visual cortex; instead, modulation is seen further down the ventral pathway in inferior temporal cortex, and in areas that might be plausibly related to the motor responses such as motor cortex and cerebellum. Selective attention to visual conjunctions acts at multiple points between presentation of a visual stimulus and production of a response to influence neural processing. This implies that theoretical attempts to distinguish between 'early' and 'late' selection by attention of incoming stimuli for further processing may be doomed to failure, and might account for some of the confusion in the existing literature (19). Moreover our results are consistent with theoretical approaches that emphasise the role of selective attention in terms of preparation for action (20).

The most important theoretical aspect of our results is the demonstration of physiologically distinct types of modulation in functionally segregated cortical areas. These correspond to our analysis of the effects of attention developed earlier in the paper, and demonstrate the credibility and empirical validity of our approach. To our knowledge, this distinction between the theoretically plausible modulatory effects of attention has not previously been demonstrated. We suggest that the distinction between the different response characteristics of functionally segregated cortical areas may allow a distinction to be made between the site and source of attentional effects. Establishing causality in a functional imaging study that shows correlations between cortical activity and experimental manipulation must proceed cautiously. Nevertheless, the analysis we developed based on our pretheoretic intuitions may allow us to make steps in this direction. In a theoretical model of attention such as the biased competition model, one would anticipate that manipulation of the rate of presentation of items to be attended to would have differential effects on the site and source of attentional manipulation, if the effects of attention are constant over time. In the specific context of our experimental design, we would expect that the source of attentional effects would show no effect of rate of presentation, but an effect of attentional task alone. By contrast, and as we have argued, the structures subserving stimulus categorisation and response production will show an effect of rate of presentation of items. The site of attentional modulation will be within these areas, as we have demonstrated. The point we are making here is not the teleological one, that we have seen attentional effects within rate responsive areas (as we predicted). Here we are pointing out the existence of an isolated cortical area that shows a selective activation to visual conjunctions, but no effect of rate. Our analysis leads us to suggest that this area is the site of attentional modulation; its position in frontal cortex and the functional neuroanatomy and neurophysiology of Brodmann area 8 lends credence to this position (see (10) for further discussion of this point).

Conclusion

We believe our new methodology takes PET beyond the characterisation of functional neuroanatomy. Using this technique, it is possible to distinguish between phasic modulation and tonic modulation by attention, and to measure the amplitude of transient responses to stimuli. Moreover, our theoretical analysis and empirical data strongly suggest that it is possible to distinguish the site and source of attentional effects. By studying the modulation of these effects by processes such as attention we can begin to use PET to look at the physiological mechanisms underlying cognitive processes.

Acknowledgements

We are grateful to all our colleagues in the Functional Imaging Laboratory for their help with the conduct of these studies. The authors thank the Wellcome Trust for their essential and continuing support.

Table 1.
Areas that show significant correlation of rCBF with stimulus presentation rate

Increases		Decreases	
Area	Coordinates	Area	Coordinates
Left inferior occipital gyrus	-42 -88 -10	Anterior cingulate	-2 36 -6
Right fusiform	48 -60 -18	Right inferior frontal gyrus	22 48 16
Left cerebellum	-32 -66 -22	Right inf. temporal gyrus	62 -20 -24
Right cerebellum	2 -54 -8	Left inferior temporal gyrus	-62 -16 -18
Left precentral gyrus	-56 -2 50		
Left premotor cortex	-12 -10 64		

Table 2.
Areas that show significant modulation of activity by the conjunction task relative
to the two feature tasks

Area	Coordinates	Type of modulation
Left inferior temporal gyrus	-62 -16 -18	Tonic (type IV)
Left precuneus	12 -58 44	Phasic (type III)
Left premotor cortex	-16 -12 62	Phasic (type III)
Left cerebellum	-32 -62 -20	Phasic (type III)
Right cerebellum	28 -54 -26	Tonic (type IV)

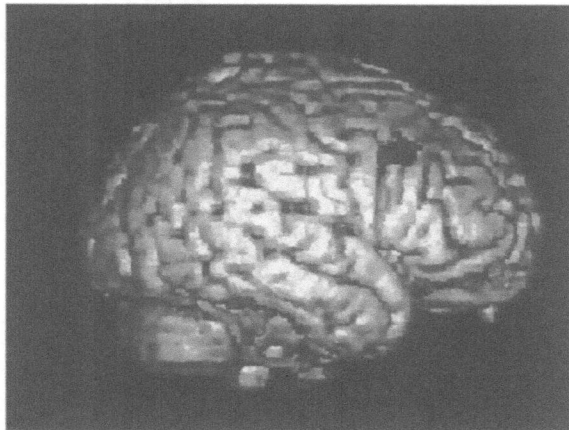

Figure 3. Images showing activation due to the conjunction task relative to the feature tasks. The image is created by overlaying the relevant statistical parametric maps onto a T1 weighted structural MRI.

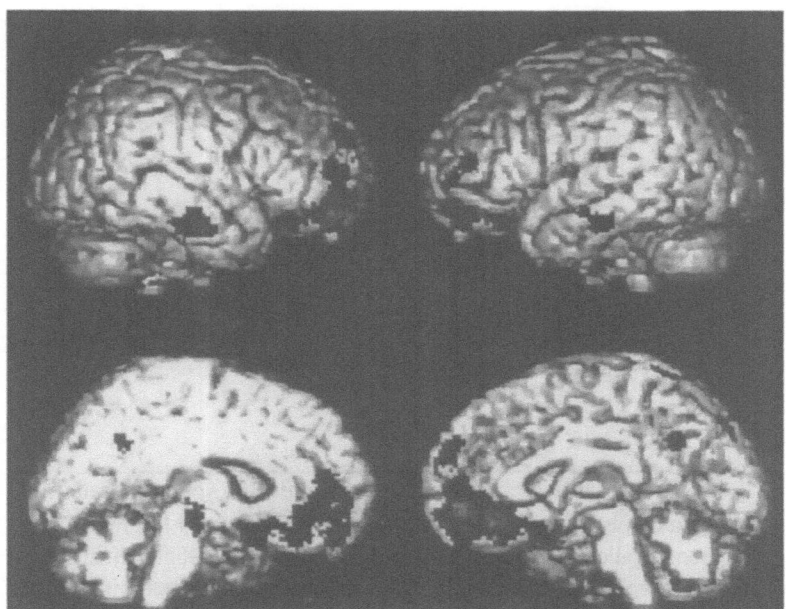

Figure 4. Images showing activations in left inferior temporal and medial prefrontal structures reflecting decreases in activity with stimulus presentation rate. The images are created by overlaying the relevant statistical parametric maps onto a T1 weighted structural MRI.

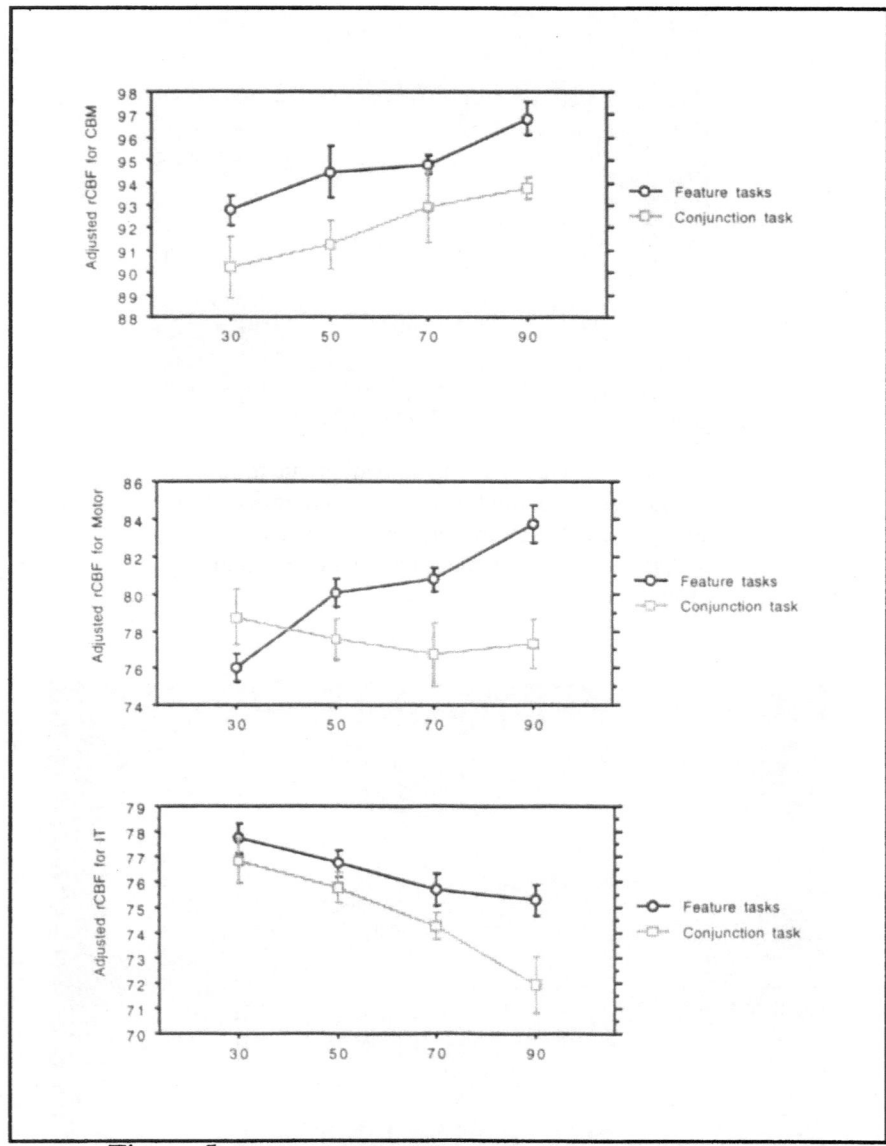

Figure 5 . Plots relating rCBF to stimulus presentation rate in right cerebellum (5a), left premotor cortex (5b) and left inferior temporal cortex (5c). Square points represent average adjusted rCBF values over all subjects for the conjunction task; circular points represent values from the feature tasks. Error bars are one standard error.

References

(1) James, W. (1890:1976) *The Principles of Psychology* , Harvard University Press, Cambridge MA

(2) M.I.Posner (1994) Attention: The mechanisms of consciousness, *Proc. Natl. Acad. Sci. USA*, 91, 7398-7403

(3) Posner, M.I. and Dehaene, S. (1994) Attentional Networks, *Trends in Neurosci* , 17, 75

(4) Corbetta, M., Miezin, F.M., Dobmeyer, S., Shulman, G.L., and Petersen, J. (1991) *J Neurosci* 11, 2383

(5) Desimone, R. and Duncan, J. (1995), Neural Mechanisms of Selective Visual Attention, *Annu. Rev. Neurosci*, 18, 193-222

(6) Fox, P.T. and Raichle, M.E. (1984) Stimulus rate dependence of regional cerebral blood flow in human striate cortex, demonstrated by positron emission tomography, *Journal of Neurophysiology* 51, 1109-20

(7) Price, C., Wise, R., Ramsay, S., Friston, K., Howard, D., Patterson, K. and Frackowiak, R. (1992) Regional response differences within the human auditory cortex when listening to words, *Neurosci Lett* 146, 179-182

(8) Frith, C.D. and Friston, K.J. (1996) The role of the thalamus in 'top-down' modulation of attention to sound. *Neuroimage* in press

(9) Corbetta, M., Miezin, F.M., Shulman, G.L. and Petersen, S.E. (1993), A PET Study of Visuospatial Attention, *J Neurosci*, 13, 1202-1226; Corbetta, M., Shulman, G.M., Miezin, F.M., Petersen, S.E. (1995) Superior Parietal Cortex Activation During Spatial Attention Shifts and Visual Feature Conjunction. *Science* 270, 802-4

(10) Wilson, F.A., Scalaidhe, S.P., Goldman-Rakic, P.S. (1993) Dissociation of object and spatial processing domains in primate prefrontal cortex. *Science*, 260, 1955-8

(11) Fuster, J.M., Bauer, R.H., Jervey, J.P. (1985), Functional interactions between inferotemporal and prefrontal cortex in a cognitive task. *Brain Res*, 330(2), 299-307

(12) Rees, G.E., Frackowiak, R.S.J., and Frith, C.D. (1996) Functional interaction between frontal and inferior temporal cortex during selective visual attention to conjunctions. Submitted

(13) SPM96, Wellcome Department of Cognitive Neurology, London. See http://www.fil.ion.ucl.ac.uk

(14) Friston, K.J. *et al.* (1994) *Human Brain Mapping* 2, 165-189; Friston, K.J. *et al*, (1995)*Human Brain Mapping* 3, 189-210

(15) Vandeberghe, R., Dupont, P., Bormans, G., Mortelmans, L. and Orban, G. (1995) Blood Flow in Human Anterior Temporal Cortex Decreases with Stimulus Familiarity, *Neuroimage* 2, 306-313

(16) Miller, E.K. and Desimone, R. (1994), Parallel neuronal mechanisms for short-term memory *Science*, 263, 520-2

(17) Friston, K.J., Price, C.J. and Buchel, C. (1996). A taxonomy of study design. Available by ftp from http://www.fil.ion.ucl.ac.uk

(18) Raichle, M. (1987), Circulatory and metabolic correlates of brain function in normal humans. In *Handbook of Physiology, The Nervous System V* (Mountcastle, V. and Plum, F., Eds.), pp. 643-674. American Physiological Society, Bethesda, MD

(19) Allport, A. (1993) Attention and Control: Have We Been Asking the Wrong Questions? A Critical Review of Twenty-Five Years. In *Attention & Performance XIV* (Meyer, D.E. and Kornblum, S. Eds), pp.183-218. MIT Press, Cambridge, MA

(20) Allport, A. (1987) in H. Heuer and A.F. Sanders (eds.), *Perspectives on Perception and Action*, Erlbaum, Hillsdale, NJ

Part six:
Appendix

PET CENTERS AND PET PROJECTS IN CENTRAL AND EASTERN EUROPE

"PET in Eastern and Central Europe" was the title and theme of an international conference held under the auspices of the Hungarian PET Foundation in Hungary's PET Center in Debrecen between 23-25 May 1996. The conference attracted over 250 participants from sixteen countries, mainly from the Central and Eastern European region, indicating a strong interest in establishing new imaging centers in this region and introducing PET as a diagnostic and research technique in the fields of medical diagnosis as well as clinical, pharmaceutical and biomedical research.

Existing and prospective PET centers in these countries face extraordinary challenges, for PET, at least at first sight, is an extremely expensive technique. It should, however, be emphasised that both the clinical and research applications of PET prove to be highly cost-effective: Diagnostic PET investigations in properly chosen patient populations, designed to answer well formulated diagnostic questions, cut significantly medical care costs by shortening the time of diagnostic procedures, eliminating the need for other expensive diagnostic techniques, improving diagnostic precision and therapeutic efficacy, by that way contributing to the patients' improved life quality, longer survival time, saver follow-up, etc. On the other hand, research applications of PET can e.g. markedly cut time-to-market in the field of drug development, resulting in significant cost reduction for both the companies and the end-users: the patients.

These advantages of the technique have been widely recognised and significant efforts have been made in several Central and Eastern European countries to establish PET centers. The following overviews, which are the shortened versions of the papers delivered at the aforementioned conference held in Debrecen in May 1996, cannot be complete as new projects sprout. They, however, can provide the reader with a first orientation in the recent situation in the field of PET in Central and Eastern Europe.

Balázs Gulyás

B. Gulyás and H.W. Müller-Gärtner (eds.),
Positron Emission Tomography: A Critical Assessment of Recent Trends, 417-428.
© 1998 *Kluwer Academic Publishers.*

THE DEBRECEN PET CENTRE

Lajos TRÓN
PET Centre, Debrecen University Medical School,
H-4027 Debrecen, Bem tér 18C, Hungary

The first PET centre in Hungary and Central Europe was inaugurated on 26th January, 1994 at the University Medical School in Debrecen (220 km east of Budapest). The Centre, which is located on the premises of Institute for Nuclear Research of the Hungarian Academy of Sciences, is equipped with a MGC 20 cyclotron (proton energy 20 MeV, intensity of the external beam is about 30 μA) and a GE 4096 Plus whole body scanner. The cyclotron is shared with the Institute for Nuclear Research.

Following initial tests and animal experiments, the first patient investigations commenced on the 28th of June 1994. Since that time, during the first two years of the Centre's activity, more than 500 human investigations have been ere performed with FDG. Patients from all over Hungary are entitled to diagnostic PET investigation, provided that the National Board of PET Investigations, appointed by the Ministry of Welfare, approves the case. The National Insurance Fund partially covers the costs of those investigations approved by the Board. Other patients were investigated within the frame of clinical research projects. Most of the above patients came from Hungary, but there were also patients from the abroad.

The patient populations included subjects with intracranial tumours (29 %), extracranial tumours (20 %), epilepsy (18 %), myocardial infarct or angina (11 %), dementia (5 %), a number of miscellaneous neurological cases, Hodgkin lymphoma, systemic lupus erythematosus, etc. In addition, normal healthy volunteers served as a control group.

At the present 9 staff members (physicists, radiochemists, mathematicians and MD's) are engaged with the PET project, while the PET Centre's activities are further supported by staff members of the Institute for Nuclear Research and the Debrecen University Medical School. There are a number of collaborating MD's from the five Hungarian medical universities on each medical project.

Future developments during the next years include the development of a new multi-headed fully automated target system, the introduction of new tracers (15O-water, 13N-ammonia, and 18F-, receptor specific ligands), a more systematic use of quantitative PET investigations, and starting joint projects with pharmaceutical companies.

A certain area of our research activity is closely related to the diagnostic investigations, another field relates to basic research; both application areas fit well into the profile of a PET program. Medical projects cover the special fields listed above. For these planned investigations concerning basic research we need "flow tracers" which can report on the regions of the brain displaying extra activity. The basis of these experiment is that while the brain performs specific tasks, extra energy is needed by those particular parts of the brain that are involved in the mental process in question. This extra need of energy results in an increased blood flow (perfusion) of the region which can be detected by labelled molecules freely diffusing through the tissues. These so called brain activation processes require a special technique allowing the simultaneous visualisation of functional maps of the brain (as it is provided by PET scans) and that of the anatomy of the brain (as it is recorded by CT or MRI). Given the possibility to render the activated regions to well defined anatomic structures a lot of information can be gained about the localisation of the information processing by the central nervous system. This program requires a strong support on behalf of programming and informatics. Fusion of images by different imaging modalities (PET, CT, MRI, etc) is a complex procedure. This problem is solved for the images of our PET camera on the one side and CT and MRI tomographs available for us on the other side using programs freely available on the internet after the necessary modifications and completion to the specific local conditions.

CONTACTPERSON

Lajos Trón
PET Centre, Debrecen University Medical School,
H-4027 Debrecen, Bem tér 18C, Hungary
Phone/Fax: +36-52-431958
Email: tron@atomki.hu

THE PET CENTER AT THE INSTITUTE OF THE HUMAN BRAIN, RUSSIAN ACADEMY OF SCIENCE, ST.PETERSBURG, RUSSIA

Sviatoslav V. MEDVEDEV
The Institute of the Human Brain of the Russian Academy of Sciences
9 Pavlov Street, St.Petersburg, 197376 Russia

The center was opened in August 1991 and is dedicated to clinical studies and fundamental research in psychophysiology.

EQUIPMENT

Cyclotron. Scanditronix MC-17 (for protons 17 MeV, for deuterons 8,6 MeV)

Radiochemistry. The radiochemical facility covers 3 hot cells and RB-86 Anatech robot. Available radiotracers are 18FDG, 11C-metionine, H215O, 13N-NH3. The general concept developmental synthetic strategy of the radiochemical department is creation of tracers fully corresponding to natural metabolites that gives opportunity of correct estimation of kinetic parameters. In the frames of this concept at the present time personnel works on the project directed to development of a new radiopharmaceutical such as pyruvic acid labeled with carbon 11. The promotion of this tracer to clinical practice will enable visualization of hypoxic areas in the brain that is of great importance for patients suffering from brain ischemia in elaboration of their adequate cure. Another project is elaboration of synthesis of L-DOPA labeled with carbon-11 that will give opportunity to determine the status of dopaminergic system in the human brain. The robotics technologies created in this department were successfully promoted in PET centers of the USA, Italy, Canada, The Netherlands, and Taiwan.

PET camera. Scanditronix PC2048-15B.

MAIN TOPICS

Oncology.
Evaluation of glucose and protein metabolism in primary and recurrent tumors.

Neurology.
1. Preoperative morphological and functional brain studies in epilepsy.
2. Assessment of viability of ischemic tissue after stroke and determination of therapeutic strategy.
3. Assessment of viability of damaged tissue after head trauma and determination of therapeutic strategy.
4. Assessment of changes of CBF after electromagnetic therapy.

Cardiology.
Differential diagnosis of myocardiopatia in small children.

Neurophysiology.
1. Brain mechanisms involved in selective attention (in collaboration with Prof.
    R.Naatanen team from Psychological Unit, Helsinki University).
2. Brain mechanisms involved in unconscious attention (in collaboration with Prof.
    R.Naatanen team from Psychological Unit, Helsinki University).
3. Brain maintenance of reading, word and text comprehension.
4. Brain maintenance of semantic processes.
5. Brain maintenance of memory.
(Main findings: The PET correlates of selective attention, mismatch negativity, reading,
    determination the physical and grammatical characteristics of words,
    comprehension of the text content.)

CONTACTPERSON

Dr. Sviatoslav V. Medvedev
Director of the Institute of the Human Brain
of the Russian Academy of Sciences
Head of the Laboratory of Positron Emission Tomography
9 Pavlova St.
St.Petersburg 197376 Russia
Tel. +7 (812) 234 5732; Fax +7 (812) 234 3247
E-mail: medvedev@brain.nw.ru

PET PROJECT IN SLOVENIA

Meta MILCINSKI[1], Budihna NATASA[1], Stanovnik ALES[2], Sket BORIS[3], Staric MARKO[2]

[1]University Medical Centre, Nuclear Medicine, [2]Jozef Stefan Institute, Elementary Physics, [3]Organic Chemistry, University of Lubljana, Slovenia

Slovenia is a small country with a population of 2 million and an area of 20.245 square kilometres. Due to adequate research funding throughout the past years, the country has a low power (250 kW) research nuclear reactor (TRIGA Mark II) in Podgorica, near the capital Ljubljana. Anhydrous ^{18}F-HF (fluoride) can be produced in this reactor by irradiation of ^{6}Li enriched Li_2CO_3 in a neutron flux of 1×10^{13} n cm^{-2}s^{-1} for 6 hours (Fajgelj et al., 1985).

In the early seventies, Professor Marjan Erjavec, head of the Department for Nuclear Medicine at the Institute of Oncology explored the possibility of using positron emitting substances for diagnostic investigations. Together with his collaborators, he collimated a Nuclear Chicago detector system and created a whole-body scanner, with which they performed bone scans using ^{18}F- (37 MBq per investigation). This technique was in use from the early seventies up to 1983.

In 1983 the group labelled glucose with fluoride and produced 18-F-3-deoxy-D-glucose (^{18}F-3-FDG). Reactor produced ^{18}F- was used to displace the trifluoromethanesulphonyl group from 1,2,5,6-diisopropylidene-3-O-triflyl-allose. Acid hydrolysis of the resulting diisopropylidene and 3-deoxy-3-fluoro-D-glucose yields ^{18}F-3-FDG. With this product the group performed experimental and clinical studies on both volunteers and patients with large un-differentiated malignant tumours. These studies confirmed earlier observations on the uptake of labelled glucose: it was intense and fast in the brain, somewhat less in the heart and the liver, and no uptake was found in the tumour tissue (Erjavec et al., 1984). The detector system has been out of function since 1985.

Whereas the detector system has been out of function since 1985, during the eighties a group at the Jozef Stefan Research institute began to develop a multiwire proportional chamber system for positron emission tomography wherein positron sensitive detectors make up two multiwire chambers with 10 micrometer diameter anode wires at 2 mm distance. The cathode planes consist of 1.5 mm wide strips with 5 mm spacing at 2 mm pitch, with anode-to-cathode spacing of only 1.5 mm. The strips are covered with a 0.1 mm thick layer of eutectic mixture of lead (Pb) and bismuth (Bi), which serves as a converter of 511 keV photons. The electron is knocked out of the Bi-Pb layer on the cathode strip either by photoeffect or by Compton scattering of the annihilation photon. The cathode strips are capacitatively coupled to delay lines so that the hit coordinates are obtained by measuring the signal times at both ends of the delay lines relative to the anode signal. The system has a good resolution of 3 mm FWHM,

In 1990 the PET system (Model 4200, Cyclotron Corporation, Berkeley 1977) from the Hannover Medical School was transferred to Ljubljana and established there with the help of Professors Hundeshagen and Gerd Meyer. The re-installation of the PET system was a joint effort University Medical Centre and the Jozef Stefan Institute. The PET system has earlier been run by a PDP 55/11 computer, more recently it is connected to the VAX of the Jozef Stefan Institute. The first images of a guinea pig skeleton were produced in 1994 with Na. Recently, further testing and phantom studies with ^{18}F- are performed. The resolution of the system is in the range of 18-19 mm.

Recent plans include the synthesis of 2-deoxy-D-glucose from reactor produced fluoride. (Maximally available ^{18}F- activity from the reactor is around 2.2 GBq (60 mCi), thus after a successful synthesis 150-220 MBq (4 to 6 mCi) FDG is expected, which will be used for clinical investigations on the available system. Furthermore, the construction of larger multiwire proportional chambers (30 x 30 cm) with improved spatial resolution is in an advanced state. These chambers are used for clinical studies in other centres (Findlay et al., 1996).

Being fully aware with the draw-backs of our old system, we work on its improvement in order to make it more adequate for clinical studies. In the next stage, together with all interested institutions in the country we plan to prepare an application for obtaining sufficient financial support for purchasing a newer PET camera system with clinically acceptable spatial resolution, a dedicated baby cyclotron and automated radiochemistry systems. Until these plans will have been materialised, we are to push the performance of the existing facility to its limits.

Our former experience with PET-systems and radiochemistry has paved the way to the establishment of a dedicated biomedical PET centre. Clearly, the future development of PET in Slovenia has a sound background as far as professional experience and human factors are concerned; the materialisation of a fully-fledged PET project depends now predominantly on governmental or private grant agencies and sponsors.

REFERENCES

Erjavec, M., Fajgelj, A., Guna, F., Novak, J., Sket, B., Zupan, M. Synthesis and clinical experiments with 18-F-deoxy-D-glucose. VIIIth International Symposium on Medical Chemistry. Abstracts. Uppsala, Sweden. 1984:65.

Fajgelj, A., Novak, J., Stegnar, P., Dimic, V., Kuznik, A. Preparation of 18-F-fluoride for labelling organic compounds using a low-power research reactor. Vestn. Slov. Kem. Drus. 21(1985):319-324.

Findlay, M., Young, H., Cunningham, D., Iveson A. Cronin B. Hickish T. Pratt B. Husband J. Flower M. Ott R., Noninvaseive monitoring of tumor metabolism using fluorodeoxyglucose and positron emission tomography in colorectal cancer liver metastases: correlation with tumor response to fluorouracil. J. Clin. Oncol. 14(1996):691-696.

Staric, M., Korbar, D., Stanovnik, A. Tests of a mini positron emission tomograph based on multiwire proportional chambers. Physica Medica 9(1993):219-223.

424

CONTACTPERSON

Metka Milcinski
University Medical Centre, Nuclear Medicine
Zaloska 7
1525 Ljubljana, Slovenia
Tel/Fax: +386 61 132 72 72
Email: META.MILCINSKI@mf.uni-lj.si

PERSPECTIVES OF PET IN POLAND

Leszek KRÓLICKI[1] and Anna GRABOWSKA[2]
[1]Department of Nuclear Medicine and Magnetic Resonance, Medical
School of Warsaw; and [2]Department of Neurophysiology, Nencki
Institute of Experimental Biology, Warsaw, Poland

In recent years PET techniques have become a very useful tool for a non-invasive imaging of biological processes in humans. In Poland, as in many other countries both medical and scientific communities undertake activities aiming at establishing a PET center. As we are not very advanced in those affords, the experience gained by Debrecen PET Center is of high interest to us. The two meetings organized by Prof. B. Gulyas in Debrecen during 1996 gave us a chance to get acquainted both with the PET technique itself and with its application in medicine and neuroscience. It also provided us with better understanding of what problems may arise at the stage of establishing such a center and during its work. Those two meetings were of high importance to us as they created a forum for exchanging ideas, discussing problems and seeking the ways of their solving.

In Poland there are two communities that form a lobby for introducing PET method. These are: Polish Society for Nuclear Medicine and Polish Neuroscience Society. Most people agree that PET will be a method of choice in future years and feel the necessity of introducing it also in our medical examinations and research. Some argue, however, that the method is very expensive and we should improve first the very poor condition of more basic nuclear medicine techniques. There are about 60 gamma-cameras in Poland (this result in an index of 1.3 gamma-cameras per 1 million people). Most of the units are old, some of them are over 10-15 years old. The total number of nuclear medicine examinations is about 130 000 per year. This means that we can perform just 4 examinations per 1000 people. Certainly, this index is not satisfactory. There are 10 MR apparatus. Thus, one apparatus serves to 4 million people. There are also 2 cyclotrons (in Cracow and Warsaw). They produce a limited number of of isotopes and radiopharmaceuticals for medical purposes. On the other hand, we have a group of well educated doctors, physicists and chemists who are ready to develop PET in our country.

Our neuroscience community is also very much interested in introducing PET technique to scientific reasearch. Neuroscience in Poland has a very good and long tradition. Polish Neuroscience Society forms a discussion platform for specialists representing various branches of neuroscience. Among them there are researchers who are familiarized withneuroimaging techniques. They gained experience in various brain research institutions all over the world.

We discussed extensively in the Ministry of Health the necessity of development of PET in Poland. As the total cost of the equipment is very high it has to be covered by several different sources: Ministry of Health, National Committee for

Scientific Research and foundations supporting Science. Another problem is the daily maintenance of the Center. According to our plans the National Committee for Scientific research will be the main supporting institution. The founds will be awarded by a special grant system supporting best projects submitted to the Committee. Participation in international programs could be another source. Commercial activity is also taken into consideration but it can be used only to a limited extend.

In summary, there is a good ground for establishing PET Center in Poland. Both medical and neuroscience communities fully understand importance of introducing that technique to achieve a progress in medicine and neuroscience. At present the atmosphere around the realization of such a project is favourable and we attracted the interest of Ministry of Health and other central institutions to this program. We hope, it will result in a concrete financial support in near future. There is a sufficient number of enthusiasts of the PET method in Poland to believe that we succeed.

CONTACTPERSON

Anna Grabowska Ph.D.
Nencki Institute of Experimental Biology
Department of Neurophysiology
e-mail: grabow@nencki.gov.pl

THE SLOVAK REPUBLIC'S CYCLOTRON FACILITY PROJECT

Árpád DUKA-ZÓLYOMI[1], Ján RUŽIČKA[1] and Izabela MAKAIOVÁ[2]
[1]*Department of Nuclear Physics, Comenius University, and* [2]*Oncological Institute of St. Elisabeth, Clinic of Nuclear Medicine, Medical Faculty, Bratislava, Slovakia*

After several failed attempts to build a cyclotron facility in the Slovak Republic, by the end of year 2000 a dream of several institutions in the country (including the Slovak Academy of Sciences, universities and health care institutions) will have been materialised.

The Slovak cyclotron project is supported by the government of the Slovak Republic and will also gain support from IAEA grants. The cyclotron laboratory will be located on the premises of the Slovak Institute of Metrology in Bratislava. This institute has already a long time experience with a battery of different radiation sources. The isochronous cyclotron will be constructed in the Joint Institute for Nuclear Research (JINR) at Dubna. The construction will be financed by the Russian Federation as a part of down-payment of its debt to Slovakia.

As the cyclotron will serve several purposes, it is designed to yield maximum beam energy and intensity and a wide diapason of accelerated ions within the limits of the existing financial and constructional possibilities.

The basic technical parameters of the cyclotron will be as follows:
- dee diameter: 200 cm
- maximal energy of the proton beam: 75 MeV with maximal intensity of 100 μA
- acceleration of protons, deuterons, etc. up to Xe ions
- availability of neutron beam for neutron therapy and metrology.

The projected main activities of the cyclotron laboratory will include:
- production of radionuclides for nuclear medicine (including positron emitters)
- production of neutrons and protons for medical therapy
- basic standards for metrology of ionising radiation
- applied research and development projects
- basic and applied research in nuclear physics and nuclear techniques
- educational programs.

In addition to the cyclotron as a basic facility, other technical facilities are expected to be installed in the laboratory, providing the isotope repertoire of nuclear medicine diagnostics with 123I and 81Rb isotopes as well as biogenic positron emitters (18F, 11C, 15O, 13N), with an eye on the future establishment of a positron emission tomography center.

CONTACTPERSONS:

Árpád Duka-Zólyomi
Department of Nuclear Physics,
Faculty of Mathematics and Physics,
Comenius University,
Mlynská dolina F1
842 15 Bratislava, Slovakia
Phone: +42 7 724000
Fax: +42 7 497 877
Email: egyutt@netlab.sk

Ján Ružička
Department of Nuclear Physics,
Faculty of Mathematics and Physics,
Comenius University,
Mlynská dolina F1
842 15 Bratislava, Slovakia
Phone: +42 7 724000
Fax: +42 7 425882
Email: jan.ruzicka@fmph.uniba.sk

Izabela Makaiova
Oncological Institute of St. Elisabeth
Clinic of Nuclear Medicine
Medical Faculty UK
Heydukova 10,
812 50 Bratislava
Phone: +42 7 3849 111
Fax: +42 7 323 711
Email: imaka@ousa.sk

PET CENTRES AROUND THE WORLD

Compiled by
Balázs GULYÁS
Department of Neuroscience. Karolinska Institute,
S-171 77 Stockholm, Sweden

EUROPE

Detailed information about the European PET Centers is available at the following web site:
http://www.neuro.ki.se/neuro/pet/europet.html

Austria

University of Vienna - AKH, Department of Cardiology

Address: Waehringer Gürtel 18-20, A-1090 Vienna
Phone: +43-1-404004616
Fax: +43-1-4081148
Contactpersons: Dr. Heinz Sochor, Dr. Gerold Porenta

Belgium

University Hospital, PET Center UZ/RUG

Address: Proeftuinstraat 86, B-9000 Ghent
Phone: +32 91 40 30 31
Fax: +32 91 40 4 991
Email: neuropet@rug.ac.be
Web site: www.admin.rug.ac.be
Contactpersons: Dr. J. De Reuck, Acting Director,
           J. Stryckmans, Secretary

PET / Biomedical Cyclotron Unit, Universite Libre de Bruxelles, Hopital Erasme

Address: 808 route de Lennik, B-1070 Brussels
Phone: +32 2 256 3111

430

Fax: +32 2 555.4701
Web site: www.ulb.ac.be/medecine/pet/
Contactperson: Serge Goldman, M.D., Acting Director
                 (phone: +32 2 555.53.72; email: sgoldman@ulb.ac.be)

Unité de Tomographie par Positrons, Université Catholique de Louvain,
Faculté de Médicine

Address: Chemin du Cyclotron 2, B-1348 Louvain-la-Neuve
Phone: +32 10 47 28 22 or +32 2 764 10 44
Fax: +32 10 45 21 83 or +32 2 764 36 97
Email: michel@topo.ucl.ac.be
Web-site: www.topo.ucl.ac.be
Contactpersons: Jacques Melin M.D., Professor,
                 Christian Michel

K.U.Leuven, U. Z. Gasthuisberg, PET Centre / Nuclear Medicine

Address: Herestraat 49, B-3000 Leuven
Phone: +32 16 34 37 14
Fax: +32 16 34 37 59
Email: Luc.Mortelmans.@uz.kuleuven.ac.be
Contactperson: Dr. Luc Mortelmans, Professor, Director

Centre Hospitalier Universitaire de Liège, Service de Médicine Nucléaire

Address: Sart Tilman, B 35, B-4000 Liège
Phone: +32 41 66 72 00
Fax: +32 41 66 82 57
Web-site: www.ulg.ac.be
Contactperson: Dr. Pierre Rigo

Denmark

Aarhus General Hospital, Positron Emission Tomography Center

Address: 44 Norrebrogade, Aarhus C, DK-8000
Phone: +45 8949 3030
Fax: +45 8949 3020
Email: albert@tiger.pet.akh.arhusamt.dk
Web site: panter.soci.aau.dk
Contactperson: Dr. Albert Gjedde, Professor

PET Units at Rigshospitalet, Copenhagen University Hospital:

Division of Neurobiology
Contactperson: Olaf B. Paulson M.D., Professor

Division of Cardiology,
Contactperson: Henning Kelbæk, M.D., Chief Physician

Division of Clinical Physiology and Nuclear Medicine,
Contactperson: Lars Friberg, M.D., Chief Physician

Address: Blegdamsvej 9, DK-2100 Copenhagen
Phone: +45 35 456 711
Fax: +45 3545 6713
Email: paulson@pet.rh.dk
Web site: neuro.pet.rh.dk
Contactperson: Olaf B. Paulson M.D., Professor

Finland

University of Helsinki, Department of Radiochemistry

Address: P.O.Box 5, FIN-00014, Helsinki
Phone: +358 9 191 49 141
Fax: +358 9 191 49 121
Contactperson: Dr. T. Jaakkola

Turku PET Center, Turku University Central Hospital

Address: FIN-20520 Turku
Phone: +358-2-2611611 or +358-2-2612772 (secretary)
Fax: +358-2-2318191
Web site: www.utu.fi/med/pet
Contactperson: Juhani Knuuti, M.D., Ph.D., Director
                 (phone: +358-2-2612842, fax: +358-2-2318191,
                 email: jknuuti@utu.fi)

France

CERMEP, Hopital Neurocardiologique

Address: 59 Boulevard Pinel, F-69003 Lyon
Phone: +33 7268 8600
Fax: +33 7268 8610
Email: comar@univ-lyon1.fr
Web-site: www.univ-lyon1.fr
Contactperson: Dominique Comar Ph.D.

Centre de Cyceron

Address: Bld. Henri Becquerel, B. P. 5229, F-14074, Caen Cedex
Phone: +33 2 31 47 02 00
Fax: +33 2 31 47 02 22
Email: derlon@cyceron.fr
Web-site: www.cyceron.fr/General/sommaire.html
Contactperson: Dr. J. M. Derlon, Director

Service Hospitalier Frederic Joliot, CEA-DSV, Hospital d'Orsay

Address: F-91406 Orsay
Phone: +33 1 69 86 77 02
Fax: +33 1 69 86 77 68
Contactperson: Dr. A. Syrota

Germany

Aachen PET Center, University of Technology, Department of Nuclear Medicine

Address: Pauwelsstrasse 30, D-52057 Aachen
Phone: +49 241 80 88 740
Fax: +49 241 88 88 424
Email: buell@nuk-gate.nukmed.rwth-aachen.de
Web-site: www.imib.rwth-aachen.de/NUKLEAR/gwag1.html
Contactpersons: U. Büll, O. Sabri, U. Cremenius, H. J. Kaiser

Herz und Diabeterzentrum NRW, Universitätsklinikum der Ruhr Universität Bochum, Institut für Molecular Biology, Radiopharmazie, Bochum

Address: Gerogstrasse 11, D-32545 Bad Oeynhausen
Phone: +49 5731 97 1865
Fax: +49 5731 97 2300
Contactpersons: Dr. G. Notohamiprodjo, Dr. Zeilster

Charité-Virchow University Hospital, Humboldt University, Klinik und Poliklinik für Nuklearmedizin, PET Center, Berlin

Address: Schumannstrasse 20/21, D-10098 Berlin
         Augustenburger Platz 1, D-13353 Berlin
Phone: +49 30 2802-3685
Fax: +49 30 2802 2735
Email: ivancev@rz.charite.hu-berlin.de
Contactpersons: D. L. Munz, M.D., Professor and Chairman,
                Dr. V. Ivancevic, M.D., Head Physician,
                Clinic for Nuclear Medicine, University Hospital Charité

Praxis Dr. Ruhlmann, Dr. Kozak, Bonn

Address: Cassius Bastei, Muenstr. 20, D-53111 Bonn
Phone: +49 228 985 280
Fax: +49 228 696 778
Contactpersons: Dr. Rujlmann, Dr. Kozak, Dr. Biersack

PET Center, Rossendorf Institut für Bioorganische Chemie, Dresden

Jointly run by:
Institute of Bioinorganic and Radiopharmaceutical Chemistry
(Director: Prof. B. Johannsen), Research Center Rossendorf and
Clinic and Polyclinic of Nuclear Medicine (Director: Prof. W.-G. Franke),
University Hospital, Dresden University of Technology

Address: Postfach 510119, D-01314 Dresden
Phone: +49 351 260 3170
Fax: +49 351 260 3232
Email: Johannsen@fz-rossendorf.de
Web site: www.fz-rossendorf.de
Contactpersons: B. Johanssen (address as above) and
                W.-G. Franke (Klinik und Polyklinik für Nuklearmedizin,
                Universitätsklinikum Carl Gustav Carus, Fetscherstr. 74,
                D-01307 Dresden, phone: +49 351 458 4160)

Department of Nuclear Medicine, Heinrich-Heine-Universität Düsseldorf

Address: Moorenstrasse 5, D-40225 Düsseldorf
Phone: +49 211 81 18 540
Web-site: www.nuk.uni-duesseldorf.de
Contactperson: Prof. Dr. Hans W. Müller-Gärtner

Röntgeninstitut Düsseldorf

Address: Kaiserwerther Strasse 89, D-40476 Düsseldorf
Phone: +49 211 49 66 500
Web-site: www.roentgeninstitut.de
Contactpersons: Dr. J. Betzner, Dr. D. Gahlen, Dr. W. Stork, Dr. A. Schomburg, Dr.
                E. Wannow, Dr. H.-M. Klein

University Clinic Essen, Clinic and Policlinic for Nuclear Medicine

Address: Hufelandstrasse 55, D-45122 Essen
Phone: +49 201 723 2032
Fax: +49 201 723 5964
Web site: www.uni-essen.de/nukmed
Contactperson: Dr. Andreas Bockisch

Universität Freiburg, Radiologische Universitätsklinik, Abteilung Nuklearmedizin

Address: Hugstetterstrasse 55, D-79106 Freiburg
Phone: +49 761 270 3916
Fax: +49 761 270 3916
Contactperson: Dr. E. Moser, Dr. Nitzsche, Dr. P. Brautigam

EURO-PET GmbH, Freiburg

Address: Schwabentorplatz 6, D-79078 Freiburg
Phone: +49 761 36 300
Fax: +49 761 34 051
Contactperson: Dr. P. Reuland, Dr. H. Weyer

Klinik für Nuklearmedizin, Klinikum der Johann Wolfgang Goethe-Universität, Frankfurt am Main

Address: Theodor-Stern-Kai 7, D-60590 Frankfurt am Main
Phone: +49 69 6301 6783
Fax: +49 69 6301 7143
Web-site: www.rz.uni-frankfurt.de/FB/fb19/nuclmed/
Contactpersons: Dr. G. Hoer, Dr. Scherer

Universitätsklinikum Hamburg-Eppendorf, Radiologische Klinik und Strahleninstitut, Hamburg

Address: Martinistr. 52, D-20251 Hamburg
Phone: +49 47 17 54 42
Fax: +49 47 17 67 75
Contactpersons: Dr. Hotze, Dr. Lubeck, Dr. Buchert

Krankenhaus St. Georg, Hamburg, Fachabteilung für Nuklearmedizin, Hamburg

Address: Lohmuhlerstr. 5, D-20099 Hamburg
Phone: +49 40 2488 3819
Fax: +49 40 2488 2275
Contacpersons: Dr. Leisner, Dr. Schulze Franke, Dr. Rust

Medical School of Hannover, Department of Nuclear Medicine and Special Biophysics

Address: Konstanty-Gutschow-Strasse 8, D-30623 Hannover
Phone: +49 511 532 2580
Fax: +49 511 532 2598
Email: Geerds-J-Meyer-MHH-PET@t-online.de
Contactperson: Prof. Dr. Geerd-J. Meyer
Phone: +49 511 532 2580

Fax: +49 511 532 2598
Email: Geerds-J-Meyer-MHH-PET@t-online.de
Contactperson: Prof. Dr. Geerd-J. Meyer

Department of Radiology, German Cancer Research Center (dkfz)

Address: Im Neuenheimer Feld 280, D-69120 Heidelberg
Phone: +49 6221 42 25 50
Fax: +49 6221 42 25 72
Web-site: www.dkfz-heidelberg.de/abt0520/
Contactpersons: Prof. Dr. W. J. Lorenz, Acting Director
               Prof. Dr. G. van Kaick, Clinical Director
               Priv.-Doz. Dr. G. Brix, Leading Physicist,
               (email: g.brix@dkfz-heidelberg.de)
               Dr. Uwe Haberkorn (email: U.Haberkorn@dkfz-heidelberg.de)

Universitötskliniken des Saarlandes (Homburg), Radioklinische Klinik, Abt.
Nuklearmedizin

Address: Oscar Orthstr., D-66421 Homburg (S.)
Phone: +49 6841 162 201
Fax: +49 6846 164 692
Contactperson: Dr Kirsch

PET Laboratory, Institut für Medizin, Forschungszentrum GmbH Jülich

Address: Leo-Brandt-Strasse, D-52428 Jülich
Phone: +49 2461 61 ext 5913
Fax: +49 2461 61 ext 2770
Email: h.herzog@fz-juelich.de
Web-site: ime-web.ime.kfa-juelich.de
Contactperson: Dr. Hans W. Müller Gärtner, Director

Kernforschunszentrum Karlsruhe

Address: Pf. 3640, D-76021 Karlsruhe
Phone: +49 721 782 2433
Fax: +49 721 782 3156
Contactpersons: Dr. H. Schweikert, Dr. V. Bechtold

Max-Planck-Institut für Neurologische Forschung

Address: Gleueler Str. 50, D-50931 Köln
Phone: +49-221-4726-220
Fax: +49-221-4726-298
E-Mail: Klaus.Wienhard@pet.mpin-koeln.mpg.de
Contactpersons: Prof. Dr. Wolf-Dieter Heiss, Prof. Dr. Klaus Wienhard

436

Nuklearmedizinische Klinik, Klinikum rechts der Isar der TU München

Address: Ismaninger Strasse 22, D-81 675 München
Phone: +49 89 4041 2971
Fax: +49 89 4041 4841
Email: tbb03as@sunmail.lrz-muenchen.de
Web-site: www.nuk.med.tu-muenchen.de
Contactpersons: Dr. Markus Schwaiger M.D., Professor, Director
                 Dr. G. Stöcklin, Dr. Avril, Dr. Bartenstein, Dr. Wolf

Klinikum der LMU, Klinik und Polyklinik für Nuklearmedizin, München

Address: Marchioni Strasse 15, D-81 377 München
Phone: +49 89 7095 4611
Fax: +49 89 7095 4648
Contactpersons: Dr. K. Hahn, Dr. K. Tatsch

Onkologische Praxis im Elisenhof, München

Address: Elisenhof, 6th floor, Prielmayerstr. 1, D-80355 München
Phone: +49 89 595 191
Fax: +49 89 550 4242
Contactperson: Dr. W. Abenhardt

Klinik und Polyklinik für Nuklearmedizin Münster

Address: Albert Schweitzerstr. 33, D-48149 Münster
Phone: +49 251 837 362
Fax: +49 251 837 383
Contactpersons: Dr. O. Schober, Dr. W. Bandau

Praxis Dr. Klott, Dr. Reinhardt, Dr. Schilling, Dr. Hanke, Stuttgart

Address: Seelbergstrasse 11, D-70372 Stuttgart-Bad Cannstatt
Phone: +49 711 553 820
Fax: +49 711 551 480
Contactpersons: Dr. C. J. Klott, Dr. Hanke, Dr. Feine

Katherinenhospital Stuttgart

Address: Kriegsbergstrasse 60, D-70174 Stuttgart
Phone: +49 711 278 4300
Fax: +49 711 278 4309
Contactperson: Dr. Bihl

University of Tübingen, Department of Nuclear Medicine

Address: Röntgenweg 113, D-7276 Tübingen
Phone: +49 7071 292 179
Fax: +49 7071 296 554
Web-site: www.uni-tuebingen.de
Contactpersons: Dr. Ulrich Feine, Dr. R. Bares, Dr. H. J. Machulla

Abteilung III (Nukearmedizin), Radiologische Klinik und Poliklinik, Universität Ulm

Address: Robert-Koch-Strasse 8, D-89070 Ulm
Phone: +49 731 502 4500
Fax: +49 731 502 4503
Email: sven.reske@medizin.uni-ulm.de
Web-site: www.uni-ulm.de/klinik/radklinik/rad3/
Contactperson: Dr. S. N. Reske

Praxis Dr. Spizt, Dr. Benz, Stadt Horst Schmidt Kliniken, Wiesbaden

Address: L. Erhardtstr. 100, D-65199 Wiesbaden
Phone: +49 611 432 323
Fax: +49 611 432 322
Contactpersons: Dr. P. Benz, Dr. J. Spitz, Dr. Jauch

Onkologisches Zentrum Wupperthal

Address: Morianstr. 27, D-42103 Wupperthal
Phone: +49 202 24 890
Fax: +49 202 24 299
Contactperson: Dr. Martin

Hungary

PET Centre, Debrecen University Medical School

Address: H-4012 Debrecen, Nagyerdei körút 98
Phone: +36 52 431 958
Fax: +36 52 431 958
Email: tron@atomki.hu, galuska@ibel.dote.hu
Web site: petindigo.atomki.hu
Contactperson: Dr. Lajos Trón, Director, Dr. László Galuska, Co-director

Italy

Instituto di Fisiologia Clinica, C.N.R.

Address: Via Savi, 8, I-56100 Pisa
Phone: +39 50 59 80 64
Fax: +39 50 50 23 54
Email: ldonato@po.ifc.pi.cnr.it
Web site: www.ifc.pi.cnr.it/
Contactperson: Luigi Donato, Professor, Acting Director

Centro Ciclotrone/PET, Cattedra di Medicina Nucleare dell'Università, Instituto di Neuroscienze e Bioimmagini del C.N.R., Instituto Scientifico H. S. Raffaele

Address: Via Olgettina, 60, I-20132 Milano
Phone: +39 2 215 3056/2643.2716
Fax: +39 2 2640390
Email: fazio@mednuc.hsr.it
Web-site: www.hsr.it/sciee.html
Contactperson: Dr. Feruccio Fazio M.D., Professor, Director

Instituto Nazionale per lo Studio e la Cura dei Tumori, Milano

Address: Via Venezian 1, I-20133 Milano
Phone: +39 2 239 0220
Fax: +39 2 236 7874
Contactpersons: Dr. Bombardieri, Dr. Flavio Crippa, Dr. Claudio Pascali

Universita di Roma "La Sapienzia", Dipartimento di Scienze Neurologiche

Address: Viale dell'Universita, 30, I-00185 Roma
Phone: +39 6 4457376
Fax: +39 6 445 4907
Contactperson: Dr. Gian Luigi Lenzi

Centro per la Medicina Nucleare, Consiglio Nazionale delle Ricerche

Address: Via S. Pansini 5, I-80131 Napoli,
Phone: +39 81 7463560
Fax :   + 39 81+ 5457081
Email: Alfanobr@unina.it
Contactperson: Bruno Alfano

University of Padova, Department of Physics

Address: Via Roma 4, I-35020 Padova
Phone: +39 49 806 8482
Fax: +39 49 64 19 25
Contactpersons: Dr. Mochini, Dr. Giron

The Netherlands

University Hospital PET Center

Address: P.O.Box 30.001, 9700 RB Groningen
Phone: +31 50 361 33 11
Fax: +31 50 361 1687
Email: w.vaalburg@pet.azg.nl
Contactperson: Willem Vaalburg

Russia

Institute of the Human Brain of the Russian Academy of Sciences

Address: 9 Pavlova St., 197376 St.Petersburg
Phone: +7 812 234 5732
Fax: +7 812 234 3247
E-mail: medvedev@brain.nw.ru
Contactperson: Dr. Sviatoslav V. Medvedev, Director

Institute of Radiology, St. Petersburg

Phone: +7 812 437 5866
Fax: +7 812 437 5600
Contactpersons: Professor Fiontine, Director of the Institute
Dr. Nickolai Konstenikov, Chief of the PET Group

Bakulev Institute of Cardiac Surgery, Russian Academy of Medical Science

Address: Leninsky prospect 8, 1117049 Moscow
Phone: +7 95 236 92 83
Fax: +7 95 237 78 56
Contactpersons: Professor Leo A. Bockeria, Director of the Institute
Dr. Irackli Aslanidis, M.D., Ph.D., Chief of the PET Group

Spain

Centro PET Complutense, Madrid
Address: Bartolome Cosi s/n Madrid
Contactperson: Dr. Carreras

Clinica Universitaria, Pamplona
Address: Avenida Pio XII 36, E-31008 Pamplona
Contactperson: Dr. Richter

Sweden

Uppsala University PET Center
Address: S-171 85 Uppsala
Phone: +46 18 18 33 81
Fax: +46 18 18 33 90
Contactperson: Dr. Bengt Långström, Professor

PET Units at the Karolinska Hospital / Karolinska Institute:

Division of Human Brain Research, Department of Neuroscience, Karolinska Institute
Address: S-171 77 Stockholm
Phone: +46 8 728 77 84
Fax: +46 8 30 90 45
Email: per.roland@neuro.ki.se, balazs.gulyas@neuro.ki.se
Web site: pet.neuro.ki.se
Contactpersons: Per E. Roland, Balázs Gulyás

Department of Psychiatry, Karolinska Hospital
Address: S-171 76 Stockholm
Phone: +46 8 5177 4445
Fax: +46 8 34 65 63
Email: larsf@psyk.ks.se, kristerh@psyk.ks.se
Web site   www.ki.se/cns/PsychiatrySection
Contactpersons: Lars Farde, Christer Halldin

Department of Neuroradiology, Karolinska Hospital
Address: S-171 76 Stockholm
Phone: +46 8 5177 4896
Fax: +46 8 33 94 12
Email: kaje@neuro.ks.se
Contactperson: Kaj Ericson

Department of Clinical Neurophysiology, Karolinska Hospital

Address:	S-171 76 Stockholm
Phone:	+46 8 5177 5134
Fax:	+46 8 33 99 53
Email:	martini@neuro.ks.se
Web site:	hem1.passagen.se/martini1
Contactperson:	Martin Ingvar

The PET Clinic, Department of Radiation Physics, University Hospital of Lund

Address:	S-22185 Lund
Phone:	+46 46 173116
Fax:	+46 46 127249
Email:	Sven-Erik.Strand@radfys.lu.se
Contactperson:	Sven-Erik Strand

Switzerland

Hopital Cantonal Universitaire, Division de Medicine Nucleaire

Address:	CH-1200 Geneve, Switzerland
Phone:	+41 22 70 27 145
Contactperson:	Alfred Donath

PSI PET-program, Paul Scherrer Institut

Address:	CH-5234 Villigen
Phone:	+41 56 310 36 83
Fax:	+41 56 310 31 32
Email:	leenders@psi.ch
Web site:	www.psi.ch/
Contactpersons:	K. L. Lenders M.D., Director
	Paul Maguire Ph.D.

Geneva University Hospital, Division of Nuclear Medicine

Address:	CH-1211 Geneva 14
Phone:	+41 22 37 27 150
Fax:	+41 22 37 27 169
Email:	slosman@cmu.unige.ch
Web-site:	dmnu-pet5.hcuge.ch
Contactperson:	Daniel O. Slosman M.D., Ph.D., Professor

442

Division of Nuclear Medicine, Department of Medical Radiology, University Hospital

Address: Rämistrasse 100, CH-8091 Zürich
Phone: +41 1 255 25 80
Fax: +41 1 255 44 14
Email: vonschulthessr@dmr.usz.ch
Web-site: www-usz.unizh.ch/PET/E/index.html
Contactperson: Prof. Dr. Gustav K. von Schulthess, Director

United Kingdom

Institute of Neurology, Wellcome Department of Cognitive Neurology, Leopold Müller Functional Imaging Laboratory

Address: Queen Square, London, WC1N 3BG
Phone: +44 171 837 3611
Fax: +44 171 813 1420
Email: r.frackowiak@fil.ion.bpmf.ac.uk
Web site: www.fil.ion.bpmf.ac.uk
Contactperson: Richard S. J. Frackowiak, Professor

The Clinical PET Centre, Guy's and St. Thomas' Hospital, Lambeth Wing

Address: London SE1 7EH
Phone: +44 171 922 8106
Fax: +44 171 620 0790
Emaila: p.marsden@uM.D.s.ac.uk,
              m.maisey@uM.D.s.ac.uk,
              margd@gerbil.uM.D.s.ac.uk
Web site: http://www-pet.uM.D.s.ac.uk
Contactpersons: M N Maisey, Director, Professor,
                 P K Marsden, Senior Scientist,
                 Margaret Dakin, manager

Department of Biomedical Physics and Bioengineering, University of Aberdeen

Address: Foresterhill, Aberdeen AB9 2ZD
Phone: +44 1224 681818 Ext 52721
Fax: +44 1224 685645
Email: a.welch@biomed.abdn.ac.uk
Web Site: www.biomed.abdn.ac.uk/Research/PET/
Contactperson: Andy Welch Ph.D.

Medical Research Council, Cyclotron Unit, Hammersmith Hospital

Address: Du Cane Road, London W12 0NN
Phone: +44 181 383 3162
Fax: +44 181 383 2029/1783
Email: tjones@cu.rpms.ac.uk
Contactpersons: Terry Jones, Professor, Head of Cyclotron Unit and
Head of PET Methodology
David Brooks, Professor, Head of PET Neurosciences
Paulo Camici, Professor, Head of PET Cardiology
Dr Pat Price, Head of PET Oncology
Dr Vic Pike, Head of PET Radiochemistry
Dr Paul Grasby, Head of PET Psychiatry

Guy's Hospital, Radiological Sciences

Address: London SE1 9RT
Phone: +44 171 955 4531
Fax: +44 171 955 4532
Contactperson: Michael N. Maisey

ICR/RMH PET Centre, The Institute of Cancer Research, Physics Department,
Royal Marsden Hospital

Address: Sutton SM2 5PT
Phone: +44 181 642 6011
Fax: +44 181 643 3812
Email: bob@icr.ac.uk
Web site: icr.ac.uk
Contactperson: Robert (Bob) J. Ott

Wolfson Brain Imaging Centre, University of Cambridge, School of Clinical Medicine

Address: University of Cambridge, Box 65, Addenbrooke's Hospital,
Cambridge, UK, CB2 2QAQ
Phone: +44 1223 331 823
Fax: +44 1223 331 826
Web-site: www.wbic.cam.ac.uk
Contactpersons: Dr. L. Hall, Dr. John Clark, Dr. John Pickard

Positron Imaging Centre, University of Birmingham

Phone: +44 121 414 72 1311
Web-site: www.birmingham.ac.uk/physics/ap/pet.html

ASIA

Japan

National Institute of Radiological Sciences,
Division of Advanced Technology for Medical Imaging

Address: 4-9-1 Anagaw, Inage-ku, Chiba-shi, Chiba-ken 263
Phone: +81 43 251 2111
Fax: +81 43 251 7147
Email: suhara@nirs.go.jp (Tetsuya Suhara)
Contactpersons: Yasuhito Sasaki (Director,M.D.),
               Kazutoshi Suzuki (Section Head,Ph.D. chemist),
               Katuya Yoshida (section Head, M.D. cardiologist),
               Tetsuya Suhara (senior researcher, M.D. psychiatrist)

Gunma University Hospital

Address: Maebashi, Gunma 371
Phone: +81 27 220 8400
Fax: +81 27 220 8409
Email: endo@sb.gunma-u.ac.jp
Contactperson: Keigo Endo

International Medical Center of Japan

Address: 1-21-1 Toyama, Shinjuku-ku, Tokyo
Phone: +81 3 3202 7181
Fax: +81 3 3207 1038
Email: thara@t3.rim.or.jp
Contactperson: Toshihiko Hara

Akita Research Institute of Brain and Blood Vessels

Address: 6-10 Senshu-Kubota Machi, Akita 010
Phone: +81 188 33 0115
Fax: +81 188 33 2104
Email: hatazawa@akita-noken.go.jp
Web site: www.akita-noken.go.jp
Contactperson: Jun Hatazawa

Kanazawa Cardiovascular Hospital

Address: 16 Ha, Tnakmachi, Kanazawa-shi, Ishikawa-ken
Phone: +81 762 53 8000
Fax: +81 762 53 0008
Contactperson: K. Hayakawa

Tohoku University Cyclotron and Radioisotope Center

Address: Aramaki Aoba-ku, Sendai 980-77
Phone: +81 22 217 7797
Fax: +81 22 263 5358
Contactperson: Tatsuo Ido

National Cardiovascular Center

Address: 5-7-1 Fujishiro-dai, Suita, Osaka 565
Phone: +81 6 833 5012
Fax: +81 6 872 7486
Email: yishida@hsp.ncvc.go.jp
Contactperson: Yoshio Ishida

Biomedical Imaging Research Center, Fukui Medical School

Address: 23 Shimoaizuki, Matsuoka-cho, Yoshida-gun, Fukui 910-11
Phone: +81 776 61 3111
Fax: +81 776 61 8137
Email: yonekura@fmsrsa.fukui-med.ac.jp
Contactperson: Yoshiharu Yonekura, M.D., Ph.D., Professor

Nagoya University Hospital

Address: 65 Tsurumaicho, Showaku, Nagoya 466
Phone: +81 52 744 2442
Fax: +81 52 744 2444
Email: katsuki@tsuru.med.nagoya-u.ac.jp
Contactpersons: Katsuki Ito M.D., Ph.D. (address as above),
                 Takeo Ishigaki M.D., Ph.D. (Professor and Chairman of
                 Department of Radiology, School of Medicine,
                 Nagoya University) (phone: +81-52-744-2327,
                 fax: +81-52-744-2335, email: i45265a@nucc.cc.nagoya-u.ac.jp)

National Institute of Longevity Science

Address: 36-3, Gengo, Moriokacho, Oobu, Aichi 474
Phone: +81-562-46-2311
Fax: +81-562-48-2373
E-mail: kito@nils.go.jp
Contactperson: Kengo Ito M.D., Ph.D.

Nagoya Rehabilitation Center

Address: 1-2 Mikanyama, Yatomi-cho, Mizuho-ku, Nagoya
Phone: +81 52 835 3811

Fax: +81 52 835 3745
Contactpersons: Noriyuki Kato or Akihiko Iida

Kyoto University Faculty of Medicine,
Department of Nuclear Medicine and Diagnostic Imaging

Address: 54 Kawaharachou, Shougoin, Sakyou-ku, Kyoto 606
Phone: +81 75 751 3760
Fax: +81 75 771 9709
Email: jkonishi@kuhp.kyoto-u.ac.jp
Web site: http://www-dnm.kuhp.kyoto-u.ac.jp/
Contactperson: Junji Konishi

National Center Hospital for Mental, Nervous and Muscular Disorders,
National Center of Neurology and Psychiatry

Address: 4-1-1 Ogawahigishichou, Kodaila-shi, Tokyo 187
Phone: +81 423 41 2271
Fax: +81 423 44 6745
Email: VX8H-MTD@asahi-net.or.jp
Contactperson: H. Matsuda

Nikko Memorial Hospital Clinical PET Center

Address: 1-5-13 Shintomicho, Muroran, Hokkaido 051
Phone: +81 143 24 1331
Fax: +81 143 24 1064
Email: nhskenky@nikkomhp.dp.u-netsurf.or.jp
Contactperson: Akio Nishimura, CEO/President

Osaka City University Medacal School, Division of Nuclear Medicine

Address: I, Asahimachi, Abenoku, Osaka 545
Phone: +81 6 645 2196
Fax: +81 6 646 0686
Contactperson: Hironobu Ochi M.D., Professor

University of Tokyo Faculty of Medicine, Department of Radiology

Address: 7-3-1 Hongo Bunkyo-ku, Tokyo 113
Phone: +81 3 3815 5411
Fax: +81 5689 8218
Contactperson: Yasuhito Sasaki

Tokyo Metropolitan Institute of Gerontology, Positron Medical Center

Address: 35-2 Sakaechou, Itabashi-ku, Tokyo 173
Phone: +81 3 3964 3241 / ext. 3500
Fax: +81 3 3579 4776
Email: senda@pet.tmig.or.jp
Web site: www.tmig.or.jp/PET/pet_E.html
Contactperson: Michio Senda

Cyclotron Research Center, Iwate Medical Univercity (CRC) & Nishina Memorial
Cyclotron Center, Japan Radioisotope Association (NMCC)

Address: 348 Tomegamori, Takizawa-mura, Iwate-ken 020-01
Phone: +81 19 688 6071
Fax: +81 19 688 6072
Email: DHF23535@pcvan.or.jp
Contactpersons: K.Sera (CRC) and S. Futatsugawa (NMCC)

Chiba University Hospital, Department of Radiology

Address: 1-8-1 Inohana, Chuou-ku, Chiba-shi, Chiba-ken 280
Phone: +81 43 222 7171
Fax: +81 43 226 2101
Contactperson: K. Uno

Nishijin Hospital, Gotuji Agaru, Shichihonmatu-dohri

Address: Kamigyo-shi, 602 Japan
Phone: +81 75 461 8800
Fax: +81 75 461 5514
Contactperson: T. Yagyu

Himedic Imaging Center at Lake Yamanaka

Address: 562-12 Aza Yanaghihara, Hirano, Yamanakakomura, Minamitsuru-gun,
             Yamanshi-ken 401 95
Phone: +81 555 65 9135
Fax: +81 555 20 3007
Contactperson: Akira Shohtsu

Saudi Arabia

King Faisal Specialist Hospital and Research Centre
Bioanalytical and Drug Development Laboratory, Biological and Medical Research

Address: MBC-03, P.O.Box 3354, Riyadh 11211
Phone: +966 1 442 7859
Fax: +996 1 442 7858
Email: enein@kfshrc.edu.sa
Web site: www.fhshrc.edu.sa
Contactpersons: Dr. Sultan T.Al-Sedairy, Executive Director
          Hassan Y. Aboul-Enein, Professor, Pricipal Scientist

South Korea

Korea Cancer Centre Hospital, Korea Advanced Energy Research Institute

Address: 215-4 Gongneug-dong, Dobong-ku, Seoul
Contactperson: Yun Taik Koo

Taiwan

Veterans General Hospital, Department of Nuclear Medicine

Address: Taipei 11217
Phone: +886 2 871 5849
Fax: +886 2 872 6035
Contactperson: Peter Shin-Hwa Yeh

AUSTRALIA

Austin & Repatriation Medical Centre,
Department of Nuclear Medicine & Centre for PET

Address: Studley Road, Heidelberg, VIC 3084
Phone: +61 3 9496 5669
Fax: +61 3 9457 6605
Email: sub@austin.unimelb.edu.au
Web site: www.austin.unimelb.edu.au
Contactpersons: Dr W.John McKay, Director
          Dr Andrew M. Scott, Head of Clinical Services
          Dr Salvatore U. Berlangieri, Head of Clinical PET Program

Royal Prince Alfred Hospital, PET Department, Level A7, Bldg 63

Address: Camperdown NSW 2050, Sydney
Phone: +61 2 95157908
Fax: +61 2 95158690
Email: mfulham@pet.nucmed.rpa.cs.nsw.gov.au
Web site: www.cs.nsw.gov.au/rpa/pet
Contactperson: Dr Michael J. Fulham, Director of PET, Senior Staff Neurologist

NORTH AMERICA

Canada

University of Ottawa Heart Institute, Cardiac PET Centre, Division of Cardiology

Address: 1053 Carling Avenue, Ottawa, Ontario K1Y 4E9
Phone: +1 613 761 5296
Fax: +1 613 761 4690
Contactperson: Dr. Bob Beanlands, Director, Cardiac PET Centre
                      Email: rbeanlan@heartinst.on.ca
                      Dr. Terry Ruddy, Head, Nuclear Cardiology Research
                      Email: truddy@heartinst.on.ca
                      Dr. R. deKemp, PET Physicist
                      Email: rdekemp@heartinst.on.ca

VM Rakoff PET Centre, Clarke Institute of Psychiatry

Address: 250 College Street, Toronto, Ontario M5T 1R8
Phone: +1 416 979 4651
Fax: +1 416 979 4656
Email: shoule@clarke-inst.on.ca
Web site: research.clarke-inst.on.ca/pet/
Contactperson: Sylvian Houle M.D.,Ph.D., Director

McConnell Brain Imaging Centre, Montreal Neurological Institute

Address: 3801 University St, Montreal, Quebec H3A 2B4
Phone: +1 514 398 8925
Fax: +1 514 398 8948
E-mail: alan@bic..mni.mcgill.ca
Web site: www.bic.mni.mcgill.ca
Contactperson: Alan C. Evans, Ph.D., Professor, Director

United States

Good Samaritan Medical Center, Nuclear Medicine, Samaritan PET Center

Address: 1111 East McDowell Road, Box 2989, Phoenix, AZ 85006
Phone: +1 602 239 4229
Fax: +1 602 239 3073
Contactperson: Michael Lawson

St. Joseph's Hospital and Medical Center, Barrow Neurological Institute, Department of Radiology

Address: 350 West Thomas Road, Phoenix, AZ 85013
Phone: +1 602 406 3974
Fax: +1 602 406 6396
Contactperson: Burton P. Drayer

University of Arizona, University Medical Center, Nuclear Medicine Department

Address: 1501 North Campbell Avenue, Tucson, AZ 85724
Phone: +1 602 626 7709
Fax: +1 602 694 2412
Email: dpatton@radiology.arizona.edu
Web site: www.radiology.arizona.edu/~nucmed/title.htm
Contactperson: Dennis Patton M.D.

Northern California PET Imaging Center

Address: 3195 Folsom Blvd., Sacramento, CA 95816
Phone: +1 916 737 3200
Fax: +1 916 737 6203
Contactperson: Ruth Dean Tesar

U. C. Irvine Brain Imaging Center,

Address: Room 182, Irvine Hall, UCI-COM, Irvine, CA 92697
Phone: +1 714 824 7867
Fax: +1 714 856 2230
Email: jcwu@uci.edu
Contactperson: Joseph Wu and Steven Potkin

Division of Nuclear Medicine, UCLA Clinical PET Center, UCLA School of Medicine

Address: CHS AR-175, 10833 Le Conte Avenue, Los Angeles, CA 90095-6942
Phone: +1 310 206 9896
Fax: +1 310 206 4899

Email: jmaddahi@mail.nuc.ucla.edu
Web-site: www.nuc.ucla.edu
Contactperson: Jamshid Maddahi

Medical Research and Biophysics, University of California at Berkeley

Address: 1 Cyclotron Road, M. S. 55-121, Berkeley, CA 94720
Phone: +1 510 486 5435
Fax: +1 510 486 4768
Web site: www.cfi.lbl.gov
Contactperson: Tom Budinger

PET Imaging Science Center, Radiology and Clinical Pharmacy,
USC School of Medicine and Pharmacy

Address: 1510 San Pablo Street, Suite 350, Los Angeles, CA 90033
Phone: +1 213 342 5940
Fax: +1 213 342 5778
Email: pconti@hsc.usc.edu
Contacperson: Peter S. Conti M.D., Ph.D., Director

PET Center, Palo Alto VA Health Care System, V.A. Medical Center / Palo Alto
Nuclear Medicine Service - 115

Address: 3801 Miranda Avenue, Palo Alto, CA 94304
Phone: +1 415 858 3945
Fax: +1 415 852 3421
Email: gss@icon.palo-alto.med.va.gov
Contactperson: George Segall, M.D., Director

Nuclear Medicine, West Los Angeles VA Medical Center,
Nuclear Medicine Service 115

Address: 11301 Wilshire Blvd., Los Angeles, CA 90073
Phone: +1 310 268 3587
Fax: +1 310 268 4916
Email: Blahd.William@West-L.A.VA.gov
Contactperson: William H. Blahd M.D.

Yale University- VA Positron Imaging Laboratory

Address: 950 Campbell Avenue - 115A, West Haven, CT 06516
Phone: +1 203 937 3882
Fax: +1 203 937 4509
Email: Soufer.Robert@West-Haven.VA.GOV
Contactperson: Robert Soufer M.D., Director

Nuclear Medicine and PET, Baptist Hospital of Miami,
Nuclear Medicine and Miami Vascular Institute

Address: 8900 N. Kendall Drive, Miami, FL 33176
Phone: +1 305 596 1960 / ext. 4474
Fax: +1 305 273 2315
Email: JZiffer@bhssf.org
Contactperson: Jack Ziffer, Ph.D., M.D., Medical Director, or
                Stacy Morris, Administrative Director

Radiology, Broward General Medical Center

Address: 1600 S. Andrews Avenue, Ft. Lauderdale, FL 33316
Phone: +1 305 355 5500
Fax: +1 503 5517
Contactperson: Kevin Fusen

Memorial Medical Center of Jacksonville

Address: 3627 University Bldv. S, #135, Jacksonville, FL 32216
Phone: +1 904 391 1111
Fax: +1 904 391 1450
Contactperson: Mitch Lyle

Nuclear Cardiology, Crawford Long Hospital of Emory University,
Department of Medicine and Radiology

Address: 550 Peachtree Street, NE, Rm. 4323, Atlanta, GA 30365
Phone: +1 404 686 8203
Fax: +1 404 686 4957
Contactperson: Randolph E. Patterson

Emory Center for PET, Emory University Hospital

Address: 1364 Clifton Road, N.E., Rm E 163, Atlanta, GA 30322
Phone: +1 404 712 7422
Fax: +1 404 712 7961
Email: erno@petone.eushc.prg
Website: www.cc.emory.edu/RADIOLOGY/pet.html
Contactperson: Ernest Garcia

Cardiac Services, Kennestone PET Center

Address: 840 Church Street, Bldg. G, Suite 1., Marietta, GA 30060
Phone: +1 770 793 5800
Fax: +1 793 7928
Contactperson: Claudia Ceminsky

PET Lab, St. Joseph's Hospital of Atlanta

Address: 5665 Peachtree Dunwood Road, Atlanta GA 30342
Phone: +1 404 851 7919
Fax: +1 404 851 5521
Email: LSTepps@PO1.SJHA.ORG
Contactpersons: Lisa Stepps, OM/DB, Coordinator
              Thom Lane, ARRT (N), Technical Supervisor
              Jayne King, Director, Cardiac Services

The Methodist Medical Center of Illinois, Downstate Clinical PET Center

Address: 112 Crescent Avenue, Peoria, IL 61636
Phone   +1 309 672 4190
Fax   +1 309 671 2166
Email: cbingham@ccgate.mmci.org
Contactperson: Cathy Bingham, R.N., Coordinator
              Carter S. Young, D.O., Medical Director

Rush PET Center for Metabolic Imaging, Rush-Presbyterian-St. Luke's Medical Center

Address: 1653 West Congress Parkway, Chicago, IL 60612
Phone: +1 312 942-5757
Fax: +1 312 942-5320
email: aali@rad.rpslmc.edu
Web site: www.rad.rpslmc.edu
Contactperson: Amjad Ali, M.D., Medical Director

PET, University of Chicago

Address: 5841 S. Maryland Avenue, FMI MC, 1037, Chicago, IL 60637
Phone: +1 312 702 6282
Fax: +1 312 702 5986
Contactperson: Malcolm Cooper

Indiana University, VA Rodenbush PET Facility, Indiana University School of Medicine

Address: Long Clinical Building, Rm CL-120, Indianapolis, IN 46202
Phone: +1 317 274 3687
Fax: +1 317 274 4074
Email: ghutchins@xray.indyrad.iupui.edu
Web site: www.indyrad.iupui.edu
Contactperson: Gary D. Hutchins, Director, Division of Imaging Science

University of Iowa PET Imaging Center, Department of Radiology and Nuclear Medicine, University of Iowa

Address: 200 Hawkins Drive, Iowa City, IA 52242
Phone: +1 319 356 4101, 356-4101, also 6-4102, 6-4103 or 6-4104
Fax: +1319 356 2220
Email: peter-kirchner@uiowa.edu
Contactpersons: Peter T. Kirchner M.D., Professor of Radiology and Medicine,
                Director of Nuclear Medicine
                (address as above, phone: +1 319-356-4302, email as above)
                Richard D. Hichwa Ph.D., Associate Professor of Radiology,
                Director of PET Centre
                (address: as above, phone: +1 319-356-4104,
                email: richard-hichwa@uiowa.edu)

PET Imaging Center, Biomedical Research Institute of NW Louisiana

Address: P.O.Box 38050, Shreveport, LA 71133-8050
Phone: +1 318 675 4000
Fax: +1 318 675 4120
Email: mtorma@mail-sh.lsumc.edu
Web site: www.biomed.org/
Contactperson: Michael J. Torma, M.D. (phone:+1 318 657 4102)

PET Center, Our Lady of the Lake Regional Medical Center,
Nuclear Medicine Department

Address: 5000 Nehhesy Blvd., Baton Rouge, LA 70809
Phone: +1 504 765 7776
Fax: +1 504 766 5645
Contactperson: Edie Kearley

PET Department, NIH Clinical Center

Address: Bldg. 10., ACRF 1C499, 9000 Rockville Pike, Bethesda, M.D. 20892
Phone: +1 301 496 6455
Fax: +1 301 402 3521
Email: Eckelman@NM.D.PET.CC.NIH.GOV
Web site: www.cc.nih.gov/pet/
Contactperson: William Eckelman, Chief, PET Department

PET Center, The Johns Hopkins Medical Institutions, Division of Nuclear Medicine

Address: 615 North Wolfe St, Rm 2001, Baltimore, M.D. 21205
Phone: +1 410 955 2916
Fax: +1 410 955 0691
Email: robert_dannals@tracer.nm.jhu.edu
Contactperson: Robert Dannals

PET Center, Massachusetts General Hospital

Address: 32 Fruit St., Boston; MA 02114
Phone: +1 617 726 8353
Fax: +1 617 726 6165
Email: Fischman@PETW6.MGH.Harvard.edu
Contactperson: Alan Fishman M.D., Ph.D., Director MGH PET facility

PET Imaging Service, Bio-Metabolic Imaging

Address: 30781 Stephenson Hwy, Madison Heights, MI 48071
Phone: +1 810 585 5115
Contactperson: Ram Gunabalam

PET Center, Children's Hospital of Michigan, Wayne State University

Address: 3901 Baubien Blvd., Detroit, MI 48201
Phone: +1313 993 2867
Fax: +1 313 993 3845
Email: HChugani@PET.wayne.edu
Contactperson: Harry Chugani

PET Center, Division of Nuclear Medicine, University of Michigan Medical Center

Address: 1500 E. Medical Center Drive, Ann Arbor, MI 48109-0028
Phone: +1313 936 5388
Fax: +1 313 936 8182
Email: DKuhl@emich.edu
Contactperson: David E. Kuhl M.D., Professor, Director

Positron Diagnostic Center, William Beaumont Hospital,
Department of Nuclear Medicine

Address: 3601 West Thirteen Mile Rd., Royal Oak, MI 48073
Phone: +1 810 551 8131
Fax: +1 810 551 0768
Email: jjuni@beaumont.edu
Web site: www.beaumont.edu
Contactperson: Jack Juni M.D., Director of the Center

PET Imaging Service (11P), V.A. Medical Center / Minneapolis

Address: One Veterans Drive, Minneapolis, MN 55417
Phone: +1 612 725 2230
Fax: +1 612 725 2068
Contactperson: David A. Rottenberg

Division of Nuclear Medicine, Mallinckrodt Institute of Radiology,
Washington University Medical Center, Division of Nuclear Medicine

Address: 510 S. Kingshighway Blvd., St. Louis, MO 63110-1076
Phone: +1 314 362 2809
Fax: +1 314 362 2806
Web site: ibc.wustl.edu:70/1/mir
Contactperson: Barry A. Siegel

St. Louis University PET Imaging Center, St. Louis University Medical Center,
Division of Nuclear Medicine

Address: 3635 Vista Ave, P.O. Box 15250, St. Louis, MO 63110-0250
Phone: +1 314 577-8801   Fax +1 314 268-5486
Email: lowe@nucmed.slu.edu
Contactperson: Penny Yost, Chief PET Technologist
                 Val J. Lowe, M.D., PET Center Director:

Alegent Health Bergan Mercy PET Center, Department of Nuclear Medicine
Bergan Mercy Medical CenterBergab Mercy Hospital

Address: 7500 Mercy Road, Omaha, NE 68124
Phone: +1402 398 6984
Fax: +1 402 697.0116
Email: SamMehr@sprintmail.com
Contactperson: Sam Mehr, M.D., Director

Creighton Center for Metabolic Imaging, University of Nebraska Medical Center,
Department of Nuclear Medicine

Address: 600 South 42nd Street, Omaha, NE 68198-1045
Phone: +1 402 559 4341
Fax: +1 402 559 1011
Email: hanneman@creighton.edu
Contactperson: Lisa Gobar M.D.

Positron Emission Tomography Laboratory, Division of Cardiology,
Beth Israel Medical Center

Address: First Avenue at 16th Street, 8th Floor Dazian Building,
           New York, NY 10003
Phone: +1 212 420 4634
Fax: +1 212 420 4172
Contactpersons: Steven F. Horowitz M.D.
                 Dahlia Garza M.D.

Brookhaven Center for Imaging and Neurosciences, Brookhaven National Laboratory

Address: Upton, NY 11973-500
Phone: +1 516 344 4397
Fax: +1 516 344 7902
Email: loisc@bnl.gov
Web site: www.chemistry.bnl.gov/chemistry.html
Contactpersons: J. S. Fowler, Head
                 Alfred P. Wolf, Co-Head

State University New York at Buffalo, Center for Positron Emission Tomogrpahy,
Department of Nuclear Medicine

Address: 105 Parker Hall, 3435 Main Street, Buffalo, NY 14214-3007
Phone: +1 716 838 5889
Fax: +1 716 838 4918
Email: fred@nucmed.buffalo.edu
Web sites: Dept of Nuclear Medicine: prometheus.nucmed.buffalo.edu/
            PET Center: prometheus.nucmed.buffalo.edu/cpethm.htm
Contactpersons: Robert Ackerhalt, Ph.D., PET Center Director and Acting Chairman
                 Fred Covelli, MBA, Center Administrator and Fiscal Officer
Imaging Suite 1:
Veteran Affairs Medical Center WNY Healthcare System

Address: 3495 Bailey Avenue, Buffalo, N.Y. 14215
Phone: +1 716 862 3450
Fax: +1 716 862 3462
Web site: Same as above
Contactpersons: Alan Lockwood, M.D., Director of PET Operations
                 Jayakumari Gona, M.D., Chief of Nuclear Medicine and PET
Imaging Suite 2:
Millard Fillmore Hospital, Lucy Dent Imaging Center, Dent Neurological Institute

Address: 3 Gates Circle, Buffalo, N.Y. 14209
Phone: +1 716 887 4435
Fax: +1 716 887 4440
Contactperson: Robert Miletich, M.D., Ph.D., Chief, PET

The Center for Positron Emission Tomography,
Buffalo Cardiology and Pulmonary Associates, P.C.

Address: 5305 Main Sreet, Buffalo, NY 14221
Phone: +1 716 634 5100
Fax: +1 716 634 5134
Email: heartbcpa @aol.com
Web site: www.bcpa.com
Contactperson: Michael Merhige, Chief Operating Officer

Kreitchman PET Center - Columbia University, Columbia Presbyterian Medical Center

Address: Milstein Bldg., Suite 2-132, 177 Fort Washington Avenue,
             New York, NY 10032
Phone: +1 212 305 9670
Fax: +1 212 305 9694
Contactperson: Ernest J. DeSalvo M.D., PET Clinical Coordinator

Laurent and Alberta Gerschel PET Center, Nuclear Medicine Service,
Department of Radiology

Address: 1275 York Avenue, New York, NY. 10021
Phone: +1 212 639 7373
Fax: +1 212 717 3263
Email :   larsons@mskcc.org
Contactperson: Steven M. Larson, M.D., Director

Mount Sinai Medical Center / New York, Department of Psychiatry

Address: One Gustave Levy Place, New York, NY 10029
Phone: +1 212 241 5294
Fax: +1 212 423 0819
Contactperson: Monte Buchsbaum

NSUH Cyclotron/PET Center, North Shore University Hospital

Address: 300 Community Drive, Manhasset, NY 11030
Phone: +1 516 562 0100
Fax: +1 516 562 1608
Email: donaldm@nshs.edu
Contactperson: Donald Margouleff M.D., Medical Director

Radiological Diagnostic Imaging, P.C.

Address: 100 Lafayette Drive, Syosset, NY 11791
Phone: +1 516 364 4600
Fax: +1 516 364 4646
Contactperson: Azad K. Anand

PET Center, Bowman Gray School of Medicine

Address: Medical Center Blvd., Winston-Salem, NC 27157
Phone: +1 910 716 7461
Fax: +1 910 716 5639
Email: jwk@pet.bgsm.edu
Web site: www.rad.bgsm.edu/pet/Web_page/main.htm
Contactperson: John W. Keyes Jr., M.D., Director

Carolinas Medical Center, Department of Nuclear Medicine

Address: P.O.Box 32861, Charlotte, NC 28232-2861
Phone: +1 704 355 5856
Fax: +1 704 355 5910
Contactperson: David M. Coates

Duke University Medical Center, Department of Radiology

Address: North, Box 3949, Durham, NC 27710
Phone: +1 919 684 7245
Fax: +1 919 684 7135
Email: colem010@mc.duke.edu
Contactperson: R. Edward Coleman

Nuclear Medicine Department, Cleveland Clinic Foundation

Address: 9500 Euclid Avenue, Cleveland, OH 44195
Phone: +1 216 444 2665
Fax: +1 216 444 3943
Contactperson: Raymundo T. Go

Nuclear Medicine, Kettering Medical Center, Nuclear Medicine / PET

Address: 535 Southern Boulevard, Ketterin, OH 45429
Phone: +1 513 296 7211
Fax: +1 513 296 4265
Contactperson: Joseph Mantil

The Christ Hospital, Metabolic Imaging Center

Address: 2139 Auburn Avenue, Cincinnati, Ohio 45219
Phone: +1 513 369 8833
Fax: +1 513 369 8835
Email: KramerB@healthallcom
Web site: Intranet.healthall.com
Contactperson: Stephen J. Pomeranz, M.D., Medical Director, Advanced Imaging
              Thomas Kemme, B.S., CNMT, Manager, Nuclear Medicine
              /Metabolic Imaging

PET Facility, University Hospitals of Cleveland

Address: 11100 Euclid Ave., Cleveland OH 44106-5056
Phone: +1 216 844 3107
Fax: +1 216 844 3106
Email: leisure@uhrad.com
Web site: www.uhrad.com
Contactperson: Floro Miraldi, M.D., ScD, Director PET Facility

University of Pennsylvania PET Center
Division of Nuclear Medicine, Department of Radiology
Hospital of the University of Pennsylvania

Address: 3400 Spruce St., Philadelphia, PA.19104
Phone: +1 215 662 3069
Fax: +1 215 349 5843
E-mail: alavi@darius.pet.upenn.edu
Contactperson: Abass Alavi M.D., Medical Director

PET Facility, University of Pittsburgh Medical Center

Address: 200 Lothrop Street, Pittsburgh, PA 15213
Phone: +1 412 647 0736
Fax: +1 412 647 0300
Email: mintun@tippi.pet.upmc.edu
Contactperson: Mark A. Mintun

PET Suite, Baptist Hospital

Address: 2000 Church Street, Nashville, TN 37236
Phone: +1 615 329 4774
Contactperson: Andrew B. Carlsen, M. D., Medical Director
              (300 20th Avenue, North, Suite 702, Nashville, TN 37203,
              phone: 615-329 0203, Fax: 615 329 4039)

University of Tennessee Medical Center

Address: 1924 Alcoa Highway, Knoxville, TN 37920-6999
Phone: +1 423 544 9662
Fax: +1 423 544 8883
Web site: www.mc.utk.edu/docs/radiology/pet/pet.html
Contactperson: Karl Hubner

PET Center, Vanderbilt University, Department of Radiology

Address: 1251 Medical research Bldg., 222 Pierce Avenue,
              Nashville, TN 37232-6531
Phone: +1 615 343 7511
Fax: +1 615 343 6531
Email: martin.sandler@mcmail.vanderbilt.edu
Contactperson: Dominique Delbeke M.D., Ph.D., Clinical Director of PET

South Texas Veterans Health Care System (STVHCS)

Address:	7400 Merton Minter Blvd, San Antonio, TX 78284
Phone:	+1 210 617 5117
Fax:	+1 210 617 5198
Email:	CHAUDHURI,TUHIN@SAN ANTONIO.VA.GOV
	tuhin@flash.net

Contactperson:	Tuhin K. Chaudhuri, M.D., FACNP
	Chief, Nuclear Medicine, STVHCS,
	Professor of Radiology, UTHSCSA

Columbia / HCA, Medical City Dallas, PET Center

Address:	7777 Forest Lane, Bldg. A / Medical City Dallas, Dallas, TX 75230
Phone:	+1 214 788 6649
Fax:	+1 214 788 6648
Contactperson:	Roy E. Aldridge

Anderson Cancer Center, University of Texas, Division of Diagnostic Imaging

Address:	Box 59, 1515 Holcombe Blvd., Houston, TX 77030
Phone:	+1 713 794 1052
Fax:	+1 713 794 5456
Email:	edmund_kim@diag_imaging.M.D.a.uth.tmc.edu
Contactperson:	E. Edmund Kim, Director, Metabolic Imaging
	Chief, Section of Experimental Nuclear Medicine

PET Imaging Center,
University of Texas Health Science Center-Houston Medical School

Address:	6431 Fannin, MSB 4.256 MSB, Houston, TX 77030
Phone:	+1 713-500-6611
Fax:	+1 713-500-6615
Email:	gould@heart.med.uth.tmc.edu
Contactperson:	K. Lance Gould, M.D., Medical Director

Research Imaging Center, Medical School,
The University of Texas Health Science Center at San Antonio

Address:	7703 Floyd Curl Drive, San Antonio, TX 78284-6240
Phone:	+1 210 567 8100
Fax:	+1 210 567 8152
Email:	fox@uthscsa.edu
Web site:	ric.uthscsa.edu
Contactperson:	Peter T. Fox M.D., Professor, Director

University of Washington Medical Center, Department of Radiology
Division of Nuclear Medicine RC-70

Address: 1959 Pacific Street, N. E., Seattle, WA 98195
Phone: +1 206 548 4240
Fax: +1 206 548 4496
Contactperson: Michael M. Graham

Department of Radiology, West Virginia University

Address: P.O.Box 9235, Morgantown, W·V 26506
Phone: +1 304 293 7798
Fax: +1 304 293 7142
Contactperson: Naresh C. Gupta

University of Wisconsin PET Imaging Center,
University of Wisconsin Hospital and Clinics, Department of Radiology

Address: E3/311 CSC, 600 Highland Avenue, Madison, WI 53792-3252
Phone: +1 608 265 8731
Fax: +1 608 265 8737
Email: sperlman@Facstaff.wisc.edu, jmhanson@facstaff.wisc.edu
Contactperson: Scott Perlman, Director, UW PET Imaging Center
                 Doug Brown, Associate Director, UW PET Imaging Center

***

Further information on the www is available at the home page of the Institute for Clinical PET: **www.icppet.org**

LIST OF CONTRIBUTORS

C. Aine
Biophysics Group, Los Alamos
National Laboratory, Los Alamos,
New Mexico 87545, USA

Stanovnik Ales
Jozef Stefan Institute,
Department of Elementary Physics,
University of Ljubljaba,
1525 Ljubljana, Slovenia

Katrin Amunts
Department of Anatomy and C. and O.
Vogt-Institut for Brain Research,
Heinrich-Heine University Düsseldorf,
D-40225 Düsseldorf, Germany

Dale Bailey
MRC Cyclotron Unit,
Royal Postgraduate Medical School,
Hammersmith Hospital,
Du Cane Road, London W12 0NN, U.K.

László Balkay
DOTE PET Centre, H-4012 Debrecen,
Nagyerdei körút 98, Hungary

Ervin Berényi
Diagnostic Centre,
Pannon Agricultural University,
Kaposvár, Hungary

Katalin Borbély
National Institute of Neurosurgery
Amerikai út 57, H-1145 Budapest,
Hungary

Sket Boris
Department of Organic Chemistry,
University of Ljubljaba,
1525 Ljubljana, Slovenia

István Boros
DOTE PET Centre, H-4012 Debrecen,
Nagyerdei körút 98, Hungary

David Brooks
MRC Cyclotron Unit,
Royal Postgraduate Medical School,
Hammersmith Hospital,
Du Cane Road, London W12 0NN, U.K.

Christian Buechel
The Wellcome Department of
Cognitive Neurology, Leopold Muller
Functional Imaging Laboratory,
Institute of Neurology,
12 Queen Square, London WC1N 3BG,
U.K.

László Csiba
Department of Neurology
Debrecen University Medical School
H-4012 Debrecen,
Nagyerdei körút 98, Hungary

Maurizio Corbetta
Mallinckrodt Institute of Radiology
Washington University Medical Center
Division of Radiation Sciences
4525 Scott Avenue, Campus Box 8225
Saint Louis, Missouri 63110, USA

Magnus Dahlbom
Division of Nuclear Medicine and
Biophysics, Department of
Radiological Sciences,
UCLA School of Medicine,
10833 Le Conte Ave,
Los Angeles, CA, 90024-1721, USA

Árpád Duka Zólyomi
Department of Nuclear Physics,
Faculty of Mathematics and Physics,
Comenius University
Mlynská dolina F1, 842 15 Bratislava,
Slovakia

Miklós Emri
DOTE PET Centre, H-4012 Debrecen,
Nagyerdei körút 98, Hungary

464

Olga Ésik
National Institute of Oncology
Ráth. Gy. u. 7-9.
H-1122 Budapest, Hungary

Lars Farde
Department of Psychiatry,
Karolinska Hospital
S-171 76 Stockholm, Sweden

Peter T. Fox
Research Imaging Center
The University of Texas Health
Science Center at San Antonio
7703 Floyd Curl Drive
San Antonio, TX 78284-6240, USA

Richard S. J. Frackowiak
The Wellcome Department of
Cognitive Neurology, Leopold Muller
Functional Imaging Laboratory,
Institute of Neurology,
12 Queen Square, London WC1N 3BG,
U.K.

Karl Friston
The Wellcome Department of
Cognitive Neurology, Leopold Muller
Functional Imaging Laboratory,
Institute of Neurology,
12 Queen Square, London WC1N 3BG,
U.K.

Chris D. Frith
The Wellcome Department of
Cognitive Neurology, Leopold Muller
Functional Imaging Laboratory,
Institute of Neurology,
12 Queen Square, London WC1N 3BG,
U.K.

Tamás Galambos
Research Group of the Hungarian
Academy of Sciences,
DOTE PET Centre, H-4012 Debrecen,
Nagyerdei körút 98, Hungary

László Galuska
Department of Nuclear Medicine,
Debrecen University Medical School
H-4012 Debrecen,
Nagyerdei körút 98, Hungary

J. George
Biophysics Group, Los Alamos
National Laboratory, Los Alamos,
New Mexico 87545, USA

Nathalie Ginovart
Department of Psychiatry,
Karolinska Hospital
S-171 76 Stockholm, Sweden

Anna Grabowska
Department of Neurophysiology,
Nencki Institute of Experimental
Biology, Warsaw, Poland

Balázs Gulyás
Department of Neuroscience
Karolinska Institute
S-171 77 Stockholm, Sweden

Ferenc Gyulai
Department of Anesthesiology and
Critical Care Medicine, University of
Pittsburgh Medical Center,
200 Lothrop Street, Room C-207
Pittsburgh, PA15213-2582, USA

Håkan Hall
Department of Psychiatry,
Karolinska Hospital
S-171 76 Stockholm, Sweden

Christer Halldin
Department of Psychiatry,
Karolinska Hospital
S-171 76 Stockholm, Sweden

Andy Holley
CTI Europe
15 Knowles Avenue, Crowthorne
Berkshire RG45 6DU, U.K.

Hiroshi Ito
Department of Psychiatry,
Karolinska Hospital
S-171 76 Stockholm, Sweden

Levente Kerényi
Department of Neurology
Debrecen University Medical School
H-4012 Debrecen,
Nagyerdei körút 98, Hungary

György Kövér
Diagnostic Centre,
Pannon Agricultural University,
Kaposvár, Hungary

J. B. Krause
Institut für Medizin,
Forschungszentrum Jülich GmbH
D-52425 Jülich, Germany

Leszek Królicki
Department of Nuclear Medicine and
Magnetic Resonance,
Medical School of Warsaw,
Warsaw, Poland

Jack L. Lancaster
Research Imaging Center
The University of Texas Health
 Science Center at San Antonio
7703 Floyd Curl Drive
San Antonio, TX 78284-6240, USA

Szabolcs Lehel
DOTE PET Centre, H-4012 Debrecen,
Nagyerdei körút 98, Hungary

Stig Lindbäck
GEMS PET Systems AB,
Husbyborg, S-752 29 Uppsala,
Sweden

József Lövey
National Institute of Oncology
Ráth. Gy. u. 7-9.
H-1122 Budapest, Hungary

Camilla Lundkvist
Department of Psychiatry,
Karolinska Hospital
S-171 76 Stockholm, Sweden

E.T. MacKenzie
University of Caen, CNRS UMR 6551,
Cyceron, Cyclotron Biomedical Unit,
Bd H. Becquerel, BP 5229,
F-14074 Caen Cedex, France

Izabella Makaiova
Oncological Institute of St. Elisabeth
Clinic of Nuclear Medicine,
Medical Faculty, Heydukova 10,
812 50 Bratislava, Slovakia

Teréz Márián
DOTE PET Centre, H-4012 Debrecen,
Nagyerdei körút 98, Hungary

Staric Marko
Jozef Stefan Institute,
Department of Elementary Physics,
University of Ljubljaba,
1525 Ljubljana, Slovenia

Julie A. McCarron
MRC Clinical Sciences Center
Cyclotron Unit
Royal Postgraduate Medical School
Hammersmith Hospital
Du Cane Road, London W12 0NN,
U.K.

Sviatoslav Medvedev
Institute of the Human Brain of the
Russian Academy of Sciences,
Laboratory for Positron Emission
Tomography, 9 Pavlova St.,
St. Petersburg, 197376 Russia

Gerd-J. Meyer
Department of Nuclear Medicine
Medical School of Hannover
Konstanty-Gutschow-Strasse 8,
D-30623 Hannover, Germany

Metka Milcinski
University Medical Center,
Department of Nuclear Medicine,
University of Ljubljaba,
1525 Ljubljana, Slovenia

Hans W. Müller-Gärtner
Institut für Medizin,
Forschungszentrum Jülich GmbH
D-52425 Jülich, Germany

Tamás Molnár
DOTE PET Centre, H-4012 Debrecen,
Nagyerdei körút 98, Hungary

Budihna Natasa
University Medical Center,
Department of Nuclear Medicine,
University of Ljubljaba,
1525 Ljubljana, Slovenia

466

Ferenc Németh
Research Group of the Hungarian
Academy of Sciences,
DOTE PET Centre, H-4012 Debrecen,
Nagyerdei körút 98, Hungary

Anna-Lena Nordström
Department of Psychiatry,
Karolinska Hospital
S-171 76 Stockholm, Sweden

Svante Nyberg
Department of Psychiatry,
Karolinska Hospital
S-171 76 Stockholm, Sweden

A. M. J. Paans
PET Center,
University Hospital Groningen
P.O.Box 30001, 9700 RB Groningen,
The Netherlands

Lawrence M. Parssons
Research Imaging Center
The University of Texas Health
Science Center at San Antonio
7703 Floyd Curl Drive
San Antonio, TX 78284-6240, USA

Victor W. Pike
MRC Clinical Sciences Center
Cyclotron Unit
Royal Postgraduate Medical School
Hammersmith Hospital
Du Cane Road, London W12 0NN,
U.K.

Cathy J. Price
The Wellcome Department of
Cognitive Neurology, Leopold Muller
Functional Imaging Laboratory,
Institute of Neurology,
12 Queen Square, London WC1N 3BG,
U.K.

J. Pruim
PET Center,
University Hospital Groningen
P.O.Box 30001, 9700 RB Groningen,
The Netherlands

D. Ranken
Biophysics Group, Los Alamos
National Laboratory, Los Alamos,
New Mexico 87545, USA

Geraint E. Rees
The Wellcome Department of
Cognitive Neurology, Leopold Muller
Functional Imaging Laboratory,
Institute of Neurology,
12 Queen Square, London WC1N 3BG,
U.K.

Ján Ruzicka
Department of Nuclear Physics,
Faculty of Mathematics and Physics,
Comenius University,
Mlynská dolina F1, 842 15 Bratislava,
Slovakia

Ivanka Savic
Department of Neuroscience
Karolinska Institute
S-171 77 Stockholm, Sweden

G. Schlaug
Department of Neurology,
Heinrich-Heine-University Düsseldorf
Moorenstrasse 5, D-40225 Düsseldorf,
Germany

H. Schlitt
Institut für Medizin,
Forschungszentrum Jülich GmbH
D-52425 Jülich, Germany

Pascale Schumann
University of Caen, CNRS UMR 6551,
Cyceron, Cyclotron Biomedical Unit,
Bd H. Becquerel, BP 5229,
F-14074 Caen Cedex, France

M. F. Schüller
Department of Neurology,
Heinrich-Heine-University Düsseldorf
Moorenstrasse 5, D-40225 Düsseldorf,
Germany

Göran Sedvall
Department of Psychiatry,
Karolinska Hospital
S-171 76 Stockholm, Sweden

Rüdiger J. Seitz
Department of Neurology,
Heinrich-Heine-University Düsseldorf
Moorenstrasse 5, D-40225 Düsseldorf,
Germany

N. J. Shah
Institut für Medizin,
Forschungszentrum Jülich GmbH
D-52425 Jülich, Germany

Paul D. Shreve
University of Michigan,
Faculty of Medicine
B1J505/UH/Box 0028
1500 East Medical Centre Drive
Ann Arbor, Michigan 48109-0028
USA

Hans Steinert
Departement of Medical Radiology and
Nuclear Medicine,
Universitätsspital Zürich
Rämistrasse 100
CH-8091 Zürich, Switzerland

Carl-Gustav Swahn
Department of Psychiatry,
Karolinska Hospital
S-171 76 Stockholm, Sweden

Gerhard Stöcklin
Nuklearmedizinische Klinik,
Klinikum v. d. I. der TU München
Ismaninger Str. 22.
D-81675 München, Germany

Zsolt Szabó
Divisions of Nuclear Medicine and
Radiology, The Johns Hopkins
Medical Institutions
600 N Wolfe Street, B1-130 Tower,
Baltimore, MD 21205, USA

Szabolcs Szakáll Jr.
DOTE PET Centre, H-4012 Debrecen,
Nagyerdei körút 98, Hungary

Arthur Toga
UCLA School of Medicine
Reed Neurological Institute
Department of Neurology
Division of Brain Mapping
710 Westwood Plaza, Rm. 4238
Los Angeles, CA 90024-1769, USA

Emese Tóth
National Institute of Oncology
Ráth. Gy. u. 7-9.
H-1122 Budapest, Hungary

Lajos Trón
DOTE PET Centre, H-4012 Debrecen,
Nagyerdei körút 98, Hungary

Willem Vaalburg
PET Center,
University Hospital Groningen
P.O.Box 30001, 9700 RB Groningen,
The Netherlands

A. Van Waarde
PET Center,
University Hospital Groningen
P.O.Box 30001, 9700 RB Groningen,
The Netherlands

Gábor Veress
DOTE PET Centre, H-4012 Debrecen,
Nagyerdei körút 98, Hungary

Hans-Jürgen Wester
Nuklearmedizinische Klinik,
Klinikum v. d. I. der TU München
Ismaninger Str. 22.
D-81675 München, Germany

A. T. M. Willemsen
PET Center,
University Hospital Groningen
P.O.Box 30001, 9700 RB Groningen,
The Netherlands

INDEX

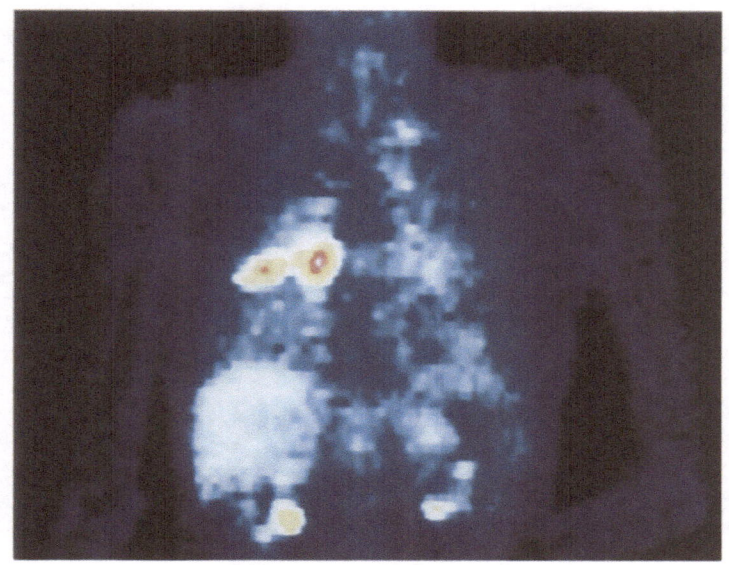

Figure 1 from page 201.

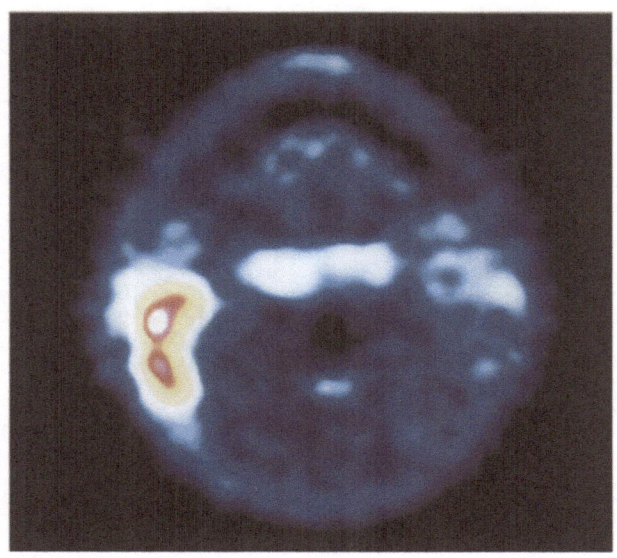

Figure 2 from page 201.

476

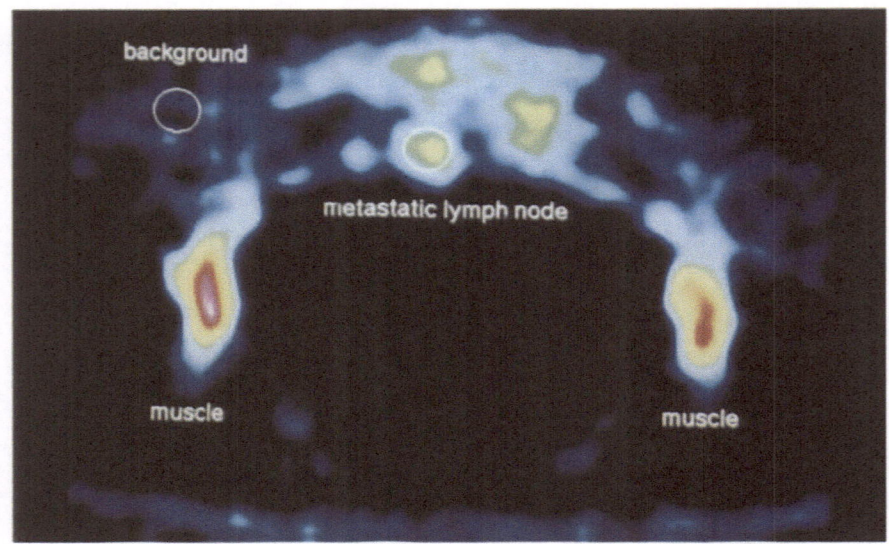

Figure 7 from page 209.

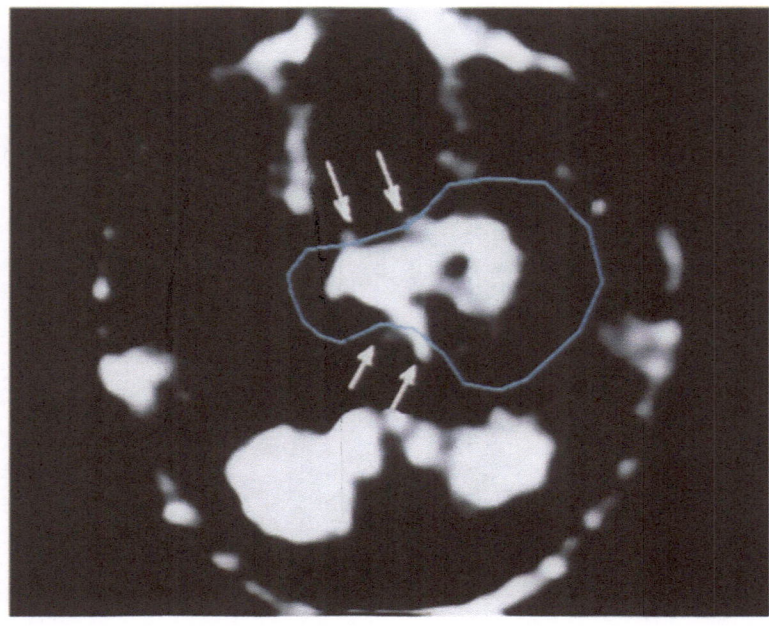

Figure 8 from page 209.

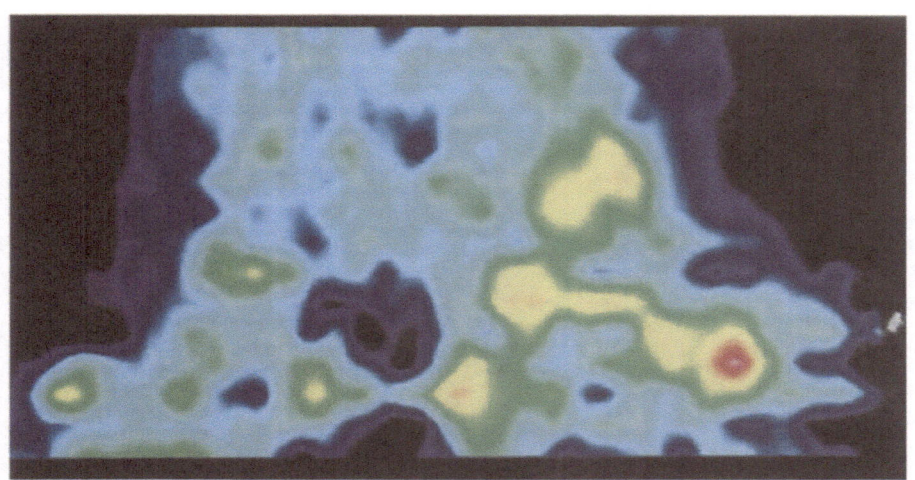

Figure 1 from page 226.

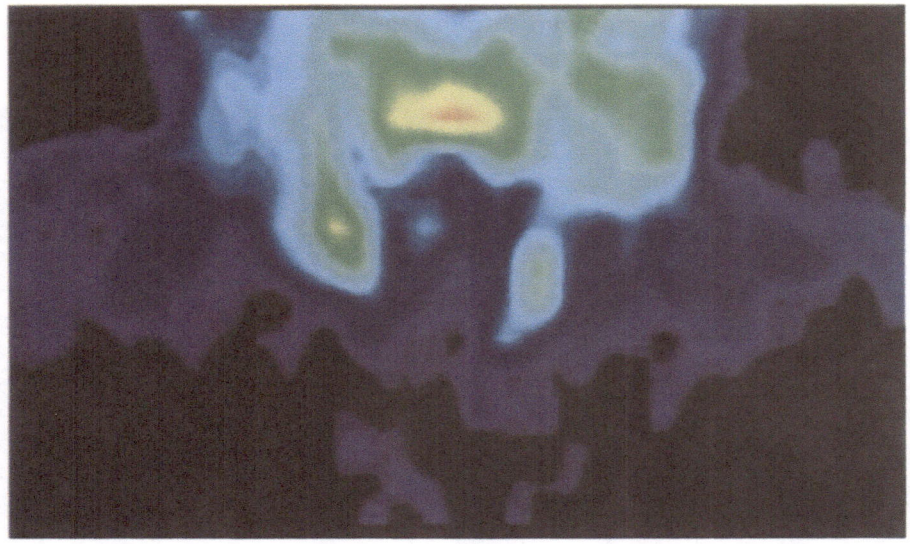

Figure 2 from page 226.

478

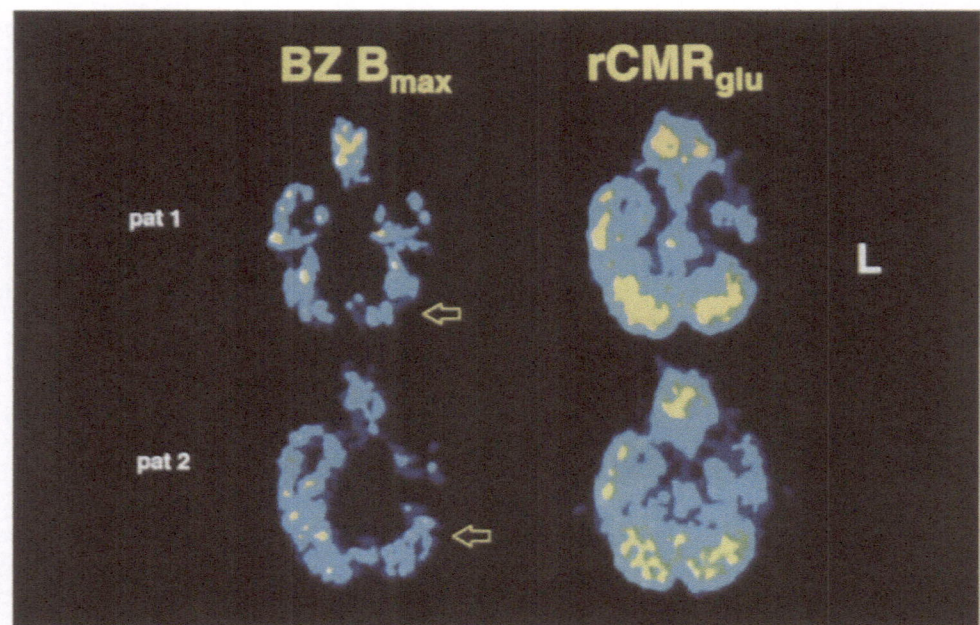

Figure 2 from page 265.

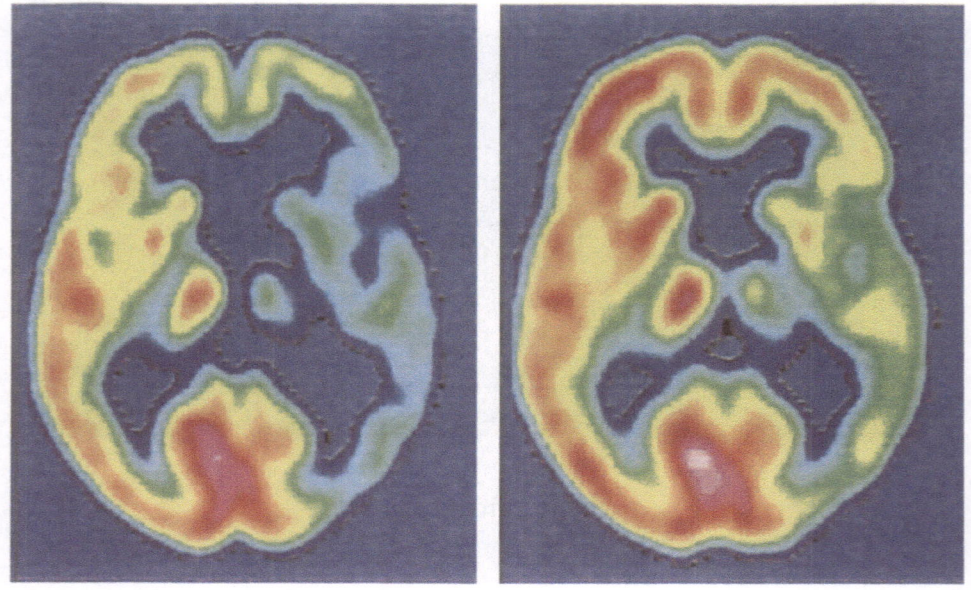

Figure 3 from page 301.

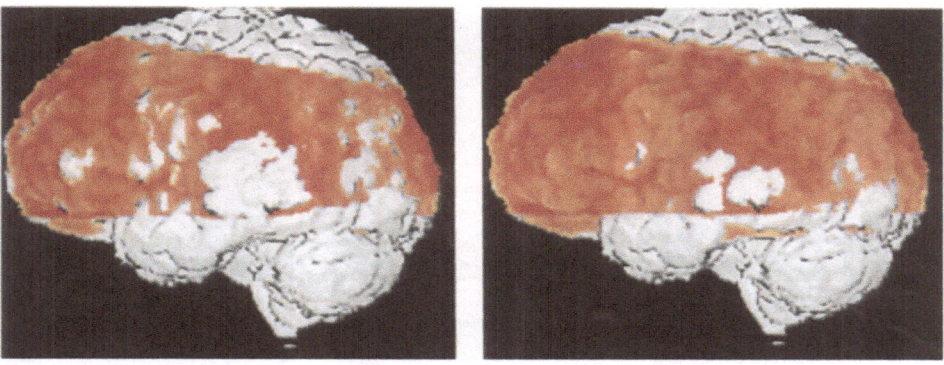

Figure 4 from page 302.

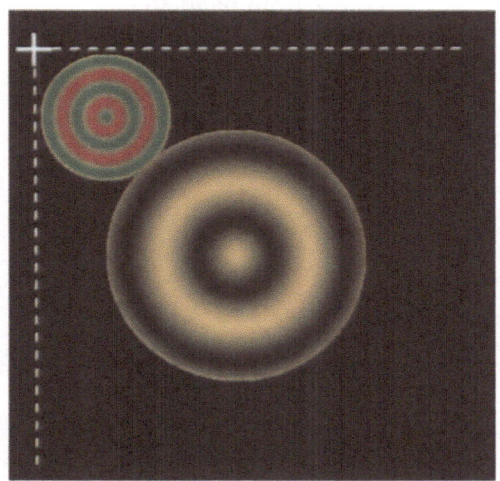

Figure 1 from page 362.

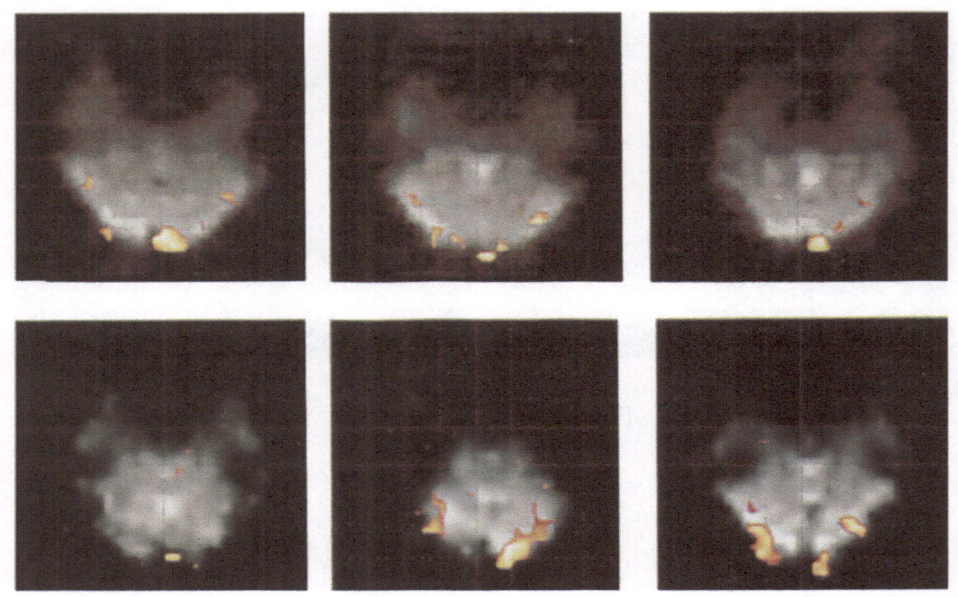

Figure 2 from page 363.

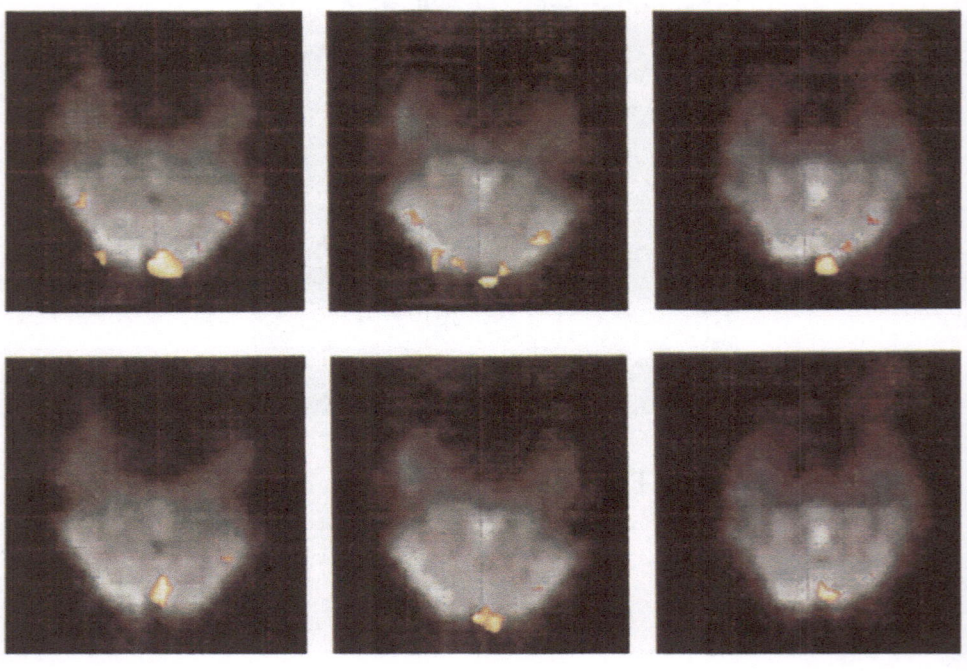

Figure 4 from page 365.

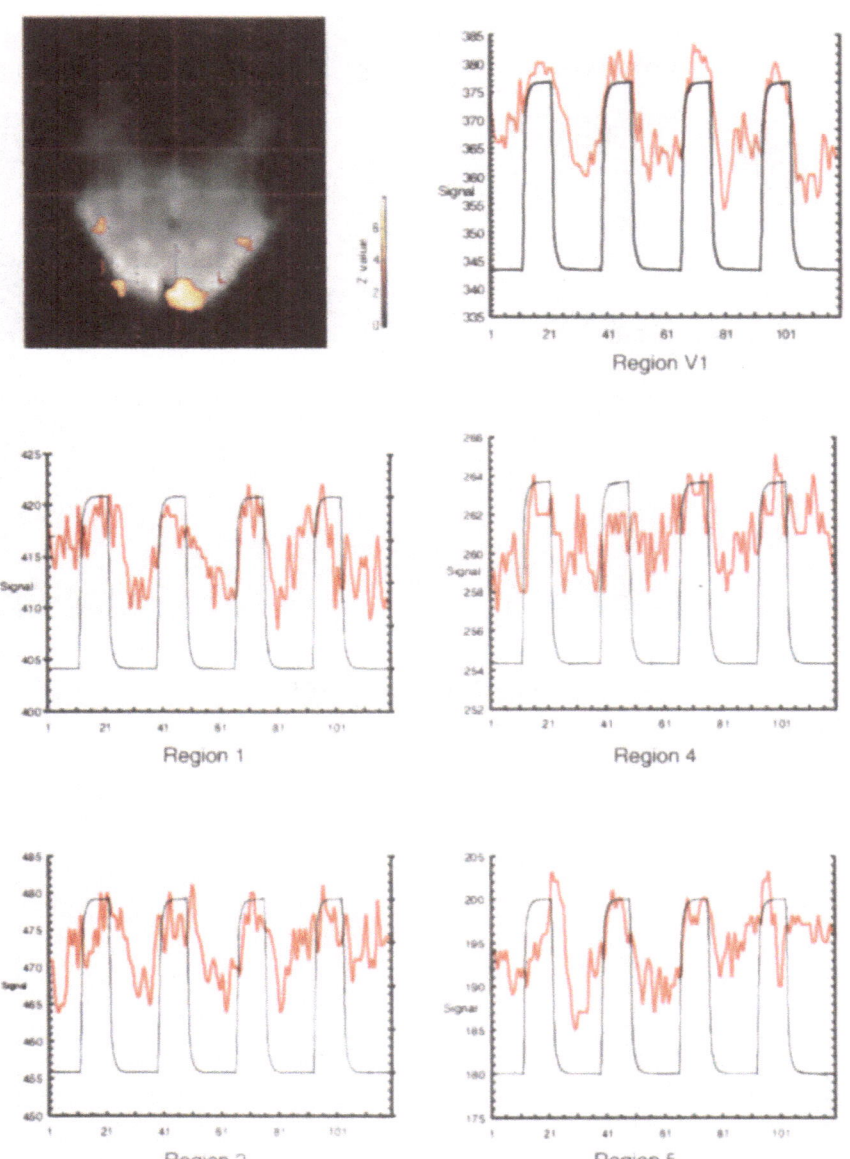

Figure 3 from page 364.

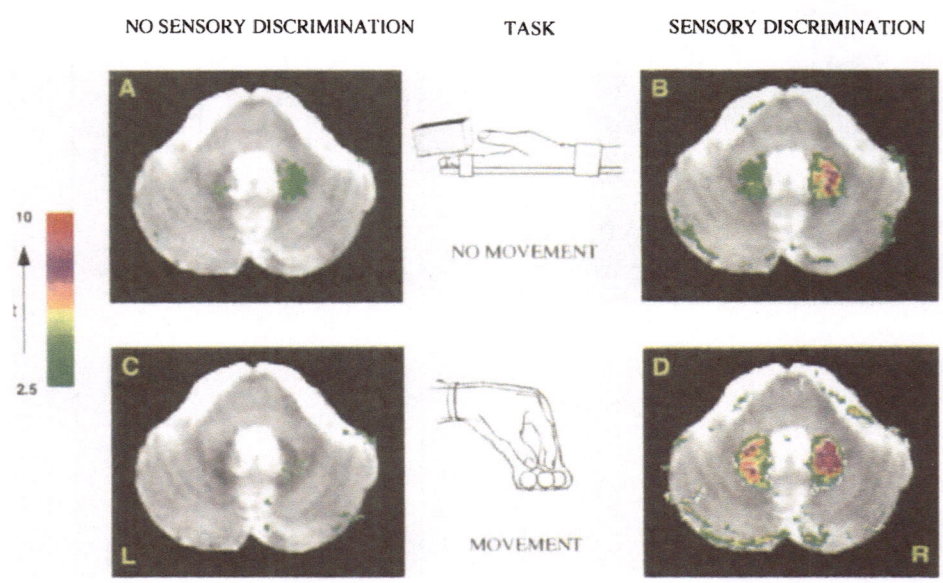

Figure 3 from page 381.